Product Formulas

$\cos\alpha \cos\beta = \dfrac{\cos(\alpha-\beta) + \cos(\alpha+\beta)}{2}$

$\sin\alpha \sin\beta = \dfrac{\cos(\alpha-\beta) - \cos(\alpha+\beta)}{2}$

$\sin\alpha \cos\beta = \dfrac{\sin(\alpha+\beta) + \sin(\alpha-\beta)}{2}$

$2\cos\alpha \sin\beta = \sin(\alpha+\beta) - \sin(\alpha-\beta)$

Sum Form.

$\cos x + \cos y = 2\cos\left(\dfrac{x+y}{2}\right)\cos\left(\dfrac{x-y}{2}\right)$

$\cos x - \cos y = 2\sin\left(\dfrac{x+y}{2}\right)\sin\left(\dfrac{x-y}{2}\right)$

$\sin x + \sin y = 2\sin\left(\dfrac{x+y}{2}\right)\cos\left(\dfrac{x-y}{2}\right)$

$\sin x - \sin y = 2\sin\left(\dfrac{x-y}{2}\right)\cos\left(\dfrac{x+y}{2}\right)$

[2.1]	a^n	the nth power of a, or a to the nth power; $a \cdot a \cdot a \cdots a$ (n factors)
	$P(x), D(y)$, etc.	P of x, D of y, etc.
[3.1]	a^0	1 ($a \neq 0$)
	a^{-n}	the reciprocal of a^n, $\dfrac{1}{a^n}$ ($a \neq 0$)
[3.2]	$a^{1/n}$	the nth real root of a for $a \in R$, or the positive one if there are two
	$a^{m/n}$	$(a^{1/n})^m$
[3.3]	$\sqrt[n]{a}$	the nth real root of a for $a \in R$, or the positive one if there are two
[4.6]	(a, b)	interval of real numbers between a and b
	$[a, b]$	interval of real numbers between a and b, and a and b
[5.1]	(a, b)	the ordered pair of numbers whose first component is a and whose second component is b
	$R \times R$, or R^2	the Cartesian product of R and R
	f, g, h, F, etc.	symbols denoting functions
	$f(x)$	f of x, or the value of f at x
[5.2]	d	the distance between two points
	m	the slope of a line
[6.1]	$[x]$	the greatest integer not greater than x
[6.2]	f^{-1}	the inverse function of f
[7.1]	e	an irrational number, approximately equal to 2.7182818
[7.2]	$\log_b x$	the logarithm to the base b of x, or the logarithm of x to the base b
[7.3]	$\text{antilog}_b x$	the antilogarithm to the base b of x, or the antilogarithm of x to the base b.

Modern College Algebra and Trigonometry

Fourth Edition

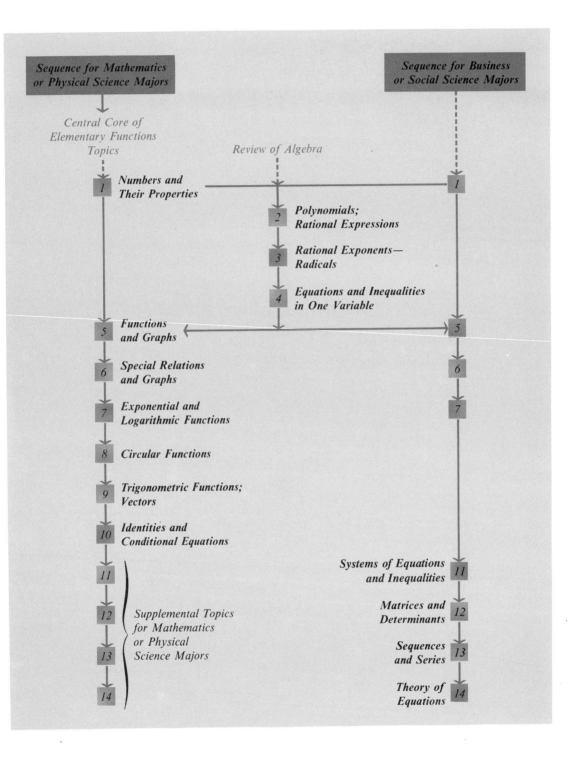

Modern College Algebra and Trigonometry
Fourth Edition

Edwin F. Beckenbach
University of California, Los Angeles

Irving Drooyan
Los Angeles Pierce College

Michael D. Grady
Loyola Marymount University

Wadsworth Publishing Company
Belmont, California

A division of Wadsworth, Inc.

Mathematics Editor: Richard Jones
Production: Greg Hubit Bookworks

© 1981 by Wadsworth, Inc.

© 1977, 1975, 1973, 1972 by Wadsworth Publishing Company, Inc. All rights reserved. No part of this book may be reproduced, stored in a retrieval system, or transcribed, in any form or by any means, electronic, mechanical, photocopying, recording, or otherwise, without the prior written permission of the publisher, Wadsworth Publishing Company, Belmont, California 94002, a division of Wadsworth, Inc.

Printed in the United States of America

1 2 3 4 5 6 7 8 9 10—85 84 83 82 81

Library of Congress Cataloging in Publication Data

Beckenbach, Edwin F
   Modern college algebra and trigonometry.

   Includes index.
   1. Algebra. 2. Trigonometry, Plane.
I. Drooyan, Irving, joint author. II. Grady, Michael D., 1946–   joint author. III. Title.
QA154.2.B43   1981   512.9   80-19882
ISBN 0-534-00871-2

Contents

1 Numbers and Their Properties — *1*

1.1 Preliminary Concepts — 1
1.2 Real Numbers — 5
1.3 Field Properties — 10
1.4 Order in R — 14
1.5 Complex Numbers — 18
Chapter Review — 23

2 Polynomials; Rational Expressions — *25*

2.1 Definitions; Sums of Polynomials — 25
2.2 Products of Polynomials — 31
2.3 Factoring Polynomials — 34
2.4 Quotients of Polynomials — 37
2.5 Synthetic Division — 41
2.6 Equivalent Fractions — 44
2.7 Sums and Differences of Rational Expressions — 48
2.8 Products and Quotients of Rational Expressions — 51
2.9 Partial Fractions — 55
Chapter Review — 61

3 Rational Exponents—Radicals — *63*

3.1 Powers with Integral Exponents — 63
3.2 Powers with Rational Exponents — 68
3.3 Radical Expressions — 73
3.4 Operations on Radical Expressions — 77
Chapter Review — 80

4 Equations and Inequalities in One Variable — 82

- 4.1 Equivalent Equations; First-Degree Equations — 82
- 4.2 Second-Degree Equations — 87
- 4.3 The Quadratic Formula — 93
- 4.4 Equations Involving Radicals — 95
- 4.5 Substitution in Solving Equations — 97
- 4.6 Solution of Linear Inequalities — 99
- 4.7 Solution of Nonlinear Inequalities — 103
- 4.8 Equations and Inequalities Involving Absolute Value — 109
- 4.9 Word Problems — 112
- Chapter Review — 117

5 Functions and Graphs — 120

- 5.1 Pairings of Real Numbers — 120
- 5.2 Linear Functions — 126
- 5.3 Forms for Linear Equations — 132
- 5.4 Quadratic Functions — 136
- 5.5 Polynomial Functions — 141
- 5.6 Rational Functions — 145
- Chapter Review — 151

6 Special Relations and Graphs — 153

- 6.1 Special Functions — 153
- 6.2 Inverse Functions — 156
- 6.3 Variation as a Functional Relationship — 160
- 6.4 Conic Sections — 165
- 6.5 Linear and Quadratic Inequalities — 173
- Chapter Review — 176

7 Exponential and Logarithmic Functions — 178

- 7.1 Exponential Functions — 178
- 7.2 Logarithmic Functions — 180
- 7.3 Logarithms to the Base 10 and the Base e — 185
- 7.4 Applications of Logarithms — 191
- Chapter Review — 196

8 Circular Functions — 198

- 8.1 The Functions Cosine and Sine — 198
- 8.2 Special Function Values of Cosine and Sine — 203

8.3	Function Values $\cos x$ and $\sin x$, $x \in R$	208
8.4	Graphs of $y = \cos x$ and $y = \sin x$	214
8.5	Tangent Function	222
8.6	Other Circular Functions	227
8.7	Inverse Circular Functions	231
	Chapter Review	238

9 Trigonometric Functions; Vectors — 241

9.1	Angles and Their Measure	241
9.2	Functions of Angles	247
9.3	Right Triangles	255
9.4	The Law of Sines	260
9.5	The Law of Cosines	266
9.6	Geometric Vectors	269
	Chapter Review	275

10 Identities and Conditional Equations — 278

10.1	Basic Identities	278
10.2	Special Formulas for the Cosine Function	283
10.3	Special Formulas for the Sine Function	287
10.4	Special Formulas for the Tangent Function	291
10.5	Conditional Equations	295
10.6	Conditional Equations for Multiples	299
	Chapter Review	301

11 Systems of Equations and Inequalities — 303

11.1	Systems of Linear Equations in Two Variables	303
11.2	Systems of Linear Equations in Three Variables	310
11.3	Systems of Nonlinear Equations	316
11.4	Systems of Inequalities	322
11.5	Convex Sets—Polygonal Regions	324
11.6	Linear Programming	327
	Chapter Review	329

12 Matrices and Determinants — 331

12.1	Definitions; Matrix Addition	331
12.2	Matrix Multiplication	336
12.3	Solution of Linear Systems by Using Row-Equivalent Matrices	342
12.4	The Determinant Function	349

12.5	Properties of Determinants	355
12.6	The Inverse of a Square Matrix	363
12.7	Solution of Linear Systems Using Inverses of Matrices	369
12.8	Cramer's Rule	372
	Chapter Review	377

13 *Sequences and Series* — *379*

13.1	Sequences	379
13.2	Series	384
13.3	Limits of Sequences and Series	390
13.4	The Binomial Theorem	396
13.5	Mathematical Induction	403
13.6	Basic Counting Principles; Permutations	407
13.7	Combinations	413
	Chapter Review	417

14 *Theory of Equations* — *419*

14.1	Synthetic Division and the Factor Theorem Over C	419
14.2	Complex Zeros of Polynomial Functions	421
14.3	Real Zeros of Polynomial Functions	424
14.4	Rational Zeros of Polynomial Functions	429
	Chapter Review	432

Appendices — *435*

A *De Moivre's Theorem; Polar Coordinates* — *436*

A.1	Trigonometric Form of Complex Numbers	436
A.2	De Moivre's Theorem—Powers and Roots	441
A.3	Polar Coordinates	444

B *Mathematical Structure* — *450*

Postulates for Real Numbers	450
Properties of Numbers	451
Mathematical Systems	463

C *Tables* — *465*

Table I	Common Logarithms	466
Table II	Exponential Functions	468

Contents

Table III Natural Logarithms of Numbers	469
Table IV Squares, Square Roots, and Prime Factors	470
Table V Values of Circular Functions	471
Table VI Values of Trigonometric Functions	475

Odd-Numbered Answers — *484*

Index — *535*

Preface

This fourth edition of *Modern College Algebra and Trigonometry*, like its predecessors, is designed for students who have completed from one and one-half to two years of high school algebra and a year of geometry, but who need additional preparation for the study of calculus or mathematically oriented courses in fields such as business or biology. The organization of the material permits considerable flexibility in the kind of course for which the book is appropriate. Specific suggestions are charted on page ii (facing the title page).

Users of the third edition will find numerous changes in this edition. Among the more significant are:

> Parts of Chapters 1 through 4 have been rewritten; the exercises in these chapters have been reviewed and revised in the light of classroom experience. An optional section on partial fractions has been added to Chapter 2.
>
> Chapters 5 and 6 have been reorganized so that polynomial functions and rational functions are now introduced in Chapter 5. The material on the theory of equations that was contained in Chapter 6 of the third edition is now included in Chapter 14.
>
> Chapter 8 has been reorganized so that reference arcs rather than reduction formulas are now used to compute the values of the circular functions. The material on the inverse circular functions that was contained in Chapter 10 of the third edition is now included in Chapter 8. Chapters 9 and 10 have been rewritten to accommodate the changes made in Chapter 8.
>
> Chapter 11, which is devoted to solving systems of equations and systems of inequalities, is new.
>
> Appendix B, which is new, includes the basic substance of the text in compact form and is a handy guide for following the development of the course. It includes a brief treatment of mathematical systems as related to the number systems considered in the text.

Provision has been made for the optional use of a hand calculator. *Either* tables or a hand calculator can be used in the work with exponential, logarithmic, circular, and trigonometric functions. Table V is graduated in hundredths of a radian, and Table VI is graduated in tenths of a degree, so that the entries are compatible with the trigonometric function values obtained using a calculator. Uses of both tables and a calculator are shown in the examples.

The first chapter, which introduces the student to the real number system, is followed by three chapters—on polynomials, rational exponents, and equations and inequalities in one variable—that may be optional, if two years of high school algebra is a prerequisite. Beginning with Chapter 5, discussions generally center around the function concept. Polynomial and rational functions (and their graphs) are covered in detail; the logarithmic function is developed from a consideration of the exponential function; determinants are presented as functions of matrices; and sequences are treated as functions having sets of positive integers as domain.

A section at the end of the book provides answers and graphs for odd-numbered problems at the end of each section; answers are also provided for all the exercises in the chapter reviews. For additional assistance, a solutions manual providing detailed solutions to the even-numbered problems is available to the student.

As in the third edition, a second color is used functionally to highlight key procedures in routine manipulations and to focus attention on key elements of figures.

Edwin F. Beckenbach
Irving Drooyan
Michael D. Grady

Modern College Algebra and Trigonometry

Fourth Edition

1 Numbers and Their Properties

In earlier mathematics courses, you explored the basic properties of the set of real numbers and saw how some of their properties apply to real-life situations. In elementary algebra, you manipulated not only numbers but symbols which represent numbers. Such manipulations will continue in this course, but the symbols may represent objects which belong to systems other than real numbers. The purpose of this first chapter is to review, briefly, some facts about the real numbers and to introduce a more advanced system, the complex numbers. The chapter begins with a review of the basic concepts and terminology of set theory in order to provide a convenient mathematical language and notation.

1.1 Preliminary Concepts

A **set** is simply a collection of some kind. In algebra, we are interested in sets of numbers of various sorts and in their relations to sets of points or lines in a plane or in space. Any one of the collection of things in a set is called a **member** or **element** of the set. For example, the counting numbers 1, 2, 3, ... (where the three dots indicate that the sequence continues indefinitely) are the elements of the set we call the set of **natural numbers**.

Set notation Sets are usually designated by means of capital letters, A, B, C, etc. They are identified by means of **braces**, { }, with the members either listed or described. For example, the set N of *natural numbers* is $\{1, 2, 3, \ldots\}$.

Using the undefined notion of set membership, we can be more specific about some other terms we shall be using.

Definition 1.1 *Two sets A and B are **equal**, $A = B$, if and only if they have the same elements.*

The order in which the elements of a set are named is of no importance in determining its membership, nor is the fact that an element might be named more than once. Thus, if A denotes $\{1, 2, 3\}$, B denotes $\{3, 2, 1\}$, C denotes $\{2, 3, 4\}$, and D denotes {natural numbers between 1 and 5}, then $A = B$ and $C = D$. The phrase "if and only if" used in this definition is simply the mathematician's way of making two statements at once. Definition 1.1 means: "Two sets are equal if they have the same elements. Two sets are equal only if they have the same elements." The second of these statements is logically equivalent to: "Two sets have the same elements if they are equal."

Definition 1.2 *If every element of a set A is an element of a set B, then A is a **subset** of B.*

Subset symbol The symbol $\subset$ (read "is a subset of" or "is contained in") will be used to denote the subset relationship. Thus

$$\{1, 2, 3\} \subset \{1, 2, 3, 4\} \quad \text{and} \quad \{1, 2, 3\} \subset \{1, 2, 3\}.$$

Notice that, by definition, every set is a subset of itself.

The set that contains no elements is called the **empty set**, or **null set**, and is denoted by the symbol $\emptyset$ (read "the empty set" or "the null set"); $\emptyset$ is a subset of every set. If a set S is the null set or contains exactly n elements for some fixed natural number n, then S is said to be **finite**. A set that is not finite is said to be **infinite**. For example, the set of *all* natural numbers, $\{1, 2, 3, \ldots\}$, is an infinite set.

Definition 1.3 *The **union** of two sets A and B is the set of all elements that belong either to A or to B or to both.*

Set-union symbol The symbol $\cup$ is used to denote the union of sets. Thus $A \cup B$ (read "the union of A and B") is the set of all elements that are in A or B or both. For example, if $A = \{1, 2, 3, 4\}$ and $B = \{3, 4, 5\}$, then

$$A \cup B = \{1, 2, 3, 4, 5\}.$$

Notice that each element in $A \cup B$ is listed only once in this example, since repetition would be redundant.

Definition 1.4 *The **intersection** of two sets A and B is the set of all elements common to both A and B.*

Set-intersection symbol The symbol $\cap$ is used to denote the intersection of sets. Thus $A \cap B$ (read "the intersection of A and B") is the set of all elements that are in both A and B. For example, if $A = \{1, 2, 3, 4\}$ and $B = \{3, 4, 5\}$, then

$$A \cap B = \{3, 4\}.$$

If two sets A and B contain no element in common, A and B are said to be **disjoint**. That is, A and B are disjoint if and only if $A \cap B = \emptyset$. For example, if $A = \{1, 2, 3\}$ and $B = \{5, 6, 7\}$, then A and B are disjoint.

When discussing sets, it is often helpful to have in mind some general set from which the elements of all sets under consideration are drawn. For example, if we wish to talk about sets of integers, we may use as a general set the set of all integers; or taking a larger view we might use the set of all real numbers, or any one of many possible general sets. Such a general set is called the **universal set** and is usually denoted simply by the capital letter U.

Definition 1.5 The **complement** *of a set A in a universal set U is the set of all elements in U not in A.*

The symbol A' denotes the complement of A. Thus if $U = \{1, 2, 3, 4, 5, 6\}$ and $A = \{1, 2, 3, 4\}$, then

$$A' = \{5, 6\}.$$

Set-membership notation

The symbol $\in$ (read "is a member of" or "is an element of") is used to denote membership in a set. Thus,

$$2 \in \{1, 2, 3\}.$$

Note that we write

$$\{2\} \subset \{1, 2, 3\} \quad \text{and} \quad 2 \in \{1, 2, 3\},$$

since $\{2\}$ is a *subset*, whereas 2 is an *element*, of $\{1, 2, 3\}$.

When discussing an individual but unspecified element of a set containing more than one member, we usually denote the element by a lowercase italic letter (for example, a, d, s, x), or sometimes by a letter from the Greek alphabet: α (alpha), β (beta), γ (gamma), and so on. Symbols used in this way are called *variables*.

Definition 1.6 *A **variable** is a symbol representing an unspecified element of a given set containing more than one element.*

The given set is called the **replacement set,** or **domain,** of the variable. For example,

$$x \in A$$

means that the variable x represents an (unspecified) element of the set A. The elements of the replacement set are called the **values** of the variable. A symbol with just one possible value is called a **constant.**

Negation symbol

The slant bar, /, drawn through certain symbols of relation, is used to indicate negation. Thus $\neq$ is read "is not equal to," $\not\subset$ is read "is not a subset of," and $\notin$ is

Set-builder notation

read "is not an element of." For example,

$$\{1, 2\} \neq \{1, 2, 3\}, \quad \{1, 2, 3\} \not\subset \{1, 2\}, \quad \text{and} \quad 3 \notin \{1, 2\}.$$

Another symbolism useful in discussing sets is $\{x \mid x \text{ has a certain property}\}$; for example,

$$\{x \mid x \in A \quad \text{and} \quad x \notin B\}$$

(read "the set of all x such that x is a member of A and is not a member of B"). This symbolism, called **set-builder notation,** is used extensively in this book. What it does is specify a variable (in this case, x) and, at the same time, state a condition on the variable (in this case, that x is contained in the set A and is not contained in the set B).

Exercise 1.1

Designate each of the following sets using braces and listing the members.

Example   {natural numbers less than 7}

Solution $\{1, 2, 3, 4, 5, 6\}$

1. {natural numbers between 3 and 11 inclusive}
2. {natural numbers between 31 and 35 inclusive}
3. {letters in the word "mathematics"}
4. {letters in the words "college algebra"}
5. {days in the week}
6. {months in the year}

In Exercises 7–36, let $U = \{1, 2, 3, 4, 5, 6, 7, 8\}$, $A = \{1, 2, 3, 4\}$, $B = \{5, 6, 7, 8\}$, $C = \{2, 4, 6, 8\}$, and $D = \{1, 3, 5, 7\}$.

Replace the asterisk $*$ with either $=$ or $\neq$ to make a true statement.

7. $A * \{12, 34\}$
8. $A * \{4, 3, 2, 1\}$
9. $A * \{1, 1, 2, 2, 3, 4\}$
10. $A * \{1234\}$

Replace the asterisk $*$ with either $\subset$ or $\not\subset$ to make a true statement.

11. $B * \{4, 5, 6, 7\}$
12. $B * \{5, 6, 7\}$
13. $B * \{5, 6, 7, 8\}$
14. $B * \{4, 5, 6, 7, 8\}$

Replace the asterisk $*$ with either $\in$ or $\notin$ to make a true statement.

15. $3 * A$
16. $3 * B$
17. $\{3\} * C$
18. $\{3\} * D$

1.2 Real Numbers

List the elements of the following sets.

19. $A \cup C$
20. $B \cup D$
21. $B \cap C$
22. $A \cap D$
23. A'
24. C'
25. $(A \cap C)'$
26. $(B \cap D)'$
27. $(A \cup C)'$
28. $(B \cup D)'$
29. $A' \cup D$
30. $B' \cup C$
31. $(A' \cup C) \cap D$
32. $(B' \cap C) \cup D$
33. $(A \cap C') \cap D$
34. $(B \cap C) \cap D$
35. $A \cup (D \cap \emptyset)$
36. $B \cap (C \cup \emptyset)$

Designate the given set in set-builder notation. Use N to denote the set of natural numbers.

Example {even natural numbers}

Solution $\{x \mid x = 2n, n \in N\}$

37. {odd natural numbers}
38. {negative integers}
39. {multiples of 3 in N}
40. {multiples of 5 in N}
41. {natural numbers less than 100}
42. {natural numbers greater than 5}

Describe the following sets in terms of A, B, and C, and unions, intersections, and complements.

Examples a. $\{x \mid x \in A$ and $x \notin B\}$ b. $\{x \mid x \in A$ or $x \in B\}$

Solutions a. $A \cap B'$ b. $A \cup B$

43. $\{x \mid x \in B$ and $x \in C\}$
44. $\{x \mid x \in A$ or $x \in C'\}$
45. $\{x \mid x \in A$ and x is an element of B or $C\}$
46. $\{x \mid x \notin A$ and $x \notin B\}$
47. $\{x \mid x \in A$ but x is not an element of either B or $C\}$
48. Let U be a set with three elements. Count all the subsets of U. Repeat for sets with four and five elements and use these results to find a formula for the number of subsets of a set with n elements.

1.2 Real Numbers

We shall frequently refer to the following five sets of numbers.

1. The set N of **natural numbers,** whose elements are the counting numbers:

$$N = \{1, 2, 3, \ldots\}.$$

2. The set J of **integers,** whose elements are the counting numbers, their negatives, and zero:
$$J = \{\ldots, -2, -1, 0, 1, 2, \ldots\}.$$

3. The set Q of **rational numbers,** whose elements are all those numbers that can be represented as the quotient of two integers $\frac{a}{b}$ (or a/b, or $a \div b$), where b is not 0. Among the elements of Q are such numbers as $-3/4$, $18/27$, $3/1$, and $-6/1$. In symbols,
$$Q = \left\{x \mid x = \frac{a}{b}, \quad a, b \in J, \quad b \neq 0\right\}.$$
Every rational number can be represented by a terminating or repeating decimal numeral, such as 3.2, 1.975, 6.3333..., and 2.171717....

4. The set H of **irrational numbers,** whose elements are the numbers with decimal representations that are nonterminating and nonrepeating. Among the elements of this set are such numbers as $\sqrt{2}$, $\sqrt{3}$, $\sqrt{5}$, $\sqrt{7}$, π, and $-\sqrt{7}$. An irrational number cannot be represented in the form a/b, where a and b are integers.

5. The set R of **real numbers,** which contains all of the elements of the set of rational numbers and all of the elements of the set of irrational numbers:
$$R = Q \cup H.$$

The foregoing sets of numbers are related as indicated in Figure 1.1. Thus we have
$$H \subset R, \quad Q \subset R, \quad \text{and} \quad N \subset J \subset Q.$$

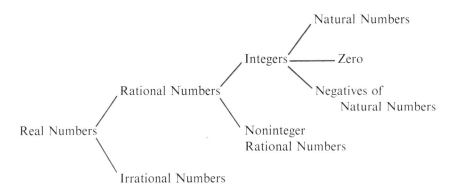

Figure 1.1

We can characterize a set by stating properties that we assume it has. In mathematics, when we make formal assumptions about members of a set or about their properties, we call the assumptions **axioms,** or **postulates.** The words **property, law,** and **principle** are sometimes used in referring to assumptions, although these words may also be applied to certain consequences thereof.

1.2 Real Numbers

Equality postulates

The first assumptions to be considered here have to do with *equality*. An **equality,** or an "is equal to" assertion, is simply a mathematical statement that two symbols or words, or two groups of symbols or words, are names for the same thing.

We postulate that, for any elements a, b, and c of any set S, the equality $(=)$ relationship satisfies the following laws:

E-1 $a = a$ *Reflexive law for equality.*
E-2   If $a = b$, then $b = a$.   *Symmetric law for equality.*
E-3   If $a = b$ and $b = c$, then $a = c$.   *Transitive law for equality.*
E-4   If $a = b$, then a may be replaced by b and b by a in any mathematical statement without altering the truth or falsity of the statement.†   *Substitution law for equality.*

The equality axioms indicate how we use symbols and warn that we must neither change the meaning of a symbol in the middle of a discussion nor use the same symbol for two different things in the same context. Other axioms, somewhat different conceptually, are used to characterize mathematical systems. They are assumptions made about the behavior of elements of sets under *binary operations*, that is, operations which involve using two elements of a set to produce a third element in the set. We express the fact that the element produced is also in a set A by saying that A is **closed** under the operation.

Addition and multiplication postulates

Since the set R of real numbers is the set of greatest interest to us at the moment, we postulate the basic laws for the elements of R in relation to the binary operations of addition and multiplication. You are probably familiar with these laws from your earlier work. They are listed on page 8 for reference. Parentheses are used in stating some of the relations to indicate that symbols within parentheses are to be viewed as representing a single entity.

If, for a given set and a given pair of binary operations (not necessarily the set of real numbers or ordinary addition and multiplication), Postulates F-1 through F-11 are satisfied by the elements of the set, then the set is called a **field** under these operations. Thus, we speak of the **field R of real numbers.**

To give meaning to expressions $a + b + c$, $a \times b \times c$, $a + b + c + d$, and so on, let us make the following agreement.

Definition 1.7   If $a, b, c, d, \ldots \in R$, then

$$a + b + c = a + (b + c), \quad a + b + c + d = a + (b + c + d), \ldots,$$

and

$$a \times b \times c = a \times (b \times c), \quad a \times b \times c \times d = a \times (b \times c \times d), \ldots.$$

The associative properties F-2 and F-7 ensure that $a + b + c$, $a \times b \times c$, etc., can actually be evaluated in any order.

---

†We have adopted a very powerful postulate in E-4, one that includes E-2 and E-3 as special cases. By wording E-4 as we have, we can eliminate a great deal of detail in later arguments, and, because E-2 and E-3 are fundamental properties of the equality relation, we have elected to retain them as postulates.

Field postulates of the set R of real numbers

Let a, b, c be arbitrary elements of R.

F-1 $a + b$ is a unique element of R. *Closure law for addition.*

F-2 $(a + b) + c = a + (b + c)$. *Associative law for addition.*

F-3 There exists an element $0 \in R$ (called the **identity element for addition**) with the property
$$a + 0 = a \quad \text{and} \quad 0 + a = a$$
for each $a \in R$. *Additive-identity law.*

F-4 For each $a \in R$, there exists an element $-a \in R$ (called the **additive inverse,** or **negative,** of a) with the property
$$a + (-a) = 0 \quad \text{and} \quad (-a) + a = 0.$$
Additive-inverse law.

F-5 $a + b = b + a$. *Commutative law for addition.*

F-6† $a \times b$ is a unique element of R. *Closure law for multiplication.*

F-7 $(a \times b) \times c = a \times (b \times c)$. *Associative law for multiplication.*

F-8 There exists an element $1 \in R$ (called the **identity element for multiplication**), $1 \neq 0$, with the property
$$a \times 1 = a \quad \text{and} \quad 1 \times a = a$$
for each $a \in R$. *Multiplicative-identity law.*

F-9 For each element $a \in R$, $a \neq 0$, there exists an element $\frac{1}{a} \in R$ (called the **multiplicative inverse,** or **reciprocal,** of a) with the property
$$a \times \frac{1}{a} = 1 \quad \text{and} \quad \frac{1}{a} \times a = 1.$$
Multiplicative-inverse law.

F-10 $a \times b = b \times a$. *Commutative law for multiplication.*

F-11 $a \times (b + c) = (a \times b) + (a \times c)$
and
$(b + c) \times a = (b \times a) + (c \times a)$. *Distributive law.*

†In all that follows, we shall adopt the customary practice of writing ab or $a \cdot b$ for $a \times b$.

1.2 Real Numbers

Inverse operations

In terms of the operations of addition and multiplication, we can define two additional operations on real numbers: **subtraction** (finding a *difference*) and **division** (finding a *quotient*).

Definition 1.8   The **difference** of elements $a \in R$ and $b \in R$, denoted by $a - b$, is given by

$$a - b = a + (-b).$$

Definition 1.9   The **quotient** of elements $a \in R$ and $b \in R$, $b \neq 0$, denoted by $\frac{a}{b}$, a/b, or $a \div b$, is given by

$$\frac{a}{b} = a \times \frac{1}{b}.$$

Observe that the quotient $\frac{a}{b}$ requires $b \neq 0$; that is, division by 0 is not defined.

Exercise 1.2

State whether each statement is true or false. The sets N, J, Q, H, and R are as given in the text.

1. $-1 \in N$
2. $0 \in Q$
3. $\sqrt{3} \in R$
4. $\frac{1}{2} \notin H$
5. $0.125 \in Q$
6. $0.1717\ldots \in Q$
7. $Q \cap H = \{0\}$
8. $Q \cap H = \emptyset$
9. $J \subset H$
10. $J \cap Q = N$
11. $N \not\subset Q$
12. $\{1, 2\} \not\subset J$

Each variable in Exercises 13–50 denotes a real number. Each of the statements 13–20 is an application of one of the Postulates E-1 through E-4. Justify the statement by citing an appropriate postulate. (There may be more than one correct justification.)

Examples

   a.   If $x + y = 8$ and $y = 3$, then $x + 3 = 8$.

   b.   If $x + y = x + z$, then $x + z = x + y$.

Solutions

   a.   Substitution law for equality, E-4.

   b.   Symmetric law for equality, E-2.

13. If $x = 3z$, then $3z = x$.
14. $3 + x = 3 + x$
15. If $a + b = b + c$ and $b + c = c + d$, then $a + b = c + d$.
16. If $x = 3y$ and $3y = 6z$, then $x = 6z$.
17. If $x = 6$ and $3x = y$, then $3 \cdot 6 = y$.

18. If $a + b = 1$ and $c - (a + b) = d$, then $c - 1 = d$.
19. $2a - (b + c) = 2a - (b + c)$
20. If $a + b = b + c$, then $b + c = a + b$.

Each statement in Exercises 21–36 is an application of one of the Postulates F-1 through F-11. Justify the statement by citing the appropriate postulate.

Examples a. $3 \cdot (x + 5) = (x + 5) \cdot 3$ b. $3(x + 5) = 3 \cdot x + 3 \cdot 5$

Solutions a. Commutative law for multiplication, F-10.   b. Distributive law, F-11.

21. $x + (y + 0) = (x + y) + 0$
22. $(3x)y = y(3x)$
23. $(p + q)(r + s) = (r + s)(p + q)$
24. $(a + 1) \cdot 1 = a + 1$
25. $3a + 0 = 3a$
26. $(a + 1) \cdot 1 = a \cdot 1 + 1 \cdot 1$
27. $(kl)(m + n) = k(l(m + n))$
28. $1 + (1 + 1) = (1 + 1) + 1$
29. $(p + q)(r + s) = p(r + s) + q(r + s)$
30. $0 + (3 + x) = 3 + x$
31. $1 \cdot (1 + 1) = 1 + 1$
32. $(5(x + y)) \cdot 2 = 5((x + y) \cdot 2)$
33. $(a + 3) + [-(a + 3)] = 0$
34. $(0 + k) \cdot 1 = 0 \cdot 1 + k \cdot 1$
35. $(a + 1)(b + 1) = (a + 1) \cdot b + (a + 1) \cdot 1$
36. $5x + (-5x) = 0$

Use the distributive laws to rewrite any product among the following as a sum, and any sum as a product.

Examples a. $3(7 + p)$ b. $8 + 4m$

Solutions a. $3(7 + p) = 3 \cdot 7 + 3p$ b. $8 + 4m = 4 \cdot 2 + 4m$
$ = 21 + 3p$ $ = 4(2 + m)$

37. $a(b + 1)$
38. $(a + 1)b$
39. $p(q + r)$
40. $ab + cb$
41. $(d + 3)d$
42. $(18 + 2)10$
43. $(de)f + gf$
44. $xy + xyz$
45. $(13 + q)r$
46. $a(2 + b)$
47. $st + su + sv$
48. $ab + (-1)b$
49. $3(x + y + z)$
50. $xyz + xy + xz$

1.3 Field Properties

The field postulates together with the postulates for equality imply other properties of the real numbers. Such implications are generally stated as **theorems.** A theorem

1.3 Field Properties

is simply an assertion of a fact that follows logically from the postulates (or axioms) and other theorems. We shall list (without proof) some of the theorems ordinarily encountered in lower-level algebra courses. We have numbered these theorems to provide an efficient way to refer to them later, and have also named those that have commonly accepted names. In most cases, illustrative examples are provided.

Addition law for equality   First, consider the following result, which reaffirms the uniqueness of the sum of two real numbers.

Theorem 1.1   If $a, b, c \in R$ and $a = b$, then
$$a + c = b + c \quad \text{and} \quad c + a = c + b.$$

Example   If $x - 3 = 5$, then $(x - 3) + 3 = 5 + 3$.

Multiplication law for equality   A theorem closely analogous to Theorem 1.1 can be stated as follows.

Theorem 1.2   If $a, b, c \in R$ and $a = b$, then
$$ac = bc \quad \text{and} \quad ca = cb.$$

Example   If $5z = 15$, then $\frac{1}{5}(5z) = \frac{1}{5}(15)$.

Additional properties   The next theorem asserts that the additive inverse of a real number and the multiplicative inverse of a nonzero real number are unique—that is, that a given real number has only one additive inverse and only one multiplicative inverse.

Theorem 1.3

I   If $a, b \in R$ and $a + b = 0$, then
$$b = -a \quad \text{and} \quad a = -b.$$

II   If $a, b \in R$ and $a \cdot b = 1$, then
$$a = \frac{1}{b} \quad \text{and} \quad b = \frac{1}{a} \quad (a, b \neq 0).$$

Examples

a.   If $x + 3 = 0$, then $x = -3$ and $3 = -x$.

b.   If $4x = 1$, then $x = \frac{1}{4}$ and $4 = \frac{1}{x}$.

The proofs of the succeeding theorems follow directly from the previous results.

Theorem 1.4   If $a, b, c \in R$ and $a + c = b + c$, then $a = b$.

Theorem 1.5   If $a, b, c \in R$, $c \neq 0$, and $ac = bc$, then $a = b$.

1 Numbers and Their Properties

Theorem 1.6   For every $a \in R$, $a \cdot 0 = 0$.

Theorem 1.7   If $a, b \in R$ and $a \cdot b = 0$, then either $a = 0$ or $b = 0$ or both.

Examples
a. If $y + 2 = 3 + 2$, then $y = 3$.   b. If $5z = 5 \cdot 3$, then $z = 3$.
c. $3 \cdot 0 = 0$ d. If $5(x + y) = 0$, then $x + y = 0$.

Combining Theorems 1.6 and 1.7, we see that for $a, b \in R$ we have $ab = 0$ if and only if at least one of the factors is 0.

The following theorem concerns the familiar "laws of signs" for operating with real numbers.

Theorem 1.8   If $a, b \in R$, then

I $-(-a) = a$,

II $(-a) + (-b) = -(a + b)$,

III $(-a)(b) = -(ab)$,

IV $(-a)(-b) = ab$,

V $\dfrac{-a}{b} = \dfrac{a}{-b} = -\dfrac{a}{b} = -\dfrac{-a}{-b}$ $(b \neq 0)$,

VI $\dfrac{-a}{-b} = \dfrac{a}{b}$ $(b \neq 0)$.

Examples
a. $-(-10) = 10$ b. $(-1) + (-3) = -(1 + 3) = -4$
c. $(-2) \cdot 3 = -6$ d. $(-2)(-4) = 2(4) = 8$
e. $\dfrac{-6}{3} = \dfrac{6}{-3} = -\dfrac{-6}{-3} = -\dfrac{6}{3} = -2$ f. $\dfrac{-8}{-2} = \dfrac{8}{2} = 4$

Theorem 1.9   If $a, b, c \in R$, then

$$\frac{a}{b} = \frac{c}{d} \text{ if and only if } ad = bc \quad (b, d \neq 0).$$

As a direct consequence of this characterization of equal quotients, we have a theorem that is sometimes referred to as the **fundamental principle of fractions**.

Theorem 1.10   If $a, b, c \in R$, then

$$\frac{ac}{bc} = \frac{a}{b} \text{ and } \frac{a}{b} = \frac{ac}{bc} \quad (b, c \neq 0).$$

Examples
a. $\dfrac{4}{12} = \dfrac{1 \cdot 4}{3 \cdot 4} = \dfrac{1}{3}$ b. $\dfrac{3}{5} = \dfrac{3 \cdot 2}{5 \cdot 2} = \dfrac{6}{10}$

Finally, let us group a number of assertions about quotients into a single theorem.

1.3 Field Properties

Theorem 1.11 If $a, b, c, d \in R$, then

I $\dfrac{1}{a} \cdot \dfrac{1}{b} = \dfrac{1}{ab}$ $(a, b \neq 0)$,

II $\dfrac{a}{b} \cdot \dfrac{c}{d} = \dfrac{ac}{bd}$ $(b, d \neq 0)$,

III $\dfrac{a}{c} + \dfrac{b}{c} = \dfrac{a+b}{c}$ $(c \neq 0)$,

IV $\dfrac{a}{b} + \dfrac{c}{d} = \dfrac{ad+bc}{bd}$ $(b, d \neq 0)$,

V $\dfrac{a}{b} - \dfrac{c}{d} = \dfrac{ad-bc}{bd}$ $(b, d \neq 0)$,

VI $\dfrac{1}{\frac{a}{b}} = \dfrac{b}{a}$ $(a, b \neq 0)$,

VII $\dfrac{\frac{a}{b}}{\frac{c}{d}} = \dfrac{a}{b} \div \dfrac{c}{d} = \dfrac{ad}{bc}$ $(b, c, d \neq 0)$.

Examples

a. $\dfrac{1}{2} \cdot \dfrac{1}{5} = \dfrac{1}{2 \cdot 5} = \dfrac{1}{10}$

b. $\dfrac{2}{3} \cdot \dfrac{5}{7} = \dfrac{2 \cdot 5}{3 \cdot 7} = \dfrac{10}{21}$

c. $\dfrac{2}{7} + \dfrac{3}{7} = \dfrac{2+3}{7} = \dfrac{5}{7}$

d. $\dfrac{1}{5} + \dfrac{3}{8} = \dfrac{1 \cdot 8 + 5 \cdot 3}{5 \cdot 8} = \dfrac{8+15}{40} = \dfrac{23}{40}$

e. $\dfrac{\frac{1}{5}}{\frac{8}{5}} = \dfrac{8}{5}$

f. $\dfrac{\frac{2}{3}}{\frac{7}{11}} = \dfrac{2}{3} \div \dfrac{7}{11} = \dfrac{2 \cdot 11}{3 \cdot 7} = \dfrac{22}{21}$

Exercise 1.3

In Exercises 1–20, each statement is justified by one part of Theorems 1.1–1.11. Cite an appropriate justification. All variables denote elements of the set R of real numbers.

Examples a. $-(x + (-y)) = -x + (-(-y))$ b. If $x - 3 = 1$, then $(x - 3) + 3 = 1 + 3$.

Solutions a. Theorem 1.8-II. b. Theorem 1.1.

1. $\dfrac{x}{3} \cdot \dfrac{y}{5} = \dfrac{xy}{15}$

2. If $5(xy) = 15$, then $xy = 3$.

3. If $x + 5 = 15$, then $x = 10$.

4. $\left(-\dfrac{1}{3}\right)\left(\dfrac{1}{5}\right) = -\dfrac{1}{15}$

5. If $3(x + y) = 0$, then $x + y = 0$.

6. If $p + \dfrac{1}{2} = 0$, then $p = -\dfrac{1}{2}$.

7. $\dfrac{2}{5} \div \dfrac{3}{7} = \dfrac{2 \cdot 7}{5 \cdot 3}$

8. $\dfrac{a}{2} + \dfrac{b}{3} = \dfrac{3a + 2b}{6}$

9. $\dfrac{xy}{xz} = \dfrac{y}{z}$

10. $x - (-y) = x + y$

11. If $3x = 5y$, then $\dfrac{x}{5} = \dfrac{y}{3}$.

12. $(-8)(-2) = 16$

13. $\dfrac{a+b}{2} + \dfrac{c}{3} = \dfrac{3(a+b) + 2c}{6}$

14. If $x - 2 = 5$, then $(x - 2) + 2 = 7$.

15. If $\dfrac{x}{2} = 5$, then $x = 10$.

16. $\dfrac{1}{x} + \dfrac{1}{y} = \dfrac{y+x}{xy}$

17. If $\dfrac{x}{5} = \dfrac{3}{10}$, then $10x = 15$.

18. If $10x = 15$, then $x = \dfrac{15}{10}$.

19. $\dfrac{2x+1}{2} + \dfrac{x}{3} = \dfrac{3(2x+1) + 2x}{6}$

20. $\dfrac{\frac{2x+1}{2}}{\frac{x}{3}} = \dfrac{(2x+1) \cdot 3}{2x}$

1.4 Order in R

There is a one-to-one correspondence between the real numbers and the points on a geometric line (to each real number there corresponds one and only one point on the line, and vice versa); this fact is formalized in the Appendix. To illustrate this, we imagine the line scaled in convenient units, with the positive direction (from 0 toward 1) denoted by an arrowhead. The line is then called a **number line**; the real number corresponding to a point on the line is called the **coordinate** of the point, and the point is called the **graph** of the number. For example, number-line representations of 1, 3, and 5 are shown in Figure 1.2. Figure 1.2 may also be interpreted as the graph of the set {1, 3, 5}.

Figure 1.2

A horizontal number-line graph directed to the right can be used to illustrate the separation of the real numbers into three disjoint subsets: {negative real numbers}, {0}, {positive real numbers}. The point associated with 0 is called the **origin**. The set of numbers whose elements are associated with the points on the right-hand side of the origin belong to the set R_+ of **positive real numbers**, and the set whose elements are associated with the points on the left-hand side belong to the set R_- of **negative real numbers**.

1.4 Order in R

Notice that the word "negative" has now been used in two ways. In one case, we refer to the *negative of a number*, as in Postulate F-4, whereas, in the other, we refer to a *negative number*, which is the negative of a positive number.

Order postulates

It is possible to categorize the set of positive real numbers without recourse to geometric considerations, though of course we shall continue to find it convenient to refer also to the number line. With this in mind, let us state two more postulates that apply to real numbers.

O-1 If a is a real number, then exactly one of the following statements is true: a is positive, a is zero, or $-a$ is positive.   *Trichotomy law.*

O-2 If a and b are positive real numbers, then $a + b$ is positive and ab is positive.   *Closure law for positive numbers.*

The first of these postulates asserts that every real number belongs to one of the sets R_+, $\{0\}$, or R_-, but to only one of them. The second asserts that the set R_+ of positive real numbers is closed with respect to the binary operations of addition and multiplication.

Since the set R of real numbers satisfies Postulates O-1 and O-2 as well as Postulates F-1 through F-11, we say that R is an **ordered field**. Similarly, the set Q of rational numbers is an ordered field.

By Postulate O-2, if a and b are positive real numbers, then ab is a positive real number. This fact, together with Parts III and IV of Theorem 1.8, is sufficient to establish that the product of a positive real number and a negative real number is a negative real number, while the product of two negative real numbers is a positive real number.

Less than and greater than

The addition of a positive real number d to a real number a can be visualized on a number-line graph as the process of locating the point corresponding to a on the line and then moving along the line d units to the right to arrive at the point corresponding to $a + d$ (Figure 1.3). With this idea in mind, we define what is meant by "less than."

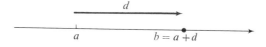

Figure 1.3

Definition 1.10 If $a, b \in R$, then a is **less than** b if and only if there exists a positive real number d such that $a + d = b$.

The fact that d is positive implies, for any two real numbers a and b, that if the graph of a lies to the left of the graph of b, then a is less than b. The inequality symbol $<$ is used to denote the phrase "is less than," and $a < b$ is read "a is less than b." The inequality symbol $>$ means "is greater than." The statements $a < b$ and $b > a$ are taken as equivalent.

Definition 1.10 and the field postulates have the following implications, which we shall not prove.

Theorem 1.12 For any $a, b, c \in R$:

I   If $a < b$ and $b < c$, then $a < c$.   II   If $a < b$, then $a + c < b + c$.

III   If $a < b$ and $c > 0$, then $ac < bc$.   IV   If $a < b$ and $c < 0$, then $ac > bc$.

Examples

a. $2 < 5$ and $5 < 7$, so $2 < 7$
b. $3 < 8$, so $3 + 5 < 8 + 5$, or $8 < 13$
c. $3 < 6$ and $2 > 0$, so $2 \cdot 3 < 2 \cdot 6$, or $6 < 12$
d. $2 < 7$ and $-2 < 0$, so $-2(2) > -2(7)$, or $-4 > -14$

As you can see, Theorem 1.12-I is comparable to the transitive law for equality. That is, we can say that "less than" is a transitive relationship.

Each of the symbols $\leq$ and $\geq$ (read "is less than or equal to" and "is greater than or equal to," respectively) is a contraction for two symbols, one of equality and one of inequality, connected by the word "or." For example, $x \leq 7$ is the statement that x is less than 7 or x equals 7.

We often write together two inequalities that express a transitive relationship. Thus $a < b$ and $b < c$ are written together as $a < b < c$ and read "a is less than b and b is less than c." Similarly, $0 < x$ and $x \leq 1$ are written together as $0 < x \leq 1$.

Absolute value The graphs of the numbers a and $-a$ on a number line lie the same distance from the origin, but on opposite sides of it. If we wish to refer to the *distance* of the graph of a number from the origin, and not to the side of the origin on which it is located, then we use the term **absolute value**. Thus, the absolute value of a and the absolute value of $-a$ are the same nonnegative number. The symbol $|a|$ is used to denote the absolute value of a. We formalize the definition as follows:

Definition 1.11 If $a \in R$, then the **absolute value** of a is given by

$$|a| = \begin{cases} a, & \text{if } a \geq 0, \\ -a, & \text{if } a < 0. \end{cases}$$

For example, $|-3| = -(-3) = 3$, $|7| = 7$, and $|0| = 0$.

We have noted in the foregoing discussion that certain algebraic statements concerning the order of real numbers can be interpreted geometrically. We summarize some of the more common correspondences in the following table, where in each case $a, b, c \in R$.

1.4 Order in R

Algebraic statement	Geometric statement
1. a is positive	1. The graph of a lies to the right of the origin.
2. a is negative	2. The graph of a lies to the left of the origin.
3. $a > b$	3. The graph of a lies to the right of the graph of b.
4. $a < b$	4. The graph of a lies to the left of the graph of b.
5. $a < c < b$	5. The graph of c is to the right of the graph of a and to the left of the graph of b.
6. $|a| < c$	6. The graph of a is less than c units from the origin.
7. $|a - b| < c$	7. The graphs of a and b are less than c units from each other.
8. $|a| < |b|$	8. The graph of a is closer to the origin than the graph of b.

Exercise 1.4

For $x, y \in R$, justify the given statement by citing one part of Theorem 1.12.

Example   If $-2x > 1$, then $x < -\dfrac{1}{2}$.

Solution   Part IV. Each member of $-2x > 1$ is multiplied by $-\dfrac{1}{2}$.

1. If $x - y > 0$, then $x > y$.
2. If $x < 5$, then $2x < 10$.
3. If $x < y$ and $y < 5$, then $x < 5$.
4. If $-x > 1$, then $x < -1$.
5. If $3x < 18$, then $x < 6$.
6. If $x + 3 < 7$, then $x < 4$.
7. If $-2 < -x$, then $x < 2$.
8. If $2 < x$ and $x < z$, then $2 < z$.

Express the given statement by means of the symbols $<$, $\leq$, $>$, and $\geq$.

Examples   a. x is at least 7   b. x is strictly between 1 and 3

Solutions   a. $x \geq 7$ b. $1 < x < 3$

9. x is greater than 3
10. x is less than 3
11. x is no greater than 3
12. x is no smaller than 3
13. x is at least as great as 3
14. x is at most 3
15. x is between 1 and 3 inclusive
16. x is not greater than 3
17. x is at least 1 and less than 3
18. x is greater than 1 but no greater than 3

For x, y ∈ R rewrite the given expression without using absolute value notation.

Examples a. |−13|
   b. |x − 1|

Solutions
a. Since −13 is negative,
   |−13| = −(−13) = 13.

b. $|x - 1| = \begin{cases} x - 1 & \text{if } x - 1 \geq 0 \\ -(x - 1) & \text{if } x - 1 < 0 \end{cases}$

so $|x - 1| = \begin{cases} x - 1 & \text{if } x \geq 1 \\ 1 - x & \text{if } x < 1 \end{cases}$

19. |−(−11)| 20. |−11| 21. |11| 22. −|−11|
23. −|11| 24. |x| 25. |−x| 26. |1 − x|
27. −|1 − x| 28. −|1 + x| 29. |1 + x| 30. |x²|

From each of the following lists of numbers, select the greatest.

31. −1, |−2|, 0 32. |−3|, |−4|, 2
33. |−3|, 4, |2| 34. 3, |−4|, 2
35. |−1|, −2, 0 36. 1, −|2|, 0

1.5 Complex Numbers

In subsequent chapters we will be interested in solving equations, but as you know some equations do not have solutions in the set of real numbers. For example, if $b > 0$ then $x^2 = -b$ has no real-number solution, because there is no real number whose square is negative. In this section, we wish to consider a set of numbers containing solutions to $x^2 = -b$ and also containing a subset that can be identified with the set of real numbers. We shall see that this new set of numbers C, called the **complex numbers**, will provide solutions for all polynomial equations with real coefficients. To construct the set of complex numbers from the real numbers, we introduce a new symbol i to represent a specific complex number and then make the following definitions.

Definition 1.12 An expression of the form $a + bi$, where $a, b \in R$, is a **complex number**.

Definition 1.13 The complex numbers $a + bi$ and $c + di$ are **equal** if and only if $a = c$ and $b = d$.

Examples The expressions

$$\frac{1}{3} + 5i, \quad 1 + \sqrt{2}\,i, \quad 0 + 1i, \quad \frac{1}{2} + 0i, \quad 5 + (-1)i$$

are all complex numbers. The equality $a + bi = 1 + 7i$ holds if and only if $a = 1$ and $b = 7$.

1.5 Complex Numbers

Complex numbers are added and multiplied according to the following rules. These definitions may appear arbitrary and unusual to start with (particularly for products) but we will see that they relate to the corresponding operations in R in a direct and useful way. As customary, we use the variable z for complex numbers.

Definition 1.14 If $z_1 = a + bi$ and $z_2 = c + di$, then

I $\quad z_1 + z_2 = (a + c) + (b + d)i$,

II $\quad z_1 \cdot z_2 = (ac - bd) + (ad + bc)i$.

Examples

a. $(2 + 3i) + (6 + 1i) = (2 + 6) + (3 + 1)i = 8 + 4i$
b. $(2 + 3i)(6 + 1i) = (2 \cdot 6 - 3 \cdot 1) + (2 \cdot 1 + 3 \cdot 6)i = 9 + 20i$

For convenience, we write

a for $a + 0i$, $\quad bi$ for $0 + bi$, $\quad a - bi$ for $a + (-b)i$, $\quad i$ for $0 + 1i$.

With a identified with $a + 0i$, every real number is contained in C. Moreover, addition and multiplication of real numbers in C are the same as they are in R. For instance,

$$3 + 5 = (3 + 0i) + (5 + 0i)$$
$$= (3 + 5) + (0 + 0)i$$
$$= 8 + 0i = 8,$$

and

$$3 \cdot 5 = (3 + 0i) \cdot (5 + 0i)$$
$$= (3 \cdot 5 - 0 \cdot 0) + (3 \cdot 0 + 0 \cdot 5)i$$
$$= 15 + 0i = 15.$$

With $0 + 1i$ written as i, the special role of this specific complex number becomes apparent. For

$$i^2 = (0 + 1i) \cdot (0 + 1i)$$
$$= (0 \cdot 0 - 1 \cdot 1) + (0 \cdot 1 + 1 \cdot 0)i$$
$$= -1 + 0i = -1,$$

that is,

$$i^2 = -1.$$

Also, for positive real numbers b,

$$(\sqrt{b}\, i)^2 = b(-1) = -b.$$

Thus negative real numbers $-b$ have square roots in our enlarged number system C. In fact you can check that $(-\sqrt{b}i)^2$ is also equal to $-b$, so each nonzero real number has two square roots in C.

Because $i^2 = -1$, i is often written as $\sqrt{-1}$; similarly for real $b > 0$,

$$\sqrt{-b} = \sqrt{b}\, i = i\sqrt{b}.$$

The symbol $\sqrt{-b}$ must be used with care, however, since certain relationships involving square roots which are valid for real numbers are not valid for complex numbers. For example,

$$\sqrt{-2} \cdot \sqrt{-3} = (i\sqrt{2})(i\sqrt{3})$$
$$= i^2\sqrt{6} = -\sqrt{6} \neq \sqrt{(-2)(-3)} = \sqrt{6}.$$

For this reason, complex numbers should be written in the form $a + bi$ before making computations.

Making use of the fact that $i^2 = -1$, we can multiply complex numbers just as we multiply real polynomial expressions.

Examples

a. $i(2 - 3i) = 2i - 3i^2$
$= 2i - 3(-1)$
$= 3 + 2i$

b. $(2 + 3i)(6 + 1i) = 12 + 18i + 2i + 3i^2$
$= 12 + 20i - 3$
$= 9 + 20i$

The complex numbers $a + bi$ with $b \neq 0$ are called **imaginary numbers**; the imaginary numbers bi with $a = 0$ are called **pure imaginary numbers**. Thus, our number systems are related as shown in Figure 1.4.

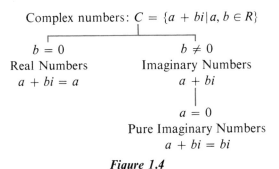

Figure 1.4

The complex numbers, with addition and multiplication defined as in Definition 1.14, satisfy all the field postulates of the set R on page 8. Hence the complex numbers are a field. Since Theorems 1.1 to 1.11 can be proved using only the field postulates, these theorems are also true for the complex numbers. The complex numbers cannot be ordered, however, in such a way as to satisfy the order properties O-1 and O-2.

The operation of subtraction for the field C can be defined just as it was for R:

$$z_1 - z_2 = z_1 + (-z_2).$$

In the usual notation,

$$(a + bi) - (c + di) = (a - c) + (b - d)i.$$

Example

$(2 + 3i) - (5 + 6i) = (2 - 5) + (3 - 6)i = -3 - 3i$

1.5 Complex Numbers

The quotient of two complex numbers can be found by using the following theorem, which is analogous to Theorem 1.10 (fundamental principle of fractions) in the set of real numbers.

Theorem 1.13 *If z_1, z_2, and z_3 are elements of C, and z_2 and z_3 are not zero, then*

$$\frac{z_1}{z_2} = \frac{z_1 z_3}{z_2 z_3}.$$

The notion of the *conjugate* of a complex number is also used in rewriting quotients with imaginary denominators.

Definition 1.15 *The **conjugate** of $z = a + bi$, denoted by $\bar{z}$, is $a - bi$.*

Examples
a. The conjugate of $2 + 3i$ is $2 - 3i$.
b. The conjugate of $-3 - i$ is $-3 + i$.

The quotient $(a + bi)/(c + di)$ can be written in standard form by using Theorem 1.13 to multiply the numerator and the denominator by $c - di$, the conjugate of the denominator.

Examples

a. $\dfrac{4+i}{2+3i} = \dfrac{(4+i)(2-3i)}{(2+3i)(2-3i)} = \dfrac{8 - 10i - 3i^2}{4 - 9i^2}$

$= \dfrac{8 - 10i + 3}{4 + 9} = \dfrac{11}{13} - \dfrac{10}{13}i$

b. $\dfrac{5}{-3-i} = \dfrac{5(-3+i)}{(-3-i)(-3+i)} = \dfrac{-15 + 5i}{9 - i^2}$

$= \dfrac{-15 + 5i}{9 + 1} = -\dfrac{15}{10} + \dfrac{5}{10}i = -\dfrac{3}{2} + \dfrac{1}{2}i$

Exercise 1.5

Write each expression in the form $a + bi$.

Examples
a. $(2 + i) + (3 - 2i)$
b. $3i - (4 + i)$

Solutions
a. $(2 + i) + (3 - 2i)$
$= (2 + 3) + (1 - 2)i$
$= 5 - i$

b. $3i - (4 + i)$
$= (0 + 3i) - (4 + 1i)$
$= (0 - 4) + (3 - 1)i$
$= -4 + 2i$

1. $(5 + i) + (-4 - 2i)$
2. $(1 + i) + i$
3. $(3 - i) + 6$
4. $13 + (3 - 10i)$
5. $\left(4 - \frac{1}{2}i\right) + 3i$
6. $(6 + i) + (1 + 6i)$
7. $(5 - i) - (5 + i)$
8. $(6 + i) - (1 + 6i)$
9. $\left(\frac{1}{3} + 2i\right) - (1 - i)$
10. $\left(\frac{1}{2} - i\right) - (2 + i)$
11. $(3 + 4i) - (5 - 2i)$
12. $(3 - i) - (7 - 2i)$

Examples

a. $(1 + 3i)(2 - i)$

b. $\dfrac{1 + 3i}{2 - i}$

Solutions

a. $(1 + 3i)(2 - i)$
$= 2 + 5i - 3i^2$
$= 2 + 5i + 3 = 5 + 5i$

b. $\dfrac{1 + 3i}{2 - i} = \dfrac{(1 + 3i)(2 + i)}{(2 - i)(2 + i)}$
$= \dfrac{2 + 7i + 3i^2}{4 - i^2}$
$= \dfrac{-1 + 7i}{5} = -\dfrac{1}{5} + \dfrac{7}{5}i$

13. $(2 + i)(3 - i)$
14. $(1 - i)(5 + i)$
15. $\dfrac{2 + i}{3 - i}$
16. $\dfrac{1 - i}{5 + i}$
17. $(4 - i)(1 + 2i)$
18. $(3 + 4i)(1 + i)$
19. $\dfrac{4 - i}{1 + 2i}$
20. $\dfrac{3 + 4i}{1 + i}$
21. $i(3 + i)$
22. $3(2 + i)$
23. $\dfrac{i}{3 + i}$
24. $\dfrac{3}{2 + i}$

Examples

a. $1 + \sqrt{-2} + \sqrt{-4}$

b. $(1 + \sqrt{-2})(2 + \sqrt{-3})$

Solutions

a. $1 + \sqrt{-2} + \sqrt{-4}$
$= 1 + \sqrt{2}i + \sqrt{4}i$
$= 1 + (\sqrt{2} + 2)i$

b. $(1 + \sqrt{-2})(2 + \sqrt{-3})$
$= (1 + \sqrt{2}i)(2 + \sqrt{3}i)$
$= 2 + (2\sqrt{2} + \sqrt{3})i + \sqrt{6}i^2$
$= (2 - \sqrt{6}) + (2\sqrt{2} + \sqrt{3})i$

25. $(1 + \sqrt{-9}) + (3 - \sqrt{-2})$
26. $(2 - \sqrt{-4}) + (1 + \sqrt{-3})$

Chapter Review

27. $(1 + \sqrt{-9})(3 - \sqrt{-2})$
28. $(2 - \sqrt{-4})(1 + \sqrt{-3})$
29. $\dfrac{1 + \sqrt{-9}}{3 - \sqrt{-2}}$
30. $\dfrac{2 - \sqrt{-4}}{1 + \sqrt{-3}}$ Answer 20
31. $\sqrt{-16} + \sqrt{-12}$
32. $\sqrt{-18} + \sqrt{-25}$

Examples   a. i^3 b. $\dfrac{1}{i^3}$

Solutions   a. $i^3 = i^2 \cdot i = -1 \cdot i = -i$ b. $\dfrac{1}{i^3} = \dfrac{1}{-i} = \dfrac{1 \cdot i}{-i \cdot i} = i$

33. i^4 34. i^5 35. i^7 36. i^6

37. $\dfrac{i}{i^5}$ 38. $\dfrac{i}{i^4}$ 39. $\dfrac{i}{i^6}$ 40. $\dfrac{i}{i^7}$

Chapter Review

[1.1] Let $A = \{2, 3, 5, 7, 11\}$, $B = \{3, 6, 9, 12\}$, and $U = \{1, 2, 3, \ldots, 11, 12\}$. *List the elements of the following sets.*

1. A'
2. $A' \cup B'$
3. $A \cup B$
4. $A \cap B'$
5. $A \cap (B \cup \emptyset)$
6. $A' \cup (B \cap \emptyset)$

State whether each of the following is true or false.

7. $A' \neq \{1, 4, 6, 8, 10\}$
8. $B \subset \{x \mid x \in U, \ x \geq 3\}$
9. $9 \in A \cap B'$
10. $\{x \mid x \in U, \ x > 1\} \subset A$
11. $3 \notin A \cup B$
12. $B = \{x \mid x \in U, \ x \text{ is a multiple of } 3\}$

[1.2] Let $A = \left\{-2, 0, 2, -\dfrac{1}{2}, \dfrac{1}{2}, \sqrt{2}, -\sqrt{2}, \sqrt{-2}, \dfrac{1}{\sqrt{2}}, \dfrac{1}{\sqrt{-2}}\right\}$. *List the elements of the following sets.*

13. {natural numbers in A}
14. {integers in A}
15. {rational numbers in A}
16. {irrational numbers in A}
17. {imaginary numbers in A}
18. {positive real numbers in A}

For $a, b, c, d \in R$, justify each statement by citing an appropriate field postulate.

19. $ab + ac = a(b + c)$
20. $ab + ac = ac + ab$
21. $ab + ac = ab + ca$
22. $a(b + c) = (b + c)a$

[1.3] For $a, b, c \in R$, justify each statement by citing a statement from Theorems 1.1–1.11.

23. If $a + 2 = 7$, then $a = 5$.
24. $(-a)(-b) = ab$
25. $2 \div \dfrac{3}{4} = \dfrac{8}{3}$
26. $-(a - 2) = -a + 2$
27. $\dfrac{a}{2} + \dfrac{b}{3} = \dfrac{3a + 2b}{6}$
28. If $(a + b)c = 1$, then $c = \dfrac{1}{a + b}$.

[1.4] For $a, b \in R$ justify each statement by citing part of Theorem 1.12.

29. If $a < b$, then $3a < 3b$.
30. If $-a < b$, then $a > -b$.
31. If $a < 5$, then $a - 1 < 4$.
32. If $a + 1 < 3b$, then $\dfrac{a + 1}{3} < b$.

Express each of the following using inequality symbols.

33. a is less than b
34. a is between 3 and 7 inclusive
35. a is not less than b
36. a is at least 2

Rewrite these expressions without using absolute values.

37. $|-5|$
38. $|x - 5|$

[1.5] Write each expression in the form $a + bi$.

39. $(3 + i) + (1 - 3i)$
40. $(3 + i) - (1 - 3i)$
41. $(3 + i) \cdot (1 - 3i)$
42. $\dfrac{3 + i}{3 - i}$
43. $(1 + i)^2$
44. $1 - \sqrt{-16}$
45. i^9
46. $\dfrac{\sqrt{-4}}{\sqrt{-9}}$

2 Polynomials; Rational Expressions

In this chapter we investigate the simplest algebraic expressions—polynomials and rational expressions. We consider their behavior under the basic arithmetic operations of addition, subtraction, multiplication, and division, and discuss simplification of the results of these operations.

2.1 Definitions; Sums of Polynomials

Any grouping of constants and variables generated by applying a finite number of the elementary operations—addition, subtraction, multiplication, division, or the extraction of roots—is called an **algebraic expression.** For example,

$$\frac{3x^2 + \sqrt{2x-1}}{3x^3 + 7} \quad \text{and} \quad xy + 3x^2z - \sqrt[5]{z}$$

are algebraic expressions. If two expressions have equal values for all values of the variables for which both expressions are defined, then we say that on the set of these values the expressions are **equivalent.**

Natural-number powers

You should recall that an expression of the form a^n is called a **power** of a, where a is the **base** of the power and n is the **exponent** of the power.

Definition 2.1 *If $n \in N$ and $a \in R$, then*
$$a^n = \underbrace{a \cdot a \cdot a \cdot \cdots \cdot a.}_{n \text{ factors}}$$

Polynomial expressions

In any algebraic expression of the form $A + B + C + \cdots$, where $A, B, C, \ldots$ are algebraic expressions, $A, B, C, \ldots$ are called **terms** of the expression. For example, in $x + (y + 3)$ the terms are x and $(y + 3)$, but in $x + y + 3$ the terms are x, y, and 3. An algebraic expression that contains only nonnegative integral powers of any variable and contains no variable in a denominator is a **polynomial**. For example,

$$5x, \quad \frac{3x^2}{2} - \frac{7x}{2}, \quad 0, \quad 2x^2 - 3x + 4, \quad \text{and} \quad \frac{x}{4} - \frac{\sqrt{7}}{4}$$

are polynomials in the variable x, whereas

$$\frac{3}{x}, \quad 3 + \sqrt{x}, \quad \text{and} \quad \frac{2\sqrt{x-1}}{2\sqrt{x+1}}$$

are algebraic expressions, but not polynomials, in the variable x.

Polynomials consisting of one, two, or three terms are also called **monomials, binomials,** and **trinomials,** respectively. Thus $3x^2y$ is a monomial, $x + 4x^2$ is a binomial, and $x + y + z$ is a trinomial.

The **degree** of a monomial is given by the exponent of the variable in the monomial. Thus, 5 is of degree zero (in Chapter 3 we shall define x^0 to be equal to 1 for all $x \in R$, $x \neq 0$), $2x$ is of the first degree, and $3x^4$ is of fourth degree; but no degree is assigned to the special monomial 0. If a monomial contains more than one variable, its degree is given by the sum of the exponents on the variables; $3x^2y^3z$ is of sixth degree in x, y, and z. It can also be described as being of second degree in x, third degree in y, fifth degree in x and y, and so on. The constant factor in the monomial, 3, is called the **coefficient** of the monomial. The **degree of a polynomial** is the same as the degree of its term of highest degree. Since no degree is assigned to the monomial 0, no degree is assigned to the polynomial 0, either.

Because $a - b$ is defined to be $a + (-b)$, we shall view the signs in any polynomial as signs denoting numbers or their negatives, and the operation involved as addition. Thus

$$3x - 5y + 4z = 3x + (-5y) + 4z,$$

and its terms are $3x$, $-5y$, and $4z$. Again, an expression such as

$$a - (bx + cx^2),$$

in which a set of parentheses is preceded by a negative sign, can be written

$$a + [-(bx + cx^2)],$$

or

$$a + [-bx - cx^2],$$

or, finally,
$$a + (-bx) + (-cx^2).$$
Also, since
$$\frac{a}{b} = a\left(\frac{1}{b}\right),$$
we can view division by a constant as multiplication by its reciprocal (multiplicative inverse), and, for example, write
$$\frac{3x^2}{4} + \frac{x}{2} \quad \text{as} \quad \frac{3}{4}x^2 + \frac{1}{2}x.$$

Accordingly, since a polynomial can be considered to involve only the operations of addition and multiplication, and since the set R of real numbers is closed with respect to these operations, it follows that, for any specific real value of x, a polynomial with real coefficients represents a real number. Therefore, the postulates for the real numbers are applicable to the terms in such polynomials and to the polynomials themselves.

Simplification of polynomials

By applying the commutative, associative, and distributive laws in various ways, we can frequently rewrite polynomials and sums of polynomials in what might be termed "simpler" forms.

Example

$$(2x^2 + 3x + 5) + 2x + (6x^2 + 7) = 2x^2 + 6x^2 + 3x + 2x + 5 + 7$$
$$= (2 + 6)x^2 + (3 + 2)x + 5 + 7$$
$$= 8x^2 + 5x + 12$$

Here we have reduced the number of terms from six to three, and the original expression is said to have been "simplified."

We shall often be concerned with polynomials in one variable. A polynomial of degree n, $n \geq 0$, in x can be represented—when its terms are rearranged, if need be—by an expression in the **standard form**

$$a_0 x^n + a_1 x^{n-1} + a_2 x^{n-2} + \cdots + a_{n-1} x + a_n \quad (a_0 \neq 0),$$

where it is understood that the a's are the (constant) coefficients of the powers of x in the polynomial.

The term $a_0 x^n$ is called the **leading term,** and the coefficient a_0 is called the **leading coefficient** in the polynomial. It is often convenient to have the leading term of the form x^n. In this case (that is, when $a_0 = 1$), the polynomial is said to be **monic.**

If the coefficients in a polynomial P are real numbers, then the polynomial is called a **polynomial over the real-number field,** or simply a **polynomial over R.** If the variable is restricted to represent only real numbers, then the polynomial is said to be a **polynomial in the real variable** x. If *both* the coefficients and the variable are restricted to real values, we say that the polynomial is a **real polynomial.**

Symbols for polynomials

Polynomials are frequently represented by symbols such as

$$P(x), \quad D(y), \quad \text{and} \quad Q(z),$$

where the symbol in parentheses designates the variable. Thus, we might write

$$P(x) = 2x^3 - 3x + 2,$$
$$D(y) = y^6 - 2y^2 + 3y - 2,$$
$$Q(z) = 8z^4 + 3z^3 - 2z^2 + z - 1.$$

Value of a polynomial

The notation $P(x)$ can be used to denote values of the polynomial for specific values of x. Thus, $P(2)$ means the value of the polynomial $P(x)$ when x is replaced by 2. For example, if

$$P(x) = x^2 - 2x + 1,$$

then

$$P(2) = 2^2 - 2(2) + 1 = 1,$$
$$P(3) = 3^2 - 2(3) + 1 = 4,$$
$$P(-4) = (-4)^2 - 2(-4) + 1 = 25.$$

In some applications, the notation $P(x)|_a^b$ denotes $P(b) - P(a)$. Thus if

$$P(x) = \frac{x^2}{2} - 4x,$$

then

$$P(x)\Big|_2^3 = P(3) - P(2) = \left[\frac{3^2}{2} - 4(3)\right] - \left[\frac{2^2}{2} - 4(2)\right] = -\frac{3}{2}.$$

Exercise 2.1

Write each polynomial in standard form.

Example $(x^2 - 3x + 1) + (2x^2 - x) - (x^2 - 3x + 4)$

Solution
$(x^2 - 3x + 1) + (2x^2 - x) - (x^2 - 3x + 4)$
$= x^2 - 3x + 1 + 2x^2 - x - x^2 + 3x - 4$
$= (x^2 + 2x^2 - x^2) + (-3x - x + 3x) + (1 - 4)$
$= 2x^2 - x - 3$

1. $(x^2 + 5x - 4) + (x^2 - 4x + 4) - (x^2 - 1)$
2. $(x^3 + 2x^2 + x) + (x^3 - x^2 + 2x) + (x^2 - 1)$
3. $(x^2 - 7x - 10) - (3x^2 - x + 5) - (2x^2 - 2x - 2)$
4. $(x^3 - x + 4) - (x^2 - 3x) - (2x^3 - 3x^2 + 4x - 5)$

2.1 Definitions; Sums of Polynomials

5. $(x^3 - x^2 + x - 1) + (3x^2 - 2x + 1) - 2x^3$
6. $(x^5 - x^3 + 1) + (x^5 - 2x^3 + 3)$
7. $\left(\dfrac{1}{2}x^4 + 2x^2 + 1\right) + \left(\dfrac{1}{3}x^3 - x\right) - \left(\dfrac{1}{4}x^4 + \dfrac{1}{6}x^3\right)$
8. $\left(\dfrac{2}{3}x^2 - x + \dfrac{3}{5}\right) - \left(\dfrac{1}{2}x^2 + x - \dfrac{1}{5}\right) + \dfrac{1}{10}$
9. $(x^2 + 3x + 2) - [(x^2 + x + 1) - (2x^2 + 6x + 3)]$
10. $(x^3 - 1) - [(x^2 + 2x + 1) - (x^3 + 3x^2 + 3x + 1)]$

Example $x - [2x - (3x - 1)]$

Solution $x - [2x - (3x - 1)] = x - [2x - 3x + 1]$
$= x - [-x + 1]$
$= x + x - 1$
$= 2x - 1$

(Note how inner grouping symbols are removed first.)

11. $2x - [3 + x - (4x - 5)]$
12. $4x - [(x + 2) - (3x + 1)]$
13. $x^2 - 1 - [x - (2x^2 - 3x + 1)]$
14. $2x^2 - x + 5 - [x^2 - x - (2x^2 - 3x + 1)]$
15. $x - (2x - (3x - (4x - 1)))$
16. $x^2 - (1 - (1 - x^2))$

Simplify each polynomial by combining like terms.

Example $(3x^2 + 2xy - y^2) - (x^2y - xy + x^2)$

Solution $(3x^2 + 2xy - y^2) - (x^2y - xy + x^2) = 3x^2 + 2xy - y^2 - x^2y + xy - x^2$
$= (3x^2 - x^2) + (2xy + xy) - y^2 - x^2y$
$= 2x^2 + 3xy - y^2 - x^2y$

17. $(2x^2y + xy^2 + y^3) - (x^3 + xy^2 - 3x^2y) - (y^3 - x^3)$
18. $(2xy + x^2 - y^2) + (xy - x^2) - (y^2 - 3xy)$
19. $(x^2yz + xy^2z) - (2x^3y + 3x^2yz) + (x^3y + xy^2z)$
20. $(x^2y + x^2z) - (xyz + 2x^2z + 3xz^2) - (x^2y - xyz + xz^2)$

Let $P(x) = x^2 - 2x - 3$, $Q(x) = 2x^2 - x + 1$, and $R(x) = -x^2 + 3x + 4$. Write each polynomial in standard form.

Example $P(x) + Q(x)$

Solution
$$P(x) + Q(x) = (x^2 - 2x - 3) + (2x^2 - x + 1)$$
$$= x^2 - 2x - 3 + 2x^2 - x + 1$$
$$= (x^2 + 2x^2) + (-2x - x) + (-3 + 1)$$
$$= 3x^2 - 3x - 2$$

21. $P(x) - Q(x)$
22. $P(x) + [Q(x) - R(x)]$
23. $P(x) - [Q(x) + R(x)]$
24. $P(x) - [Q(x) - R(x)]$
25. $[P(x) - Q(x)] - R(x)$
26. $Q(x) + Q(x)$
27. $R(x) + R(x)$
28. $[P(x) - P(x)] + [Q(x) - Q(x)]$

Example   Given $P(x) = x^2 - 5x + 4$, find $P(0)$, $P(1)$, $P(2)$, $P(x)|_0^2$.

Solution
$$P(0) = 0^2 - 5 \cdot 0 + 4 = 4$$
$$P(1) = 1^2 - 5 \cdot 1 + 4 = 0$$
$$P(2) = 2^2 - 5 \cdot 2 + 4 = -2$$
$$P(x)|_0^2 = P(2) - P(0) = -2 - 4 = -6$$

29. Given $P(x) = x^3 - x + 2$, find $P(-1)$, $P(0)$, $P(2)$, $P(x)|_{-1}^0$.
30. Given $P(x) = x^5 - 2x^3 + x$, find $P(-1)$, $P(1)$, $P(3)$, $P(x)|_1^3$.
31. Given $P(x) = x^7$, find $P(-1)$, $P(0)$, $P(1)$, $P(x)|_{-1}^1$.
32. Given $P(x) = x^{14}$, find $P(-1)$, $P(0)$, $P(1)$, $P(x)|_{-1}^1$.

Example   Given $P(x) = x^2 + x$ and $Q(x) = x - 1$, find $P(Q(2))$.

Solution $Q(2) = 2 - 1 = 1$; $P(Q(2)) = P(1) = 1 + 1 = 2$

33. Given $P(x) = x + 3$ and $Q(x) = x^2$, find $P(Q(2))$ and $Q(P(-1))$.
34. Given $P(x) = x + 2$ and $Q(x) = x^2 - 1$, find $P(Q(3))$ and $Q(P(0))$.
35. Given $P(x) = x^2 + 2x + 1$ and $Q(x) = x + 3$, find $P(Q(1))$ and $Q(P(3))$.
36. Given $P(x) = x^2 - 3x$ and $Q(x) = x^2 + 1$, find $P(Q(1))$ and $Q(P(3))$.
37. If $P(x)$ is of degree n and $Q(x)$ is of degree $n - 2$, what is the degree of $P(x) + Q(x)$? Of $P(x) - Q(x)$? Of $Q(x) + Q(x)$? Of $Q(x) - Q(x)$?
38. If $P(x)$ and $Q(x)$ are each of degree n, what can be said about the degree of $P(x) + Q(x)$?

2.2 Products of Polynomials

Laws of exponents for natural-number exponents

By Definition 2.1 (page 26), for $a \in R$ and $n \in N$, we have

$$a^n = \underbrace{a \cdot a \cdot a \cdots \cdot a}_{n \text{ factors}}.$$

The product $a^m \cdot a^n$, where m and n are natural numbers, is then given by

$$a^m \cdot a^n = \underbrace{(a \cdot a \cdot a \cdots \cdot a)}_{m \text{ factors}} \underbrace{(a \cdot a \cdot a \cdots \cdot a)}_{n \text{ factors}}$$

$$= \underbrace{a \cdot a \cdot a \cdots \cdot a}_{(m+n) \text{ factors}} = a^{m+n}.$$

We state the result formally, together with two related properties whose proofs are similar to the one above and are omitted.

Theorem 2.1 If $a \in R$ and $m, n \in N$, then

    I $a^m \cdot a^n = a^{m+n}$,

    II $(a^m)^n = a^{mn}$,

    III $(ab)^n = a^n b^n$.

This theorem justifies writing the product of two natural-number powers of the same base as a power (of the same base) with an exponent equal to the sum of the two exponents. For example,

$$x^2 x^3 = x^5,$$
$$y^3 y^4 y^2 = y^9.$$

In rewriting the product of two monomials, say

$$(3x^2 y)(2xy^2),$$

we use the commutative and associative laws and Theorem 2.1 to write

$$3 \cdot 2 \cdot x^2 \cdot x \cdot y \cdot y^2 = 6x^3 y^3.$$

Products of polynomials

The **generalized distributive law,**

$$a(b_1 + b_2 + \cdots + b_n) = ab_1 + ab_2 + \cdots + ab_n,$$

can be applied to write as a polynomial the product of a monomial and a polynomial containing more than one term. For example,

$$3x(x + y + z) = 3xx + 3xy + 3xz = 3x^2 + 3xy + 3xz.$$

The distributive law can be applied successively to the products of polynomials containing more than one term.

Example

$$(3x + 2y)(x - y) = 3x(x - y) + 2y(x - y)$$
$$= 3x^2 - 3xy + 2xy - 2y^2$$
$$= 3x^2 - xy - 2y^2$$

Of course, products of polynomials should ordinarily be simplified mentally if it is convenient to do so. The following binomial products are types so frequently encountered that you should learn to recognize them on sight:

$$(x + a)(x + b) = x^2 + (a + b)x + ab,$$
$$(x + a)^2 = x^2 + 2ax + a^2,$$
$$(x + a)(x - a) = x^2 - a^2.$$

Exercise 2.2

Simplify each product by combining all constants and all powers of each variable.

Examples   a. $(2x^2y)(5xyz^2)$ b. $(3a^k)(2a^{k+3})$

Solutions   a. $(2x^2y)(5xyz^2)$ b. $(3a^k)(2a^{k+3})$
$= 2 \cdot 5 \cdot x^2 \cdot x \cdot y \cdot y \cdot z^2$ $= 3 \cdot 2 \cdot a^k \cdot a^{k+3}$
$= 10x^3y^2z^2$ $= 6a^{k+k+3}$
     $= 6a^{2k+3}$

1. $(-x^2y)(2xy^3)$
2. $(x^4y^2)(3xy^2)$
3. $(2x^2y)(-3x^2y^2)(xy^3)$
4. $(-2x^2)(xy^3)(-3xy)$
5. $x^{n+2}(3x^n)$
6. $(2x^2)(4x)(x^n)$
7. $(-x^ny)(2x^2y^n)$
8. $(3x^2y^n)(-3x^{2n}y^n)$

Examples   a. $3(x^3 - x)$ b. $(x - 3)(x + 1)$

Solutions   a. $3(x^3 - x) = 3x^3 - 3x$ b. $(x - 3)(x + 1) = x^2 + x - 3x - 3$
     $= x^2 - 2x - 3$

9. $5(x - 2)$
10. $2(x^2 + 1)$
11. $(x + 2)(x + 3)$
12. $(x - 1)(x + 5)$
13. $(3x - 1)(x + 3)$
14. $(x + 4)(2x + 1)$
15. $3(2x - 1)(x + 1)$
16. $5(x + 2)(3x - 1)$

2.2 Products of Polynomials

Examples   a. $-(x-1)(x+2)$ b. $x(2x^2+x-1)$

Solutions a. $-(x-1)(x+2) = -[x^2+2x-x-2]$ b. $x(2x^2+x-1) = 2x^3+x^2-x$
$= -[x^2+x-2]$
$= -x^2-x+2$

17. $-(x-1)(x+1)$ 18. $-(x+1)(x+2)$
19. $-(2x-1)(x+3)$ 20. $-(x+3)(3x-1)$
21. $x(2x-4)$ 22. $x(x^2+1)$
23. $x(x^3+x+1)$ 24. $x^2(x^2+x+1)$

Example $(x+2)(x^2+3x+4)$

Solution $(x+2)(x^2+3x+4) = x(x^2+3x+4) + 2(x^2+3x+4)$
$= x^3+3x^2+4x+2x^2+6x+8$
$= x^3+5x^2+10x+8$

An alternative format:

$$\begin{array}{r} x^2+3x+4 \\ x+2 \\ \hline x^3+3x^2+4x \\ 2x^2+6x+8 \\ \hline x^3+5x^2+10x+8 \end{array}$$

← This is x^2+3x+4 times x
← This is x^2+3x+4 times 2

25. $(x-1)(x^2+x+1)$ 26. $(x+1)(x^2-x-1)$
27. $(x+2)(3x^2-x+4)$ 28. $(x+5)(x^2-2x+3)$
29. $(2x+1)(x^2+2x+1)$ 30. $(2x-1)(2x^2-x+1)$
31. $x(x+1)(x^2+x+1)$ 32. $x(x-2)(x^2-2)$
33. $(x^2+x+1)(x^2-x+2)$ 34. $(x^2+x)(x^2-2x+3)$

Example   Given $P(x) = x^2+2x+3$, find $P(a+1)$, $P(a+h)$, and $P(x)|_a^{a+h}$.

Solution
$P(a+1) = (a+1)^2 + 2(a+1) + 3$
$= a^2+2a+1+2a+2+3$
$= a^2+4a+6$

$P(a+h) = (a+h)^2 + 2(a+h) + 3$
$= a^2+2ah+h^2+2a+2h+3$

$P(x)|_a^{a+h} = P(a+h) - P(a)$
$= (a^2+2ah+h^2+2a+2h+3) - (a^2+2a+3)$
$= 2ah+h^2+2h$

35. Given $P(x) = x^2 + x - 2$, find $P(c)$, $P(c + h)$, $P(x)|_c^{c+h}$.
36. Given $P(x) = x^2 - 2x - 3$, find $P(a)$, $P(a - h)$, $P(x)|_{a-h}^a$.
37. Given $P(x) = x^2 + 1$, find $P(x + 2)$, $P(x^2)$, $(P(x))^2$.
38. Given $P(x) = x^2 + x + 3$, find $P(x + 2)$, $P(x^2)$, $(P(x))^2$.
39. If $P(x)$ and $Q(x)$ have degrees m and n, respectively, what is the degree of $P(x) \cdot Q(x)$?
40. If $P(x)$ has degree n, what is the degree of $(P(x))^2 - P(x)$?

2.3 Factoring Polynomials

What do we mean when we say that we have *factored* an integer or a polynomial? It is true, for example, that

$$2 = 4\left(\frac{1}{2}\right),$$

but we would not ordinarily say that 4 and 1/2 are factors of 2. On the other hand, since

$$10 = (2)(5),$$

we do say that 2 and 5 are factors of 10 in the domain J of integers.

Now consider the polynomial

$$2x^2 - 10,$$

which is completely factored as

$$2x^2 - 10 = 2(x^2 - 5)$$

if we are limited to *integral* coefficients—that is, coefficients that are integers; but if we consider polynomials in x having real numbers as coefficients, its complete factorization is

$$2x^2 - 10 = 2(x - \sqrt{5})(x + \sqrt{5}).$$

Thus the result depends in part on the coefficients we consider permissible. In this book, we are primarily concerned with polynomials whose coefficients are integers.

In factoring polynomials having integer coefficients, we consider as factors only polynomials having integer coefficients with no common integer factor other than 1 or -1, and we say that such a polynomial is **prime** if it is not the product of two polynomials of this sort and its leading coefficient is positive. For example, $x + 1$, $2x + 1$, and $x^2 + 1$ are prime, while $2x + 4 = 2(x + 2)$ and $x^2 - 1 = (x + 1)(x - 1)$ are not.

We say that a polynomial other than 0 or 1 is **completely factored** if it is written equivalently as a product of prime polynomials or as such a product times -1.

Example

$$6x^4 - 6 = 6(x^4 - 1)$$
$$= 2 \cdot 3(x^2 - 1)(x^2 + 1)$$
$$= 2 \cdot 3(x - 1)(x + 1)(x^2 + 1)$$

2.3 Factoring Polynomials

Thus $6x^4 - 6$ is completely factored as

$$2 \cdot 3 \cdot (x - 1)(x + 1)(x^2 + 1).$$

Factors of quadratics

One very common type of factoring is that involving quadratic (second-degree) binomials or trinomials with integer coefficients. From Section 2.2, we recall that

$$(x + a)(x + b) = x^2 + (a + b)x + ab, \tag{1}$$

$$(x + a)^2 = x^2 + 2ax + a^2, \tag{2}$$

$$(x + a)(x - a) = x^2 - a^2. \tag{3}$$

These three forms, each of which involves prime polynomial factors, are those most commonly encountered in the chapters that follow. In this section, we are interested in viewing these relationships from right to left—that is, from polynomial to factored form.

A few other polynomials occur frequently enough to justify a study of their factorization. In particular, the forms

$$(a + b)(x + y) = ax + ay + bx + by, \tag{4}$$

$$(x + a)(x^2 - ax + a^2) = x^3 + a^3, \tag{5}$$

$$(x - a)(x^2 + ax + a^2) = x^3 - a^3 \tag{6}$$

are often encountered in one or another part of mathematics. We are again interested in viewing these relationships from right to left. Expressions such as the right-hand member of form (4) are factorable by grouping. For example, to factor

$$3x^2y + 2y + 3xy^2 + 2x,$$

we write it in the form

$$3x^2y + 2x + 3xy^2 + 2y$$

and factor the common monomial x from the first group of two terms and y from the second group of two terms, obtaining

$$x(3xy + 2) + y(3xy + 2)$$

If we now factor the common binomial $(3xy + 2)$ from each term, we have

$$(3xy + 2)(x + y),$$

in which both factors are prime.

The application of forms (5) and (6) is direct.

Examples

a. $a^3 - 1 = (a - 1)(a^2 + a + 1)$

b. $8a^3 + b^3 = (2a)^3 + b^3$
$= (2a + b)[(2a)^2 - 2ab + b^2]$
$= (2a + b)[4a^2 - 2ab + b^2]$

Exercise 2.3

Factor completely into products of polynomials with integer coefficients.

Examples a. $3xy^2 + 15x^2y + 6xy$ b. $x^2 + 10x + 25$

Solutions a. $3xy^2 + 15x^2y + 6xy$ b. $x^2 + 10x + 25$
 $= 3xy(y + 5x + 2)$ $= (x + 5)^2$ using formula (2)

1. $2x^3y^2 + 8x^2y + 16x^2y^2$
2. $7y^4z^2 + 14y^3z + 7y^2z^2$
3. $x^2 - 16$
4. $2x^2 - 18$
5. $3x^2y^4 - 12x^2$
6. $x^2y^2z^2 - 4x^2z^2$
7. $2x^2 + 12x + 18$
8. $x^2 + 14x + 49$
9. $3y^3x^2 + 6y^3x + 3y^3$
10. $2x^2y + 16xy + 32y$
11. $5x^2yz + 10xyz + 15x$
12. $2x^3y^2 + 16x^2y^3 + 32x^2y^2$

Examples a. $x^2 + 5x + 6$ b. $xy + 3x + 4y + 12$

Solutions a. $x^2 + 5x + 6 = (x + 2)(x + 3)$ b. $xy + 3x + 4y + 12$
 using formula (1) $= x(y + 3) + 4(y + 3)$
 $= (x + 4)(y + 3)$
 using formula (4)

13. $x^2 + 3x - 10$
14. $x^2 - 5x - 6$
15. $x^2 - 8x + 15$
16. $2x^2 + 6x + 4$
17. $3x^2 + 12x - 15$
18. $x^3 - 2x^2 - 8x$
19. $xy^2 + 5xy - 14x$
20. $x^2y^2 + 5xy^2 - 14y^2$
21. $xy - y + 3x - 3$
22. $xy + 3y - x - 3$
23. $2xy + 6x - 4y - 12$
24. $xy^2 - 2y^2 + 2xy - 4y$

Examples a. $x^3 + 8$ b. $x^3 - 8y^3$

Solutions a. $x^3 + 8 = x^3 + 2^3$ b. $x^3 - 8y^3 = x^3 - (2y)^3$
 $= (x + 2)(x^2 - 2x + 4)$ $= (x - 2y)(x^2 + 2xy + 4y^2)$
 using formula (5)   using formula (6)

25. $27 + x^3$
26. $x^3y^3 - 8$
27. $1 + (x + 1)^3$
28. $(x + 1)^3 - x^3$
29. $64x^3 - 8y^3$
30. $64x^3 - y^3$

2.4 Quotients of Polynomials

Factor into products of polynomials with integer coefficients. Assume all exponents are natural numbers.

Examples

a. $x^{2n} + 3x^n + 2$ b. $x^{3n} - 1$

Solutions

a. $x^{2n} + 3x^n + 2$
$= (x^n)^2 + 3x^n + 2$
$= (x^n + 1)(x^n + 2)$

b. $x^{3n} - 1 = (x^n)^3 - 1$
$= (x^n - 1)(x^{2n} + x^n + 1)$

31. $x^{2n} - 1$
32. $3x^{2n} - 9x^n - 12$
33. $x^{4n} - y^{2n}$
34. $x^{3n} + 8$
35. $x^{6n} + 2x^{4n} + x^{2n}$
36. $x^{4n} - 8x^{3n} + 16x^{2n}$
37. $x^{6n} + 1$
38. $x^{6n} + 2x^{3n} + 1$

Factor completely into products of polynomials with integer coefficients.

39. $x^3 + 3x^2 + yx + 3y$
40. $x^2 y^3 - 8x^2$
41. $x^2 + 11x + 30$
42. $x^4 + 6x^2 + 8$
43. $x^2 y^2 + 3x^2 + y^2 + 3$
44. $x - xy^2$
45. $x^5 - x$
46. $1 - xy + x - y$

47. Consider the polynomial $x^4 + x^2 y^2 + 25y^4$. If $9x^2 y^2$ is both added to and subtracted from this expression, we have

$$x^4 + x^2 y^2 + 25y^4 + 9x^2 y^2 - 9x^2 y^2,$$
$$(x^4 + 10x^2 y^2 + 25y^4) - 9x^2 y^2,$$
$$(x^2 + 5y^2)^2 - (3xy)^2,$$
$$[(x^2 + 5y^2) - 3xy][(x^2 + 5y^2) + 3xy],$$
$$(x^2 - 3xy + 5y^2)(x^2 + 3xy + 5y^2).$$

By adding and subtracting an appropriate monomial, factor $x^4 + x^2 y^2 + y^4$.

48. Use the method of Exercise 47 to factor $x^4 - 3x^2 y^2 + y^4$.
49. Use the method of Exercise 47 to factor $x^4 + x^2 + 1$.
50. Use the method of Exercise 47 to factor $x^4 + 3x^2 + 4$.

2.4 Quotients of Polynomials

It is easy to show that the set of integers is not closed with respect to division; 2/3, 1/5, 8/9, and $-3/4$ are all examples of noninteger quotients of integers. Similarly, we can see that the set of polynomials is not closed with respect to division, because $1/x$ is a counterexample. That is, $1/x$ is the quotient of the polynomials 1 and x but is not, itself, a polynomial.

Let us examine some ways in which we can rewrite quotients of polynomials, even if the resulting expressions are not always, themselves, polynomials. We begin with the simplest case, that in which the polynomials are monomials and their quotient is a monomial. Consider

$$\frac{a^m}{a^n} \quad (a \neq 0, m, n \in N, \text{ and } m > n).$$

We have

$$\frac{a^m}{a^n} = a^m \cdot \frac{1}{a^n}$$

$$= (a^{m-n} \cdot a^n) \cdot \frac{1}{a^n}$$

$$= a^{m-n} \cdot \left(a^n \cdot \frac{1}{a^n}\right)$$

$$= a^{m-n} \cdot 1,$$

$$\frac{a^m}{a^n} = a^{m-n}.$$

Laws of exponents for quotients

This establishes the first part of the following result. Proof of the second part is similar to the proof of Theorem 2.1 and is omitted.

Theorem 2.2 *If $a, b \in R$ ($a, b \neq 0$), $m, n \in N$, then*

$$\text{I} \quad \frac{a^m}{a^n} = a^{m-n}, \quad m > n,$$

$$\text{II} \quad \left(\frac{a}{b}\right)^m = \frac{a^m}{b^m}.$$

This theorem enables us, for example, to write

$$\frac{12a^5b^3}{4a^2b^2} = \frac{12}{4} \cdot a^{5-2}b^{3-2} = 3a^3b \quad (a, b \neq 0).$$

Note that a and b are not permitted to take the value 0, because if they were, $3a^3b$ would represent a real number, 0, although $(12a^5b^3)/(4a^2b^2)$ would not be defined.

Quotients of polynomials

Theorem 1.11-III (together with the closure laws) permits us to rewrite quotients of polynomials whose numerators are not monomials. For example,

$$\frac{2x^3 + 4x^2 + 8x}{2x} = \frac{2x^3}{2x} + \frac{4x^2}{2x} + \frac{8x}{2x},$$

2.4 Quotients of Polynomials

and, by an application of Theorem 2.2-I, the right-hand member can then be denoted by the expression

$$x^2 + 2x + \frac{8x}{2x} \quad (x \neq 0).$$

Though Theorem 2.2-I is not applicable to the variable factors in expressions such as $(8x)/(2x)$, by Theorem 1.11-II and the fact that $x/x = 1$ for all $x \neq 0$, we have

$$\frac{8x}{2x} = \frac{8}{2} \cdot \frac{x}{x} = \frac{8}{2} \cdot 1 = 4,$$

for every $x \neq 0$. Thus,

$$\frac{2x^3 + 4x^2 + 8x}{2x} = x^2 + 2x + 4 \quad (x \neq 0).$$

As mentioned at the start of this section, however, quotients of polynomials cannot always be represented by polynomials. For example, we have

$$\frac{2x^3 + 4x + 1}{x} = \frac{2x^3}{x} + \frac{4x}{x} + \frac{1}{x}$$

$$= 2x^2 + 4 + \frac{1}{x} \quad (x \neq 0),$$

where the resulting expression is not a polynomial, but rather an expression of the form

$$Q(x) + \frac{R(x)}{D(x)}$$

where the polynomial $Q(x)$ (equal to $2x^2 + 4$) may be called a "partial quotient" and $R(x)$ (equal to 1) is called a "remainder."

If the divisor of a quotient contains more than one term, the familiar **long-division algorithm** involving successive subtractions can be used to rewrite the quotient. For example, the computation

$$\begin{array}{r}
x - 3 \\
x^2 + 2x - 1 \overline{\smash{\big)}\, x^3 - x^2 - 7x + 3} \\
\underline{x^3 + 2x^2 - x } \\
-3x^2 - 6x + 3 \\
\underline{-3x^2 - 6x + 3} \\
0
\end{array}$$

shows that, for $x^2 + 2x - 1 \neq 0$,

$$\frac{x^3 - x^2 - 7x + 3}{x^2 + 2x - 1} = x - 3.$$

It is most convenient to arrange the dividend and the divisor in descending powers of the variable before using the division algorithm, and to leave an appropriate space for any missing terms (terms with coefficient 0) in the dividend.

When the divisor is not a factor of the dividend, the division process will produce a nonzero remainder. For example, from

$$x + 3 \overline{\smash{\big)} x^4 + x^2 + 2x - 1}$$
$$\underline{x^4 + 3x^3}$$
$$-3x^3 + x^2$$
$$\underline{-3x^3 - 9x^2}$$
$$10x^2 + 2x$$
$$\underline{10x^2 + 30x}$$
$$-28x - 1$$
$$\underline{-28x - 84}$$
$$83 \text{ (remainder)}$$

we see that

$$\frac{x^4 + x^2 + 2x - 1}{x + 3} = x^3 - 3x^2 + 10x - 28 + \frac{83}{x + 3} \quad (x \neq -3).$$

Observe in the foregoing example that a space is left in the dividend for a term involving x^3, even though the dividend contains no such term.

Exercise 2.4

Write each quotient as a polynomial.

Examples

a. $\dfrac{10x^4 y^2}{2x^2 y}$

b. $\dfrac{x^3 + 3x^2 + 4x}{x}$

Solutions

a. $\dfrac{10x^4 y^2}{2x^2 y} = \dfrac{10}{2} x^{4-2} y^{2-1}$

$= 5x^2 y \quad (x, y \neq 0)$

b. $\dfrac{x^3 + 3x^2 + 4x}{x} = \dfrac{x^3}{x} + \dfrac{3x^2}{x} + \dfrac{4x}{x}$

$= x^2 + 3x + 4 \quad (x \neq 0)$

1. $\dfrac{15y^4 z^3}{3yz^2}$

2. $\dfrac{8x^4 y^2}{2xy}$

3. $\dfrac{6x^4 y^2}{xy^2}$

4. $\dfrac{3x^5 y^6 z^7}{x^2 y^4 z^6}$

5. $\dfrac{4x^4 + 2x^2}{2x}$

6. $\dfrac{x^3 + 5x^2 - 6x}{x}$

7. $\dfrac{3x^2 y - 2xy^2 + xy}{xy}$

8. $\dfrac{x^2 yz - xy^2 z + xyz^2}{xyz}$

Write each quotient $\dfrac{P(x)}{D(x)}$ either in the form $Q(x)$ or $Q(x) + \dfrac{R(x)}{D(x)}$, where $Q(x)$ and $R(x)$ are polynomials and the degree of $R(x)$ is less than that of $D(x)$.

Examples

a. $\dfrac{x^3 + 3x + 5}{x}$

b. $\dfrac{x^3 + 3x + 5}{x + 1}$

2.5 Synthetic Division

Solutions

a. $\dfrac{x^3 + 3x + 5}{x}$

$= \dfrac{x^3}{x} + \dfrac{3x}{x} + \dfrac{5}{x}$

$= x^2 + 3 + \dfrac{5}{x}$

b.
$$
\begin{array}{r}
x^2 -x + 4 \\
x+1 \overline{\smash{)}\, x^3 + 0\cdot x^2 + 3x + 5} \\
\underline{x^3 + x^2 } \\
-x^2 + 3x \\
\underline{-x^2 - x } \\
4x + 5 \\
\underline{4x + 4} \\
1
\end{array}
$$

Therefore

$\dfrac{x^3 + 3x + 5}{x + 1}$

$= x^2 - x + 4 + \dfrac{1}{x + 1}$

9. $\dfrac{x^2 + 2x + 1}{x}$

10. $\dfrac{2x^2 - 3x + 4}{x}$

11. $\dfrac{4x^3 - 2x + 6}{2x}$

12. $\dfrac{z^3 + z^2 + z + 1}{z^3}$

13. $\dfrac{x^5 + 2x^3 + x - 4}{x^2}$

14. $\dfrac{3x^5 + 8x^3 + 4}{x^3}$

15. $\dfrac{x^2 + 2x + 1}{x + 1}$

16. $\dfrac{2x^2 - 3x + 4}{x + 1}$

17. $\dfrac{4x^3 - 2x + 6}{2x - 3}$

18. $\dfrac{z^3 + z^2 + 1}{z - 1}$

19. $\dfrac{x^5 + 2x^3 + x - 4}{x - 4}$

20. $\dfrac{3x^4 + 8x^2 + 4}{x - 2}$

21. $\dfrac{x^5 + 2x^3 + x - 4}{x^3 - x}$

22. $\dfrac{x^5 - 4x^3 + x}{x^2 + x + 1}$

23. $\dfrac{4x^4 + 3x^3 + 2x^2 + x}{2x^2 + 1}$

24. $\dfrac{x^5 - 1}{x - 1}$

25. $\dfrac{x^5 + 1}{x + 1}$

26. $\dfrac{x^5}{x^2 + 1}$

2.5   Synthetic Division

If the divisor is of the form $x + c$, then the process of dividing one polynomial by another can be simplified by a process called **synthetic division.** Consider the example on page 40, where $x^4 + x^2 + 2x - 1$ is divided by $x + 3$. If we omit the variables, writing only the coefficients of the terms, and use zero for the coefficient of any missing power, we have

$$
\begin{array}{r}
1 - 3 + 10 - 28 \\
1 + 3 \overline{\smash{)}1 + 0 + 1 + 2 - 1} \\
\underline{1 + 3 } \\
-3 + (1) \\
\underline{-3 - 9 } \\
10 + (2) \\
\underline{10 + 30 } \\
-28 - (1) \\
\underline{-28 - 84} \\
83 \text{ (remainder).}
\end{array}
$$

Now, observe that the numerals shown in color are repetitions of the numerals written immediately above and are also repetitions of the coefficients of the associated variables in the quotient; the numerals in parentheses, ( ), are repetitions of the coefficients of the dividend. Therefore, the whole process can be written in compact form as

$$
\begin{array}{rl}
(1) & \underline{3\,|}\ 1 0 1 2 -1 \\
(2) & 3 -9 30 -84 \\
(3) & \overline{1 -3 10 -28 83} \quad \text{(remainder: 83)}
\end{array}
$$

where the repetitions are omitted and where 1, the coefficient of x in the divisor, has also been omitted.

The entries in line (3), which are the coefficients of the variables in the quotient and the remainder, have been obtained by *subtracting* the **detached coefficients** in line (2) from the detached coefficients of terms of the same degree in line (1). We could obtain the same result by replacing 3 with -3 in the divisor and *adding* instead of subtracting at each step, and this is what is done in the *synthetic division* process. The final form then appears:

$$
\begin{array}{rl}
(1) & \underline{-3\,|}\ 1 0 1 2 -1 \\
(2) & -3 9 -30 84 \\
(3) & \overline{1 -3 10 -28 83} \quad \text{(remainder: 83).}
\end{array}
$$

Comparing the results of using synthetic division with those obtained by the same process using long division, on page 40, we observe that the entries in line (3) are the coefficients of the polynomial $x^3 - 3x^2 + 10x - 28$, and that there is a remainder of 83.

Example Write $\dfrac{3x^3 - 4x - 1}{x - 2}$ in the form $Q(x) + \dfrac{r}{D(x)}$, where $Q(x)$ and $D(x)$ are polynomials and r is a constant.

Solution Using synthetic division, we write

$$\underline{2\,|}\ 3 0 -4 -1,$$

2.5 Synthetic Division

where 0 has been inserted in the position that would be occupied by the coefficient of a second-degree term if such a term were present in the dividend. Our divisor is the negative of -2, or 2.

$$
\begin{array}{rrrrr}
(1) & \underline{2\,|\,3} & 0 & -4 & -1 \\
(2) & & 6 & 12 & 16 \\
(3) & 3 & 6 & 8 & 15 \quad \text{(remainder: 15).}
\end{array}
$$

This process employs these steps:

1. 3 is "brought down" from line (1) to line (3).
2. 6, the product of 2 and 3, is written in the next position on line (2).
3. 6, the sum of 0 and 6, is written on line (3).
4. 12, the product of 2 and 6, is written in the next position on line (2).
5. 8, the sum of -4 and 12, is written on line (3).
6. 16, the product of 2 and 8, is written in the next position on line (2).
7. 15, the sum of -1 and 16, is written on line (3).

We can use the first three entries on line (3) as coefficients to write a polynomial of degree one less than the degree of the dividend. This polynomial is the quotient lacking the remainder. The last number is the remainder. Thus, for $x - 2 \neq 0$, the quotient when $3x^3 - 4x - 1$ is divided by $x - 2$ is $3x^2 + 6x + 8$ with a remainder of 15; that is,

$$\frac{3x^3 - 4x - 1}{x - 2} = 3x^2 + 6x + 8 + \frac{15}{x - 2} \quad (x \neq 2).$$

The foregoing example illustrates a theorem that we shall state without proof.

Theorem 2.3 *If $P(x)$ is a real polynomial and c is any real number, then there exists a unique real polynomial $Q(x)$ and a real number r, such that*

$$P(x) = (x - c)Q(x) + r.$$

Although this theorem does not directly involve the quotient $\dfrac{P(x)}{x - c}$, it does assure us that, for $x \neq c$,

$$\frac{P(x)}{x - c} = Q(x) + \frac{r}{x - c}.$$

Exercise 2.5

Use synthetic division to write each quotient $\dfrac{P(x)}{x - c}$ in the form $Q(x) + \dfrac{r}{x - c}$ where $Q(x)$ is a polynomial and r is a constant.

Examples

a. $\dfrac{x^4 + x^3 - 3x^2 - 8x + 4}{x - 2}$

b. $\dfrac{x^4 - x^3 + 3x - 7}{x + 3}$

Solutions on Overleaf

Solutions a.
$$\begin{array}{r|rrrrr} 2 & 1 & 1 & -3 & -8 & 4 \\ & & 2 & 6 & 6 & -4 \\ \hline & 1 & 3 & 3 & -2 & 0 \end{array}$$

Thus

$$\frac{x^4 + x^3 - 3x^2 - 8x + 4}{x - 2}$$

$$= x^3 + 3x^2 + 3x - 2$$
$(x \neq 2).$

b.
$$\begin{array}{r|rrrrr} -3 & 1 & -1 & 0 & 3 & -7 \\ & & -3 & 12 & -36 & 99 \\ \hline & 1 & -4 & 12 & -33 & 92 \end{array}$$

Thus

$$\frac{x^4 - x^3 + 3x - 7}{x + 3}$$

$$= x^3 - 4x^2 + 12x - 33 + \frac{92}{x + 3}$$
$(x \neq -3).$

1. $\dfrac{x^2 + 4x - 21}{x - 3}$

2. $\dfrac{3x^2 - 4x - 7}{x + 1}$

3. $\dfrac{x^2 - 30}{x + 5}$

4. $\dfrac{x^2 + 30}{x - 5}$

5. $\dfrac{2x^3 + x + 18}{x + 2}$

6. $\dfrac{2x^3 + x + 18}{x - 2}$

7. $\dfrac{3x^4 + x^3 - 2x^2 + 6}{x - 2}$

8. $\dfrac{x^5 + x^3 + 1}{x + 1}$

9. $\dfrac{x^6 - 1}{x - 1}$

10. $\dfrac{x^5 + 1}{x + 1}$

11. $\dfrac{4x^5 + x^4 - 2x^3 + x^2}{x - 3}$

12. $\dfrac{3x^3 + 4x^2 - 2x + 4}{x + 2}$

13. $\dfrac{x^3 - 3x^2 + 3x - 1}{x + 1}$

14. $\dfrac{x^3 - 3x^2 + 3x - 1}{x - 1}$

15. $\dfrac{x^4 + x - 14}{x + 3}$

16. $\dfrac{x^4 + x - 14}{x - 2}$

17. $\dfrac{x^3 + 11}{x + 5}$

18. $\dfrac{x^6 - 2x^3 + 1}{x + 1}$

19. $\dfrac{x^5 + 2x^4 + 3x^3 + 9x^2 - 7}{x + 2}$

20. $\dfrac{x^4 - 3x^3 + 5x^2 - x + 2}{x - 3}$

2.6 Equivalent Fractions

Rational expressions

A fraction is an expression denoting a quotient. If the numerator (dividend) and the denominator (divisor) are polynomials, then the fraction is said to be a **rational expression**. Trivially, any polynomial can be considered as being a rational expres-

2.6 Equivalent Fractions

sion, since it is the quotient of itself and 1. For each replacement of the variable(s) for which the numerator and denominator of a fraction represent real numbers and for which the denominator is not zero, the fraction represents a real number. Of course, for any value of the variable(s) for which the denominator vanishes (is equal to zero), the fraction does not represent a real number and its value is undefined.

Since for each replacement of the variable(s) for which its denominator is not zero, a rational expression represents a real number, the following relationships for rational expressions follow directly from Theorems 1.8, 1.9, and 1.10 and the laws of closure.

For values of the variables for which the denominators do not vanish,

I $\dfrac{P(x)}{Q(x)} = \dfrac{R(x)}{S(x)}$ *if and only if* $P(x) \cdot S(x) = Q(x) \cdot R(x)$,

II $-\dfrac{P(x)}{Q(x)} = \dfrac{-P(x)}{Q(x)} = \dfrac{P(x)}{-Q(x)} = -\dfrac{-P(x)}{-Q(x)}$,

III $\dfrac{P(x)}{Q(x)} = \dfrac{-P(x)}{-Q(x)} = -\dfrac{-P(x)}{Q(x)} = -\dfrac{P(x)}{-Q(x)}$,

IV $\dfrac{P(x) \cdot R(x)}{Q(x) \cdot R(x)} = \dfrac{P(x)}{Q(x)}$ *(fundamental principle of fractions).*

Reducing fractions

A fraction is said to be in **lowest terms** when the numerator and denominator do not contain prescribed types of factors in common. The arithmetic fraction a/b, where a and b are integers and $b \neq 0$, is in lowest terms provided a and b are relatively prime—that is, provided they contain no common positive integral factors other than 1. If the numerator and denominator of a fraction are polynomials with integral coefficients, then the fraction is said to be in lowest terms if the numerator and denominator cannot be expressed as products of polynomials with integral coefficients having a common factor other than ± 1.

To express a given fraction in lowest terms (called **reducing** the fraction), we can factor the numerator and denominator and apply the fundamental principle of fractions.

Examples

a. $\dfrac{y}{y^2} = \dfrac{1 \cdot y}{y \cdot y} = \dfrac{1}{y} \quad (y \neq 0)$

b. $\dfrac{(x+1)^3}{x+1} = \dfrac{(x+1)^2 \cdot (x+1)}{1 \cdot (x+1)}$
$= (x+1)^2 \quad (x \neq -1)$

Diagonal lines are sometimes used to abbreviate the procedure of reducing a fraction. Thus in Example a, we may write

$$\dfrac{y}{y^2} = \dfrac{\cancel{y}^{\,1}}{\cancel{y^2}_{\,y}} = \dfrac{1}{y} \quad (y \neq 0).$$

2 Polynomials; Rational Expressions

Reducing a fraction to lowest terms should be accomplished mentally whenever convenient.

To reduce fractions with polynomial numerators and denominators, you should, when possible, write them in factored form. Common factors are then evident by inspection.

Example

$$\frac{2x^2 + x - 15}{2x + 6} = \frac{(2x - 5)(x + 3)}{2(x + 3)} = \frac{2x - 5}{2} \quad (x \neq -3)$$

Building fractions

We can also change fractions to equivalent fractions in higher terms by applying the fundamental principle in the form

$$\frac{P(x)}{Q(x)} = \frac{P(x) \cdot R(x)}{Q(x) \cdot R(x)}.$$

In general, to change a fraction $P(x)/Q(x)$ to an equivalent fraction with $Q(x) \cdot R(x)$ as a denominator, we can determine the factor $R(x)$ by inspection, and then can multiply the numerator and the denominator of the original fraction by this factor.

Example

To express $\dfrac{x}{x + 1}$ as a fraction with denominator $x^2 + 3x + 2$, we observe that $x^2 + 3x + 2 = (x + 1)(x + 2)$ and write

$$\frac{x}{x + 1} = \frac{x(x + 2)}{(x + 1)(x + 2)} = \frac{x^2 + 2x}{x^2 + 3x + 2}.$$

Exercise 2.6

Reduce the fractions to lowest terms. Specify restrictions on the variables for which the reduction is not valid.

Examples

a. $\dfrac{5x^3 y}{15xy^2}$

b. $\dfrac{a + 3}{a^2 + 6a + 9}$

Solutions

a. $\dfrac{5x^3 y}{15xy^2} = \dfrac{(5xy)(x^2)}{(5xy)(3y)}$

$= \dfrac{x^2}{3y} \quad (x, y \neq 0)$

b. $\dfrac{a + 3}{a^2 + 6a + 9} = \dfrac{(a + 3) \cdot 1}{(a + 3)(a + 3)}$

$= \dfrac{1}{a + 3} \quad (a \neq -3)$

1. $\dfrac{8x^3 y^4}{2x^2 y}$

2. $\dfrac{3x^{11} y^3}{33y^2}$

3. $\dfrac{4x^2 + 2x}{2x + 1}$

4. $\dfrac{x^2 - x}{(x - 1)^2}$

5. $\dfrac{x^2 - 1}{x - 1}$

6. $\dfrac{1 - 4x^2}{2x - 1}$

2.6 Equivalent Fractions

7. $\dfrac{a^2 - b^2}{a + b}$ 8. $\dfrac{c^2 - cd}{(1 - c)(1 - d)}$ 9. $\dfrac{ab^3 - a^3b^5}{ab + a^2b^2}$

10. $\dfrac{xy^3 - xy}{xy - x}$ 11. $\dfrac{x^2 - 2x + 1}{x - 1}$ 12. $\dfrac{x^2 + 3x - 4}{x^2 - 1}$

13. $\dfrac{x^2 - 3x - 4}{x^2 - 1}$ 14. $\dfrac{x^2 + 5x + 6}{x^2 + 4x + 4}$ 15. $\dfrac{x^3 + 1}{(x + 1)^3}$

Examples a. $\dfrac{x^2 + 1}{x^2 - 1}$ b. $\dfrac{x^2 + 2xy + y^2}{(x + y)^2}$

Solutions a. Expression is in lowest terms b. $\dfrac{x^2 + 2xy + y^2}{(x + y)^2} = \dfrac{(x + y)^2}{(x + y)^2}$

$= 1 \quad (x \neq -y)$

16. $\dfrac{x^2 - 4x - 5}{x^2 + 4x + 3}$ 17. $\dfrac{x^3 - 8}{x^2 - 5x + 6}$ 18. $\dfrac{x + 5}{x^2 + 14x + 45}$

19. $\dfrac{x^4 - 1}{2x^2 + 2}$ 20. $\dfrac{3x^2 - 6x - 45}{9x^2 + 27x}$ 21. $\dfrac{x^3 - 1}{(x - 1)(x^2 + x + 1)}$

22. $\dfrac{a^4 - x^4}{x^2 - a^2}$ 23. $\dfrac{1 - 9x^2}{4 - 12x}$ 24. $\dfrac{2x - 6}{x^3 - 27}$

25. $\dfrac{x^4 - 4x^2 - 21}{x^3 + 3x}$ 26. $\dfrac{xy - 2x + 3y - 6}{x^2 + x - 6}$ 27. $\dfrac{b^2 - 4c^2}{ab + 2ac + bc + 2c^2}$

Express each fraction as an equivalent fraction with the indicated denominator. Specify the values of the variables for which the fractions are equivalent.

Examples a. $\dfrac{2}{5}, \dfrac{}{35}$ b. $\dfrac{a}{bc}, \dfrac{}{ab^2c}$

Solutions a. $35 = 5 \cdot 7$, so b. $ab^2c = (ab)(bc)$, so

$\dfrac{2}{5} = \dfrac{2 \cdot 7}{5 \cdot 7} = \dfrac{14}{35}$ $\dfrac{a}{bc} = \dfrac{a(ab)}{(bc)(ab)} = \dfrac{a^2b}{ab^2c} \quad (a, b, c \neq 0)$

28. $\dfrac{2}{9}, \dfrac{}{54}$ 29. $\dfrac{3}{7}, \dfrac{}{49}$ 30. $\dfrac{a^2}{ab^2c}, \dfrac{}{a^2b^2c^2}$

31. $\dfrac{xyz}{xz}, \dfrac{}{x^2yz^2}$ 32. $\dfrac{x}{x + 1}, \dfrac{}{x^2 + x}$ 33. $\dfrac{x + 1}{x - 1}, \dfrac{}{x^2 - 1}$

34. $\dfrac{3}{x - y}, \dfrac{}{x^2 - y^2}$ 35. $\dfrac{1}{x + 1}, \dfrac{}{x^3 + 1}$ 36. $\dfrac{x + 1}{4}, \dfrac{}{8x^2 + 8x}$

37. Is the value of the fraction $\dfrac{x(1-x)}{x^2-3x+2}$ equal to that of $\dfrac{x}{2-x}$ for all values of x? If not, for what value(s) of x does the equality fail to hold?

38. Write three equivalent forms of the fraction $\dfrac{1}{a-b}$ $(a \neq b)$ by changing the sign or signs of the numerator, denominator, or fraction itself.

39. What is the condition on a and b for the fraction $\dfrac{-1}{a-b}$ to represent a positive number? A negative number?

40. Is the fraction $\dfrac{x-2}{1+x^2}$ defined for all values of $x \in R$? For what value(s) of x does the fraction equal zero?

2.7 Sums and Differences of Rational Expressions

Since rational expressions represent real numbers for each replacement of the variable(s) for which the denominators are not zero, the following relationships are a direct consequence of Theorem 1.11-III and the laws of closure.

For values of the variable for which $Q(x) \neq 0$,

$$\frac{P(x)}{Q(x)} + \frac{R(x)}{Q(x)} = \frac{P(x) + R(x)}{Q(x)},$$

and

$$\frac{P(x)}{Q(x)} - \frac{R(x)}{Q(x)} = \frac{P(x) - R(x)}{Q(x)}.$$

Examples a. $\dfrac{3x^2}{5} + \dfrac{x}{5} = \dfrac{3x^2 + x}{5}$ b. $\dfrac{3}{y} - \dfrac{x}{y} = \dfrac{3-x}{y}$ $(y \neq 0)$

This principle, of course, extends to any number of fractions. If the fractions in a sum or difference have unlike denominators, we can replace the fractions with equivalent fractions having common denominators and then write the sum or difference as a single fraction.

For values for which denominators do not equal zero,

$$\frac{P(x)}{Q(x)} + \frac{R(x)}{S(x)} = \frac{P(x) \cdot S(x)}{Q(x) \cdot S(x)} + \frac{R(x) \cdot Q(x)}{S(x) \cdot Q(x)}$$

$$= \frac{P(x) \cdot S(x) + R(x) \cdot Q(x)}{Q(x) \cdot S(x)}$$

2.7 Sums and Differences of Rational Expressions

and
$$\frac{P(x)}{Q(x)} - \frac{R(x)}{S(x)} = \frac{P(x) \cdot S(x) - R(x) \cdot Q(x)}{Q(x) \cdot S(x)}.$$

Least common multiple

In rewriting fractions in a sum or difference so that they share a common denominator, any such denominator may be used. If the **least common multiple** of the denominators (called the **least common denominator**) is used, however, the resulting fraction will be in simpler form than if any other common denominator is employed. The least common multiple of two or more natural numbers is the least natural number that is exactly divisible (that is, each quotient is a natural number) by each of the given numbers.

Example Find the least common multiple of 24, 45, and 60.

Solution Factor   24,   45,   60

as $2^3 \cdot 3$, $3^2 \cdot 5$, $2^2 \cdot 3 \cdot 5$.

The least common multiple is $2^3 \cdot 3^2 \cdot 5 = 360$.

The notion of a least common multiple among several polynomial expressions is, in general, meaningless without further specification of what is desired. We can, however, define the least common multiple of a set of polynomials with integer coefficients to be the polynomial of lowest degree with integer coefficients yielding a polynomial quotient upon division by each of the given polynomials, and to be, among all such polynomials, the one having the least possible positive leading coefficient.

Very often, the least common multiple of a set of natural numbers or polynomials can be determined by inspection. When inspection fails us, however, we can find the least common multiple of a set of polynomials with integer coefficients as follows:

1. Express each polynomial in completely factored form.
2. Write as factors of a product each *different* factor occurring in any of the polynomials, including each factor the greatest number of times it occurs in any one of the given polynomials.

Example Find the least common multiple of $x^3 - x$, $x^3 + x^2$,   and $x^2 - 3x + 2$.

Solution Factor $x^3 - x$, $x^3 + x^2$, $x^2 - 3x + 2$

as $x(x - 1)(x + 1)$, $x^2(x + 1)$, $(x - 1)(x - 2)$.

The least common multiple is $x^2(x - 1)(x + 1)(x - 2)$.

To simplify sums of fractions having different denominators, we can ascertain the least common denominator of the fractions, determine the factor necessary to express each of the fractions as a fraction having this common denominator, write the fractions accordingly, and then express the sum as a single fraction.

2 Polynomials; Rational Expressions

Example

Write $\dfrac{1}{x} + \dfrac{-2}{x+1} + \dfrac{2x}{(x+1)^2}$ as a single fraction.

Solution

The least common denominator of the fractions is $x(x+1)^2$. Thus

$$\dfrac{1}{x} + \dfrac{-2}{x+1} + \dfrac{2x}{(x+1)^2} = \dfrac{1(x+1)^2}{x(x+1)^2} + \dfrac{-2x(x+1)}{x(x+1)^2} + \dfrac{2x \cdot x}{x(x+1)^2}$$

$$= \dfrac{x^2 + 2x + 1}{x(x+1)^2} + \dfrac{-2x^2 - 2x}{x(x+1)^2} + \dfrac{2x^2}{x(x+1)^2}$$

$$= \dfrac{x^2 + 1}{x(x+1)^2} \quad (x \neq 0, -1)$$

Exercise 2.7

Find the least common multiple.

Examples

a. $54, 20, 63$

b. $x^3,\ x^2 + x,\ x^4 + 2x^3 + x^2$

Solutions

a. Factor $54, 20, 63$ as $2 \cdot 3^3,\ 2^2 \cdot 5,\ 3^2 \cdot 7$; least common multiple is $2^2 \cdot 3^3 \cdot 5 \cdot 7 = 3780$.

b. Factor $x^3,\ x^2 + x,\ x^4 + 2x^3 + x^2$ as $x^3,\ x(x+1),\ x^2(x+1)^2$; least common multiple is $x^3(x+1)^2$.

1. $18, 27, 35$ **2.** $15, 20, 45$ **3.** $21, 56, 48$
4. $x^2 - 1,\ x^3 + x^2$ **5.** $x^4 - x^3,\ x^2 + 3x - 4$ **6.** $2x^3,\ 4x \pm 2,\ 4x^2 + 8x + 1$

Write each sum or difference as a single fraction in lowest terms. Assume no variable in a denominator takes a value for which the denominator vanishes.

Examples

a. $\dfrac{x+1}{x} + \dfrac{x-1}{x+1}$

b. $\dfrac{x}{x^2-1} + \dfrac{1}{x-1} - \dfrac{1}{x+1}$

Solutions

a. $\dfrac{x+1}{x} + \dfrac{x-1}{x+1}$

$= \dfrac{(x+1)^2}{x(x+1)} + \dfrac{x(x-1)}{x(x+1)}$

$= \dfrac{x^2 + 2x + 1 + x^2 - x}{x(x+1)}$

$= \dfrac{2x^2 + x + 1}{x(x+1)}$

b. $\dfrac{x}{x^2-1} + \dfrac{1}{x-1} - \dfrac{1}{x+1}$

$= \dfrac{x}{x^2-1} + \dfrac{1 \cdot (x+1)}{(x-1)(x+1)} - \dfrac{1 \cdot (x-1)}{(x+1)(x-1)}$

$= \dfrac{x + (x+1) - (x-1)}{x^2 - 1}$

$= \dfrac{x+2}{x^2 - 1}$

7. $\dfrac{x-1}{x} + \dfrac{1}{x^2+x}$

8. $\dfrac{1}{x^2-1} + \dfrac{x}{x-1}$

9. $\dfrac{y+3}{4y} - \dfrac{1}{2y^2}$

10. $\dfrac{x-2}{x^2} + \dfrac{1}{x}$

11. $\dfrac{1}{x-1} - \dfrac{1}{x+1}$

12. $\dfrac{y-1}{y+2} + \dfrac{y-3}{y+4}$

13. $\dfrac{1}{2-x} - \dfrac{x}{x-2}$

14. $\dfrac{2}{x-x^2} - \dfrac{1}{x^2-x}$

15. $\dfrac{1}{w^2-3w+2} + \dfrac{1}{w^2-1}$

16. $\dfrac{1}{x^2+5x+4} - \dfrac{1}{x^2-1}$

17. $\dfrac{x+1}{x^2-5x+6} - \dfrac{1}{x^2-6x+9}$

18. $\dfrac{2}{u^2+u-2} + \dfrac{u}{(u-1)^2}$

19. $\dfrac{1}{x+1} - \dfrac{1}{x^2-6x+9}$

20. $\dfrac{1}{x} - \dfrac{1}{(x+1)^2}$

21. $\dfrac{1}{x+1} - \dfrac{1}{x^2-1} + \dfrac{x}{x-1}$

22. $\dfrac{2}{x} + \dfrac{x+1}{x^2+x} + \dfrac{1}{x+1}$

23. $z - \dfrac{1}{z} - \dfrac{1}{z^2}$

24. $1 + \dfrac{x}{1+x} + \dfrac{x^2}{1+2x+x^2}$

25. $1 - \dfrac{1}{x-1} + \dfrac{x}{x^2-1}$

26. $a + 1 + \dfrac{1}{a} + \dfrac{1-a^3}{a^2+a}$

27. $\dfrac{x+3}{3x^2+7x+4} - \dfrac{x-7}{3x^2+13x+12}$

28. $\dfrac{z+4}{2z^2-5z-3} + \dfrac{2z-1}{2z^2+3z+1}$

29. $\dfrac{x}{x^2-a^2} - \dfrac{a}{x+a}$

30. $\dfrac{4ax}{x^2-a^2} + \dfrac{x+a}{x-a}$

2.8 Products and Quotients of Rational Expressions

By Parts II and VII of Theorem 1.11 and the laws of closure, we have the following result.

For values of the variables for which the denominators do not vanish,

$$\dfrac{P(x)}{Q(x)} \cdot \dfrac{R(x)}{S(x)} = \dfrac{P(x) \cdot R(x)}{Q(x) \cdot S(x)},$$

and

$$\dfrac{P(x)}{Q(x)} \div \dfrac{R(x)}{S(x)} = \dfrac{P(x) \cdot S(x)}{Q(x) \cdot R(x)}.$$

The quotient $\dfrac{P(x)}{Q(x)} \div \dfrac{R(x)}{S(x)}$ can also be denoted by

$$\dfrac{\dfrac{P(x)}{Q(x)}}{\dfrac{R(x)}{S(x)}}$$

This latter form is called a **complex fraction**; that is, it is a fraction containing a fraction in either the numerator or denominator or both.

We can use the above relationships to rewrite a product or quotient of fractions as a single fraction in lowest terms.

Example

$$\dfrac{x^2 - 4}{x^2 + x - 12} \cdot \dfrac{x - 3}{x^2 + 4x + 4} = \dfrac{(x - 2)(x + 2)}{(x + 4)(x - 3)} \cdot \dfrac{x - 3}{(x + 2)^2}$$

$$= \dfrac{(x - 2)(x + 2)(x - 3)}{(x + 4)(x - 3)(x + 2)^2}$$

$$= \dfrac{x - 2}{(x + 4)(x + 2)} = \dfrac{x - 2}{x^2 + 6x + 8} \quad (x \neq -2, 3, -4)$$

Since the factors of the numerator and denominator of the product of two fractions are just the factors of the numerators and denominators (respectively) of the fractions, we can divide common factors out of the numerators and denominators before writing the product as a single fraction. Thus, in the example above, we could write

$$\dfrac{x^2 - 4}{x^2 + x - 12} \cdot \dfrac{x - 3}{x^2 + 4x + 4} = \dfrac{(x - 2)\cancel{(x + 2)}^{1}}{(x + 4)\cancel{(x - 3)}_{1}} \cdot \dfrac{\cancel{x - 3}^{1}}{(x + 2)\cancel{(x + 2)}_{1}}$$

$$= \dfrac{x - 2}{(x + 4)(x + 2)} \quad (x \neq -2, 3, -4).$$

Example

$$\dfrac{x^2 - 5x + 4}{x^2 + 5x + 6} \div \dfrac{x - 1}{x + 2} = \dfrac{\cancel{(x - 1)}^{1}(x - 4)}{\cancel{(x + 2)}_{1}(x + 3)} \cdot \dfrac{\cancel{x + 2}^{1}}{\cancel{x - 1}_{1}}$$

$$= \dfrac{x - 4}{x + 3} \quad (x \neq -2, -3, 1)$$

Reduction of complex fractions

When the quotient of two fractions is given in the form of a complex fraction, we have a choice of procedures available to us for writing the quotient in the form of a simple (not complex) fraction.

2.8 Products and Quotients of Rational Expressions

Example Write $\dfrac{\dfrac{x+1}{4}}{x - \dfrac{2}{3}}$ as a simple fraction in lowest terms.

Solution 1 We can apply the fundamental principle of fractions to multiply numerator and denominator by the least common denominator of the simple fractions involved. Then, we have

$$\frac{\left(\dfrac{x+1}{4}\right)12}{\left(x - \dfrac{2}{3}\right)12} = \frac{3x+3}{12x-8} \quad \left(x \neq \frac{2}{3}\right).$$

Solution 2 Alternatively, we can rewrite the complex fraction as

$$\frac{\dfrac{x+1}{4}}{x - \dfrac{2}{3}} = \frac{\dfrac{x+1}{4}}{\dfrac{3x-2}{3}} = \frac{x+1}{4} \cdot \frac{3}{3x-2} = \frac{3x+3}{12x-8} \quad \left(x \neq \frac{2}{3}\right).$$

In the event we have a more complicated expression involving a complex fraction, we can rewrite the expression by simplifying small parts of it at a time.

Example Write $\dfrac{1}{x + \dfrac{1}{x + \dfrac{1}{x}}}$ as a simple fraction in lowest terms.

Solution We can begin by concentrating on the lower right-hand expression, $\dfrac{1}{x + \dfrac{1}{x}}$. We have

$$\frac{1}{x + \dfrac{1}{x}} = \frac{(1)x}{\left(x + \dfrac{1}{x}\right)x} = \frac{x}{x^2 + 1}.$$

Thus,

$$\frac{1}{x + \dfrac{1}{x + \dfrac{1}{x}}} = \frac{1}{x + \dfrac{x}{x^2 + 1}}.$$

From this point, we can apply either of the methods shown in the previous example to the right-hand member above. Using the first method, we have

$$\frac{1}{x + \dfrac{1}{x + \dfrac{1}{x}}} = \frac{1(x^2 + 1)}{\left(x + \dfrac{x}{x^2 + 1}\right)(x^2 + 1)} = \frac{x^2 + 1}{x^3 + x + x} = \frac{x^2 + 1}{x^3 + 2x} \quad (x \neq 0).$$

Exercise 2.8

Write each product or quotient as a simple fraction in lowest terms. Assume that no denominator vanishes.

Examples

a. $\dfrac{3a^2b}{2c^2} \div \dfrac{6a}{b^2c}$

b. $\dfrac{x^2 - 1}{x} \cdot \dfrac{x^2}{x - 1}$

Solutions

a. $\dfrac{3a^2b}{2c^2} \div \dfrac{6a}{b^2c} = \dfrac{3a^2b}{2c^2} \cdot \dfrac{b^2c}{6a}$

$= \dfrac{3a^2b^3c}{12ac^2} = \dfrac{ab^3}{4c}$

b. $\dfrac{x^2 - 1}{x} \cdot \dfrac{x^2}{x - 1} = \dfrac{(x - 1)(x + 1)x^2}{x(x - 1)}$

$= (x + 1) \cdot x$

1. $\dfrac{x^2 yz}{ab} \cdot \dfrac{a^2 bc}{xy^2 z}$

2. $\dfrac{xy}{a} \div \dfrac{ay}{x}$

3. $\dfrac{4xyz}{3a} \div \dfrac{6xy^2}{10ab}$

4. $\dfrac{x^2}{yz} \cdot \dfrac{2x^2 y}{z}$

5. $\dfrac{x + 1}{x} \cdot \dfrac{x^2}{x^2 - 1}$

6. $\dfrac{x^2 - 2x + 1}{x + 1} \cdot \dfrac{x^2 + 2x + 1}{x - 1}$

7. $\dfrac{x^2 - 1}{x^2 - 4} \div \dfrac{x^2 + 2x + 1}{x^2 + 4x + 4}$

8. $\dfrac{2x^2 - 2}{x + 1} \div \dfrac{x}{6x + 6}$

9. $\dfrac{a^2 b^2}{a^2} \cdot \dfrac{a^2 - ab + b^2}{a^2} \div \dfrac{a^3 + b^3}{a^4}$

10. $\dfrac{x^3 + 8}{x^2 - 9} \cdot \dfrac{x^3 - 27}{x^2 + 5x + 6} \div \dfrac{x^2 - 2x + 4}{x^2 + 6x + 9}$

11. $\left(1 - \dfrac{1}{z}\right)\left(1 + \dfrac{1}{z}\right)$

12. $\left(\dfrac{1}{z} + 1\right) \div \left(\dfrac{1}{z} - 1\right)$

Examples

a. $\left(\dfrac{1}{x} - \dfrac{1}{x + 1}\right)x$

b. $a - \dfrac{a}{a + \dfrac{1}{a}}$

Solutions

a. $\left(\dfrac{1}{x} - \dfrac{1}{x + 1}\right)x$

$= \left[\dfrac{(x + 1) - x}{x(x + 1)}\right]x$

$= \dfrac{1}{x(x + 1)} \cdot x$

$= \dfrac{1}{x + 1}$

b. $a - \dfrac{a}{a + \dfrac{1}{a}} = a - \dfrac{a}{\dfrac{a^2 + 1}{a}}$

$= a - a \cdot \dfrac{a}{a^2 + 1}$

$= a - \dfrac{a^2}{a^2 + 1}$

$= \dfrac{a^3 + a - a^2}{a^2 + 1}$

$= \dfrac{a^3 - a^2 + a}{a^2 + 1}$

2.9 Partial Fractions

13. $\left[\dfrac{5}{x+5} - \dfrac{4}{x+4}\right] \cdot \dfrac{x+4}{x}$

14. $\left[\dfrac{x}{x^2+1} - \dfrac{1}{x+1}\right] \cdot \dfrac{x+1}{x-1}$

15. $\left[\dfrac{5}{x^2-9} + 1\right] \div \dfrac{x+2}{x-3}$

16. $\left[\dfrac{y}{y+1} + \dfrac{1}{y-1}\right] \div \dfrac{y^2+1}{y+1}$

17. $\dfrac{\dfrac{1}{x} + \dfrac{1}{2x}}{1 + \dfrac{1}{x}}$

18. $\dfrac{1 - \dfrac{2}{x+1}}{\dfrac{x-1}{2}}$

19. $1 + \dfrac{2}{x - \dfrac{1}{x}}$

20. $\dfrac{a + \dfrac{2}{a+3}}{a + 3 - \dfrac{1}{a+3}}$

21. $\dfrac{a + 3 + \dfrac{12}{a-5}}{a + 5 + \dfrac{16}{a-5}}$

22. $\dfrac{x - \dfrac{5}{x+1}}{x + 2 - \dfrac{3}{x-1}}$

23. $\dfrac{x - 7 + \dfrac{36}{x+6}}{x - 1 + \dfrac{12}{x+6}}$

24. $\dfrac{x - 2 + \dfrac{4}{x+3}}{x^2 - 3x + 2}$

25. $\dfrac{1 + \dfrac{1}{1 - \dfrac{x}{y}}}{1 - \dfrac{3}{1 + \dfrac{x}{y}}}$

26. $\dfrac{1 - \dfrac{1}{\dfrac{x}{y} + 2}}{1 + \dfrac{3}{\dfrac{x}{y} + 2}}$

2.9 Partial Fractions

Recall from Section 2.7 that the sum $\dfrac{P(x)}{Q(x)} + \dfrac{R(x)}{S(x)}$ can be rewritten as the single fraction

$$\dfrac{P(x) \cdot S(x) + Q(x) \cdot R(x)}{Q(x) \cdot S(x)}.$$

It is frequently useful to be able to reverse this process in the case of certain kinds of fractions—in particular, for rational expressions in which the numerator is a polynomial of lesser degree than the denominator. Such rational expressions are customarily referred to as "proper" fractions.

Notice first that by the long-division algorithm, any rational expression can be written as a polynomial or the sum of a polynomial and a proper fraction. For example, by the long-division algorithm we find that

$$\frac{x^3 + 2x^2 + 2x + 3}{x^2 - 1} = x + 2 + \frac{3x + 5}{x^2 - 1}.$$

Theorem 2.4 *If $a, b, r_1, r_2 \in R$ are given, with $r_1 \neq r_2$, then there exist constants $c_1, c_2 \in R$ such that for $x \neq r_1, r_2$,*

$$\frac{ax + b}{(x - r_1)(x - r_2)} = \frac{c_1}{x - r_1} + \frac{c_2}{x - r_2}.$$

Proof Suppose first that there are such constants $c_1, c_2 \in R$. Multiplying each member of the equation by $(x - r_1)(x - r_2)$, we get

$$ax + b = c_1(x - r_2) + c_2(x - r_1).$$

Since we wish this equation to hold for *every value of x* other than r_1 and r_2, it must hold also for $x = r_1$ and $x = r_2$. Substituting r_1 and r_2 for x in turn, we get

$$ar_1 + b = c_1(r_1 - r_2) + c_2(r_1 - r_1) \quad \text{and} \quad ar_2 + b = c_1(r_2 - r_2) + c_2(r_2 - r_1),$$

from which, since $r_1 \neq r_2$,

$$c_1 = \frac{ar_1 + b}{r_1 - r_2} \quad \text{and} \quad c_2 = \frac{ar_2 + b}{r_2 - r_1}.$$

That these values satisfy the given fractional equation can be verified by direct substitution. Thus not only do c_1 and c_2 exist, but they have the values shown.

While proof of the foregoing theorem produces formulas for c_1 and c_2, it is preferable from the standpoint of efficiency to rely on the method used in the proof to obtain the numbers c_1 and c_2 in any particular problem. The technique is known as the method of **partial fractions.**

Example Apply the method of partial fractions to express $\dfrac{3x - 2}{x^2 - x}$ as a sum of fractions.

Solution We first observe that

$$\frac{3x - 2}{x^2 - x} = \frac{3x - 2}{x(x - 1)}. \tag{1}$$

Then by Theorem 2.4, we know there exist numbers $c_1, c_2 \in R$, such that

$$\frac{3x - 2}{x(x - 1)} = \frac{c_1}{x} + \frac{c_2}{x - 1}. \tag{2}$$

2.9 Partial Fractions

Multiplying each member of the equation by $x(x - 1)$, we obtain

$$3x - 2 = c_1(x - 1) + c_2(x). \qquad (3)$$

We wish the members of Equation (2) to be equal for all values of x except 0 and 1. However, if this is to be the fact, then the members of (3) must be equal for all values of x, *including* 0 and 1. Hence, while we can arbitrarily select *any* values for x and solve for c_1 and c_2 in (3), let us take 0 and 1 because these values of x yield the values of c_1 and c_2 by inspection. Thus, if $x = 0$, we have from Equation (3)

$$3(0) - 2 = c_1(0 - 1) + c_2(0),$$

from which

$$c_1 = 2.$$

If $x = 1$, Equation (3) becomes

$$3(1) - 2 = c_1(1 - 1) + c_2(1),$$

$$c_2 = 1.$$

Therefore,

$$\frac{3x - 2}{x(x - 1)} = \frac{2}{x} + \frac{1}{x - 1}.$$

Theorem 2.4 generalizes to the following form, although the proof is omitted.

Theorem 2.5 *If $P(x)$ and $Q(x)$ are real polynomials, with $P(x)$ of degree less than $Q(x)$, and if $Q(x) = (x - r_1)(x - r_2) \cdots (x - r_n)$, where no two factors are identical, then there exist constants $c_1, c_2, \ldots, c_n \in R$ such that*

$$\frac{P(x)}{Q(x)} = \frac{c_1}{x - r_1} + \frac{c_2}{x - r_2} + \cdots + \frac{c_n}{x - r_n}.$$

If the denominator of a proper fraction can be factored into *equal* linear factors, then we need the following result, which is stated without proof.

Theorem 2.6 *If $P(x)$ and $Q(x)$ are real polynomials with $P(x)$ of degree less than $Q(x)$, and if $Q(x) = (x - r_1)^n$, then there exist constants $c_1, c_2, \ldots, c_n \in R$ such that*

$$\frac{P(x)}{Q(x)} = \frac{c_1}{(x - r_1)} + \frac{c_2}{(x - r_1)^2} + \cdots + \frac{c_n}{(x - r_1)^n}.$$

A further extension of Theorems 2.5 and 2.6 can be used to apply the method of partial fractions to rational expressions in which the denominator has one or more repeated factors. Each repeated factor contributes a sum such as the one in Theorem 2.6.

Example Express $\dfrac{5x^2 + 2}{x^3 - 2x^2 + x}$ as a sum of fractions.

Solution We first observe that $\dfrac{5x^2 + 2}{x^3 - 2x^2 + x} = \dfrac{5x^2 + 2}{x(x - 1)(x - 1)}.$

Solution Continued on Overleaf

Then we write
$$\frac{5x^2 + 2}{x(x-1)(x-1)} = \frac{c_1}{x} + \frac{c_2}{x-1} + \frac{c_3}{(x-1)^2}.$$

Multiplying each member of the equation by $x(x-1)^2$, we obtain
$$5x^2 + 2 = c_1(x-1)^2 + c_2 x(x-1) + c_3 x.$$

Arbitrarily selecting any values of x, we can solve for c_1, c_2, and c_3. Let us use 0, 1, and 2 because these numbers yield simple computations. Substituting 0 for x gives
$$5(0)^2 + 2 = c_1(-1)^2 + 0 + 0,$$
$$c_1 = 2.$$

Substituting 1 for x gives
$$5(1)^2 + 2 = 0 + 0 + c_3(1),$$
$$c_3 = 7.$$

Substituting 2 for x gives
$$5(2)^2 + 2 = c_1(1)^2 + c_2(2)(1) + c_3(2).$$

Now, since $c_1 = 2$ and $c_3 = 7$, we have
$$22 = 2 + 2c_2 + 14,$$
$$2c_2 = 6,$$
$$c_2 = 3.$$

Hence,
$$\frac{5x^2 + 2}{x^3 - 2x^2 + x} = \frac{2}{x} + \frac{3}{x-1} + \frac{7}{(x-1)^2}.$$

Note that Theorems 2.5 and 2.6 apply to fractions whose denominators can be factored into linear factors. Similar theorems apply to fractions whose denominators cannot be factored completely into linear factors. In particular, the numerator of each term of the partial-fraction expansion whose denominator is a nonfactorable quadratic expression will be a linear expression of the form $c_1 x + c_2$. However, we shall not consider such fractions here.

Exercise 2.9

Use the method of partial fractions to express each fraction as a sum of fractions with powers of linear polynomials as denominators. The numerators may be rational numbers.

Example
$$\frac{2}{x^2 + x}$$

Solution Factor $x^2 + x = x(x+1)$ and write
$$\frac{2}{x^2 + x} = \frac{a}{x} + \frac{b}{x+1} = \frac{a(x+1) + bx}{x^2 + x}.$$

2.9 Partial Fractions

Thus $2 = a(x + 1) + bx$.

Substitute $x = -1$ and $x = 0$ to obtain

$$2 = a(-1 + 1) + b(-1) = -b,$$
$$2 = a(0 + 1) + b \cdot 0 = a.$$

Therefore $a = 2$, $b = -2$ and

$$\frac{2}{x^2 + x} = \frac{2}{x} - \frac{2}{x + 1}.$$

1. $\dfrac{4}{x^2 + x}$
2. $\dfrac{x + 2}{x^2 + x}$
3. $\dfrac{x}{(x + 1)(x + 2)}$
4. $\dfrac{3}{x^2 + 4x + 3}$
5. $\dfrac{x - 2}{x^2 + 5x + 6}$
6. $\dfrac{2x - 1}{x^2 + 5x - 14}$

Example $\dfrac{x + 1}{x(x + 2)(x + 4)}$

Solution Write $\dfrac{x + 1}{x(x + 2)(x + 4)} = \dfrac{a}{x} + \dfrac{b}{x + 2} + \dfrac{c}{x + 4}$

$$= \frac{a(x + 2)(x + 4) + bx(x + 4) + cx(x + 2)}{x(x + 2)(x + 4)}.$$

Thus $x + 1 = a(x + 2)(x + 4) + bx(x + 4) + cx(x + 2)$. Substitute $x = 0, -2, -4$ to obtain

$$1 = a \cdot 2 \cdot 4 + b \cdot 0 + c \cdot 0 = 8a$$
$$-1 = a \cdot 0 \cdot 2 + b(-2) \cdot 2 + c(-2) \cdot 0 = -4b$$
$$-3 = a \cdot (-2) \cdot 0 + b(-4) \cdot 0 + c(-4)(-2) = 8c.$$

Therefore $a = \dfrac{1}{8}$, $b = \dfrac{1}{4}$, $c = \dfrac{-3}{8}$ so that

$$\frac{x + 1}{x(x + 2)(x + 4)} = \frac{\frac{1}{8}}{x} + \frac{\frac{1}{4}}{x + 2} - \frac{\frac{3}{8}}{x + 4}.$$

7. $\dfrac{2x + 1}{x(x + 1)(x + 2)}$
8. $\dfrac{x}{(x - 1)(x + 1)(x + 3)}$
9. $\dfrac{x^2 - x + 1}{x(x + 1)(x + 2)}$
10. $\dfrac{x^2 + 1}{x(x - 1)(x + 1)}$

11. $\dfrac{1}{x^3 + 3x^2 + 2x}$

12. $\dfrac{x - 1}{(x + 1)(x^2 - 2x)}$

Example $\dfrac{x + 2}{x(x + 1)^2}$

Solution Write
$$\dfrac{x + 2}{x(x + 1)^2} = \dfrac{a}{x} + \dfrac{b}{x + 1} + \dfrac{c}{(x + 1)^2}$$
$$= \dfrac{a(x + 1)^2 + bx(x + 1) + cx}{x(x + 1)^2}.$$

Thus $x + 2 = a(x + 1)^2 + bx(x + 1) + cx$. Substitute $x = 0, -1$ to obtain
$$2 = a \cdot 1 + b \cdot 0 + c \cdot 0 = a$$
$$1 = a \cdot 0 + b \cdot 0 + c(-1) = -c.$$

Substitute any other value for x to find b; here we will use $x = 1$ to obtain
$$3 = a \cdot 4 + b \cdot 2 + c = 2 \cdot 4 + 2 \cdot b - 1$$
so that $3 = 7 + 2b$ or $b = -2$. Therefore
$$\dfrac{x + 2}{x(x + 1)^2} = \dfrac{2}{x} - \dfrac{2}{x + 1} - \dfrac{1}{(x + 1)^2}.$$

13. $\dfrac{x + 1}{x^2(x - 2)}$

14. $\dfrac{x + 3}{(x + 1)(x + 2)^2}$

15. $\dfrac{x^3}{(x - 1)^2(x + 1)^2}$

16. $\dfrac{x^2 + 1}{x^4 - 2x^2 + 1}$

17. $\dfrac{1}{x^3 - x^2}$

18. $\dfrac{1}{x^4 - x^2}$

Use long division and the method of partial fractions to express each fraction as the sum of a polynomial and fractions with powers of linear polynomials for denominators.

Example $\dfrac{x^3 + 2x^2 + x + 2}{x^2 + x}$

Solution Long division yields
$$\dfrac{x^3 + 2x^2 + x + 2}{x^2 + x} = x + 1 + \dfrac{2}{x^2 + x}.$$

The method of partial fractions yields
$$\dfrac{2}{x^2 + x} = \dfrac{2}{x} - \dfrac{2}{x + 1}.$$

Thus
$$\frac{x^3 + 2x^2 + x + 2}{x^2 + x} = x + 1 + \frac{2}{x} - \frac{2}{x+1}.$$

19. $\dfrac{x^4 + x^3 + x + 2}{x^2 + x}$

20. $\dfrac{2x^2 + 9x + 10}{x^2 + 5x + 6}$

21. $\dfrac{x^3}{(x-1)^2}$

22. $\dfrac{x^3}{x^2 - 1}$

Chapter Review

[2.1] *Simplify.*

1. $(x^2 - 5x + 1) - (2 - 3x - x^2)$
2. $3\{x^2 + 1 - [x^2 - 2(x^2 - 1)] + 1\}$

Given $P(x) = 2x^2 - 3x + 4$, find the following.

3. $P(3)$
4. $P(x)|_{-1}^{2}$

[2.2] *Simplify.*

5. $2(x + 1)(3x + 2)$
6. $(x + 2)(3x^2 + x + 1)$

[2.3] *Factor completely.*

7. $y^2 - 8y + 15$
8. $x^3 + 4x^2 + 4x$
9. $x^4 - 16$
10. $z^3 - 64$

[2.4] *Write each quotient as a polynomial.*

11. $\dfrac{30x^2 y^3 z^4}{5xyz^3}$
12. $\dfrac{18y^3 + 21y^2 + 3y}{3y}$
13. $\dfrac{2x^2 + x - 15}{x + 3}$
14. $\dfrac{6x^2 + x - 2}{2x - 1}$

[2.5] *Use synthetic division to write each quotient* $\dfrac{P(x)}{D(x)}$ *in the form* $Q(x)$ *or in the form* $Q(x) + \dfrac{r}{D(x)}$, *where* $Q(x)$ *is a polynomial and r is a constant.*

15. $\dfrac{x^4 - 5x^2 + 7x - 10}{x - 2}$
16. $\dfrac{3x^3 + 6x^2 - 4x - 5}{x + 2}$

[2.6] Reduce to lowest terms and state all restrictions on the variables.

17. $\dfrac{6y^2z + 33yz^2}{3yz}$

18. $\dfrac{x^2 + 10x + 21}{x + 3}$

19. $\dfrac{x^2 + 4x - 21}{x^2 - 9}$

20. $\dfrac{3x^2 + 12x + 12}{3x^2 - 12}$

[2.7] Simplify.

21. $\dfrac{x}{3} + \dfrac{x + 1}{4}$

22. $\dfrac{x + 1}{x - 1} - \dfrac{x - 2}{x + 2}$

23. $\dfrac{x^2 + 3}{x + 1} - 2$

24. $\dfrac{y + 2}{y^2 - y - 2} + \dfrac{y - 3}{y + 1}$

[2.8] Simplify.

25. $\dfrac{x^3y}{z} \cdot \dfrac{xz^2}{y^3}$

26. $\dfrac{x^2 + 2x + 1}{x - 3} \cdot \dfrac{x^2 + 4x - 21}{x^2 + 6x + 5}$

27. $\dfrac{x^2 - 1}{x + 2} \div \dfrac{x^2 + 4x + 3}{x^2 + 4x + 4}$

28. $\dfrac{x - \dfrac{1}{x + 1}}{x + 4 + \dfrac{11}{x - 3}}$

[2.9] Use the method of partial fractions to express each fraction as a sum of fractions with powers of linear polynomials for denominators.

29. $\dfrac{5}{2x^2 - x - 3}$

30. $\dfrac{x^2 + 1}{x^3 - x}$

3 Rational Exponents—Radicals

In Chapter 2, powers of real numbers were defined for natural-number exponents, and some simple properties of products and quotients of powers were examined. In this chapter, powers with integral and rational exponents are defined in a manner consistent with these properties, and the basic rules for manipulating such powers are developed.

3.1 Powers with Integral Exponents

Reason for defining a^0 to be 1

In Section 2.4 we saw that for $a \in R$, $a \neq 0$, and $m, n \in N$, $m > n$, we have

$$\frac{a^m}{a^n} = a^{m-n}. \tag{1}$$

If (1) is to hold also for $m = n$, then we must have

$$\frac{a^n}{a^n} = a^{n-n} = a^0.$$

Since $a^n/a^n = 1$ for $a \neq 0$, we can state the following definition.

63

3 Rational Exponents—Radicals

Definition 3.1   If $a \in R$, $a \neq 0$, then
$$a^0 = 1.$$

Examples   a. $2^0 = 1$ b. $(x - 3)^0 = 1$ for $x \neq 3$

Reason for defining a^{-n} to be $1/a^n$

In a similar way, if (1) is to hold for $m = 0$, then we must have
$$\frac{a^0}{a^n} = a^{0-n} = a^{-n}.$$

Since $a^0 = 1$, we can make the following definition.

Definition 3.2   If $a \in R$, $a \neq 0$, and $n \in N$, then
$$a^{-n} = \frac{1}{a^n}.$$

Examples   a. $3^{-2} = \dfrac{1}{3^2}$ b. $(x + 1)^{-1} = \dfrac{1}{x + 1}$ for $x \neq -1$

Laws of exponents for integral exponents

Now for m and n nonnegative integers, the equation in Theorem 2.1-I, with $-n$ in place of n, is
$$a^m \cdot a^{-n} = a^{m+(-n)} = a^{m-n} \quad (a \in R, a \neq 0),$$
which is true by Definition 3.2 and Theorem 2.2-I (extended to include $m = 0$ or $n = 0$ or both; for a complete discussion, several cases must be considered). Again,
$$a^{-m} \cdot a^{-n} = \frac{1}{a^m} \cdot \frac{1}{a^n} = \frac{1}{a^{m+n}} = a^{-(m+n)} = a^{-m-n}.$$

Thus the equation in Theorem 2.1-I is true for all $a \in R$, $a \neq 0$, and all integers m, n.

By Definitions 3.1 and 3.2 and Theorem 2.1-I, we can show that, for $m, n \in J$,
$$\frac{a^m}{a^n} = a^m \cdot \frac{1}{a^n} = a^m \cdot a^{-n} = a^{m-n},$$
which extends Theorem 2.2-I to $m, n \in J$. The other parts of Theorem 2.1 and 2.2, appropriately reworded for integers, also follow from Definitions 3.1 and 3.2.

We now restate these theorems for integral exponents $m, n \in J$.

Theorem 3.1   If $a, b \in R$ and $m, n \in J$, then

I $a^m \cdot a^n = a^{m+n}$,

II $\dfrac{a^m}{a^n} = a^{m-n}$, $a \neq 0$

III $(a^m)^n = a^{mn}$,

IV $(ab)^n = a^n b^n$,

V $\left(\dfrac{a}{b}\right)^n = \dfrac{a^n}{b^n}$, $b \neq 0$

3.1 Powers with Integral Exponents

Some examples of applications of this theorem should prove enlightening.

Examples

a. $(x^2)^3 = x^{2 \cdot 3} = x^6$ By Theorem 3.1-III.

b. $(xy)^3 = x^3 y^3$ By Theorem 3.1-IV.

c. $\left(\dfrac{x^2}{y}\right)^{-4} = \dfrac{(x^2)^{-4}}{(y)^{-4}}$ By Theorem 3.1-V.

$= \dfrac{x^{-8}}{y^{-4}}$ By Theorem 3.1-III.

$= \dfrac{1/x^8}{1/y^4}$ By Definition 3.2

$= \dfrac{y^4}{x^8} \quad (x, y \neq 0)$ By Theorem 1.11-VII.

Although we shall not do anything about proving it here, Parts III, IV, and V of Theorem 3.1 can be shown to apply to expressions involving more than two factors.

Examples

a. $\left(\dfrac{a^3 b}{3c^2}\right)^{-2} = \dfrac{a^{-6} b^{-2}}{3^{-2} c^{-4}}$

$= \dfrac{3^2 c^4}{a^6 b^2}$

$= \dfrac{9c^4}{a^6 b^2} \quad (a, b, c \neq 0)$

b. $\dfrac{x^{-1} - 1}{x^{-2} y} = \dfrac{\dfrac{1}{x} - 1}{\dfrac{y}{x^2}}$

$= \left(\dfrac{1}{x} - 1\right) \cdot \dfrac{x^2}{y}$

$= \dfrac{1 - x}{x} \cdot \dfrac{x^2}{y}$

$= \dfrac{(1 - x) \cdot x}{y} \quad (x, y \neq 0)$

Integral exponents are used in an exponential form for numbers called scientific notation. A number is represented in *scientific notation* if it is expressed as a product of a number between 1 and 10 (excluding 10) and a power of 10.

Examples

a. $8102 = 8.102 \times 1000 = 8.102 \times 10^3$

b. $0.2108 = 2.108 \times \dfrac{1}{10} = 2.108 \times 10^{-1}$

In order to avoid the necessity of constantly noting exceptions, we shall assume that in the exercises and examples in this chapter the variables are restricted so that no denominator vanishes.

3 Rational Exponents—Radicals

Exercise 3.1

For each of the following expressions, write an equivalent basic numeral—that is, a numeral with exponent 1.

Examples

a. $4 \cdot 3^{-2}$ b. $\dfrac{3^{-2}}{6^{-1}}$ c. $2^{-1} + 2^0 + 2^1$

Solutions

a. $4 \cdot 3^{-2} = 4 \cdot \dfrac{1}{3^2}$ b. $\dfrac{3^{-2}}{6^{-1}} = \dfrac{\frac{1}{3^2}}{\frac{1}{6}}$ c. $2^{-1} + 2^0 + 2^1 = \dfrac{1}{2} + 1 + 2$

$= \dfrac{4}{9}$ $= \dfrac{1}{9} \cdot \dfrac{6}{1}$ $= \dfrac{7}{2}$

$= \dfrac{6}{9} = \dfrac{2}{3}$

1. 3^{-1} 2. 4^{-2} 3. $(-2)^{-3}$ 4. $(-2)^{-4}$

5. $\dfrac{1}{2^{-2}}$ 6. $\dfrac{1}{3^{-1}}$ 7. $\dfrac{2^0 \cdot 3^{-1}}{4}$ 8. $\dfrac{3^0}{4^{-1}5^{-2}}$

9. $\left(\dfrac{1}{3}\right)^{-1}$ 10. $\left(\dfrac{2}{5}\right)^{-2}$ 11. $4^{-1} + 2^{-2}$ 12. $3^{-2} - 7^0$

Write each expression as a product or quotient in which each variable occurs at most once in the expression and all exponents are positive.

Examples

a. $x^4 x^{-5}$ b. $(x^{-2} y^3)^{-2}$ c. $\left(\dfrac{x^{-1} y z^0}{x y^{-1} z}\right)^{-1}$

Solutions

a. $x^4 x^{-5} = x^{-1}$ b. $(x^{-2} y^3)^{-2} = x^4 y^{-6}$ c. $\left(\dfrac{x^{-1} y z^0}{x y^{-1} z}\right)^{-1} = \dfrac{x^1 y^{-1} z^0}{x^{-1} y z^{-1}}$

$= \dfrac{1}{x}$ $= \dfrac{x^4}{y^6}$ $= \dfrac{x \cdot x \cdot z}{y \cdot y}$

$= \dfrac{x^2 z}{y^2}$

13. $x^8 x^{-5}$ 14. $y^{-3} y^2$ 15. $\dfrac{y^{-1}}{x^2}$ 16. $\dfrac{x^3}{y^{-2}}$

17. $(x^2 y)^5$ 18. $(xy^{-3})^2$ 19. $\left(\dfrac{x^2}{2y}\right)^3$ 20. $\left(\dfrac{2x}{y^2}\right)^2 \cdot \dfrac{y}{x^2}$

21. $\left(\dfrac{x^{-1} y^2}{2z}\right)^{-2}$ 22. $\left(\dfrac{x^0 y^2}{z^2}\right)^{-1} \cdot \dfrac{x}{y^{-1}}$

3.1 Powers with Integral Exponents

23. $\left(\dfrac{xy^2}{z^3}\right)^{-2} \cdot \left(\dfrac{xy^0}{z^2}\right)^{-1}$

24. $\left(\dfrac{2x^{-1}}{y^2}\right)^{-1} \cdot \dfrac{x^2}{y}$

Represent each expression as a single fraction involving positive exponents only.

Examples   a. $x^{-1} + y^{-1}$ b. $\dfrac{x^{-1}}{(x^{-1}+y^{-2})^{-1}}$

Solutions   a. $x^{-1}+y^{-1} = \dfrac{1}{x}+\dfrac{1}{y}$

$= \dfrac{y+x}{xy}$

b. $\dfrac{x^{-1}}{(x^{-1}+y^{-2})^{-1}} = \dfrac{\dfrac{1}{x}}{\left(\dfrac{1}{x}+\dfrac{1}{y^2}\right)^{-1}}$

$= \dfrac{1}{x}\left(\dfrac{1}{x}+\dfrac{1}{y^2}\right)$

$= \dfrac{1}{x}\left(\dfrac{y^2+x}{xy^2}\right)$

$= \dfrac{y^2+x}{x^2y^2}$

25. $x^{-1}+y^{-2}$ 26. $x^{-2}-y^{-1}$ 27. $xy^{-1}-x^{-1}y$

28. $\dfrac{x}{y^{-1}}+\dfrac{y^{-1}}{x}$ 29. $\dfrac{x^2}{y}-\dfrac{x^{-1}}{y^2}$ 30. $x(x+1)^{-1}$

31. $xy(x^{-1}+y^{-1})$ 32. $x^{-1}(x^{-1}+1)$ 33. $\dfrac{x^{-1}+y^{-1}}{x^{-1}y^{-1}}$

34. $\dfrac{x^{-1}-y^{-2}}{(x^2y)^0}$ 35. $(x^{-2}+y^{-1})^{-2}$ 36. $\dfrac{(x^{-1}+y^{-1})^{-1}}{x^{-1}+y^{-1}}$

For each expression, write an equivalent product in which each variable occurs only once.

Examples   a. $\dfrac{x^k x^{k+3}}{x^{k-1}}$ b. $\dfrac{(x^{k-1})^2}{x^k}$

Solutions   a. $\dfrac{x^k x^{k+3}}{x^{k-1}} = x^{k+(k+3)-(k-1)}$

$= x^{k+4}$

b. $\dfrac{(x^{k-1})^2}{x^k} = x^{2(k-1)-k}$

$= x^{k-2}$

37. $x^n x^{n-3}$ 38. $x^n x^{n+1}$ 39. $y^{k+1} y^{-k+1}$

40. $\dfrac{x^{n+1} x^{n+2}}{x^{n+3}}$ 41. $\dfrac{y^{3n} y^{2n-1}}{y^{-2}}$ 42. $\dfrac{x^2}{x^{n-2}}$

43. $\left(\dfrac{x^n}{x^{-1}}\right)^2$ 44. $\dfrac{(x^2)^n}{(x^n)^{-1}}$ 45. $\dfrac{x^n y^{n+2}}{(xy)^{-1}}$

68   3 Rational Exponents—Radicals

46. $\dfrac{x^{2n+1}y^{2n-1}}{x^n y^n}$ **47.** $\left(\dfrac{xy}{x^n}\right)^{-1}$ **48.** $\left(\dfrac{x^{n+1}y}{xy^{n+1}}\right)^{-1}$

Write each expression as a rational expression, that is, as the quotient of two polynomials.

Examples   a. $(x + 3x^{-1})(x^{-1} + 1)$ b. $x^{-2}y + x^2y^{-3}$

Solutions a. $(x + 3x^{-1})(x^{-1} + 1)$ b. $x^{-2}y + x^2y^{-3} = x^{-2}y^{-3}(y^4 + x^4)$

$= 1 + 3x^{-2} + x + 3x^{-1}$ $= \dfrac{y^4 + x^4}{x^2 y^3}$

$= 1 + \dfrac{3}{x^2} + x + \dfrac{3}{x}$

(Observe how the least power of each variable was factored out.)

$= \dfrac{x^2 + 3 + x^3 + 3x}{x^2}$

$= \dfrac{x^3 + x^2 + 3x + 3}{x^2}$

49. $(x^{-1} + 1)(x^{-1} - 1)$ **50.** $x^{-1}(x + 2x^2 + 3x^3)$

51. $(x^{-1} + y^{-1})(x + y)$ **52.** $(3y + y^{-1})(y^{-1} + 3)$

53. $(2x - x^{-1})(x^{-2} + x)$ **54.** $(3y + y^{-1})(y^{-1} - 2)$

55. $x^{-3}y^2 + 4xy^{-1}$ **56.** $xy^{-1} + x^{-1}y^{-2}$

57. $3x^{-3}y^{-4} + 4x^{-4}y^{-3}$ **58.** $4y^{-1} + x^2y^{-3}$

59. $xy^2z^3 - x^2y^{-1}z^{-3}$ **60.** $x^{-1}z^{-3} + 2y^{-2}z$

Write each number in scientific notation.

Examples   a. 13400   b. 0.00532

Solutions a. $13400 = 1.34 \times 10^4$ b. $0.00532 = 5.32 \times 10^{-3}$

61. 19711   **62.** 36   **63.** 1976   **64.** 6140

65. 0.0586   **66.** 0.0002   **67.** 0.01001   **68.** 0.9

3.2 Powers with Rational Exponents

In Section 3.1, we defined powers of real numbers with 0 and negative-integer exponents so that Theorem 3.1-I would be consistent with Theorem 2.1-I. It followed that the remaining laws of exponents hold for these exponents also. Now

3.2 Powers with Rational Exponents

we want to give meaning to powers of real numbers with rational numbers as exponents. As before, we shall want any such definition to be consistent with all of the laws of exponents for integers.

We observed in Chapter 2 that for $a \in R$ and $m, n \in N$,

$$(a^m)^n = a^{mn}. \tag{1}$$

If (1) is to hold for $m = 1/n$, and $a^{1/n}$ is a real number, then we must have

$$(a^{1/n})^n = a^{(1/n)(n)} = a^{n/n} = a^1 = a,$$

so that the nth power of $a^{1/n}$ must be a. A number having a as its nth power is called an **nth root** of a. In particular, for n equal to 2 or 3, respectively, an nth root is called a **square root** or a **cube root.**

Number of roots

For n odd, each $a \in R$ has just one real nth root. Thus

$$(-2)^3 = -8 \quad \text{and} \quad 2^3 = 8,$$

so that -2 is the cube root of -8, and 2 is the cube root of 8.

For n even and $a > 0$, a has two real nth roots. Thus,

$$(-2)^4 = 16 \quad \text{and} \quad 2^4 = 16,$$

so that -2 and 2 are both fourth roots of 16.

For n even and $a < 0$, a has no real nth root. Thus -1 has no real square root, since the square of each real number is nonnegative.

If $a = 0$, then a has exactly one real nth root, namely 0.

We therefore make the following definition.

Definition 3.3 *If $a \in R$, $n \in N$, then $a^{1/n}$ is the real number if one exists, and is the positive real number if two exist, such that*

$$(a^{1/n})^n = a.$$

Examples

a. $25^{1/2} = 5$ b. $-25^{1/2} = -5$ c. $(-25)^{1/2}$ is not a real number.
d. $27^{1/3} = 3$ e. $-27^{1/3} = -3$ f. $(-27)^{1/3} = -3$

Notice in parts b and e that $-25^{1/2}$ and $-27^{1/3}$ denote $-(25^{1/2})$ and $-(27^{1/3})$, respectively.

Powers $a^{m/n}$

To generalize from rational exponents of the form $1/n$, for $n \in N$, to rational exponents of the form m/n, for $m \in J$, $n \in N$, we need the results expressed in the following two theorems, which will be stated without proof.

Theorem 3.2 *If $a^{1/n} \in R$, $m \in J$, and $n \in N$, with $a \neq 0$ for $m \leq 0$, then*

$$(a^{1/n})^m = (a^m)^{1/n}.$$

Observe that Theorem 3.2 requires that $a^{1/n}$ be a real number. This requirement is not satisfied if n is even and a is negative.

Consider $((-3)^{1/2})^2$ and $((-3)^2)^{1/2}$. The expression $((-3)^2)^{1/2} = 9^{1/2} = 3$. But $(-3)^{1/2}$ does not exist in R, and neither does $((-3)^{1/2})^2$. Therefore the equality in Theorem 3.2 does not hold here, because $(-3)^{1/2} \notin R$.

Examples a. $9^{3/2} = (9^{1/2})^3 = 3^3 = 27$ b. $8^{-2/3} = (8^{1/3})^{-2} = 2^{-2} = \dfrac{1}{4}$

Theorem 3.3   If $a^{1/np} \in R$, $m \in J$, and $n, p \in N$, with $a \neq 0$ for $m \leq 0$, then
$$(a^{1/np})^{mp} = (a^{1/n})^m.$$

Theorems 3.2 and 3.3 show that if $a^{1/np}$ is a real number, then $(a^{1/n})^m$, $(a^m)^{1/n}$, $(a^{1/np})^{mp}$, and $(a^{mp})^{1/np}$ all represent the same number.

Examples a. $(16^{1/2})^3 = 4^3 = 64$ b. $(16^3)^{1/2} = 4096^{1/2} = 64$
c. $(16^{1/4})^6 = 2^6 = 64$ d. $(16^6)^{1/4} = 16{,}777{,}216^{1/4} = 64$

We make the following definition.

Definition 3.4   If $a^{1/n} \in R$, $m \in J$, and $n \in N$, then
$$a^{m/n} = (a^{1/n})^m.$$

Observe that requiring $n \in N$ does not alter the fact that m/n can represent every rational number, since all that is done is to restrict the denominator of the fraction representing the rational number to be positive, which is always possible in light of Theorems 1.8-V and 1.8-VI

Examples a. $8^{2/3} = (8^{1/3})^2 = 2^2 = 4$ b. $16^{-3/4} = (16^{1/4})^{-3} = 2^{-3} = \dfrac{1}{8}$

Because we define $a^{1/n}$ to be the positive nth root of a for a positive and n an even natural number, and since a^m is positive for a negative and m an even natural number, it follows that, for m and n even natural numbers and a any real number,
$$(a^m)^{1/n} = |a|^{m/n}.$$

For the special case $m = n$ (m and n even),
$$(a^n)^{1/n} = |a|.$$

Thus, $[(-2)^2]^{1/2} = 4^{1/2} = 2 = |-2|$.

Powers with rational exponents have the same fundamental properties as powers with integral exponents, as long as the base is positive. Theorem 3.1 can be invoked to rewrite exponential expressions involving rational exponents whenever $a, b > 0$. If the base is negative, however, considerable care must be exercised in dealing with any exponents m/n where m and n are even.

Examples a. $\dfrac{x^{3/5}}{x^{2/5}} = x^{3/5 - 2/5} = x^{1/5}$ b. $(x^6 y^3)^{1/3} = (x^6)^{1/3}(y^3)^{1/3} = x^2 y$
c. $(a^2)^{1/2} = |a|^{2/2} = |a|$ d. $(a^6)^{1/2} = |a|^{6/2} = |a|^3$

3.2 Powers with Rational Exponents

Observe that in the last two examples it was necessary to use absolute value notation because the expressions had to be positive. For example, if $a = -2$ then

$$((-2)^2)^{1/2} = 4^{1/2} = 2 \neq (-2)^1$$

and

$$((-2)^6)^{1/2} = 64^{1/2} = 8 \neq (-2)^3.$$

Exercise 3.2

For each expression, write an equivalent basic numeral—that is, a numeral not involving exponents.

Examples

a. $36^{1/2}$ b. $(-8)^{5/3}$ c. $\left(\dfrac{1}{27}\right)^{-2/3}$

Solutions

a. $36^{1/2} = 6$

b. $(-8)^{5/3} = ((-8)^{1/3})^5$
$\phantom{(-8)^{5/3}} = (-2)^5$
$\phantom{(-8)^{5/3}} = -32$

c. $\left(\dfrac{1}{27}\right)^{-2/3} = \left(\left(\dfrac{1}{27}\right)^{1/3}\right)^{-2}$
$\phantom{\left(\dfrac{1}{27}\right)^{-2/3}} = \left(\dfrac{1}{3}\right)^{-2} = \dfrac{1}{\left(\dfrac{1}{3}\right)^2}$
$\phantom{\left(\dfrac{1}{27}\right)^{-2/3}} = \dfrac{1}{\dfrac{1}{9}} = 9$

1. $9^{1/2}$
2. $9^{-1/2}$
3. $4^{-3/2}$
4. $4^{4/2}$
5. $8^{2/3}$
6. $8^{-2/3}$
7. $(-8)^{2/3}$
8. $(-8)^{-2/3}$
9. $(-32)^{-2/5}$
10. $(-32)^{2/5}$
11. $\left(\dfrac{9}{4}\right)^{3/2}$
12. $\left(\dfrac{9}{4}\right)^{-3/2}$

Write each expression as a product or quotient in which each variable occurs only once and all exponents are positive. Assume all variable bases are positive and all variables in exponents are natural numbers.

Examples

a. $\dfrac{x^{1/2}}{x^{1/3}}$ b. $\dfrac{(x^{1/2}y)^2}{(xy^{1/3})^3}$ c. $(x^n \cdot x^{(n+1)/2})^3$

Solutions

a. $\dfrac{x^{1/2}}{x^{1/3}} = x^{1/2 - 1/3}$
$\phantom{\dfrac{x^{1/2}}{x^{1/3}}} = x^{1/6}$

b. $\dfrac{(x^{1/2}y)^2}{(xy^{1/3})^3} = \dfrac{xy^2}{x^3 y}$
$\phantom{\dfrac{(x^{1/2}y)^2}{(xy^{1/3})^3}} = \dfrac{y}{x^2}$

c. $(x^n \cdot x^{(n+1)/2})^3$
$= (x^{n + [(n+1)/2]})^3$
$= (x^{(3n+1)/2})^3$
$= x^{(9n+3)/2}$

13. $x^{1/2} \cdot x^{1/3}$
14. $x^{-1/2} \cdot x^{2/3}$
15. $y \cdot y^{1/3}$

16. $y^{-1/3} \cdot y^2$

17. $\dfrac{y^{1/2}}{y^{1/6}}$

18. $\dfrac{y^{3/5}}{y^{2/5}}$

19. $\dfrac{(4x)^{1/2}}{(16x)^{-1/4}}$

20. $\dfrac{(8x)^{1/3}}{(8x)^{-2/3}}$

21. $\left(\dfrac{x^3 y^{1/2}}{x^{1/3}}\right)^6$

22. $\left(\dfrac{x^{1/2} y^{1/3}}{x^{1/4}}\right)^2$

23. $\left(\dfrac{125 x^3 y^4}{27 x^{-6} y}\right)^{1/3}$

24. $\left(\dfrac{8x^3 y^2}{x^{1/2}}\right)^{1/3}$

25. $\left(\dfrac{3 x^{1/2} y^2}{x^{-1/2}}\right)^2$

26. $(x^{3n})^2 (y^4)^{n/2}$

27. $(x^{n/2} y^{n+1})^3$

28. $\left(\dfrac{a^n}{b^{2n}}\right)^{1/2}$

29. $\dfrac{(a^n)^3}{a^{3-2n}}$

30. $(x^{2n} y^{-n})^{-1/n}$

Apply the distributive law to write each product as a sum.

Examples

a. $x^{1/4}(x^{3/4} - x)$
b. $(x^{1/2} - x)(x^{1/2} + x)$

Solutions

a. $x^{1/4} x^{3/4} - x^{1/4} x$
   $= x - x^{5/4}$

b. $x^{1/2} x^{1/2} - x^2$
   $= x - x^2$

31. $x(x^{1/2} + x)$

32. $x^{1/2}(1 + x)$

33. $x^{1/2}(x^{1/2} - 1)$

34. $x^{1/3}(x^{1/3} + 2)$

35. $y^{2/3}(y - y^{1/3})$

36. $y^{1/2}(y^2 - y^{1/2})$

37. $(x^{1/2} - y^{1/2})(x^{1/2} + y^{1/2})$

38. $(x^{-1/2} + y^{1/2})(x^{-1/2} - y^{1/2})$

39. $(x + y)^{1/2}[(x + y)^{1/2} - (x + y)]$

40. $(x - y)^{2/3}[(x - y)^{-1/3} + (x - y)]$

Factor as indicated. (Observe that in each expression the least power of the variable is factored out.)

Examples

a. $y^{-1/2} + y^{1/2} = y^{-1/2}(?)$
b. $x^{3/2} - x^{-1/2} = x^{-1/2}(?)$

Solutions

a. $y^{-1/2}(1 + y)$
b. $x^{-1/2}(x^2 - 1)$

41. $x^{3/2} + x = x(?)$

42. $y - y^{2/3} = y^{2/3}(?)$

43. $x^{-3/2} + x^{-1/2} = x^{-3/2}(?)$

44. $y^{3/4} - y^{-1/4} = y^{-1/4}(?)$

45. $(x + 1)^{1/2} - (x + 1)^{-1/2} = (x + 1)^{-1/2}(?)$

46. $(y + 2)^{1/5} - (y + 2)^{-4/5} = (y + 2)^{-4/5}(?)$

In the previous exercises the variables were restricted to positive numbers. In the remaining exercises, consider each variable base to be any real number. Simplify.

Examples

a. $[(-5)^2]^{1/2}$
b. $(u^2)^{1/2}$

Solutions

a. $[(-5)^2]^{1/2} = 25^{1/2} = 5$
b. $(u^2)^{1/2} = |u|$

3.3 Radical Expressions

47. $((-3)^6)^{1/3}$ **48.** $((-2)^8)^{1/4}$ **49.** $(9x^2)^{1/2}$

50. $(x^2(x^2+1))^{1/2}$ **51.** $(x^4(1-x))^{-1/2}$ **52.** $(x^{-2}(x+1))^{-1/2}$

3.3 Radical Expressions

Powers of real numbers with rational numbers for exponents are frequently denoted by symbols involving the use of the radical sign, $\sqrt{}$.

Definition 3.5   *If $a^{1/n} \in R$ and $n \in N$, then*

$$\sqrt[n]{a} = a^{1/n}.$$

Naturally, the radical expression on the left is not defined if the power on the right is not. In the symbolism $\sqrt[n]{a}$, a is called the **radicand** and n the **index** of the radical, and the expression is called a **radical expression of order** n. If no index is shown with a radical expression, as, for example, in the case $\sqrt{a}$, then the index 2 is understood to apply. The symbol $\sqrt{a}$ denotes the nonnegative square root of a, where, of course, a cannot be negative if $\sqrt{a}$ is to be real. The symbol $\sqrt{x^2}$, where $x \in R$, therefore provides us with an alternative means of writing $|x|$. That is, $\sqrt{x^2} = |x|$.

Examples   a. $\sqrt{5^2} = |5| = 5$ b. $\sqrt{(-3)^2} = |-3| = 3$

Properties of radicals

An immediate consequence of the foregoing definition and the theorems pertaining to exponents is the following.

Theorem 3.4   *For real values of a and b for which all the radical expressions denote real numbers,*

     I_A $\sqrt[n]{a^n} = a$ (n an odd natural number),

     I_B $\sqrt[n]{a^n} = |a|$ (n an even natural number),

     II $\sqrt[n]{a^m} = (\sqrt[n]{a})^m$ ($n \in N, m \in J$),

     III $\sqrt[n]{a} \cdot \sqrt[n]{b} = \sqrt[n]{ab}$ ($n \in N$),

     IV $\dfrac{\sqrt[n]{a}}{\sqrt[n]{b}} = \sqrt[n]{\dfrac{a}{b}}$ ($b \neq 0, n \in N$),

     V $\sqrt[cn]{a^{cm}} = \sqrt[n]{a^m}$ ($m \in J, n, c \in N$).

It might be noted in III and IV that if $a, b < 0$ and n is even, then the radicals in the left-hand members are not defined, even though the radical in the right-hand member is.

The several parts of this theorem can be used to rewrite radical expressions in various ways, and, in particular, to write them in what is called **standard form**. A radical expression is said to be in standard form if:

a. the radicand contains no polynomial factor raised to a power equal to or greater than the index of the radical,
b. the radicand contains no fractions,
c. no radical expressions are contained in denominators of fractions, and
d. the index of the radical is as small as possible.

Examples

a. $\sqrt[3]{x^4} = \sqrt[3]{x^3}\sqrt[3]{x} = x\sqrt[3]{x}$

b. $\sqrt[3]{\dfrac{1}{x^3}} = \dfrac{\sqrt[3]{1}}{\sqrt[3]{x^3}} = \dfrac{1}{x}$

c. $\dfrac{1}{\sqrt{x}} = \dfrac{1\sqrt{x}}{\sqrt{x}\sqrt{x}} = \dfrac{\sqrt{x}}{x}$

d. $\sqrt[6]{a^3} = \sqrt[3 \cdot 2]{a^{3 \cdot 1}} = \sqrt{a}$

The process employed in simplifying the expressions in parts b and c in the foregoing examples is called "rationalizing the denominator," because the result is a fraction with the denominator free of radicals. This does not exclude the possibility that the denominator is an irrational number.

It should be noted, though, that it is not *always* preferable, in working with fractions, to have their denominators rationalized. Sometimes, in fact, it is desirable to rationalize the numerator. Thus, for example,

$$\dfrac{\sqrt[3]{2}}{\sqrt{3}} = \dfrac{\sqrt[3]{2}\sqrt[3]{4}}{\sqrt{3}\sqrt[3]{4}} = \dfrac{2}{\sqrt{3}\sqrt[3]{4}}.$$

Exercise 3.3

(*Assume that all variables represent positive real numbers and that all radicands are positive.*)

Write in radical form.

Examples

a. $4^{1/3}$
b. $x^{1/2}y$
c. $(x^2 + y^2)^{-1/2}$

Solutions

a. $4^{1/3} = \sqrt[3]{4}$
b. $x^{1/2}y = \sqrt{x}y = y\sqrt{x}$
c. $(x^2 + y^2)^{-1/2} = \dfrac{1}{\sqrt{x^2 + y^2}}$

1. $5^{1/2}$
2. $2^{1/4}$
3. $2x^{1/3}$
4. $(2x)^{1/3}$
5. $(9y)^{1/2}$
6. $9y^{1/2}$
7. $x^{3/5}$
8. $x^{5/4}$
9. $xy^{1/3}$
10. $x^{1/3}y$
11. $(xy)^{1/3}$
12. $(x + y)^{1/3}$
13. $(x^2 + y^2)^{1/2}$
14. $(1 + 4x^2)^{-1/2}$
15. $(x^2 - y^2)^{-3/4}$
16. $(x^3 + y^3)^{2/3}$

3.3 Radical Expressions

Write in exponential form.

Examples
a. $\sqrt{y^5}$
b. $\sqrt[3]{3x^2}$
c. $\sqrt[5]{x^2+y^2}$

Solutions
a. $\sqrt{y^5} = (y^5)^{1/2}$
$= y^{5/2}$

b. $\sqrt[3]{3x^2} = (3x^2)^{1/3}$
$= 3^{1/3}x^{2/3}$

c. $\sqrt[5]{x^2+y^2} = (x^2+y^2)^{1/5}$

17. $\sqrt[3]{8x^4}$
18. $8\sqrt[3]{x^4}$
19. $\sqrt[4]{x^2y^3}$
20. $\sqrt{4y^4}$

21. $\sqrt{(x+y)^3}$
22. $\sqrt[5]{x^6y^4}$
23. $\sqrt[3]{x^2+1}$
24. $\sqrt{1-x^2}$

Find the indicated root.

Examples
a. $\sqrt[3]{8}$
b. $\sqrt{x^4}$
c. $\sqrt[3]{8x^6}$

Solutions
a. $\sqrt[3]{8} = 2$

b. $\sqrt{x^4} = x^{4/2}$
$= x^2$

c. $\sqrt[3]{8x^6} = \sqrt[3]{8}\sqrt[3]{x^6}$
$= 2x^{6/3}$
$= 2x^2$

25. $\sqrt[3]{-27}$
26. $\sqrt[4]{81}$
27. $\sqrt[5]{-32}$
28. $\sqrt{100}$

29. $\sqrt{x^6y^2}$
30. $\sqrt[3]{x^6y^3}$
31. $\sqrt{9x^2y^4}$
32. $\sqrt[3]{\dfrac{125x^6}{8y^3}}$

Write in standard form.

Examples
a. $\sqrt{8x^5}$
b. $\sqrt{3xy}\sqrt{6xy^3}$

Solutions
a. $\sqrt{8x^5} = \sqrt{4x^4}\sqrt{2x}$
$= 2x^2\sqrt{2x}$

b. $\sqrt{3xy}\sqrt{6xy^3} = \sqrt{18x^2y^4}$
$= \sqrt{9x^2y^4}\sqrt{2}$
$= 3xy^2\sqrt{2}$

33. $\sqrt{x^7}$
34. $\sqrt{9x^5}$
35. $\sqrt[3]{-27x^4}$

36. $\sqrt[3]{2x^3y^4}$
37. $\sqrt{x}\sqrt{xy}$
38. $\sqrt[3]{xy^2}\sqrt[3]{x^2y}$

39. $\sqrt[3]{4x^2}\sqrt[3]{16x^{10}}$
40. $\sqrt{5xy^3}\sqrt{20x}$
41. $\dfrac{\sqrt{xy}\sqrt{x}}{\sqrt{y}}$

42. $\dfrac{\sqrt[3]{xy}\sqrt[3]{x^4y}}{\sqrt[3]{x^2y^5}}$
43. $\dfrac{\sqrt[3]{2x^4}\sqrt[3]{16y}}{\sqrt[3]{4x}}$
44. $\dfrac{\sqrt{3y}\sqrt{6x}}{\sqrt{2x^3y^3}}$

Rationalize the denominator of each expression.

Examples a. $\dfrac{1}{\sqrt{3}}$ b. $\dfrac{x}{\sqrt{y}}$ c. $\dfrac{1}{\sqrt[3]{x}}$

Solutions a. $\dfrac{1}{\sqrt{3}} = \dfrac{1\sqrt{3}}{\sqrt{3}\sqrt{3}} = \dfrac{\sqrt{3}}{3}$

b. $\dfrac{x}{\sqrt{y}} = \dfrac{x\sqrt{y}}{\sqrt{y}\sqrt{y}} = \dfrac{x\sqrt{y}}{y}$

c. $\dfrac{1}{\sqrt[3]{x}} = \dfrac{1\sqrt[3]{x^2}}{\sqrt[3]{x}\sqrt[3]{x^2}} = \dfrac{\sqrt[3]{x^2}}{\sqrt[3]{x^3}} = \dfrac{\sqrt[3]{x^2}}{x}$

45. $\dfrac{1}{\sqrt{2}}$ 46. $\dfrac{6}{\sqrt{3}}$ 47. $\dfrac{4}{\sqrt{2x}}$ 48. $\dfrac{\sqrt{x}}{\sqrt{2y}}$

49. $\dfrac{1}{\sqrt[3]{4}}$ 50. $\dfrac{3}{\sqrt[3]{9}}$ 51. $\dfrac{1}{\sqrt[3]{y^2}}$ 52. $\dfrac{1}{\sqrt[3]{xy^2}}$

Rationalize the numerator of each expression.

Examples a. $\dfrac{\sqrt{x}}{2}$ b. $\dfrac{\sqrt[3]{x}}{4}$

Solutions a. $\dfrac{\sqrt{x}}{2} = \dfrac{\sqrt{x}\sqrt{x}}{2\sqrt{x}} = \dfrac{x}{2\sqrt{x}}$

b. $\dfrac{\sqrt[3]{x}}{4} = \dfrac{\sqrt[3]{x}\sqrt[3]{x^2}}{4\sqrt[3]{x^2}} = \dfrac{x}{4\sqrt[3]{x^2}}$

53. $\dfrac{\sqrt{2}}{2}$ 54. $\dfrac{\sqrt[3]{2}}{2}$ 55. $\dfrac{\sqrt[3]{x}}{y}$ 56. $\dfrac{\sqrt{xy}}{x}$

Reduce the order of each radical.

Examples a. $\sqrt[4]{5^2}$ b. $\sqrt[12]{81}$ c. $\sqrt[4]{x^2 y^2}$

Solutions a. $\sqrt[4]{5^2}$ b. $\sqrt[12]{81}$ c. $\sqrt[4]{x^2 y^2}$

$= \sqrt[2\cdot 2]{5^{1\cdot 2}}$ $= \sqrt[3\cdot 4]{3^{1\cdot 4}}$ $= \sqrt[2\cdot 2]{x^{1\cdot 2} y^{1\cdot 2}}$

$= \sqrt{5}$ $= \sqrt[3]{3}$ $= \sqrt{xy}$

57. $\sqrt[6]{81}$ 58. $\sqrt[10]{32}$ 59. $\sqrt[4]{16x^2}$ 60. $\sqrt[9]{8a^3}$

61. $\sqrt[6]{8x^3}$ 62. $\sqrt[6]{125z^3}$ 63. $\sqrt{(x-1)^4}$ 64. $\sqrt[12]{(x-2y)^4}$

3.4 Operations on Radical Expressions

We have defined radical expressions so that they represent real numbers, and therefore the properties of the real numbers can be applied. For example, the distributive law permits us to express certain sums as products.

Examples
a. $2\sqrt{5} + 4\sqrt{5} = (2 + 4)\sqrt{5} = 6\sqrt{5}$
b. $5\sqrt{2} - 9\sqrt{2} = (5 - 9)\sqrt{2} = -4\sqrt{2}$
c. $5\sqrt{2} + \sqrt{75} = 5\sqrt{2} + 5\sqrt{3} = 5(\sqrt{2} + \sqrt{3})$

The distributive law in the form $c(a + b) = ca + cb$ permits us to write certain products as sums or differences.

Examples
a. $2(\sqrt{2} - 7) = 2\sqrt{2} - 14$
b. $(\sqrt{x} - 3)(\sqrt{x} + 3) = \sqrt{x}(\sqrt{x} + 3) - 3(\sqrt{x} + 3)$
$= x + 3\sqrt{x} - 3\sqrt{x} - 9$
$= x - 9$
c. $(\sqrt{x} + 2)(\sqrt{x} - 1) = \sqrt{x}(\sqrt{x} - 1) + 2(\sqrt{x} - 1)$
$= x - \sqrt{x} + 2\sqrt{x} - 2$
$= x + \sqrt{x} - 2$

Binomial denominators

The distributive law also provides us with a means of rationalizing denominators of fractions in which radicals occur in one or both of two terms. To accomplish this, we first observe that

$$(a - \sqrt{b})(a + \sqrt{b}) = a^2 - b,$$

where the expression in the right-hand member contains no radical term. Each of the two factors of a product exhibiting this property is said to be the **conjugate** of the other. Now consider a fraction of the form

$$\frac{a}{b + \sqrt{c}},$$

where c is positive and $b \neq -\sqrt{c}$. If we multiply the numerator and denominator of this fraction by the conjugate of the denominator, then the denominator of the resulting fraction will contain no term involving $\sqrt{c}$ and hence will be free of radicals. That is,

$$\frac{a}{b + \sqrt{c}} = \frac{a(b - \sqrt{c})}{(b + \sqrt{c})(b - \sqrt{c})} = \frac{ab - a\sqrt{c}}{b^2 - c},$$

where the denominator has been rationalized. This process is equally applicable to radical fractions of the form

$$\frac{a}{\sqrt{b} + \sqrt{c}},$$

since

$$\frac{a}{\sqrt{b} + \sqrt{c}} = \frac{a(\sqrt{b} - \sqrt{c})}{(\sqrt{b} + \sqrt{c})(\sqrt{b} - \sqrt{c})} = \frac{a\sqrt{b} - a\sqrt{c}}{b - c}.$$

As noted on page 74, it is sometimes preferable, in working with fractions, to have their numerators rationalized. Thus, for example,

$$\frac{\sqrt{b} + \sqrt{c}}{a} = \frac{(\sqrt{b} + \sqrt{c})(\sqrt{b} - \sqrt{c})}{a(\sqrt{b} - \sqrt{c})} = \frac{b - c}{a(\sqrt{b} - \sqrt{c})}.$$

Exercise 3.4

Write each sum as a product.

Examples a. $2\sqrt{3} + 4\sqrt{3} - \sqrt{3}$ b. $3\sqrt{2} + \sqrt{8}$

Solutions a. $2\sqrt{3} + 4\sqrt{3} - \sqrt{3} = (2 + 4 - 1)\sqrt{3}$ b. $3\sqrt{2} + \sqrt{8} = 3\sqrt{2} + 2\sqrt{2}$
$$= 5\sqrt{3} \qquad\qquad\qquad\qquad = 5\sqrt{2}$$

1. $3\sqrt{5} + 4\sqrt{5}$
2. $\sqrt{6} - 2\sqrt{6}$
3. $4\sqrt{3} + 3\sqrt{3} - 2\sqrt{3}$
4. $4\sqrt{2} - \sqrt{2} + 3\sqrt{2}$
5. $5\sqrt{3} - 3\sqrt{12}$
6. $\sqrt{18} + \sqrt{2}$
7. $\sqrt{8} + \sqrt{98} - \sqrt{2}$
8. $\sqrt{75} - \sqrt{27} + \sqrt{3}$
9. $\sqrt[3]{2} + 3\sqrt[3]{16}$
10. $2\sqrt[3]{54} - \sqrt[3]{250}$
11. $\sqrt[4]{32} - 3\sqrt[4]{2}$
12. $6\sqrt[4]{3} + \sqrt[4]{48}$

Multiply factors and write all radicals in the result in simplest form. Assume all variables represent positive real numbers and that all radicands are positive.

Examples a. $\sqrt{2x}(\sqrt{x} + \sqrt{2x})$ b. $(\sqrt{x} + \sqrt{y})(\sqrt{x} + 2\sqrt{y})$

Solutions a. $\sqrt{2x}(\sqrt{x} + \sqrt{2x})$ b. $(\sqrt{x} + \sqrt{y})(\sqrt{x} + 2\sqrt{y})$
$$= \sqrt{2x \cdot x} + \sqrt{2x \cdot 2x} \qquad\qquad = (\sqrt{x})^2 + 3\sqrt{x}\sqrt{y} + 2(\sqrt{y})^2$$
$$= x\sqrt{2} + 2x \qquad\qquad\qquad\qquad = x + 2y + 3\sqrt{xy}$$

13. $3(2 - \sqrt{3})$
14. $\sqrt{5}(3 + 3\sqrt{5})$
15. $(\sqrt{3} - 1)(\sqrt{3} + 1)$
16. $(\sqrt{2} + \sqrt{3})(2\sqrt{2} - \sqrt{3})$
17. $(\sqrt{5} + 2)(1 - \sqrt{5})$
18. $(3 - \sqrt{2})(3 + \sqrt{2})$

3.4 Operations on Radical Expressions

19. $\sqrt{x}(\sqrt{2x} + \sqrt{8x})$
20. $\sqrt{2x}(\sqrt{3x} + \sqrt{x})$
21. $(\sqrt{x} - 1)(2\sqrt{x} + 3)$
22. $(\sqrt{2} + \sqrt{x})(1 - \sqrt{x})$
23. $(\sqrt{3} - \sqrt{x})(\sqrt{3} + \sqrt{x})$
24. $(2\sqrt{3} + \sqrt{x})(\sqrt{x} - \sqrt{3})$

Rationalize the denominators.

Examples

a. $\dfrac{1}{\sqrt{2} + 1}$

b. $\dfrac{\sqrt{x}}{\sqrt{x} + \sqrt{x-2}}$

Solutions

a. $\dfrac{1}{\sqrt{2} + 1} = \dfrac{1}{(\sqrt{2} + 1)} \dfrac{(\sqrt{2} - 1)}{(\sqrt{2} - 1)}$

$= \dfrac{\sqrt{2} - 1}{2 - 1}$

$= \sqrt{2} - 1$

b. $\dfrac{\sqrt{x}}{\sqrt{x} + \sqrt{x-2}}$

$= \dfrac{\sqrt{x}}{(\sqrt{x} + \sqrt{x-2})} \dfrac{(\sqrt{x} - \sqrt{x-2})}{(\sqrt{x} - \sqrt{x-2})}$

$= \dfrac{x - \sqrt{x(x-2)}}{x - (x-2)}$

$= \dfrac{x - \sqrt{x^2 - 2x}}{2}$

25. $\dfrac{-1}{1 - \sqrt{3}}$
26. $\dfrac{2}{\sqrt{2} + 2}$
27. $\dfrac{x}{1 - \sqrt{x}}$
28. $\dfrac{x}{\sqrt{x} + 1}$
29. $\dfrac{3x}{\sqrt{x} - 2}$
30. $\dfrac{\sqrt{x}}{\sqrt{2x} + 3}$
31. $\dfrac{1}{\sqrt{x+1} - 1}$
32. $\dfrac{\sqrt{x+3}}{3 - \sqrt{x+3}}$
33. $\dfrac{\sqrt{y}}{\sqrt{x} - \sqrt{y}}$
34. $\dfrac{4\sqrt{x} - \sqrt{3x+1}}{\sqrt{x} + \sqrt{3x+1}}$
35. $\dfrac{\sqrt{x+1} + \sqrt{x}}{\sqrt{x+1} - \sqrt{x}}$
36. $\dfrac{\sqrt{2x-1}}{\sqrt{x} + \sqrt{2x-1}}$

Rationalize the numerators.

Examples

a. $\dfrac{1 - \sqrt{2}}{2}$

b. $\dfrac{1 - \sqrt{x+1}}{\sqrt{x+1}}$

Solutions

a. $\dfrac{1 - \sqrt{2}}{2} = \dfrac{(1 - \sqrt{2})(1 + \sqrt{2})}{2 \ (1 + \sqrt{2})}$

$= \dfrac{1 - 2}{2 + 2\sqrt{2}}$

$= \dfrac{-1}{2 + 2\sqrt{2}}$

b. $\dfrac{1 - \sqrt{x+1}}{\sqrt{x+1}} = \dfrac{(1 - \sqrt{x+1})(1 + \sqrt{x+1})}{\sqrt{x+1} \ (1 + \sqrt{x+1})}$

$= \dfrac{1 - (x+1)}{\sqrt{x+1} + x + 1}$

$= \dfrac{-x}{\sqrt{x+1} + x + 1}$

37. $\dfrac{\sqrt{3}+1}{2}$ 38. $\dfrac{\sqrt{3}-\sqrt{2}}{3}$ 39. $\dfrac{\sqrt{2}+\sqrt{x+1}}{x}$

40. $\dfrac{\sqrt{x-1}-1}{\sqrt{x-1}}$ 41. $\dfrac{\sqrt{x}+\sqrt{y}}{\sqrt{x}-\sqrt{y}}$ 42. $\dfrac{\sqrt{x}}{\sqrt{x}+1}$

Chapter Review

[3.1] Write each expression as a product or quotient in which each variable occurs at most once in the expression and all exponents are positive.

1. $(x^3 y^2)^4$
2. $\dfrac{x^{-1} y^2}{x^2 y^{-1}}$
3. $\dfrac{(xy)^{-2}}{x^0 y^{-3}}$
4. $\dfrac{(x^{-1} y^0)^{-1}}{x^{-2}}$

Represent each expression as a single fraction involving positive exponents only.

5. $x^{-1} + x^{-3}$
6. $\dfrac{x^{-1} + y^{-2}}{(xy)^{-2}}$

Write each expression as an equivalent expression in which each variable occurs only once.

7. $\dfrac{x^{n+2}}{x^n x^3}$
8. $\dfrac{x^{n+1} y^2}{xy^n}$

Write each number in scientific notation.

9. 35100
10. 0.00018

[3.2] Write each expression as a product or quotient of powers in which each variable occurs only once and all exponents are positive. Assume that all variable bases are positive and all variables in exponents are natural numbers.

11. $x^{1/2} \cdot x^{2/3}$
12. $\left(\dfrac{x^3}{y^6}\right)^{-1/6}$
13. $(x^{2n} y^{4n})^{-1/2}$
14. $\left(\dfrac{y^{2n}}{x^n y^{3n}}\right)^{1/n}$

Apply the distributive law to write each product as a sum.

15. $y^{1/3}(y + y^{2/3})$
16. $(2y + y^{1/2})(2y - y^{1/2})$

Factor as indicated.

17. $x^{-1/4} + x^{1/2} = x^{-1/4}(?)$
18. $x^{2/3} - x^{-2/3} = x^{-2/3}(?)$

Chapter Review

[3.3] *Write in simplest form.*

19. $\sqrt{4x^5y^3}$
20. $\sqrt{3x} \cdot \sqrt{12xy}$
21. $\dfrac{\sqrt{2x}\sqrt{3xy}}{\sqrt{2y}}$
22. $\dfrac{\sqrt[3]{6x}\sqrt[3]{9x^2y}}{\sqrt[3]{2y}}$

23. Rationalize the denominator: $\sqrt[3]{\dfrac{x}{y}}$

24. Rationalize the numerator: $\dfrac{\sqrt{3y}}{6y}$

Reduce the order of each radical.

25. $\sqrt[4]{25}$
26. $\sqrt[6]{27x^3y^3}$

[3.4] *Write each sum as a product.*

27. $3\sqrt{8} - 2\sqrt{50} + \sqrt{2}$
28. $3\sqrt[3]{40} + 2\sqrt[3]{5}$

Multiply factors and write all radicals in the result in simplest form.

29. $(\sqrt{3} - \sqrt{2})(\sqrt{3} + \sqrt{2})$
30. $(2\sqrt{x} - 5)(\sqrt{x} + 1)$

Rationalize the denominators.

31. $\dfrac{1}{3 - \sqrt{2}}$
32. $\dfrac{\sqrt{x} + 1}{2\sqrt{x} - 1}$

Rationalize the numerators.

33. $\dfrac{\sqrt{3} + 1}{2}$
34. $\dfrac{\sqrt{x} - 1}{\sqrt{x} + 2}$

4 Equations and Inequalities in One Variable

Equations and inequalities that involve variables are called **open sentences**. Equations and inequalities involving only constants are referred to as **statements**. For example,

$$x + 2 = 7, \quad x^2 - y \geq 4, \quad \text{and} \quad |x - 3| < 2$$

are open sentences, whereas

$$3 + 2 = 5, \quad 7 < 10 - 1, \quad \text{and} \quad |3 + 2| > 0$$

are statements. Although a statement can be adjudged true or false, no such judgment is possible in the case of an open sentence. Thus, $3 - 2 = 1$ is a true statement, and $3 - 2 = 2$ is false, but we cannot assert that $x - 2 = 1$ is either true or false until we know something more about x. Open sentences can be looked upon as set selectors. Given any set of numbers, an equation such as $x - 2 = 1$ or an inequality such as $x - 2 > 1$ will serve to select certain numbers from the set and reject others, depending on whether the numbers make the resulting statement true or false.

In this chapter, we shall be concerned with open sentences (both equations and inequalities) in one variable, where the replacement set is the set of real numbers.

4.1 Equivalent Equations; First-Degree Equations

If we replace the variable x in $P(x) = Q(x)$ with an element from its replacement set U, and if the resulting statement is true, the element is a **solution**, or **root**, of the equation and is said to **satisfy** the equation. Thus, if $x \in J$, then 2 is a solution of $x + 3 = 5$, because $2 + 3 = 5$ is a true statement. On the other hand, 3 is not a solution of the equation, because $3 + 3 = 5$ is false. The subset of U consisting of all solutions of an equation is said to be the **solution set** of the equation. In the example we have been using here, $x + 3 = 5$ and $x \in J$, the solution set is $\{2\}$.

4.1 Equivalent Equations; First-Degree Equations

Equations that have the same solution set are called **equivalent equations**. For example, if $x \in J$, the equations

$$x + 3 = -3 \quad \text{and} \quad x = -6$$

are equivalent, because the solution set of each is $\{-6\}$.

Elementary transformations

To solve an equation over a given set, we usually either determine the members of the solution set by inspection or else generate a sequence of equivalent equations until we arrive at one with an obvious solution set. The following theorem, which is a direct consequence of the addition and multiplication laws for real numbers, is frequently used in generating equivalent equations over the set of real numbers.

Theorem 4.1 *If $P(x)$, $Q(x)$, and $R(x)$ are expressions, then for all values of x for which $P(x)$, $Q(x)$ and $R(x)$ are real numbers, the sentence*

$$P(x) = Q(x)$$

is equivalent to each of the following sentences:

$$\begin{aligned} &\text{I} \quad P(x) + R(x) = Q(x) + R(x), \\ &\text{II}^\dagger \quad P(x) - R(x) = Q(x) - R(x), \\ &\text{III} \quad P(x) \cdot R(x) = Q(x) \cdot R(x) \\ &\text{IV}^\dagger \quad \frac{P(x)}{R(x)} = \frac{Q(x)}{R(x)} \end{aligned} \quad \text{for } x \in \{x \mid R(x) \neq 0\}.$$

Theorem 4.1 allows us to manipulate an equation $P(x) = Q(x)$ by adding, subtracting, multiplying or dividing by an expression $R(x)$. As long as we do not multiply or divide by zero, the result will be an equivalent equation. The goal of these manipulations is generally to isolate the variable.

Example Solve $\dfrac{y + 8}{y} = 5$.

Solution We generate the following sequence of equivalent equations.

$$\frac{y+8}{y} = 5$$
$$y + 8 = 5y \quad (y \neq 0)$$
$$8 = 5y - y$$
$$8 = 4y$$
$$2 = y$$

The equation $2 = y$ is equivalent to the original for $y \neq 0$; therefore the solution set is $\{2\}$.

† Parts II and IV can be considered special cases of Parts I and III, respectively.

Any application of any part of Theorem 4.1 is called an **elementary transformation**. An elementary transformation *always* produces an equivalent equation. Care must be exercised in the application of Parts III and IV, however, for we have specifically excluded multiplication or division by zero. For example, to solve the equation

$$\frac{x}{x-3} = \frac{3}{x-3} + 2 \qquad (x \in R), \qquad (1)$$

we might first multiply each member by $(x-3)$ to find an equation that is free of fractions. We have

$$(x-3)\frac{x}{x-3} = (x-3)\frac{3}{x-3} + (x-3)2,$$

or

$$x = 3 + 2x - 6, \qquad (2)$$

from which

$$x = 3.$$

Thus 3 is a solution of Equation (2). But, upon substituting 3 for x in Equation (1), we have

$$\frac{3}{3-3} = \frac{3}{3-3} + 2 \quad \text{or} \quad \frac{3}{0} = \frac{3}{0} + 2,$$

and neither member is defined. In obtaining Equation (2), each member of Equation (1) was multiplied by $(x-3)$; but if x is 3, then $(x-3)$ is zero, and Theorem 4.1-III is not applicable. Equation (2) is *not* equivalent to Equation (1), and in fact Equation (1) has no solution.

We can always ascertain whether what we think is a solution of an equation is such in reality by substituting the suggested solution in the original equation and determining whether or not the resulting statement is true. If each equation in a sequence is obtained by means of an elementary transformation, the sole purpose for such checking is to detect arithmetic errors. We shall dispense with checking solution sets in the examples that follow unless we apply what may be a nonelementary transformation—that is, unless we multiply or divide by an expression that vanishes for some value or values of the variable.

Solution of a linear equation

The equation

$$ax + b = 0, \qquad (3)$$

where $a, b \in R$ and $a \neq 0$, is a **first-degree**, or **linear**, **equation**. Any equation that can be reduced to this form by elementary transformations, therefore, is equivalent to a first-degree equation. We can show that such an equation always has one and only one solution. By Theorem 4.1-I,

$$ax = -b \qquad (4)$$

is equivalent to Equation (3); further, by Theorem 4.1-III, Equation (4) is equivalent to

$$x = -\frac{b}{a}, \qquad (5)$$

4.1 Equivalent Equations; First-Degree Equations

which, of course, has the unique solution $-b/a$. Since Equations (3), (4), and (5) are equivalent, Equation (3) has the unique solution $-b/a$.

An equation containing more than one variable, or containing symbols such as a, b, and c, representing constants, can often be solved for one of the symbols in terms of the remaining symbols by applying elementary transformations until the desired symbol is obtained by itself as one member of an equation.

Example Solve $ay = b + y$, for y.

Solution We generate the following sequence of equivalent equations.
$$ay - y = b$$
$$y(a - 1) = b$$
$$y = \frac{b}{a - 1} \quad (a \neq 1)$$

Exercise 4.1

Solve. Consider R to be the replacement set of the variable.

Example $6y - 2(2y + 5) = 6(5 + y)$

Solution
$$6y - 4y - 10 = 30 + 6y$$
$$2y - 10 = 30 + 6y$$
$$2y - 40 = 6y$$
$$-40 = 4y$$
$$-10 = y$$

The solution set is $\{-10\}$.

1. $3x + 2(8 - x) = 3(x - 2)$
2. $5(x + 1) = 4(x + 2) - 3$
3. $-2[x - (x + 3)] = x + 1$
4. $-[2x - (3 - x)] = 4x - 11$
5. $x^2 - 3 = 1 + (x + 1)(x - 2)$
6. $2 - x - x^2 = 1 - (x - 1)^2$

Example $\dfrac{3x + 4}{5} = \dfrac{7x + 6}{10}$

Solution
$$10 \left(\frac{3x + 4}{5} \right) = 10 \left(\frac{7x + 6}{10} \right)$$
$$2(3x + 4) = 7x + 6$$
$$6x + 8 = 7x + 6$$
$$2 = x$$

The solution set is $\{2\}$.

7. $\dfrac{x-5}{4} = 1 + \dfrac{x-9}{12}$

8. $\dfrac{2x+3}{6} = \dfrac{x+1}{9}$

9. $\dfrac{x}{3} - 1 = \dfrac{2x-3}{3}$

10. $\dfrac{x}{3} + \dfrac{x+1}{2} = 3$

11. $\dfrac{2}{x+1} = \dfrac{x}{x+1} + 1$

12. $\dfrac{3}{x-2} = \dfrac{1}{2} + \dfrac{2x-7}{2x-4}$

Example

$$\dfrac{4}{2x+3} + \dfrac{4x}{4x^2-9} = \dfrac{1}{2x+3}$$

Solution

$$\left(\dfrac{4}{2x+3} + \dfrac{4x}{4x^2-9}\right)(4x^2-9) = \dfrac{1}{2x+3}(4x^2-9)$$

$$4(2x-3) + 4x = 2x - 3$$

$$10x = 9$$

$$x = \dfrac{9}{10}$$

The solution set is $\left\{\dfrac{9}{10}\right\}$.

13. $\dfrac{1}{x-1} + \dfrac{2}{x+1} = \dfrac{3x-1}{x^2-1}$

14. $\dfrac{4}{x+2} - \dfrac{1}{x} = \dfrac{2x-1}{x^2+2x}$

15. $\dfrac{3}{x+1} + \dfrac{x-4}{x^2-x-2} = \dfrac{-10}{x-2}$

16. $\dfrac{1}{x} + \dfrac{1}{x^2+x} = \dfrac{3}{x+1}$

Solve for the indicated variable. Assume that all constants are real numbers and that the replacement set of all variables is R. Leave the results in the form of an equation equivalent to the given equation. Indicate any restrictions on the variables.

Example Solve $l = a + (n-1)d$, for d.

Solution Add $-a$ to each member.

$$l - a = (n-1)d$$

Multiply each member by $\dfrac{1}{n-1}$ $(n \neq 1)$.

$$\dfrac{l-a}{n-1} = d$$

$$d = \dfrac{l-a}{n-1} \quad (n \neq 1)$$

4.2 Second-Degree Equations

17. $v = k + gt$, for k
18. $v = k + gt$, for t
19. $A = \dfrac{h}{2}(b + c)$, for c
20. $S = \dfrac{a}{1 - r}$, for r
21. $l = a + (n - 1)d$, for n
22. $\dfrac{1}{r} = \dfrac{1}{r_1} + \dfrac{1}{r_2}$, for r

Example   Solve $x'(x - 3) + 4 = 5(x + x') - 9$, for x.

Solution
$$x'(x - 3) + 4 = 5(x + x') - 9$$
$$x'x - 3x' + 4 = 5x + 5x' - 9$$
$$x'x - 5x = 5x' - 9 + 3x' - 4$$
$$x(x' - 5) = 8x' - 13$$
$$x = \dfrac{8x' - 13}{x' - 5} \quad (x' \neq 5)$$

23. $x^2 y' - 3x - 2y^3 y' = 1$, for y'
24. $2xy' - 3y' + x^2 = 0$, for y'
25. $x_1 x_2 - 2x_1 x_3 = x_4$, for x_1
26. $3x_1 x_3 + x_1 x_2 = x_4$, for x_1
27. $\dfrac{y - y_1}{x - x_1} = 6$, for y
28. $\dfrac{y - y_1}{x - x_1} = 2$, for x

29. For what value of k does the equation $2x - 3 = \dfrac{4 + x}{k}$ have as its solution set $\{-1\}$?

30. Find a value of k so that the equation $3x - 1 = k$ is equivalent to the equation $2x + 5 = 1$.

4.2 Second-Degree Equations

If a, b, and c are real numbers, then the equation

$$ax^2 + bx + c = 0 \quad (a \neq 0)$$

is a **second-degree**, or **quadratic**, **equation**. Any equation that can be reduced to this form by elementary transformations is therefore equivalent to a quadratic equation. We shall designate the form shown above as the **standard form** for such equations.

Solution by factoring   The following theorem, which is a consequence of Theorems 1.6 and 1.7, will prove useful to us in solving quadratic equations.

Theorem 4.2   If $a, b \in R$, then $ab = 0$ if and only if

$$a = 0 \quad \text{or} \quad b = 0 \quad \text{or both}.$$

Example Find the solution set of $x^2 + 2x - 15 = 0$.

Solution The equation
$$x^2 + 2x - 15 = 0$$
is equivalent to
$$(x + 5)(x - 3) = 0.$$
Since $(x + 5)(x - 3) = 0$ is true if and only if
$$x + 5 = 0 \quad \text{or} \quad x - 3 = 0,$$
we can see by inspection that the only values of x that satisfy the original equation are -5 and 3. Hence the solution set is $\{-5, 3\}$.

Number of solutions Ordinarily the solution set of a quadratic equation over the real numbers contains two elements. The equation in the foregoing example has two real solutions. However, consider the equation
$$x^2 - 2x + 1 = 0.$$
Since $x^2 - 2x + 1 = 0$ is equivalent to
$$(x - 1)^2 = 0,$$
and since the only value of x for which $(x - 1)^2 = 0$ is 1, this is the only member of the solution set of $x^2 - 2x + 1 = 0$. For reasons of convenience, we wish to consider every quadratic equation to have two roots. Accordingly, we say that the solution of any quadratic equation having only one solution is of **multiplicity two;** that is, we count it twice as a solution.

If we cannot readily factor a quadratic equation, we must use other methods of solution. Let us first consider the special case of the quadratic equation,
$$x^2 - a = 0.$$
Since $x^2 - a = 0$ is equivalent to $x^2 = a$, and since $x^2 = a$ implies that x must be a square root of a, we have as the solution set $\{\sqrt{a}, -\sqrt{a}\}$, where the members are real if $a \geq 0$ and imaginary if $a < 0$. This method of solving a quadratic equation is sometimes called **extraction of roots.**

General solution Next, quadratic equations of the form
$$(x - a)^2 = b,$$
can be solved by observing that $x - a$ must be one of the square roots of b. That is, either
$$x - a = \sqrt{b} \quad \text{or} \quad x - a = -\sqrt{b},$$
and conversely. Thus the solution set of $(x - a)^2 = b$ is $\{a + \sqrt{b}, a - \sqrt{b}\}$.

Example Solve $(x - 3)^2 = -4$.

Solution By extraction of roots, we have
$$x - 3 = \sqrt{-4} \quad \text{or} \quad x - 3 = -\sqrt{-4},$$

4.2 Second-Degree Equations

from which
$$x - 3 = 2i \quad \text{or} \quad x - 3 = -2i.$$
Then, $x = 3 + 2i$ or $x = 3 - 2i$, and the solution set is $\{3 + 2i, 3 - 2i\}$.

Being able to find solution sets for quadratic equations of the form $(x - a)^2 = b$ enables us to find the solution set of any quadratic equation. Let us first consider the general quadratic equation in standard form
$$ax^2 + bx + c = 0,$$
for the special case in which $a = 1$, that is,
$$x^2 + bx + c = 0. \tag{1}$$
We can write this equation in the equivalent form
$$(x - p)^2 = q,$$
which we can solve as above. We begin the process by adding $-c$ to each member of Equation (1), which yields
$$x^2 + bx = -c. \tag{2}$$
If we then add $(b/2)^2$ to each member of Equation (2), we obtain
$$x^2 + bx + \left(\frac{b}{2}\right)^2 = -c + \left(\frac{b}{2}\right)^2, \tag{3}$$
in which the left-hand member is equal to $(x + b/2)^2$, and we have
$$\left(x + \frac{b}{2}\right)^2 = -c + \frac{b^2}{4}. \tag{4}$$

Since we have performed only elementary transformations, Equation (4) is equivalent to (1), and we can solve (4) by the method of extraction of roots.

The technique used to obtain Equations (3) and (4) is called **completing the square**. We can determine the term necessary to complete the square in (2) by dividing the coefficient b of the first-degree term by the number 2 and squaring the result. The expression obtained, $x^2 + bx + (b/2)^2$, is called a **perfect square** and may be written in the form $(x + b/2)^2$. The process of completing the square has other applications in addition to solving quadratic equations. Some are indicated in Exercises 31–48.

Exercise 4.2

Solve by factoring.

Example $x^2 + x = 12$

Solution Write an equivalent equation in standard form and then factor the left-hand member.
$$x^2 + x - 12 = 0$$
$$(x + 4)(x - 3) = 0$$

Solution Continued on Overleaf

Determine solutions by inspection, or set each factor equal to zero and solve.

$$x + 4 = 0 \qquad x - 3 = 0$$
$$x = -4 \qquad x = 3$$

The solution set is $\{-4, 3\}$.

1. $x^2 + x = 2$
2. $x^2 = 6 - x$
3. $x^2 - 5x = 0$
4. $x^2 - 5x - 10 = 4$
5. $x^2 - x = -1 + x$
6. $x(2x - 1) = 1$
7. $6x^2 + 5x + 1 = 0$
8. $3x^2 = 16x - 5$
9. $x(2x + 9) = -9$
10. $5 = \dfrac{6}{x^2} - \dfrac{7}{x}$
11. $(x - 5)(x + 1) = -8$
12. $3 = \dfrac{-2}{x + 2} + \dfrac{2}{x + 5}$

Solve by the extraction of roots.

Example $(x - 2)^2 = 5$

Solution Set $x - 2$ equal to each square root of 5.

$$x - 2 = \sqrt{5} \qquad x - 2 = -\sqrt{5}$$
$$x = 2 + \sqrt{5} \qquad x = 2 - \sqrt{5}$$

The solution set is $\{2 + \sqrt{5}, 2 - \sqrt{5}\}$

13. $(x + 1)^2 = 4$
14. $(2x - 3)^2 = 16$
15. $x^2 = -9$
16. $(x + 2)^2 = 7$
17. $(x - 1)^2 = -1$
18. $(x - 3)^2 = -5$

Solve by completing the square.

Example $2x^2 - 6x - 3 = 0$

Solution Write an equivalent equation with the coefficient of x^2 equal to 1 and the constant term as the right-hand member.

$$x^2 - 3x = \dfrac{3}{2}$$

Add the square of half of the coefficient of the first degree term to each member.

$$x^2 - 3x + \left(-\dfrac{3}{2}\right)^2 = \dfrac{3}{2} + \left(-\dfrac{3}{2}\right)^2$$

Rewrite the left-hand member as the square of an expression.

$$\left(x - \dfrac{3}{2}\right)^2 = \dfrac{15}{4}$$

4.2 Second-Degree Equations

Extract the roots.

$$x - \frac{3}{2} = \frac{\sqrt{15}}{2} \qquad x - \frac{3}{2} = -\frac{\sqrt{15}}{2}$$

$$x = \frac{3}{2} + \frac{\sqrt{15}}{2} \qquad x = \frac{3}{2} - \frac{\sqrt{15}}{2}$$

The solution set is $\left\{ \frac{3}{2} + \frac{\sqrt{15}}{2}, \frac{3}{2} - \frac{\sqrt{15}}{2} \right\}$.

19. $x^2 + 8x - 9 = 0$ 20. $x^2 - 4x + 4 = 0$ 21. $x^2 + 11x + 30 = 0$
22. $4x^2 - 3x - 1 = 0$ 23. $3x^2 - 5x - 2 = 0$ 24. $5x^2 + 8x = 4$
25. $x^2 - x - 4 = 0$ 26. $x^2 - 2x - 2 = 0$ 27. $x^2 - 3x + 3 = 0$
28. $3x^2 + x + 7 = 0$ 29. $4x^2 + 9 = 0$ 30. $9x^2 + 2 = 0$

Reduce each equation to the form $(x - h)^2 + (y - k)^2 = r^2$ by completing the squares in x and y.

Example

$x^2 + y^2 - 4x + 6y = 5$

Solution

Write an equivalent equation in the form

$$[x^2 - 4x + (\ )] + [y^2 + 6y + (\ )] = 5 + (\ ) + (\ ).$$

Complete the squares in x and y.

$$[x^2 - 4x + 4] + [y^2 + 6y + 9] = 5 + 4 + 9$$

Write each expression in brackets as the square of a binomial.

$$(x - 2)^2 + (y + 3)^2 = 18 \quad \text{or} \quad (x - 2)^2 + [y - (-3)]^2 = (\sqrt{18})^2$$

31. $x^2 + y^2 - 4x - 4y - 17 = 0$ 32. $x^2 + y^2 + 6x - 6y + 18 = 0$
33. $x^2 + y^2 + 6x - 2y + 6 = 0$ 34. $x^2 + y^2 - 2x + 4y + 2 = 0$
35. $4x^2 + 4y^2 - 4x + 8y = 11$ 36. $16x^2 + 16y^2 - 8x + 16y = 59$

Reduce each of the following equations to the form $y = (x - a)^2 + b$, where a and b are constants, by completing the square in x.

Example

$y = x^2 + 4x + 3$

Solution

Write the equation in the indicated form.

$$y = [x^2 + 4x + (\ )] + 3$$

Solution Continued on Overleaf

Complete the square in x; add $4 + (-4)$ to the right-hand member.
$$y = [x^2 + 4x + (4)] + 3 + (-4)$$
$$y = (x + 2)^2 - 1, \quad \text{or} \quad y = [x - (-2)]^2 + (-1)$$

37. $y = x^2 - 2x + 5$ 38. $y = x^2 + 6x + 3$ 39. $y = x^2 + 8x + 4$
40. $y = x^2 - 4x + 7$ 41. $y = x^2 - 3x + 5$ 42. $y = x^2 + 5x - 2$

In each of the following equations, find the real value for x which makes the corresponding value of y as small as possible.

Example $y = x^2 + 4x + 3$

Solution Rewrite in the form $y = (x - a)^2 + b$ as in the previous example, obtaining
$$y = (x + 2)^2 - 1.$$
Now observe that $(x + 2)^2 \geq 0$ so the right-hand member is always greater than or equal to -1. When $x = -2$, $x + 2 = 0$, and the right-hand member is equal to -1. This is the smallest possible value for y.

43. $y = x^2 - 2x + 5$ 44. $y = x^2 + 6x + 3$ 45. $y = x^2 + 8x + 4$
46. $y = x^2 - 4x + 7$ 47. $y = x^2 - 3x + 5$ 48. $y = x^2 + 5x - 2$

Since r_1 and r_2 are solutions of the quadratic equation $(x - r_1)(x - r_2) = 0$, it follows that $(x - r_1)(x - r_2) = 0$, or $x^2 - (r_1 + r_2)x + r_1 r_2 = 0$, is a quadratic equation having solutions r_1 and r_2. Given the solutions of a quadratic equation in Exercises 49–52, write the equation in standard form having integral coefficients, with no common integer factors other than 1 and -1 and with leading coefficient positive.

Example $\dfrac{1}{3}$ and 5.

Solution A quadratic equation having $\dfrac{1}{3}$ and 5 as roots is
$$\left(x - \frac{1}{3}\right)(x - 5) = 0$$
or equivalently
$$x^2 - \frac{16}{3}x + \frac{5}{3} = 0.$$
To obtain an equivalent equation in the desired form, multiply each member by 3:
$$3x^2 - 16x + 5 = 0.$$

49. 3 and 2 50. $-\dfrac{1}{2}$ and 3 51. $-\dfrac{2}{3}$ and $\dfrac{1}{2}$ 52. $\dfrac{2}{9}$ and $-\dfrac{2}{9}$

4.3 The Quadratic Formula

Because the general quadratic equation
$$ax^2 + bx + c = 0 \quad (a \neq 0)$$
can be written equivalently in the form
$$x^2 + \frac{b}{a}x + \frac{c}{a} = 0,$$
the process of completing the square that we considered in Section 4.2 can be applied to obtain the **quadratic formula**,
$$x = \frac{-b \pm \sqrt{b^2 - 4ac}}{2a},$$
where the solutions, or roots, of the general quadratic equation are expressed in terms of the coefficients (see Exercise 28 of this exercise section). The symbol $\pm$ is used to condense the two equations
$$x = \frac{-b + \sqrt{b^2 - 4ac}}{2a} \quad \text{or} \quad x = \frac{-b - \sqrt{b^2 - 4ac}}{2a}$$
into a single equation. We need only substitute the coefficients a, b, and c of a given quadratic equation in the formula to find the solution set for the equation.

Determination of number of solutions

An examination of the quadratic formula,
$$x = \frac{-b \pm \sqrt{b^2 - 4ac}}{2a}.$$
shows that, if $ax^2 + bx + c = 0$, $a, b, c \in R$, is to have a nonempty solution set in the set of real numbers, then $\sqrt{b^2 - 4ac}$ must be real. This, in turn, implies that only those quadratic equations for which $b^2 - 4ac \geq 0$ have real solutions. The number represented by $b^2 - 4ac$ is called the **discriminant** of the quadratic equation $ax^2 + bx + c = 0$. It yields the following information about the nature of the solution set of the equation for $a, b, c \in R$.

1. If $b^2 - 4ac = 0$, then there is precisely one real solution (multiplicity two).
2. If $b^2 - 4ac < 0$, then there are two imaginary solutions.
3. If $b^2 - 4ac > 0$, then there are two real solutions.

Exercise 4.3

Solve for x by using the quadratic formula.

Example $x^2 - 5x + 5 = 0$

Solution on Overleaf

Solution Substitute 1 for a, -5 for b, and 5 for c in the quadratic formula and simplify

$$x = \frac{5 \pm \sqrt{25 - 4 \cdot 1 \cdot 5}}{2} = \frac{5 \pm \sqrt{5}}{2}.$$

The solution set is $\left\{\frac{5 + \sqrt{5}}{2}, \frac{5 - \sqrt{5}}{2}\right\}$.

Example $x^2 = 2x - 2$

Solution First write the equation in the standard form

$$x^2 - 2x + 2 = 0.$$

Substitute 1 for a, -2 for b, and 2 for c in the quadratic formula and simplify.

$$x = \frac{2 \pm \sqrt{4 - 4 \cdot 1 \cdot 2}}{2} = \frac{2 \pm \sqrt{-4}}{2}$$

$$= \frac{2 \pm 2\sqrt{-1}}{2} = 1 \pm \sqrt{-1} = 1 \pm i.$$

The solution set is $\{1 + i, 1 - i\}$.

1. $x^2 - 7x + 12 = 0$
2. $x^2 - 9x + 20 = 0$
3. $4x^2 + 5x - 6 = 0$
4. $6x^2 + 7x - 3 = 0$
5. $6x^2 + 9x + 3 = 0$
6. $3x^2 + 11x - 4 = 0$
7. $x^2 + x - 3 = 0$
8. $x^2 + 3x - 1 = 0$
9. $2x^2 + 3x + 1 = 0$
10. $x^2 - x - 1 = 0$
11. $x^2 - 3x + \frac{3}{2} = 0$
12. $x^2 + \frac{5}{3}x + \frac{2}{3} = 0$
13. $\frac{2}{3}x^2 = x - \frac{1}{3}$
14. $7x^2 + 2 = 5x$
15. $x^2 + 8x + 17 = 0$
16. $x^2 + 6x + 13 = 0$
17. $2x^2 + 6x + 5 = 0$
18. $2x^2 + 10x + 17 = 0$
19. $x^2 + 3x + 5 = 0$
20. $x^2 + x - 1 = 0$
21. $-2x^2 + x = 1$
22. $x^2 = 10x - 28$
23. Show that if r_1 and r_2 are roots of the quadratic equation $ax^2 + bx + c = 0$, then $r_1 + r_2 = -b/a$ and $r_1 r_2 = c/a$.

Using the formulas in Exercise 23, find the sums and products of the roots in 24–27.

24. $x^2 - 3x + 2 = 0$
25. $2x^2 + 3x - 6 = 0$

26. $x^2 + 2x - 3 = 0$ 27. $3x^2 - 9x - 5 = 0$

28. Show that the equation $ax^2 + bx + c = 0$ ($a \neq 0$) can be written equivalently as

$$x = \frac{-b \pm \sqrt{b^2 - 4ac}}{2a}.$$

4.4 Equations Involving Radicals

Equality of like powers

In order to find solution sets for equations containing radical expressions, we shall need the following result.

Theorem 4.3 *If $U(x)$ and $V(x)$ are algebraic expressions in x, then the solution set of $U(x) = V(x)$ is a subset of the solution set of $[U(x)]^n = [V(x)]^n$, for each natural number n.*

This theorem, which follows simply from the fact that products of equal numbers are equal numbers, permits us to raise both members of an equation to the same natural-number power with the assurance that we do not lose any solutions of the original equation in the process. On the other hand, it does not assert that the resulting equation will be equivalent to the original equation, and indeed it will not always be so. The equation

$$[U(x)]^n = [V(x)]^n$$

may have additional solutions (called **extraneous solutions**) that are not solutions of $U(x) = V(x)$. Thus, if $a = b$, then $a^4 = b^4$, but the converse does not necessarily hold. That is, a^4 and b^4 may be equal, but $a \neq b$. For example, $(3)^4 = (-3)^4$, but $3 \neq -3$. The solution set of the equation $x^4 = 81$, obtained from $x = 3$ by raising each member to the fourth power, contains -3 as an extraneous real solution, since -3 does not satisfy the original equation even though it does satisfy $x^4 = 81$.

Necessity of checking solutions

Because the result of applying the foregoing process is not always an equivalent equation, each solution obtained through its use *must* be substituted for the variable in the original equation to check its validity. An application of this theorem is not an elementary transformation.

Example Find the solution set of $\sqrt[3]{x-1} = -1$.

Solution If we raise each member of $\sqrt[3]{x-1} = -1$ to the third power, we obtain

$$(\sqrt[3]{x-1})^3 = (-1)^3,$$
$$x - 1 = -1,$$

which is equivalent to

$$x = 0.$$

Solution Continued on Overleaf

Since $\sqrt[3]{0-1} = -1$, a solution of the original equation is 0. Moreover, 0 is the only real solution, since Theorem 4.3 guarantees that the solution set of the equation $\sqrt[3]{x-1} = -1$ is a subset of the solution set of $x = 0$.

Example Find the solution set of $\sqrt{x+2} + 4 = x$.

Solution We first write the equivalent equation

$$\sqrt{x+2} = x - 4$$

and then apply Theorem 4.3. We obtain

$$(\sqrt{x+2})^2 = (x-4)^2,$$

or

$$x + 2 = x^2 - 8x + 16.$$

This last equation is equivalent to

$$x^2 - 9x + 14 = 0,$$

or

$$(x-2)(x-7) = 0,$$

which clearly has solutions 2 and 7. Upon replacing x with 2 in the original equation, however, we obtain

$$\sqrt{2+2} + 4 = 2,$$

or

$$6 = 2,$$

which is false. Hence, 2 is not a solution of the original equation; it is an extraneous root. On the other hand, 7 does satisfy the original equation, so the solution set we seek is $\{7\}$.

It is sometimes necessary to apply Theorem 4.3 more than once in solving certain equations.

Example Find the solution set of $\sqrt{x+4} + \sqrt{9-x} = 5$.

Solution It is helpful if this equation is first transformed so that each member of the equivalent equation contains only one of the radical expressions in each member. Thus, by adding $-\sqrt{9-x}$ to each member, we obtain

$$\sqrt{x+4} = 5 - \sqrt{9-x}.$$

An application of Theorem 4.3 leads to

$$(\sqrt{x+4})^2 = (5 - \sqrt{9-x})^2,$$
$$x + 4 = 25 - 10\sqrt{9-x} + 9 - x,$$
$$2x - 30 = -10\sqrt{9-x},$$
$$x - 15 = -5\sqrt{9-x}.$$

4.5 Substitution in Solving Equations

Applying Theorem 4.3 again, we obtain

$$(x - 15)^2 = (-5\sqrt{9 - x})^2,$$
$$x^2 - 30x + 225 = 25(9 - x),$$
$$x^2 - 5x = 0,$$
$$x(x - 5) = 0.$$

It is clear that this last equation has 0 and 5 as solutions. Since both of these satisfy the original equation, the solution set we seek is $\{0, 5\}$.

Exercise 4.4

Solve and check. If there is no solution, so state.

1. $\sqrt{x} = 8$
2. $4\sqrt{x} = 9$
3. $\sqrt{y + 8} = 1$
4. $\sqrt{y + 3} = 2$
5. $x - 1 = \sqrt{2x + 1}$
6. $2y + 3 = \sqrt{y + 2}$
7. $\sqrt[3]{2 - y} = 3$
8. $\sqrt[5]{7 - x} = 2$
9. $\sqrt{x}\sqrt{x + 9} = 20$
10. $\sqrt{x + 3}\sqrt{x - 9} = 8$
11. $\sqrt{y^2 - y} = \sqrt{y^2 + 2y - 3}$
12. $x - 2 = \sqrt{2x^2 - 3x + 2}$
13. $\sqrt{y + 4} = \sqrt{y + 20} - 2$
14. $\sqrt{x} + \sqrt{2} = \sqrt{x + 2}$
15. $\sqrt{5 + \sqrt{x}} = \sqrt{x} - 1$
16. $\sqrt{13 + \sqrt{x}} = \sqrt{x} + 1$
17. $(5 + x)^{1/2} + x^{1/2} = 5$
18. $(y + 7)^{1/2} + (y + 4)^{1/2} = 3$
19. $(y^2 - 3y + 5)^{1/2} - (y + 2)^{1/2} = 0$
20. $(z - 3)^{1/2} + (z + 5)^{1/2} = 4$

Solve for the indicated variable. Leave the results in the form of an equation. Assume that denominators are not zero.

21. $r = \sqrt{\dfrac{A}{\pi}}$, for A
22. $t = \sqrt{\dfrac{2v}{g}}$, for g
23. $x\sqrt{xy} = 1$, for y
24. $P = \pi\sqrt{\dfrac{l}{g}}$, for g
25. $x = \sqrt{a^2 - y^2}$, for y
26. $y = \dfrac{1}{\sqrt{1 - x}}$, for x

4.5 Substitution in Solving Equations

Some equations that are not polynomial equations can nevertheless be solved by means of related polynomial equations. For example, though $y + 2\sqrt{y} - 8 = 0$ is not a polynomial equation, if the variable p is substituted for the radical expression $\sqrt{y}$, we then have $p^2 + 2p - 8 = 0$, which is a polynomial equation in p.

Example Find the solution set of $y + 2\sqrt{y} - 8 = 0$.

Solution If we set $p = \sqrt{y}$ and substitute in the given equation, we have

$$p^2 + 2p - 8 = 0,$$
$$(p + 4)(p - 2) = 0,$$
$$(p - 2)(p + 4) = 0,$$

which has 2 and -4 as solutions. The value $p = 2$ leads to $\sqrt{y} = 2$ or $y = 4$. The value $p = -4$ leads to $\sqrt{y} = -4$, which has no solution since $\sqrt{y}$ is always nonnegative when it is real. The solution set is therefore $\{4\}$.

The technique of substituting one variable for another—or, more generally, a variable for an expression—is not limited to cases involving radicals, but is useful in any situation in which an equation is polynomial in form. For example, in the equation

$$\left(x + \frac{1}{x}\right)^{-2} + 6\left(x + \frac{1}{x}\right)^{-1} + 8 = 0,$$

we would set

$$p = \left(x + \frac{1}{x}\right)^{-1}.$$

Similarly, in the equation

$$(y + 3)^{1/2} - 4(y + 3)^{1/4} + 4 = 0,$$

we would set

$$p = (y + 3)^{1/4}.$$

Exercise 4.5

Solve.

Example $x^4 - 6x^2 + 5 = 0$

Solution Set $x^2 = p$ so that $x^4 = p^2$. Rewrite and solve for p.

$$p^2 - 6p + 5 = 0$$
$$(p - 5)(p - 1) = 0$$
$$p = 1, \text{ or } p = 5$$

For each value of p, solve $x^2 = p$.

$$x^2 = 1 \qquad\qquad x^2 = 5$$
$$x = 1, \text{ or } x = -1 \qquad\qquad x = \sqrt{5}, \text{ or } x = -\sqrt{5}$$

The solution set is $\{1, -1, \sqrt{5}, -\sqrt{5}\}$.

1. $x^4 - 10x^2 + 9 = 0$
2. $y^4 - 4y^2 - 77 = 0$
3. $z^4 - 7z^2 + 12 = 0$
4. $x^4 - 8x^2 + 7 = 0$
5. $(y + 1)^2 + 5(y + 1) - 24 = 0$
6. $(z - 1)^2 + 11(z - 1) + 30 = 0$
7. $x + 6\sqrt{x} - 7 = 0$
8. $y - 12y^{1/2} + 35 = 0$
9. $z^{2/3} - 3z^{1/3} - 18 = 0$
10. $x^{2/3} - x^{1/3} - 6 = 0$
11. $\dfrac{1}{y^2} - \dfrac{7}{y} - 18 = 0$
12. $z^{-2} - 5z^{-1} - 14 = 0$
13. $\dfrac{x^2}{(x + 1)^2} + \dfrac{x}{x + 1} = 30$
14. $\left(1 + \dfrac{1}{y}\right)^2 + 3\left(1 + \dfrac{1}{y}\right) = 40$
15. $\sqrt{x} - 6\sqrt[4]{x} + 8 = 0$
16. $\sqrt{x} - 4 = 0$
17. $\sqrt{x - 6} + 3\sqrt[4]{x - 6} - 18 = 0$
18. $\sqrt[3]{x^2} - 12\sqrt[3]{x} + 20 = 0$
19. $(x + 3)^{-2} + 4(x + 3)^{-1} = 32$
20. $z^{-4} - 8z^{-2} + 15 = 0$
21. $(x^2 + 2x + 1)^2 - 3(x + 1)^2 + 2 = 0$
22. $x^4 - 2x^2 + 1 + 2(x^2 - 1) + 1 = 0$

4.6 Solution of Linear Inequalities

Open sentences such as

$$x + 3 \geq 10 \tag{1}$$

and

$$\dfrac{-2y - 3}{3} < 5 \tag{2}$$

are called **inequalities**. For appropriate values of the variable, one member of an inequality represents a real number that is less than ($<$), less than or equal to ($\leq$), greater than or equal to ($\geq$), or greater than ($>$) the real number represented by the other member.

Any element of the replacement set of the variable for which an inequality is valid is called a **solution**, and the set of all solutions of an inequality is called the **solution set** of the inequality. Inequalities that are true for every element in the replacement set of the variable—such as $x^2 + 1 > 0$, $x \in R$—are called **absolute** or **unconditional inequalities**. Inequalities that are not true for every element of the replacement set are called **conditional inequalities**—for example, inequalities (1) and (2) above.

Elementary transformations

As in the case with equations, we shall solve a given inequality by generating a series of **equivalent inequalities** (inequalities having the same solution set) until we arrive at one for which the solution set is obvious. To do this, we shall need the following theorem applicable to inequalities. The proof of this theorem follows directly from the properties of order for the set of real numbers.

Theorem 4.4 *If $P(x)$, $Q(x)$, and $R(x)$ are expressions, then for all values of x for which $P(x)$, $Q(x)$, and $R(x)$ are real numbers, the sentence*

$$P(x) < Q(x)$$

is equivalent to each of the following statements.

  I $P(x) + R(x) < Q(x) + R(x)$,
  II† $P(x) - R(x) < Q(x) - R(x)$,
  III $P(x) \cdot R(x) < Q(x) \cdot R(x)$
  IV† $\dfrac{P(x)}{R(x)} < \dfrac{Q(x)}{R(x)}$ $\Bigg\}$ *for* $x \in \{x \mid R(x) > 0\}$,

  V $P(x) \cdot R(x) > Q(x) \cdot R(x)$
  VI† $\dfrac{P(x)}{R(x)} > \dfrac{Q(x)}{R(x)}$ $\Bigg\}$ *for* $x \in \{x \mid R(x) < 0\}$.

Similarly, the sentence

$$P(x) \leq Q(x)$$

is equivalent to sentences of the form I-VI, with $<$ (or $>$) replaced by $\leq$ (or $\geq$) under the same conditions as above

The theorem can be interpreted as follows:

I and II. The addition or subtraction of the same expression to or from each member of an inequality produces an equivalent inequality in the same sense.

III and IV. If each member of an inequality is multiplied or divided by the same expression representing a *positive* number, the result is an equivalent inequality.

V and VI. If each member of an inequality is multiplied or divided by the same expression representing a *negative* number, and the direction of the inequality is reversed, the result is an equivalent inequality.

Note that Theorem 4.4 does not permit multiplying or dividing by zero, and variables in multipliers and divisors are restricted from values for which the expression vanishes. The result of applying any part of this theorem is an **elementary transformation.**

Solution of a linear inequality

Theorem 4.4 can be applied to solve inequalities in the same way that the theorems of equality are applied to solve equations.

† Parts II, IV, and VI can be considered special cases of Parts I, III, and V, respectively.

4.6 Solution of Linear Inequalities

Example Find the solution set of $\dfrac{-x+3}{4} > -\dfrac{2}{3}$, $x \in R$.

Solution Multiplying each member by 12 yields
$$3(-x+3) > -8,$$
$$-3x + 9 > -8.$$
Adding -9 to each member, we have
$$-3x > -17.$$
Finally, dividing each member by -3, we obtain
$$x < \frac{17}{3},$$
and the solution set is written
$$S = \left\{ x \mid x < \frac{17}{3} \right\}.$$

Graphical representation The solution set in the foregoing example can be pictured on a line graph as shown in Figure 4.1. The colored line indicates points with coordinates in the solution set. Note that the open dot on the right-hand endpoint indicates that $\dfrac{17}{3}$ is *not* a member of the solution set.

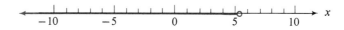

Figure 4.1

Inequalities sometimes appear in a form such as
$$-6 < 3x \le 15, \qquad (3)$$
where an expression is bracketed between two inequality symbols. As observed in Section 1.4, this means $-6 < 3x$ and $3x \le 15$. The solution set of such an inequality is obtained in the same manner as the solution set of any other inequality. In (3) above, each expression may be divided by 3 to obtain
$$-2 < x \le 5.$$
The solution set,
$$S = \{x \mid -2 < x \le 5\},$$
is shown on a line graph in Figure 4.2. The open dot at the left-hand endpoint of the interval indicates that -2 *is not* a member of the solution set, whereas the solid dot at the other ends shows that 5 *is* a member of the solution set.

Figure 4.2

Interval notation

A notation called **interval notation** is sometimes used to denote sets such as $\{x \mid -2 < x \leq 5\}$, which can be called *intervals* of real numbers. Such notation involves the use of a parentheses to denote an open endpoint and a bracket to denote a closed endpoint. Thus,

$$(-2, 5] = \{x \mid -2 < x \leq 5\}.$$

For an infinite interval such as $\left\{x \mid x < \frac{17}{3}\right\}$ we would write $\left(-\infty, \frac{17}{3}\right)$, where the symbol $-\infty$ denotes the inclusion of all real numbers less than $\frac{17}{3}$ in the interval. Similarly, $\{x \mid x \geq 4\}$ can be written as $[4, +\infty)$.

Exercise 4.6

Solve each inequality. Write the solution set in interval notation and represent it on a line graph.

Examples

a. $\dfrac{3x+1}{5} < x - 2$

b. $-1 < 2x + 3 \leq 8$

Solutions

Generate the following sequences of equivalent inequalities.

a. $3x + 1 < 5x - 10$
$11 < 2x$
$\dfrac{11}{2} < x$
$\left(\dfrac{11}{2}, +\infty\right)$

b. $-4 < 2x \leq 5$
$-2 < x \leq \dfrac{5}{2}$
$\left(-2, \dfrac{5}{2}\right]$

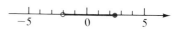

1. $x + 11 > 5$
2. $x - 3 < -2$
3. $3x + 1 \leq 10$
4. $5x - 4 \leq 11$
5. $2x + 7 \geq -5$
6. $3 - x \leq -1$
7. $x + 6 \leq 3x + 1$
8. $2x + 7 \leq x + 4$
9. $1 - 4x > x + 6$
10. $4x \geq -5x - 1$
11. $-(x + 3) \leq -1$
12. $-(2x + 4) \geq 3$
13. $-1 \leq x + 5 \leq 6$
14. $0 \leq 2x - 3 < 1$
15. $-4 < \dfrac{3x + 2}{5} \leq -2$
16. $4 \leq \dfrac{3x - 1}{2} \leq 10$
17. $5 \leq \dfrac{x + 1}{-3} \leq 8$
18. $-2 \leq \dfrac{x + 5}{2} \leq 0$

4.7 Solution of Nonlinear Inequalities

Graph each of the following sets on a number line. Describe each set as a single interval.

Example $(1, 4] \cap [-3, 2)$

Solution Graph each set separately as shown. The values for which the graphs overlap is the intersection.

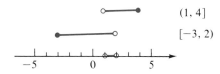

The set is $(1, 2)$.

19. $(-7, 4] \cap (0, 5)$
20. $[-5, 2) \cap [-1, 4]$
21. $(-\infty, 1] \cap (0, 2)$
22. $[1, 3) \cap [0, +\infty)$
23. $[-1, 2) \cap [2, 3]$
24. $(-\infty, -1) \cap (0, 2)$

Graph each of the following sets on a number line.

Example $[-2, 3) \cup (4, +\infty)$

Solution Graph each set on the same line. The entire region graphed is the union.

25. $(-1, 1) \cup [3, 5]$
26. $(-1, +\infty) \cup [-4, -1)$
27. $[-3, 2) \cup (2, 3]$
28. $[1, +\infty) \cup (-\infty, 4)$
29. $[-4, -2) \cup [0, 1) \cup [3, 4]$
30. $(-\infty, 1] \cup [2, 3) \cup [4, +\infty)$

4.7 Solution of Nonlinear Inequalities

As in the case with first-degree inequalities, we can generate equivalent second-degree inequalities by applying any part of Theorem 4.4. Additional procedures are necessary, however, to obtain the solution sets of such inequalities. For example, consider the inequality

$$x^2 + 4x < 5.$$

To determine values of x for which this condition holds, we might first rewrite the inequality equivalently as

$$x^2 + 4x - 5 < 0,$$

and then as

$$(x + 5)(x - 1) < 0.$$

It is clear here that only those values of x for which the factors $x + 5$ and $x - 1$ are opposite in sign will be in the solution set. These can be determined analytically by noting that $(x + 5)(x - 1) < 0$ implies either

Case I: $x + 5 < 0$ and $x - 1 > 0$

or else

Case II: $x + 5 > 0$ and $x - 1 < 0.$

Each of these two cases can be considered separately.

First, for Case I,

$$x + 5 < 0 \quad \text{and} \quad x - 1 > 0$$

imply

$$x < -5 \quad \text{and} \quad x > 1,$$

a condition which is not satisfied by any values of x. The solution set of Case I is $\emptyset$. The inequalities for Case II,

$$x + 5 > 0 \quad \text{and} \quad x - 1 < 0$$

imply

$$x > -5 \quad \text{and} \quad x < 1,$$

which lead to the solution set

$$\{x \mid -5 < x < 1\}.$$

Thus the complete solution set is

$$S = \{x \mid -5 < x < 1\} \cup \emptyset = \{x \mid -5 < x < 1\}.$$

Let us now consider an alternate method of solving quadratic inequalities. Notice in the foregoing solution process that the numbers -5 and 1 separate the set of real numbers into the three intervals shown in Figure 4.3.

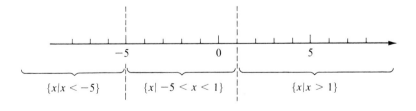

Figure 4.3

4.7 Solution of Nonlinear Inequalities

Each of these intervals either is or is not a part of the solution set of the inequality $x^2 + 4x < 5$. To determine which, we need only substitute an arbitrarily selected number from each interval and test it in the inequality. Let us use -6, 0, and 2.

$$(-6)^2 + 4(-6) \stackrel{?}{<} 5 \qquad 0^2 + 4(0) \stackrel{?}{<} 5 \qquad 2^2 + 4(2) \stackrel{?}{<} 5$$
$$36 - 24 \stackrel{?}{<} 5 \qquad 0 + 0 \stackrel{?}{<} 5 \qquad 4 + 8 \stackrel{?}{<} 5$$
$$12 \stackrel{?}{<} 5 \qquad 0 \stackrel{?}{<} 5 \qquad 12 \stackrel{?}{<} 5$$
$$\text{No.} \qquad\qquad \text{Yes.} \qquad\qquad \text{No.}$$

Clearly, the only interval involved that is in the solution set is $\{x \mid -5 < x < 1\}$, and hence this interval is the solution set. The graph is shown in Figure 4.4. If the inequality in this example were $x^2 + 4x \leq 5$, the endpoints on the graph would be shown as closed dots.

Figure 4.4

In an inequality of the form $Q(x) < 0$ or $Q(x) > 0$, numbers for which either $Q(x) = 0$ or else $Q(x)$ is undefined are called **critical numbers**. For example, critical numbers in the preceding example are -5 and 1. Again, for $2x^2 - x - 1 > 0$, the critical numbers are $-1/2$ and 1, because

$$2x^2 - x - 1 = (2x + 1)(x - 1) = 0$$

for these values; and critical numbers for $\dfrac{1}{x^2 - 4} < 0$ are 2 and -2, because $\dfrac{1}{x^2 - 4}$, or $\dfrac{1}{(x - 2)(x + 2)}$, is not defined for either number.

Let us use the notion of critical numbers to solve another inequality,

$$x^2 - 3x - 4 \geq 0. \tag{1}$$

Factoring the left-hand member yields

$$(x + 1)(x - 4) \geq 0,$$

and we observe that -1 and 4 are critical numbers because $(x + 1)(x - 4) = 0$ for these values. Hence, we wish to check the intervals shown on the number line in Figure 4.5.

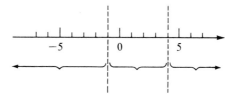

Figure 4.5

We now substitute selected arbitrary values in each interval, say -2, 0, and 5 for the variable in (1).

$$(-2)^2 - 3(-2) - 4 \overset{?}{\geq} 0 \qquad (0)^2 - 3(0) - 4 \overset{?}{\geq} 0 \qquad (5)^2 - 3(5) - 4 \overset{?}{\geq} 0$$
$$\text{Yes.} \qquad\qquad\qquad \text{No.} \qquad\qquad\qquad \text{Yes.}$$

Hence, the solution set is

$$\{x \mid x \leq -1\} \cup \{x \mid x \geq 4\}.$$

The graph is shown in Figure 4.6.

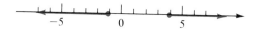

Figure 4.6

Inequalities involving fractions have to be approached with care if any fraction contains a variable in the denominator. If each member of such an inequality is multiplied by an expression containing the variable, we have to be careful either to distinguish between those values of the variable for which the expression denotes a positive and negative number, respectively, or to make sure that the expression by which we multiply is always positive. However, we can use the notion of critical numbers to avoid these complications. For example, consider the inequality

$$\frac{x}{x-2} \geq 5. \tag{2}$$

We first write (2) equivalently as

$$\frac{x}{x-2} - 5 \geq 0,$$

from which

$$\frac{x - 5(x-2)}{x-2} \geq 0,$$

$$\frac{-4x + 10}{x-2} \geq 0. \tag{2'}$$

In this case, the critical numbers are $5/2$ and 2, because $-4x + 10$ and hence $\dfrac{-4x+10}{x-2}$ equal zero for $x = 5/2$, and $\dfrac{-4x+10}{x-2}$ is undefined for $x = 2$. Thus we want to check the intervals shown on the number line in Figure 4.7.

Figure 4.7

4.7 Solution of Nonlinear Inequalities

Substituting arbitrary values for the variable in (2) or (2') in each of the three intervals, say 0, 9/4, and 3, we can identify the solution set

$$\left\{x \mid 2 < x \le \frac{5}{2}\right\},$$

as we did in the previous example. The graph is shown in Figure 4.8. Note that the left-hand endpoint is an open dot (2 is not a member of the solution set) because the left-hand member of (2) is undefined for $x = 2$.

Figure 4.8

Exercise 4.7

Solve each inequality. Write the solution set in interval notation.

Example $(x + 4)(x - 3) \le 0$

Solution The critical numbers are -4 and 3 because $(x + 4)(x - 3) = 0$ for these values. Check the intervals shown on the number line in the figure by substituting an arbitrary value from each interval for the variable in the inequality. Here we use $-5, 0$ and 5.

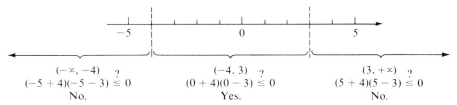

Next check the critical numbers themselves; here both are solutions. Therefore the solution set is $[-4, 3]$.

1. $(x + 1)(x - 5) > 0$
2. $(x + 6)(x + 2) < 0$
3. $(2x + 1)(3x - 8) \le 0$
4. $\left(x - \frac{1}{3}\right)(3x - 10) \ge 0$
5. $x^2 + x - 6 > 0$
6. $x^2 + 3x - 4 \le 0$
7. $x^2 < 1$
8. $x^2 + 1 \ge 0$

Example $x^2 - 2x - 1 \le 0$

Solution Find the critical numbers by the quadratic formula:

$$x = \frac{2 \pm \sqrt{4 + 4}}{2} = \frac{2 \pm 2\sqrt{2}}{2} = 1 \pm \sqrt{2}$$

Solution Continued on Overleaf

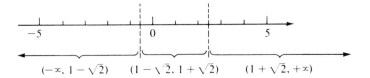

Check each interval by testing an arbitrary point in it, say -4, 1, and 4.

$$(-4)^2 - 2(-4) - 1 \stackrel{?}{<} 0 \qquad 1^2 - 2(1) - 1 \stackrel{?}{<} 0 \qquad 4^2 - 2(4) - 1 \stackrel{?}{<} 0$$
No.   Yes.   No.

Test the critical numbers themselves; since they make the expression 0, they are solutions. The solution set is therefore $[1 - \sqrt{2}, 1 + \sqrt{2}]$.

9. $x^2 - 4x + 1 \geq 0$ 10. $x^2 + 4x + 2 \leq 0$
11. $x^2 - x + 4 < 0$ 12. $-x^2 + x + 1 > 0$

Example $\dfrac{x+1}{x} \geq 2$

Solution   Write the equivalent inequalities

$$\frac{x+1}{x} - 2 \geq 0$$

$$\frac{x+1-2x}{x} \geq 0$$

$$\frac{1-x}{x} \geq 0$$

The critical numbers are 0 and 1 because $\dfrac{1-x}{x}$ is zero when $x = 1$ and undefined when $x = 0$.

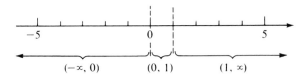

Check each interval by testing an arbitrary point in it, say $-1, \dfrac{1}{2}, 2$.

$$\frac{-1+1}{-1} \stackrel{?}{\geq} 2 \qquad \frac{\frac{1}{2}+1}{\frac{1}{2}} \stackrel{?}{\geq} 2 \qquad \frac{2+1}{2} \stackrel{?}{\geq} 2$$
No.   Yes.   No.

Test the critical numbers themselves; 1 is a solution but 0 is not. Therefore the solution set is $(0, 1]$.

4.8 Equations and Inequalities Involving Absolute Value

13. $\dfrac{x-1}{x} \le 3$

14. $\dfrac{x}{x-1} \ge 5$

15. $\dfrac{x+1}{x-1} < 1$

16. $\dfrac{1}{x-1} \le 1$

17. $\dfrac{2}{2+x} > 1$

18. $\dfrac{x}{3-x} - 3 \le 6$

Example $\dfrac{(x-1)(x+2)}{x-3} \le 0$

Solution The critical numbers are 1, −2, and 3.

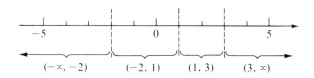

Check each interval by testing an arbitrary point in it, say −3, 0, 2, and 4.

$$\dfrac{(-3-1)(-3+2)}{-3-3} \overset{?}{\le} 0 \qquad \dfrac{(0-1)(0+2)}{0-3} \overset{?}{\le} 0$$

Yes. No.

$$\dfrac{(2-1)(2+2)}{2-3} \overset{?}{\le} 0 \qquad \dfrac{(4-1)(4+2)}{4-3} \overset{?}{\le} 0$$

Yes. No.

Test the critical numbers themselves; 1 and −2 are solutions but 3 is not. Therefore the solution set is $(-\infty, -2] \cup [1, 3)$.

19. $\dfrac{(x+1)}{(x-1)(x-3)} > 0$

20. $\dfrac{(x-1)(x+1)}{x-5} < 0$

21. $\dfrac{(x-2)(x+1)}{x} < 0$

22. $\dfrac{x}{(x+2)(x-4)} \ge 0$

23. $(x-1)(x+1)(x+3) \ge 0$

24. $(x+1)(x-4)x < 0$

25. $\dfrac{(x+2)(x-2)}{x(x+4)} > 0$

26. $x(x+2)(x-2)(x-4) \ge 0$

4.8 Equations and Inequalities Involving Absolute Value

In Section 1.4 the absolute value of a real number was defined by

$$|x| = \begin{cases} x, & \text{if } x \ge 0, \\ -x, & \text{if } x < 0, \end{cases}$$

and $|x|$ was interpreted as the distance from x to 0 on the number line. For example, $|-5| = 5$ and $|7| = 7$. Since it is also true that $|5| = 5$, the equation $|x| = 5$ has two solutions, 5 and -5. This in no way contradicts our earlier discussion concerning the number of roots of a first-degree equation. Equations involving absolute values are not polynomial equations and so cannot be assigned a degree. In fact, simple equations involving absolute values are contractions of two equations without absolute values. The same is true of inequalities. The relationships are indicated in Figure 4.9.

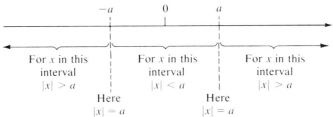

Figure 4.9

To summarize the relationships, note that for $a > 0$,

I $|x| = a$ is equivalent to $x = a$ or $x = -a$;
II $|x| < a$ is equivalent to $-a < x$ and $x < a$, i.e. $-a < x < a$;
III $|x| > a$ is equivalent to $x < -a$ or $x > a$.

To solve equations or inequalities involving absolute value, we translate into the two equivalent statements without absolute value and solve these two statements. We then recall that the everyday word "and" corresponds to set intersection and "or" (the nonexclusive "or," meaning "one or the other, or both") corresponds to set union. Our solution set will be the intersection of two sets for type II problems, the union of two sets for problems of type I or III.

Example $|x - 3| = 5$

Solution   This equality is equivalent to

$$x - 3 = 5 \quad \text{or} \quad x - 3 = -5 \quad \text{by Statement I above.}$$

The solution to the first equality is $x = 8$, the solution to the second $x = -2$. The solution set is the union, $\{-2, 8\}$.

Recall also from Section 1.4 that $|a - b| = c$ can be interpreted geometrically to mean that the distance between a and b is c. In the example we are given that the distance from x to 3 is 5. The values of x are then readily seen geometrically to be -2 and 8.

Example   Solve $|x - 3| < 5$.

Solution   This inequality is equivalent to

$$-5 < x - 3 \quad \text{and} \quad x - 3 < 5 \quad \text{by Statement II above.}$$

Solve the first inequality to obtain $-2 < x$ so that the solution set is $(-2, +\infty)$.

4.8 Equations and Inequalities Involving Absolute Value

Solve the second to obtain $x < 8$ and solution set $(-\infty, 8)$. The solution to the given inequality is therefore the intersection

$$(-2, +\infty) \cap (-\infty, 8) = (-2, 8).$$

Example Solve $|x - 3| > 5$.

Solution This inequality is equivalent to

$$x - 3 < -5 \quad \text{or} \quad x - 3 > 5 \quad \text{by Statement III above.}$$

The first inequality reduces to $x < -2$ so the solution set is $(-\infty, -2)$. The second reduces to $x > 8$ and has solution set $(8, \infty)$. The solution to the given inequality is therefore the union $(-\infty, -2) \cup (8, \infty)$.

Inequalities involving absolute values and $\leq$ or $\geq$ are treated similarly. These relationships are apparent in Figure 4.9.

IV $|x| \leq a$ is equivalent to $-a \leq x$ and $x \leq a$, i.e. $-a \leq x \leq a$;

V $|x| \geq a$ is equivalent to $x \leq -a$ or $x \geq a$.

Exercise 4.8

Solve.

Example $|x - 2| \leq 5$

Solution Write the equivalent inequalities from Statement IV above.

$$-5 \leq x - 2 \quad \text{and} \quad x - 2 \leq 5$$

Solve each inequality.

$$-3 \leq x \quad \text{and} \quad x \leq 7$$

The solution set is $[-3, +\infty) \cap (-\infty, 7] = [-3, 7]$.

1. $|x + 2| = 3$
2. $|4 - x| = 1$
3. $|3 - x| = 4$
4. $|x + 5| = 1$
5. $|2x + 1| = 3$
6. $|3x - 2| = 4$
7. $|x - 2| < 5$
8. $|x + 3| < 2$
9. $\left|x - \dfrac{1}{2}\right| < \dfrac{2}{3}$
10. $\left|x + \dfrac{3}{4}\right| < \dfrac{1}{2}$
11. $|x - 4| > 1$
12. $|x + 2| > 2$
13. $|3 + x| > 7$
14. $|2x + 1| > 3$
15. $|x - 4| \leq 1$
16. $|x + 2| \leq 5$
17. $|2x - 3| \leq 3$
18. $|3x + 4| \leq \dfrac{1}{2}$
19. $\left|x - \dfrac{1}{2}\right| \geq 2$
20. $|x + 1| \geq \dfrac{1}{2}$
21. $|3x - 1| \geq 3$

22. $\left|\dfrac{1}{2}x - \dfrac{1}{4}\right| \geq \dfrac{1}{4}$ 23. $|x| \leq 3$ 24. $|x| \geq 1$

Rewrite each of the following using absolute-value notation.

Examples a. $-1 < x < 1$ b. $-3 < x < 1$

Solutions a. $|x| < 1$ b. $-2 < x + 1 < 2$
$|x + 1| < 2$

25. $-3 \leq x \leq 3$ 26. $-4 \leq x \leq 4$
27. $-2 < x < 6$ 28. $0 \leq x \leq 8$
29. $-8 \leq x \leq -2$ 30. $-10 < x < 0$
31. $-\dfrac{4}{3} < x < \dfrac{2}{3}$ 32. $\dfrac{1}{2} < x < \dfrac{9}{2}$

4.9 Word Problems

Equations and inequalities can be used to express quantitative relations in word problems symbolically. The problem may be explicitly concerned with numbers, or it may be concerned with numerical measures of physical quantities. In either event, we seek the set of numbers (the solution set) for which the stated relationship holds. The following suggestions are frequently helpful in expressing the conditions of the problem symbolically:

i. Determine the quantities asked for and represent them by symbols. Since at this time we are using one variable only, all relevant quantities should be represented in terms of this variable.

ii. Where applicable, draw a sketch and label all known quantities thereon; label the unknown quantities in terms of symbols.

iii. Find in the problem a quantity that can be represented in two different ways and write this representation as an equation. The equation may derive from:
   a. the problem itself, which may state a relationship explicitly; for example, "What number added to 4 gives 7?" produces the equation $4 + x = 7$;
   b. formulas or relationships that are part of your general mathematical background; for example, $A = \pi r^2$, $d = rt$.

iv. Solve the resulting equation.

v. Check the results against the original problem. It is not sufficient to check the result in the equation, because the equation itself may be in error or have a root irrelevant to the problem.

In some cases, the mathematical model we obtain for a physical situation is a quadratic equation that has two real solutions. It may be that one but not both of

4.9 Word Problems

the solutions fits the physical situation. For example, if we were asked to find two consecutive *natural numbers* of which the product is 72, we would write the equation

$$x(x + 1) = 72$$

as our model. Solving this equation, we have

$$x^2 + x - 72 = 0,$$
$$(x + 9)(x - 8) = 0,$$

with solution set $\{8, -9\}$. Since -9 is not a natural number, we must reject it as a possible answer to our original question; the solution 8, however, leads to the consecutive natural numbers 8 and 9. As additional examples, observe that we would not accept -6 feet as the height of a man, or $27/4$ for the number of persons in a room.

A quadratic equation used as a model for a physical situation may have two, one, or no meaningful solutions—meaningful, that is, in a physical sense. Answers to word problems should always be checked against the set of meaningful numbers for the original problem.

Exercise 4.9

Solve the following word problems.

In Exercises 1–12, use a mathematical model in the form of a first-degree equation in one variable.

Example A collection of coins consisting of dimes and quarters has a value of $12.75. If there are 33 more dimes than quarters, how many of each are in the collection?

Solution We follow the suggestions i–v, omitting ii as not applicable.

i. Represent the unknown quantities symbolically. Let x represent the number of quarters; then $x + 33$ represents the number of dimes.

iii. Write an equation relating the values of quarters and dimes to the total value.

$$\begin{bmatrix} \text{value of} \\ \text{quarters} \\ \text{in cents} \end{bmatrix} + \begin{bmatrix} \text{value of} \\ \text{dimes} \\ \text{in cents} \end{bmatrix} = \begin{bmatrix} \text{total} \\ \text{value} \\ \text{in cents} \end{bmatrix}$$

$$25x \quad + \quad 10(x + 33) \quad = \quad 1275$$

iv. Solve for x.

$$25x + 10x + 330 = 1275$$
$$35x = 945$$
$$x = 27$$

Therefore $x + 33 = 60$ and there are 27 quarters and 60 dimes in the collection.

v. Check: do 27 quarters and 60 dimes have a value of $12.75? Yes.

1. A man has $2.35 in change consisting of two more nickels than dimes. How many dimes and how many nickels does he have?

2. The admission at a baseball game was $2.00 for adults and $1.25 for children. The receipts were $542.50 for 350 paid admissions. How many adults, and how many children, attended the game?

3. How many pounds of an alloy containing 32% silver must be melted with 25 pounds of an alloy containing 48% silver to obtain an alloy containing 42% silver?

4. How much water should be added to 6 gallons of pure acid to obtain a 15% solution?

Example

A man has an annual income of $6500 from two investments. He has $15,000 more invested at 6% than he has invested at 8%. How much does he have invested at each rate?

Solution

i. Represent the amount invested at each rate symbolically. Let A represent the amount in dollars invested at 8%; then $A + 15,000$ represents the amounts invested at 6%.

iii. Write an equation relating the interest from each investment and the total interest.

$$\begin{bmatrix} \text{interest from} \\ 8\% \text{ investment} \end{bmatrix} + \begin{bmatrix} \text{interest from} \\ 6\% \text{ investment} \end{bmatrix} = [\text{total interest}]$$

$$0.08A + 0.06(A + 15,000) = 6500$$

iv. Solve for A.

$$8A + 6A + 90,000 = 650,000$$
$$14A = 560,000$$
$$A = 40,000; \text{ therefore } A + 15,000 = 55,000$$

The man had $40,000 invested at 8% and $55,000 invested at 6%.

v. Check: Does 8% of $40,000 ($3200) added to 6% of $55,000 ($3300) equal $6500? Yes.

5. A sum of $2000 is invested, part at 6% and the remainder at 7%. Find the amount invested at each rate if the yearly income from the two investments is $128.

6. A sum of $2700 is invested, part at 5% and the remainder at $7\frac{1}{2}$%. Find the total yearly interest if the interest on each investment is the same.

7. A man has three times as much money invested in 6% bonds as he has in stocks paying 4%. How much does he have invested in each if his yearly income from the investments is $2420?

8. A man has $1000 more invested at 5% than he has invested at 4%. If his annual income from the two investments is $698, how much does he have invested at each rate?

4.9 Word Problems

Example An express train travels 150 miles in the same time that a freight train travels 100 miles. If the express goes 20 miles per hour faster than the freight, find each rate.

Solution
i. Represent the unknown quantities symbolically. Let r represent a rate for the freight train; then $r + 20$ represents the rate of the express train.

iii. The fact that the times are equal is the significant equality in the problem.

$$(t \text{ of freight}) = (t \text{ of express})$$

Express the time of each train in terms of r (time = distance/rate).

$$\frac{100}{r} = \frac{150}{r + 20}$$

iv. Solve for r.

$$(r + 20)100 = (r)150$$
$$100r + 2000 = 150r$$
$$-50r = -2000$$
$$r = 40; \text{ therefore } r + 20 = 60$$

The freight train's rate is 40 miles per hour; the express train's rate is 60 miles per hour.

v. Check: Does the time of the freight train (100/40) equal the time of the express train (150/60)? Yes.

9. An airplane travels 1260 miles in the same time that an automobile travels 420 miles. If the rate of the airplane is 120 miles per hour greater than the rate of the automobile, find the rate of each.

10. Two cars start together and travel in the same direction, one going twice as fast as the other. At the end of 3 hours they are 96 miles apart. How fast is each traveling?

11. A freight train leaves town A for town B, traveling at an average rate of 50 miles per hour. Three hours later a passenger train also leaves town A for town B, on a parallel track, traveling at an average rate of 80 miles per hour. How far from town A does the passenger train pass the freight train?

12. A boy walked to his friend's house at the rate of 4 miles per hour and he ran back home at the rate of 6 miles per hour. How far apart are the two houses if the round trip took 20 minutes?

In Exercises 13–23, use a mathematical model in the form of a second-degree equation in one variable.

Example Find two numbers whose sum is 27 and whose product is 180.

Solution
i. Let x represent one of the numbers so that $27 - x$ represents the other.

iii. The equation is

$$x(27 - x) = 180.$$

Solution Continued on Overleaf

iv. Solve for x.

$$27x - x^2 = 180$$
$$x^2 - 27x + 180 = 0$$
$$x = \frac{27 \pm \sqrt{729 - 720}}{2} = \frac{27 \pm 3}{2}$$
$$x = 15, \text{ or } x = 12$$

When $x = 15$, $27 - x = 12$; when $x = 12$, $27 - x = 15$. The solutions therefore are 12 and 15.

v. Check: $12 + 15 = 27$; $12 \cdot 15 = 180$.

13. Find two numbers whose sum is 25 and whose product is 154.

14. Find two numbers whose sum is 5 and whose product is -24.

15. Find two consecutive integers whose product is 132.

16. Find two consecutive natural numbers such that the sum of their squares is 365.

17. Two airplanes with lines of flight at right angles to each other pass each other (at slightly different altitudes) at noon. One is flying at 140 miles per hour and the other is flying at 180 miles per hour. How far apart are they at 12:30 PM? *Hint:* Use the Pythagorean Theorem.

18. A box without a top is to be made from a square piece of tin by cutting a two-centimeter square from each corner and folding up the sides. If the box is to hold 128 cubic centimeters, what should be the length of each side of the original square?

19. A ball thrown vertically upward reaches a height h in feet given by the equation $h = 56t - 16t^2$, where t is the time in seconds after the throw. How long will it take the ball to reach a height of 24 feet on its way up? How long after the throw will the ball return to the height from which it was thrown?

20. The distance s a body falls in a vacuum is given by $s = v_0 t + \frac{1}{2} g t^2$, where s is measured in feet, t is measured in seconds, v_0 is the initial velocity in feet per second, and g is the constant of acceleration due to gravity (approximately 32 ft/sec/sec). How long will it take a body to fall 150 feet if v_0 is 20 feet per second? How long will it take if the body starts from rest?

21. A man and his son working together can paint their house in four days. The man can do the job alone in six days less than the son can do it. How long would it take each of them to paint the house alone? *Hint:* What part of the job could each of them do in one day?

22. A theater that is rectangular in shape seats 720 people. The number of rows needed to seat the people would be four fewer if each row held six more seats. How many seats would then be in each row?

Chapter Review

In Exercises 23 and 24, use a mathematical model in the form of an inequality in one variable.

Example A student must have an average of 80% to 90% inclusive on five tests in a course to receive a *B*. His grades on the first four tests were 98%, 76%, 86%, and 92%. What grade on the fifth test would qualify him for a *B* in the course?

Solution
i. Let x represent a grade (in percent) on the last test.

iii. Write an inequality expressing the word sentence.
$$80 \leq \frac{98 + 76 + 86 + 92 + x}{5} \leq 90$$

iv. Solve for x.
$$400 \leq 352 + x \leq 450$$
$$48 \leq x \leq 98$$

Any grade equal to or greater than 48 and less than or equal to 98 will qualify the student for a *B*.

23. In the preceding example, what grade on the fifth test would qualify the student for a *B* if his grades on the first four tests were 78%, 64%, 88%, and 76%?

24. The Fahrenheit and centigrade temperatures are related by $C = \frac{5}{9}(F - 32)$. Within what range must the temperature be in Fahrenheit degrees for the temperature in centigrade degrees to lie between $-10°$ and $20°$?

Chapter Review

[4.1] *Solve each equation.*

1. $3 + \dfrac{x}{5} = \dfrac{7}{10}$

2. $\dfrac{3x}{4} - \dfrac{5x - 1}{8} = \dfrac{1}{2}$

3. $\dfrac{x}{x + 1} + \dfrac{1}{3} = 2$

4. $1 - \dfrac{y + 1}{y - 1} = \dfrac{3}{y}$

5. Solve $\dfrac{x + y}{5} = \dfrac{x - y}{3}$ for y in terms of x.

6. Solve $\dfrac{x + y}{5} = \dfrac{x - y}{3}$ for x in terms of y.

[4.2] *Solve by factoring.*

7. $(x - 2)(x + 1) = 4$

8. $x(x + 4) = 21$

Solve by extraction of roots.

9. $6x^2 = 30$
10. $(x - 5)^2 = 2$

Solve by completing the square.

11. $x^2 + 5x - 2 = 0$
12. $2x^2 + x - 1 = 0$
13. Write $x^2 + y^2 - 8x + 6y = 12$ in the form $(x - h)^2 + (y - k)^2 = r^2$.
14. Write $y = x^2 - bx + 2$ in the form $y = (x - a)^2 + c$.

[4.3] *Solve for x by using the quadratic formula.*

15. $2x^2 - x + 2 = 0$
16. $x^2 = 4 - 2x$
17. $kx^2 - 3x + 1 = 0$
18. $x^2 + kx - 4 = 0$

[4.4] *Solve each equation.*

19. $x - 5\sqrt{x} + 6 = 0$
20. $\sqrt{x + 1} + \sqrt{x + 8} = 7$

[4.5] *Solve each equation.*

21. $y^4 - 3y^2 + 2 = 0$
22. $y^{-2} - y^{-1} - 42 = 0$

[4.6] *Solve each inequality.*

23. $\dfrac{x - 3}{5} \geq 7$
24. $2(x + 1) < \dfrac{1}{3}x$

Graph each of the following set intersections.

25. $\{x \mid x - 5 > 2\} \cap \{x \mid 3x - 2 < 22\}$
26. $\left\{ y \mid \dfrac{y + 4}{3} \leq 6 + y \right\} \cap \left\{ y \mid \dfrac{2y - 1}{3} < 1 \right\}$

[4.7] *Solve each inequality.*

27. $x^2 + 3x - 10 < 0$
28. $\dfrac{2}{1 - x} > 3$

[4.8] *Solve each equation.*

29. $|3x + 1| = 5$
30. $\left| x - \dfrac{2}{3} \right| = \dfrac{5}{3}$

31. Solve $|x - 3| > 4$ and graph the solution set on a number line.
32. Write $-5 \leq x \leq 3$ using a single inequality involving an absolute-value symbol.

[4.9] 33. In a recent election, the winning candidate received 150 votes more than his opponent. How many votes did each candidate receive if there were 4376 votes cast?

34. When the length of each side of a square is increased by five centimeters, the area is increased by 85 square centimeters. Find the length of a side of the original square.

35. A man sailed a boat across a lake and back in two and a half hours. If his rate returning was two miles per hour less than his rate going, and if the distance each way was six miles, find his rate each way.

5 Functions and Graphs

In Chapter 4 the problems with which we dealt involve only one variable. In Chapter 5 we begin to concern ourselves with problems that involve two variables.

5.1 Pairings of Real Numbers

When we work with expressions involving one variable, the replacement sets are sets of numbers. But when we deal with an expression involving two variables, each replacement for the variables must be a pair of numbers, one number to replace each variable.

Ordered pairs When the order in which the numbers of a number pair are to be considered is specified, the pair is called an **ordered pair**, and the pair is denoted by a symbol such as $(3, 2), (2, 3), (-1, 5)$, or $(0, 3)$. Each of the two numbers in an ordered pair is called a **component** of the ordered pair; in the ordered pair (x, y), x is called the **first component** and y is called the **second component.**

$R \times R$, or R^2, and the geometric plane The most important set of ordered pairs with which we shall be concerned is the set of all ordered pairs of real numbers. This set is called $R \times R$ (read "R cross R") and is often denoted by R^2. The fact that each member of R^2 corresponds to a point in the geometric plane, and that the coordinates of each point in the geometric plane are the components of a member of R^2, is the basis for all plane graphing. As you probably recall from your earlier study of algebra, the correspondence between points in the plane and ordered pairs of real numbers is usually established through a **Cartesian** (or **rectangular**) **coordinate system**, as shown in Figure 5.1. The first component of an ordered pair is called the **abscissa** and the second component the **ordinate** of the corresponding point.

5.1 Pairings of Real Numbers

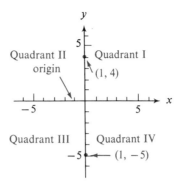

Figure 5.1

Solutions of an equation in two variables

Equations in two variables, such as

$$3x + 2y = 12, \quad x^2y + 3x = y^5, \quad \text{and} \quad xy = y^2 - 5,$$

where x and y represent real numbers, have ordered pairs of numbers as solutions. For example, if the components of $(2, 3)$ are substituted for the variables x and y, in that order, in the equation

$$3x + 2y = 12, \tag{1}$$

the result is

$$3(2) + 2(3) = 12,$$

which is true. Hence, $(2, 3)$ is a solution of Equation (1). Since many equations (and inequalities) in two variables have an infinite number of solutions, we shall sometimes use the set-builder notation

$$\{(x, y) \mid \text{condition on } x \text{ and } y\}$$

to represent the set of all solutions.

Relations

In mathematics and applications of mathematics, we are often concerned with relationships that involve the pairing of numbers. Such relationships as Fahrenheit and Celsius temperatures, standard height and weight tables, and square root tables are common examples. The following definition gives us a basis for considering such relationships mathematically.

Definition 5.1   A **relation** is a set of ordered pairs.

Relations determined by sets of ordered pairs

It follows from Definition 5.1 that any subset of $R \times R$ is a relation. The set of all first components of the pairs in a relation is called the **domain** of the relation, and the set of all second components is called the **range** of the relation. For example,

$$\{(1, 5), (2, 10), (3, 15)\} \quad \begin{array}{l}\text{elements in the domain} \\ \text{elements in the range}\end{array}$$

is a relation with domain $\{1, 2, 3\}$ and range $\{5, 10, 15\}$.

Relations determined by rules

Relations are often defined by equations or inequalities. Such an equation or inequality can be thought of as a "rule" for finding the value or values in the range of a relation paired with a given value in the domain. For brevity, we shall simply refer to equations or inequalities such as

$$y = \frac{1}{x-2}, \quad \text{or} \quad y = \sqrt{x-2}, \quad \text{or} \quad y > 2x + 1$$

as relations.

If a relation is defined by an equation with variables x and y, and the domain (set of values for x) is not specified, we shall understand that *the domain is the set of all real numbers for which a real number exists in the range*.

Examples

Find the domain of each of the following relations.

a. $y = \dfrac{1}{x-2}$ b. $y = \sqrt{x-2}$

Solutions

a. Since $1/(x-2)$ is a real number for every value x except 2, the domain of the relation is the set of all real numbers x, $x \neq 2$ or $\{x \mid x \neq 2\}$.

b. Since $\sqrt{x-2}$ is a real number only when $x - 2 \geq 0$, the domain of the relation is the set of all real numbers x, $x \geq 2$ or $\{x \mid x \geq 2\}$.

Functions

From the definition of a relation, each element in the domain is said to be *paired* with one or more elements in the range. A relation may be a pairing in which one or more elements in the domain are associated with *more than one element in the range*, as shown in Figure 5.2.

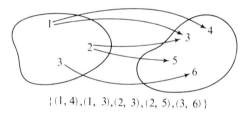

Figure 5.2

A relation may also be a pairing in which each element in the domain is associated with *only one element in the range*, as illustrated in Figures 5.3-a and 5.3-b. Such a relation is called a *function*.

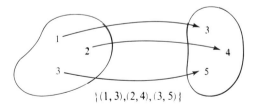

a

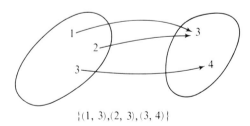

b

Figure 5.3

5.1 Pairings of Real Numbers

Definition 5.2 *A **function** is a relation in which no two elements in the range are paired with the same element in the domain.*

For example, both

$$\{(3, 4), (4, 5), (6, 8)\} \qquad (2)$$

and

$$\{(3, 4), (3, 5), (6, 8)\} \qquad (3)$$

are relations. However, only set (2) is a function. Set (3) is not a function because two elements, 4 and 5, in the range are paired with the same element, 3, in the domain.

Function notation

In general, functions are denoted by single symbols; for example, f, g, h, and F might designate functions. Furthermore, a symbol such as $f(x)$, read "f of x" or "the value of f at x," represents the value in the range of a function f associated with the value x in the domain. This is the same use we made of $P(x)$ in Section 2.1 when we discussed algebraic expressions, and in particular, polynomial expressions.

Examples

If $f(x) = x^2 + 3$, find the elements in the range associated with the given domain elements.

   a. 2   b. 3   c. 5

Solutions

Substituting the given elements in the domain for x, we have

a. $f(2) = (2)^2 + 3$ b. $f(3) = (3)^2 + 3$ c. $f(5) = (5)^2 + 3$
           $= 7$ $= 12$ $= 28$

In some applications it is useful to know the difference $f(b) - f(a)$ between two specific range values. We introduce the notation "$f(x) \,|_a^b$" for that difference. Thus,

$$f(x) \,|_a^b = f(b) - f(a).$$

Examples

For the given function f and values a and b, find $f(x) \,|_a^b$.

   a. $f(x) = x^2$; $a = 1, b = 2$ b. $f(x) = 2x - 3$; $a = 0, b = 1$

Solutions

a. $x^2 \,|_1^2 = (2)^2 - (1)^2$ b. $(2x - 3) \,|_0^1 = (2(1) - 3) - (2(0) - 3)$
            $= 4 - 1$ $= -1 - (-3)$
            $= 3$ $= 2$

Note that $f(x)$ is associated with elements in the range of a function and hence plays the same role as y in equations involving the variables x and y. For example,

$$y = x + 3 \quad \text{and} \quad f(x) = x + 3$$

define the same function.

Exercise 5.1

Supply the missing components so that the ordered pairs are solutions of the given equations.

(a) (0,)   (b) (1,)   (c) (2,)   (d) (−3,)   (e) $\left(\frac{2}{3}, \right)$

1. $2x + y = 6$
2. $y = 9 - x^2$
3. $y = \dfrac{3x}{x^2 - 2}$
4. $y = 0$
5. $y = \sqrt{3x + 11}$
6. $y = |x - 1|$

(a) Specify the domain of each relation.
(b) State whether or not each relation is a function.

Example $\{(3, 5), (4, 8), (4, 9), (5, 10)\}$

Solution
a. Domain (the set of first components): $\{3, 4, 5\}$

b. The relation is not a function, because two ordered pairs, (4, 8) and (4, 9), have the same first components.

7. $\{(2, 3), (5, 7), (7, 8)\}$
8. $\{(-1, 6), (0, 2), (3, 3)\}$
9. $\{(2, -1), (3, 4), (3, 6)\}$
10. $\{(-4, 7), (-4, 8), (3, 2)\}$
11. $\{(5, 5), (6, 6), (7, 7)\}$
12. $\{(0, 0), (2, 4), (4, 2)\}$

Specify the domain of the relation.

Examples   a. $y = \sqrt{16 - x^2}$ b. $y = \dfrac{1}{x(x + 2)}$

Solutions
a. For what values of x is $16 - x^2 \geq 0$?
The domain is $\{x \mid -4 \leq x \leq 4\}$.

b. For what values of x is $x(x + 2) \neq 0$?
The domain is $\{x \mid x \neq 0, -2\}$.

13. $y = x + 7$
14. $y = 2x - 3$
15. $y = x^2$
16. $y = \dfrac{1}{x}$
17. $y = \dfrac{1}{x - 2}$
18. $y = \dfrac{1}{x^2 + 1}$
19. $y = \sqrt{x}$
20. $y = \sqrt{4 - x}$
21. $y = \sqrt{4 - x^2}$
22. $y = \sqrt{x^2 - 9}$
23. $y = \dfrac{4}{x(x - 1)}$
24. $y = \dfrac{x}{(x - 1)(x + 2)}$

If $f(x) = x + 2$, find the given element in the range.

5.1 Pairings of Real Numbers

Examples a. $f(3)$ b. $f(a - 4)$

Solutions a. Substituting 3 for x, we have

$$f(3) = 3 + 2 = 5$$

The element is 5.

b. Substituting $a - 4$ for x, we have

$$f(a - 4) = (a - 4) + 2 = a - 2$$

The element is $a - 2$.

25. $f(0)$ 26. $f(1)$ 27. $f(-3)$ 28. $f(-2)$
29. $f(a)$ 30. $f(-a)$ 31. $f(2a)$ 32. $f(a + 2)$

If $g(x) = x^2 - 2x + 1$, find the given element in the range.

33. $g(-2)$ 34. $g(0)$ 35. $g(-1)$ 36. $g(3)$
37. $g(a - 1)$ 38. $g(a + 1)$ 39. $g(a/2)$ 40. $g(-2a)$

If $f(x) = x + 2$, find the element in the domain of f associated with the given element in the range.

Examples a. $f(x) = 5$ b. $f(x) = 2a + 1$

Solutions a. Replacing $f(x)$ with 5, we have

$$5 = x + 2, \quad \text{or} \quad x = 3$$

The element is 3.

b. Replacing $f(x)$ with $2a + 1$, we have

$$2a + 1 = x + 2, \quad \text{or} \quad x = 2a - 1$$

The element is $2a - 1$.

41. $f(x) = 3$ 42. $f(x) = -2$ 43. $f(x) = a$ 44. $f(x) = a + 2$

If $g(x) = x^2 - 1$, find all elements in the domain of g associated with the given element in the range.

45. $g(x) = 0$ 46. $g(x) = 3$ 47. $g(x) = 8$ 48. $g(x) = 5$
49. Suppose $f(x) = x + 2$ and $g(x) = x - 2$. Find each of the following.
   a. $f(0)$ b. $g(2)$ c. $f(-2)$ d. $g(-10)$
50. If $f(x) = x^2 - x + 1$, find each of the following.
   a. $f(x + h) - f(x)$ b. $\dfrac{f(x + h) - f(x)}{h}$

Compute $f(x) \vert_a^b$ for the given function and values of a and b.

Examples a. $f(x) = x^3$, $a = 0$, $b = 1$ b. $f(x) = \sqrt{x}$; $a = 1$, $b = 4$

Solutions a. $x^3 \vert_0^1 = (1)^3 - (0)^3 = 1$ b. $\sqrt{x} \vert_1^4 = \sqrt{4} - \sqrt{1} = 1$

51. $f(x) = x^2 - 4$, $a = -1$, $b = 2$

52. $f(x) = x^4$, $a = -1$, $b = 1$
53. $f(x) = \sqrt{x}$, $a = 4$, $b = 9$
54. $f(x) = \sqrt{x + 1}$, $a = 15$, $b = 24$

State whether or not the given relation is a function and explain why or why not.

Examples

a. $x^2 y = 3$

b. $x^2 + y^2 = 36$

Solutions

Solve explicitly for y.

a. $y = \dfrac{3}{x^2}$

Yes. There is only one value of y associated with each value of x ($x \neq 0$).

b. $y = \pm\sqrt{36 - x^2}$

No. There are two values of y associated with values of x satisfying $|x| < 6$.

55. $y^2 = x^3$

56. $y = x^3 + 4$

*Any function satisfying the condition that $f(-x) = f(x)$ for all x in the domain is called an **even function**. Any function satisfying the condition that $f(-x) = -f(x)$ for all x in the domain is an **odd function**. Which of the following functions are even and which are odd?*

57. **a.** $f(x) = x^2$ **b.** $f(x) = x^3$
58. **a.** $f(x) = x^4 - x^2$ **b.** $f(x) = x^3 - x$

5.2 Linear Functions

A **first-degree equation**, or **linear equation**, in x and y is an equation that can be written equivalently in the form

$$Ax + By + C = 0 \quad (A \text{ and } B \text{ not both } 0). \tag{1}$$

Graphs of first-degree equations

We shall call (1) the **standard form** for a linear equation. The graph of any such equation (technically, of its solution set) in R^2 is a straight line, although we do not prove this here.

Since two distinct points determine a straight line, it is evident that we need find only two solutions of such an equation to determine its graph—that is, the graph of the solution set of the equation. In practice, the two solutions easiest to find are usually those with first and second components, respectively, equal to zero—that is, the solutions $(0, y_1)$ and $(x_1, 0)$.

The numbers x_1 and y_1 are called the ***x*- and *y*-intercepts** of the graph. As an example, consider the equation

$$3x + 4y = 12 \tag{2}$$

5.2 Linear Functions

If $y = 0$, we have $x = 4$, and the x-intercept is 4. If $x = 0$, then $y = 3$, and the y-intercept is 3. Thus the graph of (2) appears as in Figure 5.4.

If the graph intersects both axes at or near the origin, then the intercepts either do not represent two separate points, or the points are too close together to be of much use in drawing the graph. It is then necessary to plot at least one other point at a distance far enough removed from the origin to establish the line with pictorial accuracy.

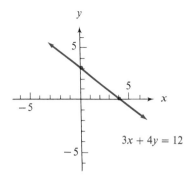

Figure 5.4

Equations of horizontal lines

Two special cases of linear equations are worth noting. First, an equation such as

$$y = 4$$

may be considered an equation in two variables,

$$0x + y = 4.$$

For each x, this equation assigns $y = 4$. That is, any ordered pair of the form $(x, 4)$ is a solution of the equation. For instance,

$$(1,4), \quad (2,4), \quad (3,4), \quad \text{and} \quad (4,4),$$

are all solutions of the equation. If we graph these points and connect them with a straight line, we have the graph shown in Figure 5.5-a.

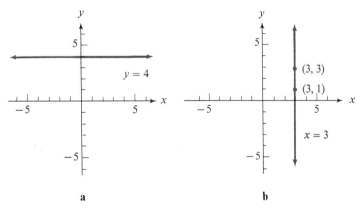

Figure 5.5

Since the equation

$$y = 4$$

assigns to each x the same value for y, the function defined by this equation is called a **constant function**.

Equations of vertical lines

The other special case of the linear equation is of the type

$$x = 3,$$

which may be looked upon as

$$x + 0y = 3.$$

Here, only one value is permissible for x, namely 3, whereas any value may be assigned to y. That is, any ordered pair of the form $(3, y)$ is a solution of this equation. If we choose two solutions, say $(3, 1)$ and $(3, 3)$, we can draw the graph shown in Figure 5.5-b. It is clear that this equation does *not* define a function with x as its first component.

If, however, $B \neq 0$ in the first-degree equation, $Ax + By + C = 0$, then this equation defines a function. Such a function is called a **linear function**. In particular, every constant function is a linear function.

Directed distances

Any two distinct points in a plane can be looked upon as the endpoints of a line segment. Two fundamental properties of a line segment are its **length** and its **inclination** with respect to the x-axis.

Let P_1, with coordinates (x_1, y_1), and P_2, with coordinates (x_2, y_2), be endpoints of a line segment. If we construct through P_2 a line parallel to the y-axis, and through P_1 a line parallel to the x-axis, the lines will meet at a point P_3, as shown in Figure 5.6. The x-coordinate of P_3 is evidently the same as the x-coordinate of P_2, and the

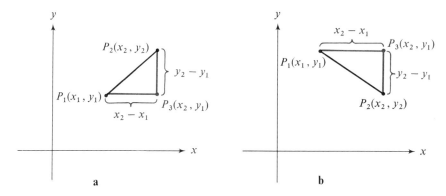

Figure 5.6

y-coordinate of P_3 is the same as that of P_1; hence the coordinates of P_3 are (x_2, y_1). By inspection, we observe that the distance from P_2 to P_3 is simply the absolute value of the difference in the y-coordinates of the two points, $|y_2 - y_1|$, and the distance between P_1 and P_3 is the absolute value of the difference of the x-coordinates of these points, $|x_2 - x_1|$.

In general, since $y_2 - y_1$ is positive or negative as $y_2 > y_1$ or $y_2 < y_1$, respectively, and $x_2 - x_1$ is positive or negative as $x_2 > x_1$ or $x_2 < x_1$, respectively, it is also convenient to designate the distances represented by $x_2 - x_1$ and $y_2 - y_1$ as positive or negative. Such distances are called **directed distances**.

5.2 Linear Functions

Distance between two points

The Pythagorean theorem can be used to find the length, d, of the line segment joining P_1 and P_2. This theorem asserts that the square on the hypotenuse of any right triangle is equal to the sum of the squares on the legs. Thus, we have

$$d^2 = (x_2 - x_1)^2 + (y_2 - y_1)^2,$$

and by considering only the positive (or nonnegative) square root of the right-hand member, we obtain

$$d = \sqrt{(x_2 - x_1)^2 + (y_2 - y_1)^2}. \tag{3}$$

Since the distances $x_2 - x_1$ and $y_2 - y_1$ are squared, it makes no difference here whether they are positive or negative—the result is the same. Equation (3) is a formula for the **distance** between any two points in the plane in terms of the coordinates of the points. The distance is always taken as positive—or 0 if the points coincide. If the points P_1 and P_2 lie on the same horizontal line, we have observed that the directed distance between them is

$$x_2 - x_1,$$

and if they lie on the same vertical line, then the directed distance is

$$y_2 - y_1.$$

If we are concerned only with distance and not direction, then these become

$$d = |x_2 - x_1| \quad \text{and} \quad d = |y_2 - y_1|.$$

Slope of a line segment

The second useful property of the line segment joining two points, its inclination, can be measured by comparing the *rise* of the segment with a given *run*, as shown in Figure 5.7.

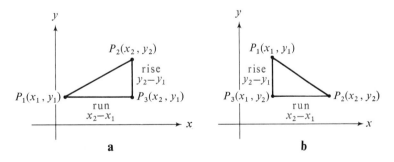

Figure 5.7

The ratio of *rise* to *run* is called the **slope** of the line segment and is designated by the letter m. Since the rise is simply $y_2 - y_1$ and the run is $x_2 - x_1$, the slope of the line segment joining P_1 and P_2 is given by

$$m = \frac{y_2 - y_1}{x_2 - x_1} \quad x_2 \neq x_1.$$

If P_2 is to the right of P_1, $x_2 - x_1$ will necessarily be positive, and the slope will be positive or negative as $y_2 - y_1$ is positive or negative. Thus positive slope indicates that a line rises to the right; negative slope indicates that it falls to the right. Since

$$\frac{y_2 - y_1}{x_2 - x_1} = \frac{-(y_1 - y_2)}{-(x_1 - x_2)} = \frac{y_1 - y_2}{x_1 - x_2},$$

the restriction that P_2 be to the right of P_1 is not necessary, and the order in which the points are considered is immaterial in determining slope.

If a line segment is parallel to the x-axis, then $y_2 - y_1 = 0$, and the line has slope 0; but if it is parallel to the y-axis, then $x_2 - x_1 = 0$, and its slope is not defined. These two special cases are shown in Figure 5.8.

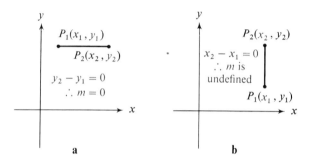

Figure 5.8

In the next section, we shall see how the slope concept is applied in discussing linear functions and their graphs.

Exercise 5.2

Graph.

Example $3x + 4y = 24$

Solution   Determine the intercepts.

If $x = 0$, then $y = 6$;
if $y = 0$, then $x = 8$.

Sketch the line through $(0, 6)$ and $(8, 0)$, as shown.

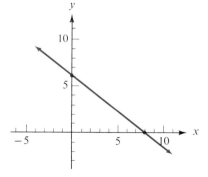

1. $y = 3x + 1$ **2.** $y = x - 5$ **3.** $y = -2x$ **4.** $2x + y = 3$

5.2 Linear Functions   131

5. $3x - y = -2$ 6. $3x = 2y$ 7. $2x + 3y = 6$ 8. $3x - 2y = 8$
9. $2x + 5y = 10$ 10. $y = 5$ 11. $x = -2$ 12. $x = -3$

Graph $f(x) = x - 1$. Represent $f(5)$ and $f(3)$ by drawing line segments from $(5, 0)$ to $(5, f(5))$ and from $(3, 0)$ to $(3, f(3))$.

Example

Solution

$f(5) = 5 - 1 = 4$

    The ordinate at $x = 5$ is 4.

$f(3) = 3 - 1 = 2$

    The ordinate at $x = 3$ is 2.

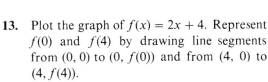

13. Plot the graph of $f(x) = 2x + 4$. Represent $f(0)$ and $f(4)$ by drawing line segments from $(0, 0)$ to $(0, f(0))$ and from $(4, 0)$ to $(4, f(4))$.

14. Plot the graph of $f(x) = 2x + 1$. Represent $f(3)$ and $f(-2)$ by drawing line segments from $(3, 0)$ to $(3, f(3))$ and from $(-2, 0)$ to $(-2, f(-2))$.

15. Suppose the function defined by $y = 2x + 1$ has as domain the set of real numbers between and including -1 and 1. Plot the graph of the function on a rectangular coordinate system. What is the range of the function?

16. Graph $x + y = 6$ and $5x - y = 0$ on the same set of axes. Estimate the coordinates of the point of intersection. What can you say about the coordinates of this point in relation to the two linear equations?

Find the distance between each of the given pairs of points, and find the slope of the line segment joining them.

Example $(3, -5), (2, 4)$

Solution Consider $(3, -5)$ as P_1 and $(2, 4)$ as P_2.

$$d = \sqrt{(x_2 - x_1)^2 + (y_2 - y_1)^2} \qquad m = \frac{y_2 - y_1}{x_2 - x_1}$$

$$= \sqrt{[2 - 3]^2 + [4 - (-5)]^2} \qquad = \frac{4 - (-5)}{2 - 3}$$

$$= \sqrt{1 + 81} \qquad = \frac{9}{-1}$$

Distance, $\sqrt{82}$; slope -9

17. $(1, 1), (4, 5)$ 18. $(-1, 1), (5, 9)$ 19. $(-3, 2), (2, 14)$
20. $(-4, -3), (1, 9)$ 21. $(2, 1), (1, 0)$ 22. $(-3, 2), (0, 0)$

23. (5, 4), (−1, 1)   24. (2, −3), (−2, −1)   25. (3, 5), (−2, 5)
26. (2, 0), (−2, 0)   27. (0, 5), (0, −5)   28. (−2, −5), (−2, 3)

Find the lengths of the sides of the triangle having vertices as given.

29. (10, 1), (3, 1), (5, 9)   30. (0, 6), (9, −6), (−3, 0)
31. (5, 6), (11, −2), (−10, −2)   32. (−1, 5), (8, −7), (4, 1)
33. Show that the triangle described in Exercise 30 is a right triangle. *Hint:* Use the converse of the Pythagorean theorem.
34. The two line segments with endpoints at (0, −7), (8, −5) and (5, 7), (8, −5) are perpendicular. Find the slope of each line segment. Compare the slopes. Do the same for the perpendicular line segments with endpoints at (8, 0), (6, 6) and (−3, 3), (6, 6). Can you make a conjecture about the slopes of perpendicular line segments?
35. The graph of a linear function contains the points (2, −3) and (6, −1). Find an equation that defines the function.
36. Determine algebraically whether or not the points lie on the same line.
    a. (2, 7), (−2, −5), (0, 1)   **b.** (9, 5), (−3, −1), (0, 1)
37. Show by similar triangles that the coordinates of the midpoint of the line segment joining the points $P_1(x_1, y_1)$ and $P_2(x_2, y_2)$ are given by

$$x = \frac{x_1 + x_2}{2} \quad \text{and} \quad y = \frac{y_1 + y_2}{2}.$$

38. Using the results of Exercise 37, find the coordinates of the midpoint of the line segment joining:
    a. (2, 4) and (6, 8)   **b.** (−4, 6) and (6, −10)

5.3 Forms for Linear Equations

Point-slope form

Assuming that the slope of the line segment joining any two points on a line does not depend on the points (as can be shown by considering similar triangles), consider a line in the plane with given slope m that passes through a given point (x_1, y_1), as shown in Figure 5.9. If we choose any other point on the line and assign to it the coordinates (x, y), it is evident that the slope of the line is given by

$$\frac{y - y_1}{x - x_1} = m,$$

from which

$$y - y_1 = m(x - x_1). \qquad (1)$$

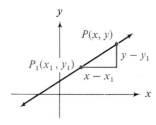

Figure 5.9

5.3 Forms for Linear Equations

Note that (1) is satisfied also by $(x, y) = (x_1, y_1)$. Since now x and y are the coordinates of *any* point on the line, (1) is an equation of the line passing through (x_1, y_1) with slope m. This is called the **point-slope form** for a linear equation.

Slope-intercept form

Now consider the equation of the line with slope m passing through a given point on the y-axis having coordinates $(0, b)$ as shown in Figure 5.10. Substituting the components of $(0, b)$ in the point-slope form of a linear equation,

$$y - y_1 = m(x - x_1),$$

we obtain

$$y - b = m(x - 0),$$

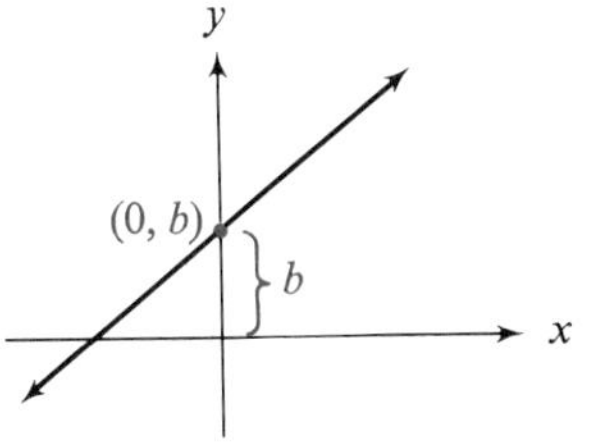

from which

$$y = mx + b. \qquad (2)$$

Figure 5.10

Equation (2) is called the **slope-intercept form** for a linear equation. Any linear equation in standard form can be written equivalently in the slope-intercept form by solving for y in terms of x if $B \neq 0$. For example,

$$2x + 3y - 6 = 0$$

can be written equivalently as

$$y = -\frac{2}{3}x + 2.$$

The slope of the line, $-2/3$, and the y-intercept, 2, can now be read directly from the last form of the equation.

Intercept form

If the x- and y-intercepts of the graph of

$$y = mx + b \qquad (3)$$

are a and b ($a, b \neq 0$), respectively, as shown in Figure 5.11, then the slope m is clearly equal to $-b/a$. Replacing m in Equation (3) with $-b/a$, we have

$$y = -\frac{b}{a}x + b,$$

$$ay = -bx + ab,$$

$$bx + ay = ab,$$

and multiplying each member by $\dfrac{1}{ab}$ produces

$$\frac{x}{a} + \frac{y}{b} = 1.$$

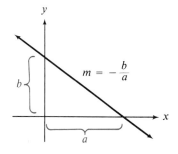

Figure 5.11

This latter form is called the **intercept form** for a linear equation. Any of the three forms discussed in this section may be used in working with linear functions and their graphs.

Exercise 5.3

Find the equation, in standard form, of the line passing through each of the given points and having the given slope.

Example $(3, -5)$, $m = -2$

Solution Substitute given values in the point-slope form of the linear equation.

$$y - y_1 = m(x - x_1)$$
$$y - (-5) = -2(x - 3)$$
$$y + 5 = -2x + 6$$
$$2x + y - 1 = 0$$

1. $(2, 1)$, $m = 4$
2. $(-2, 3)$, $m = 5$
3. $(5, 5)$, $m = -1$
4. $(-3, -2)$, $m = \dfrac{1}{2}$
5. $(0, 0)$, $m = 3$
6. $(-1, 0)$, $m = 1$
7. $(0, -1)$, $m = -\dfrac{1}{2}$
8. $(2, -1)$, $m = \dfrac{3}{4}$
9. $(-2, -2)$, $m = -\dfrac{3}{4}$
10. $(2, -3)$, $m = 0$
11. $(-4, 2)$, $m = 0$
12. $(-1, -2)$, parallel to y-axis

Write each equation in slope-intercept form; specify the slope of the line and the y-intercept.

Example $2x - 3y = 5$

Solution Solve explicitly for y.

$$-3y = 5 - 2x$$
$$3y = 2x - 5$$
$$y = \dfrac{2}{3}x - \dfrac{5}{3}$$

Compare with the general slope-intercept form $y = mx + b$.
Slope, $2/3$; y-intercept, $-5/3$.

13. $x + y = 3$
14. $2x + y = -1$
15. $3x + 2y = 1$
16. $3x - y = 7$
17. $x - 3y = 2$
18. $2x - 3y = 0$

5.3 Forms for Linear Equations

Find the equation, in standard form, of the line with the given intercepts.

Example $x = 3; \quad y = -1/2$

Solution Substitute 3 and $-1/2$ for a and b, respectively, in the intercept form $x/a + y/b = 1$.

$$\frac{x}{3} + \frac{y}{-1/2} = 1$$

$$x - 6y - 3 = 0$$

19. $x = 2; \quad y = 3$
20. $x = 4; \quad y = -1$
21. $x = -2; \quad y = -5$
22. $x = -1; \quad y = 7$
23. $x = -\frac{1}{2}; \quad y = \frac{3}{2}$
24. $x = \frac{2}{3}; \quad y = -\frac{3}{4}$

Write the equation, in standard form, of the line passing through the given point and parallel to the graph of the given equation. Hint: Parallel lines have the same slope.

25. $(2, 1); \quad y = 3x + 4$
26. $(-3, 2); \quad y = -4x + 9$
27. $(-2, 5); \quad 2x - 3y = 6$
28. $(-5, -1); \quad 5x - 4y = 7$
29. $(2, 3); \quad x = 6$
30. $(5, -4); \quad y = 2$

31. Show that, for $x_2 \neq x_1$,

$$y - y_1 = \left(\frac{y_2 - y_1}{x_2 - x_1}\right)(x - x_1)$$

is an equation of the line joining the points (x_1, y_1) and (x_2, y_2). This is the **two-point form** of the linear equation.

Use the form given in Exercise 31 to find the equation of the line through the given points.

32. $(2, 1)$ and $(-1, 3)$
33. $(3, 0)$ and $(5, 0)$
34. $(-2, 1)$ and $(3, -2)$
35. $(-1, -1)$ and $(1, 1)$
36. Consider the linear function

$$y = F(x).$$

Assuming that $(2, 3)$ and $(-1, 4)$ are known to be in F, find $F(x)$ in terms of x.

5.4 Quadratic Functions

Graph of a quadratic function

Consider the quadratic function given by the equation

$$y = x^2 - 4. \tag{1}$$

As with linear equations in two variables, solutions of this equation must be ordered pairs (x, y). We need replacements for both x and y in order to obtain a statement we may decide to be true or false. As before, such ordered pairs can be found by arbitrarily assigning values to x and computing related values for y. For instance, assigning the value -3 to x in Equation (1), we obtain

$$y = (-3)^2 - 4,$$
$$y = 5,$$

and $(-3, 5)$ is a solution. Similarly, we find that

$$(-2, 0), \quad (-1, -3), \quad (0, -4), \quad (1, -3), \quad (2, 0), \quad \text{and} \quad (3, 5)$$

are also solutions of (1). Locating the corresponding points on the plane, we have the graph in Figure 5.12-a. Clearly, these points do not lie on a straight line, and we

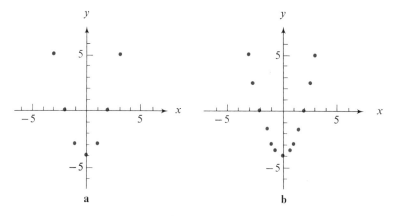

Figure 5.12

might reasonably inquire whether the graph of the solution set of (1),

$$S = \{(x, y) \mid y = x^2 - 4\},$$

forms any kind of a meaningful pattern on the plane. By graphing additional solutions of (1)—solutions with x-components between those already found—we may be able to obtain a clearer picture. Accordingly, we find the solutions

$$\left(\frac{-5}{2}, \frac{9}{4}\right), \quad \left(\frac{-3}{2}, \frac{-7}{4}\right), \quad \left(\frac{-1}{2}, \frac{-15}{4}\right), \quad \left(\frac{1}{2}, \frac{-15}{4}\right), \quad \left(\frac{3}{2}, \frac{-7}{4}\right), \quad \left(\frac{5}{2}, \frac{9}{4}\right),$$

and by graphing these points in addition to those found earlier, we have the graph in Figure 5.12-b. It now appears reasonable to connect these points in sequence, say from left to right, by a smooth curve as in Figure 5.13 and to assume that the resulting curve is a good approximation to the graph of (1). (We should realize, of course, that regardless of how many individual points are plotted, we have no absolute assurance that the smooth curve is a good approximation to the true graph; more information is needed.) This curve is an example of a **parabola**.

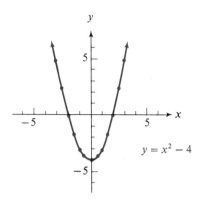

Figure 5.13

More generally, the graph of the solution set of any second-degree equation in two variables of the form

$$y = ax^2 + bx + c, \qquad (2)$$

where a, b, and c are real and $a \neq 0$, is a parabola. Since for each x an equation of the form (2) will determine only one y, such an equation defines a function (called a **quadratic function**) having as domain the entire set of real numbers and as range some subset of the real numbers. For example, we observe from the graph in Figure 5.13 that the range of the function defined by (1) is the set of real numbers

$$\{y \mid y \geq -4\}.$$

The parabola that is the graph of an equation of the form (2) will have a lowest (minimum) point (will open upward) or a highest (maximum) point (will open downward), depending on whether $a > 0$ or $a < 0$, respectively. Such a point is called the **vertex** of the parabola. The line through the vertex and parallel to the y-axis is called the **axis of symmetry,** or simply the **axis,** of the parabola; it separates the parabola into two parts, each the mirror image of the other in the axis. If we observe that the graphs of

$$y = ax^2 + bx + c \qquad (3)$$

and

$$y = ax^2 + bx \qquad (4)$$

have the same axis (Figure 5.14), we can find an equation for the axis of (3) by inspecting Equation (4). Factoring the right-hand member of $y = ax^2 + bx$ yields

$$y = x(ax + b),$$

and thus we can see that 0 and $-b/a$ are the x-intercepts of the graph. Since the axis of symmetry bisects the segment with these endpoints, an equation for the axis of symmetry is

$$x = -\frac{b}{2a}.$$

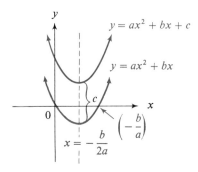

Figure 5.14

Example Find an equation for the axis of symmetry of the graph of

$$y = 2x^2 - 5x + 7.$$

Solution By comparing the given equation to $y = ax^2 + bx + c$, we see that $a = 2$ and $b = -5$. Hence, we have

$$x = -\frac{b}{2a} = -\frac{-5}{2(2)} = \frac{5}{4},$$

and $x = 5/4$ is the desired equation for the axis.

Graphing parabolas Since the vertex of a parabola lies on its axis, obtaining an equation for this axis will give us the x-coordinate of the vertex. The y-coordinate can then easily be obtained by substitution in the equation for the parabola.

When graphing a quadratic function, it is desirable first to select components for the ordered pairs that ensure that the more significant parts of the graph are displayed. For a parabola, these parts include the intercepts, if they exist, and the maximum or minimum point on the curve.

Example Graph $y = x^2 - 3x - 4$.

Solution By inspection, the y-intercept is -4. Setting $y = 0$, we have

$$0 = x^2 - 3x - 4 = (x - 4)(x + 1),$$

and the x-intercepts are 4 and -1. Since $a > 0$, the curve opens upward, and the x-coordinate of the minimum point is

5.4 Quadratic Functions

$$x = -\frac{b}{2a} = -\frac{-3}{2(1)} = \frac{3}{2}.$$

By substituting $3/2$ for x in $y = x^2 - 3x - 4$, we obtain

$$y = \left(\frac{3}{2}\right)^2 - 3\left(\frac{3}{2}\right) - 4 = \frac{9}{4} - \frac{9}{2} - 4 = -\frac{25}{4}.$$

Graphing the intercepts and the coordinates of the minimum point and then sketching the curve produce the graph shown.

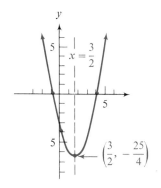

Solutions, zeros, and x-intercepts

Consider the graph of the function

$$f(x) = ax^2 + bx + c, \tag{5}$$

for $a \neq 0$, and the solution set of the equation

$$ax^2 + bx + c = 0. \tag{6}$$

Any value of x for which $f(x) = 0$ in (5) will be a solution of (6). Since any point on the x-axis has y-coordinate zero [that is, $f(x) = 0$], the x-intercepts of the graph of (5) are the real solutions of (6). Values of x for which $f(x) = 0$ are called the **zeros of the function**. Thus we have three different names for a single idea:

1. The *elements in R of the solution set* of the equation $ax^2 + bx + c = 0$. These are called the *solutions* or *roots* of the equation.

2. The *zeros in R of the function* defined by $f(x) = ax^2 + bx + c$.

3. The *x-intercepts* of the graph of the equation $f(x) = ax^2 + bx + c$.

We recall from Chapter 4 that a quadratic equation may have no real solution, one real solution, or two real solutions. If the equation has no real solution, we find that the graph of the related quadratic equation in two variables does not touch the x-axis; if there is one real solution, the graph is tangent to the x-axis; if there are two real solutions, the graph crosses the x-axis in two distinct points. These cases are illustrated in Figure 5.15.

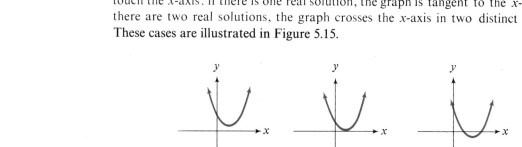

Figure 5.15

Exercise 5.4

Graph. (Obtain analytically the intercepts and the maximum or minimum point, and then sketch the curve.)

Example $y = x^2 - 7x + 6$

Solution The y-intercept is clearly 6. Since the solutions of

$$x^2 - 7x + 6 = (x - 1)(x - 6) = 0$$

are 1 and 6, these are the x-intercepts. The axis of symmetry has equation

$$x = -\frac{b}{2a} = \frac{7}{2}.$$

Writing $7/2$ for x in $y = x^2 - 7x + 6$, we obtain

$$y = -\frac{25}{4}.$$

Since the curve opens upward, the minimum point is $(7/2, -25/4)$. Using these points, you can sketch the graph as shown.

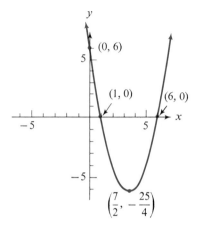

1. $y = x^2 - 5x + 4$
2. $g(x) = x^2 - 3x + 2$
3. $y = x^2 - 6x - 7$
4. $y = x^2 - 3x + 2$
5. $f(x) = -x^2 + 5x - 4$
6. $f(x) = -x^2 - 8x + 9$
7. $g(x) = \frac{1}{2}x^2 + 2$
8. $y = -\frac{1}{2}x^2 - 2$

9. Graph $f(x) = x^2 + 1$. Represent $f(0)$ and $f(4)$ by drawing line segments from $(0, 0)$ to $(0, f(0))$ and from $(4, 0)$ to $(4, f(4))$.

10. Graph $g(x) = x^2 + 1$. Represent $g(-3)$ and $g(2)$ by drawing line segments from $(-3, 0)$ to $(-3, g(-3))$ and from $(2, 0)$ to $(2, g(2))$.

Solve Exercises 11 and 12 by completing the square.

11. Find two numbers having sum 8 and product as great as possible.
12. Find the maximum possible area of a rectangle with perimeter 100 inches.
13. On a single set of axes, sketch the family of four curves that are the graphs of

$$y = x^2 + k \quad (k = -2, 0, 2, 4).$$

What effect does varying k have on the graph?

14. On a single set of axes, sketch the family of six curves that are the graphs of

$$y = kx^2 \quad \left(k = \frac{1}{2}, 1, 2, -\frac{1}{2}, -1, -2\right).$$

What effect does varying k have on the graph?

15. Graph the relation $x = y^2$.
    a. What kind of curve is the graph?
    b. Is the given relation a function? Why or why not?
    c. Does the graph of this relation have a maximum or minimum point?

16. Graph the relation $x = y^2 - 2y$.
    a. What kind of curve is the graph?
    b. Is the given relation a function? Why or why not?
    c. Does the graph of this relation have a maximum or minimum point?

Graph each of the following relations. The graphs are parabolas.

17. $x = y^2 - 4$
18. $x = y^2 - 2y - 3$
19. $x = y^2 - 4y + 4$
20. $x = 2y^2 + 3y - 2$

5.5 Polynomial Functions

Graphs of polynomial functions

In Section 5.2, we graphed linear functions

$$f(x) = a_0 x + a_1;$$

and in Section 5.4, we graphed quadratic functions

$$f(x) = a_0 x^2 + a_1 x + a_2.$$

We can graph any real polynomial function

$$P(x) = a_0 x^n + a_1 x^{n-1} + \cdots + a_n$$

by similar methods—that is, by combining the plotting of points with a consideration of certain general properties of the defining equations. In the case of the general polynomial equation, we shall lean more heavily on the use of plotted points. There is one fact about polynomial equations, however, that can be useful. This involves *turning points*, or local maximum and minimum values of y. Thus, for example, in Figure 5.16-a there is one local maximum as well as one local minimum, or a total of two turning points, and in Figure 5.16-c are one local maximum and two local minima, for a total of three turning points. In general, we have the following result.

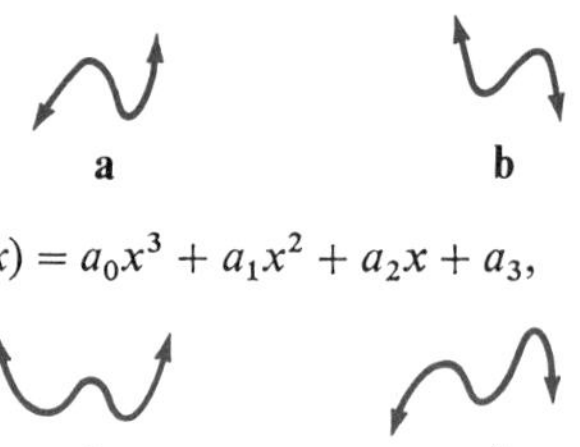

$a_0 > 0$ **a**   **b** $a_0 < 0$

$$P(x) = a_0x^3 + a_1x^2 + a_2x + a_3, \quad a_0 \neq 0$$

$a_0 > 0$ **c**   **d** $a_0 < 0$

$$P(x) = a_0x^4 + a_1x^3 + a_2x^2 + a_3x + a_4, \quad a_0 \neq 0$$

Figure 5.16

Theorem 5.1 *If $P(x) = a_0x^n + a_1x^{n-1} + \cdots + a_n$ is a real polynomial of degree n, then the graph of*

$$P(x) = a_0x^n + a_1x^{n-1} + \cdots + a_n,$$

is a smooth curve that has at most $n - 1$ turning points.

Since the proof of this theorem involves ideas we have not discussed herein, it is omitted. As a consequence of this theorem the graphs of third- and fourth-degree polynomial functions might appear as in Figure 5.16-a–5.16-d.

General form of a polynomial graph

To ascertain whether a graph ultimately goes up to the right, taking the general form (a) or (c) of Figure 5.16 rather than (b) or (d) of the figure, we can examine the leading coefficient, a_0, of the right-hand member of the defining equation; if $a_0 > 0$, then we can look for a form similar to (a) or (c), whereas if $a_0 < 0$, we can expect something similar to (b) or (d). The graph ultimately goes up or down to the right according as $a_0 > 0$ or $a_0 < 0$. If $a_0 > 0$, then it ultimately goes down to the left, as in (a) and (d), or up to the left, as in (b) and (c), according as n is odd or even. In each case, then, the leading term $a_0 x^n$ governs the behavior of the graph of the polynomial function for large values of $|x|$.

For the actual graphing process, we can obtain ordered pairs $(x, f(x))$ for any polynomial function by direct substitution, which we used in previous sections, or by using the following.

Theorem 5.2 *If $P(x)$ is a real polynomial, then for every real number c there exists a unique real polynomial $Q(x)$ such that*

$$P(x) = (x - c)Q(x) + P(c).$$

Proof From Theorem 2.3, we know that there exists a real polynomial $Q(x)$ and a real number r such that

$$P(x) = (x - c)Q(x) + r.$$

5.5 Polynomial Functions

Since this is true for all $x \in R$, then it is true for $x = c$, and we have

$$P(c) = (c - c)Q(c) + r,$$
$$P(c) = 0 \cdot Q(c) + r,$$
$$P(c) = r,$$

and the theorem is proved.

Use of the remainder theorem to find $P(c)$

This theorem is called the **remainder theorem** because it asserts that the remainder, when $P(x)$ is divided by $(x - c)$, is the value of P at c, that is, $P(c)$. Since synthetic division offers a quick means of obtaining this remainder, we can usually find values $P(c)$ more rapidly by synthetic division than by direct substitution.

Example

Given $P(x) = x^3 - x^2 + 3$, find $P(3)$ by means of the remainder theorem.

Solution

Synthetically dividing $x^3 - x^2 + 3$ by $x - 3$, we have

$$\begin{array}{r|rrrr} 3 & 1 & -1 & 0 & 3 \\ & & 3 & 6 & 18 \\ \hline & 1 & 2 & 6 & 21 \end{array}$$

and, by inspection, $r = P(3) = 21$.

Example

Graph $P(x) = 2x^3 + 13x^2 + 6x$.

Solution

Since P is defined by a cubic polynomial with positive leading coefficient, we expect to find a graph of form similar to that in Figure 5.16-a. Now, to find points $(x, P(x))$ lying on the graph, we shall use the process of synthetic division and find $P(x)$ by the remainder theorem. Since we do not know where we should look for turning points, let us start with $x = 0$. By inspection, we have $P(0) = 0$, so that the graph includes the origin. For $x = 1$, we have

$$\begin{array}{r|rrrr} 1 & 2 & 13 & 6 & 0 \\ & & 2 & 15 & 21 \\ \hline & 2 & 15 & 21 & 21 \end{array}$$

and $(1, 21)$ is on the graph. For $x = 2$, we get

$$\begin{array}{r|rrrr} 2 & 2 & 13 & 6 & 0 \\ & & 4 & 34 & 80 \\ \hline & 2 & 17 & 40 & 80 \end{array}$$

and $(2, 80)$ is on the graph. Since the signs involved at each step in the last row of the division process here are positive, it is evident that for values $x > 2$, $P(x)$ will grow increasingly large; consequently, let us turn our attention to negative values of x. For $x = -1$, we have

$$\begin{array}{r|rrrr} -1 & 2 & 13 & 6 & 0 \\ & & -2 & -11 & 5 \\ \hline & 2 & 11 & -5 & 5 \end{array}$$

and $P(-1) = 5$, so that $(-1, 5)$ is on the graph. In similar fashion, we find that the following points lie on the graph:

$$(-2, 24), \quad (-3, 45), \quad (-4, 56), \quad (-5, 45), \quad \text{and} \quad (-6, 0).$$

The graphs of the nine ordered pairs are shown in Figure a. These points make the general appearance of the graph clear, and there remains only the question of whether or not the function has a zero between -1 and 0. If we let x have values $-3/4$, $-1/4$, and $-1/2$, we obtain the additional pairs $(-3/4, 63/32)$, $(-1/4, -23/32)$, $(-1/2, 0)$. Then for $-1/2 < x < 0$, we have $P(x) < 0$, and the graph can be sketched as shown in Figure b.

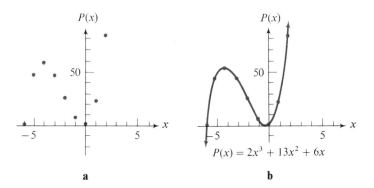

$P(x) = 2x^3 + 13x^2 + 6x$

a   b

In the particular case of the given function, we might have begun by noting that

$$\begin{aligned} P(x) &= 2x^3 + 13x^2 + 6x \\ &= x(2x^2 + 13x + 6) \\ &= x(2x + 1)(x + 6), \end{aligned}$$

and hence, observing by inspection that $P(0) = 0$, $P(-1/2) = 0$ and $P(-6) = 0$. This would have given us all three x-intercepts, and, accordingly, we could have looked for turning points between 0 and $-1/2$, and $-1/2$ and -6. However, since in general the defining equation for polynomial P will not be factorable over J, we elected to use the process we did, which is applicable to any such function.

Exercise 5.5

Use synthetic division to find values of each polynomial for the specified values of the variable.

1. $P(x) = x^3 + 2x^2 + x - 1$; $P(1)$, $P(2)$, and $P(3)$

5.6 Rational Functions

2. $P(x) = x^3 - 3x^2 - x + 3$; $P(1)$, $P(2)$, and $P(3)$
3. $P(x) = 2x^4 - 3x^3 + x + 2$; $P(-2)$, $P(2)$, and $P(4)$
4. $P(x) = 3x^4 + 3x^2 - x + 3$; $P(-2)$, $P(2)$, and $P(4)$
5. $P(x) = 3x^5 - x^3 + 2x^2 - 1$; $P(-3)$, $P(2)$, and $P(3)$
6. $P(x) = 2x^6 - x^4 + 3x^3 + 1$; $P(-3)$, $P(2)$, and $P(3)$

Graph. Use enough of the domain to include all turning points.

7. $P(x) = x^3 - 4x^2 + 3x$
8. $P(x) = x^3 - 2x^2 + 1$
9. $P(x) = 2x^3 + 9x^2 + 7x - 6$
10. $P(x) = 3x^3 + 2x^2 - x + 1$
11. $P(x) = x^4$
12. $P(x) = -x^4 + x$
13. $P(x) = x^4 - x^3 - 2x^2 + 3x - 3$
14. $P(x) = x^4 - 4x^3 - 4x + 12$

5.6 Rational Functions

So far in this chapter we have discussed polynomial functions. In this section we discuss techniques for graphing rational functions.

A function defined by an equation of the form

$$y = \frac{P(x)}{Q(x)}, \qquad (1)$$

where $P(x)$ and $Q(x)$ are polynomials in x and $Q(x)$ is not the zero polynomial, is called a **rational function**. Rational functions with real coefficients are of importance in the calculus and provide some interesting problems with respect to their graphs.

Vertical asymptotes

Since $P(x)/Q(x)$ is not defined for values of x for which $Q(x) = 0$, it is evident that we shall not be able to find points on the graph of (1) having such x-coordinates. We can, however, consider the graph for values of x as close as we please to a value, say x_0, for which $Q(x_0) = 0$, but still with $x \neq x_0$. This consideration is usually described by saying that x "approaches" x_0, "grows close" to x_0, etc., and correspondingly that $Q(x)$ approaches 0. Thus, the closer $Q(x)$ approaches 0, if $P(x)$ does not approach 0 at the same time, then the larger $|y|$ becomes in (1). For example, Figure 5.17-a shows the behavior of

$$y = \frac{2}{x-2}, \qquad (2)$$

as x approaches 2 from the right. In situations such as this, the vertical line that the curve approaches is called a **vertical asymptote**.

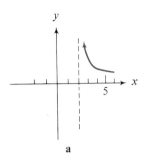
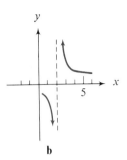

Figure 5.17

Theorem 5.3 *The graph of the function defined by $y = P(x)/Q(x)$ has a vertical asymptote $x = a$ for each value a at which $Q(x)$ vanishes and $P(x)$ does not vanish.*

In Figure 5.17-a we see the behavior of Equation (2) as x approaches 2 from the right. We are equally interested in its behavior as x approaches 2 from the left. Figure 5.17-b illustrates this. As long as $x > 2$, we have $x - 2 > 0$ and $2/(x - 2) > 0$; but if $x < 2$, we have $x - 2 < 0$ and $2/(x - 2) < 0$.

Horizontal asymptotes

The graphs of some rational functions have **horizontal asymptotes**, which can, in general, be identified by using the following theorem.

Theorem 5.4 *The graph of the rational function defined by*

$$y = R(x) = \frac{a_0 x^n + a_1 x^{n-1} + \cdots + a_n}{b_0 x^m + b_1 x^{m-1} + \cdots + b_m}, \quad (3)$$

where $a_0, b_0 \neq 0$, and n, m are nonnegative integers, has

  I *a horizontal asymptote at $y = 0$ if $n < m$,*
 II *a horizontal asymptote at $y = a_0/b_0$ if $n = m$,*
III *no horizontal asymptotes if $n > m$.*

Though we shall not give a rigorous proof of this theorem, we can certainly make the results plausible. If $n < m$, we can divide the numerator and denominator of the right-hand member of (3) by x^m to obtain, for $x \neq 0$,

$$y = R(x) = \frac{\dfrac{a_0}{x^{m-n}} + \dfrac{a_1}{x^{m-n+1}} + \cdots + \dfrac{a_n}{x^m}}{b_0 + \dfrac{b_1}{x} + \cdots + \dfrac{b_m}{x^m}}.$$

Now, as $|x|$ grows greater and greater, each term containing an x in its denominator grows closer and closer to 0, and we find the expression on the right approaching $0/b_0$, so that y approaches 0. But if, as $|x|$ increases without bound, y grows close to 0, then the graph of $y = R(x)$ approaches the line $y = 0$ asymptotically and actually must approach 0 from one of the directions shown in Figure 5.18-a. A similar

5.6 Rational Functions

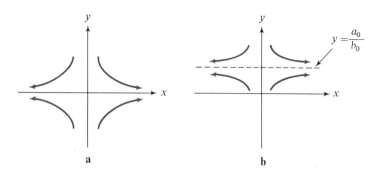

Figure 5.18

argument shows that if $n = m$, then, as $|x|$ increases without bound, the graph of $y = R(x)$ approaches the line $y = a_0/b_0$ from one of the directions shown in Figure 5.18-b. If $n > m$, then as $|x|$ becomes greater and greater, so does $|y|$. Thus the curve has no horizontal asymptote.

Example Find all horizontal and vertical asymptotes of the graph of the function

$$y = \frac{x^2 - 4}{x - 1}.$$

Solution We observe by Theorem 5.3 that $x = 1$ is a vertical asymptote and by Theorem 5.4 that there are no horizontal asymptotes.

Helpful items for graphing Identifying asymptotes is one aid to graphing a rational function. Other helpful items are the following:

1. the zeros of the function, because these give us the x-intercepts,
2. the domain and range, because these let us know where we can expect to find parts of the graph and where we cannot, and
3. some specific points on the graph, because these give us guidelines in sketching.

Example Graph $y = \dfrac{x - 1}{x - 2}$. Domain

Solution We can begin by observing that the numerator of the right-hand member will be equal to 0 when x is equal to 1. Therefore when $x = 1$ we have $y = 0$, and 1 is an x-intercept. Also, when $x = 0$ we have $y = 1/2$, so that $1/2$ is a y-intercept. Thus we can begin our graph by plotting the intercepts as shown in Figure a. By inspection, $y = (x - 1)/(x - 2)$ is defined for all x except $x = 2$, so that $\{x | x \neq 2\}$ is the domain. Similarly, if we solve the defining equation for x in terms of y, we have

$$x = \frac{2y - 1}{y - 1},$$ Range of function

which is defined for all values of y except 1. Hence $\{y \mid y \neq 1\}$ is the range of the function. From the defining equation and Theorems 5.3 and 5.4, we see that there is a vertical asymptote at $x = 2$ and a horizontal asymptote at $y = 1$. We can then add this information to our graph, as indicated in Figure a. To determine the behavior of the function near the vertical asymptote, we consider one side of the asymptote at a time. On the left side, when x is just less than 2, say $2 - p$, $0 < p < 1/10$, the denominator $x - 2$ in the expression for y is

$$2 - p - 2 = -p,$$

which is barely negative, whereas the numerator

$$x - 1 = 2 - p - 1 = 1 - p$$

is definitely positive; hence y is negative and $|y|$ large. Thus the graph goes down on the left side of the vertical asymptote. The graph appears in Figure b. To the right of

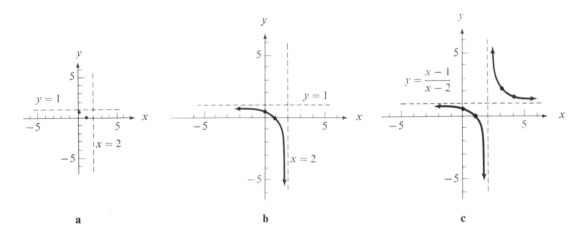

a   b   c

the vertical asymptote, we perform a similar analysis using $x = 2 + p$, $0 < p < 1/10$ to find that the graph goes up on the right of the vertical asymptote. This along with the knowledge that $y = 1$ is a horizontal asymptote leads us to the complete graph of the function $y = (x - 1)/(x - 2)$, as shown in Figure c.

Oblique asymptotes

In Theorem 5.4 if $n = m + 1$, that is, if the degree of the numerator is one greater than that of the denominator, we can argue that though the graph has no horizontal asymptote, it does have an **oblique asymptote**. We shall illustrate a special case only, but the technique involved is quite general.

Example

Find all asymptotes for the graph of the function

$$y = \frac{x^2 - 4}{x - 1}.$$

Solution

This is the same function we investigated in the example on page 147, where we found the vertical asymptote $x = 1$ and no horizontal asymptote. If we rewrite $y = (x^2 - 4)/(x - 1)$ by dividing $x^2 - 4$ by $x - 1$, we obtain

$$y = x + 1 - \frac{3}{x - 1}.$$

Now, as x grows greater and greater, $3/(x - 1)$ grows closer and closer to 0 and the graph of $y = (x^2 - 4)/(x - 1)$ approaches the graph of $y = x + 1$. Hence, the graph of $y = x + 1$, which is an oblique line, is an asymptote to the curve.

Example   Graph $y = \dfrac{x^2 - 4}{x - 1}$.

Solution   This is the same function we investigated in the preceding example, for which we found the vertical asymptote $x = 1$ and the oblique asymptote $y = x + 1$. If $x = 0$ then $y = 4$, and if $y = 0$ then $x = 2$ or $x = -2$, so that the y-intercept is 4 and the x-intercepts are 2 and -2. We therefore have the situation shown in Figure a.

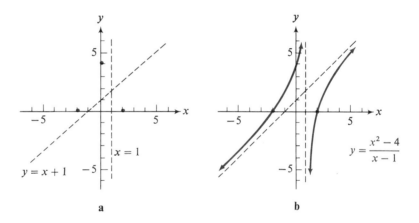

Without further information, we have reason to suspect that the graph will appear as shown in Figure b. Plotting a few check points, say $(-1, 3/2), (1/2, 15/2), (3/2, -7/2)$, and $(3, 5/2)$, would tend to confirm our conjecture.

Exercise 5.6

Determine the vertical asymptotes of the graph of each function.

Example $x^2 y - 4y = 1$

Solution   Express y explicitly in terms of x.

$$y(x^2 - 4) = 1$$

$$y = \frac{1}{x^2 - 4} = \frac{1}{(x-2)(x+2)}$$

By Theorem 5.3 there are vertical asymptotes at $x = 2$ and $x = -2$.

1. $y = \dfrac{1}{x - 3}$
2. $y = \dfrac{1}{x + 4}$
3. $y = \dfrac{4}{(x + 2)(x - 3)}$
4. $y = \dfrac{8}{(x - 1)(x + 3)}$
5. $y = \dfrac{2x - 1}{x^2 + 5x + 4}$
6. $y = \dfrac{x + 3}{2x^2 - 5x - 3}$
7. $xy + y = 4$
8. $x^2 y + xy = 3$

Graph.

9. $y = \dfrac{1}{x}$
10. $y = \dfrac{1}{x + 4}$
11. $y = \dfrac{1}{x - 3}$
12. $y = \dfrac{1}{x - 6}$
13. $y = \dfrac{4}{(x + 2)(x - 3)}$
14. $y = \dfrac{8}{(x - 1)(x + 3)}$
15. $y = \dfrac{2}{(x - 3)^2}$
16. $y = \dfrac{1}{(x + 4)^2}$

Determine any vertical, horizontal, or oblique asymptotes of the graphs of each of the following functions.

Example

$$y = \frac{6x^2 + 1}{2x^2 + 5x - 3}$$

Solution

The equation can be written equivalently as

$$y = \frac{6x^2 + 1}{(2x - 1)(x + 3)}.$$

By Theorem 5.3 there are vertical asymptotes at $x = 1/2$ and $x = -3$. By Theorem 5.4 there is a horizontal asymptote at $y = 6/2 = 3$.

17. $y = \dfrac{x}{x^2 - 4}$
18. $y = \dfrac{3x - 6}{x^2 + 3x + 2}$
19. $y = \dfrac{x^2 - 9}{x - 4}$
20. $y = \dfrac{x^3 - 27}{x^2 - 1}$
21. $y = \dfrac{x^2 - 3x + 2}{x^2 - 3x - 4}$
22. $y = \dfrac{x^2}{x^2 - x - 6}$

Chapter Review

Graph. Use information concerning the zeros of the function and concerning vertical, horizontal, and oblique asymptotes.

23. $y = \dfrac{x}{x-2}$
24. $y = \dfrac{x-1}{x+3}$
25. $y = \dfrac{2x-4}{x^2-9}$
26. $y = \dfrac{3x}{x^2-5x+4}$
27. $y = \dfrac{x^2-4}{x^3}$
28. $y = \dfrac{x-2}{x^2}$
29. $y = \dfrac{x+1}{x(x^2-4)}$
30. $y = \dfrac{x^2+x-2}{x(x^2-9)}$
31. $y = \dfrac{x^2-4x+4}{x-1}$
32. $y = \dfrac{x^2+4}{x-2}$

33. Describe the asymptotic behavior of $y = \dfrac{x^3+x^2+2}{x+1}$.

34. Sketch the graph of $y = \dfrac{x^3+x^2+2}{x+1}$.

35. Is it possible for the graph of a rational function to cross a vertical asymptote? Why or why not?

36. Find the point where the graph of $y = \dfrac{x^2+1}{x^2-2x-1}$ crosses its horizontal asymptote.

37. Find the point where the graph of $y = \dfrac{2x^3+x^2+3x}{x^2+1}$ crosses its oblique asymptote.

38. Give an example of a rational function with the property that its graph crosses its oblique asymptote more than once.

Chapter Review

[5.1] *Specify the domain of each relation.*

1. $\{(4, -1), (2, -4), (3, -5)\}$
2. $\{(1, 2), (1, 3), (1, 6)\}$
3. $y = \dfrac{1}{x+4}$
4. $y = \sqrt{x-6}$

Let $f(x) = x - 3$ and $g(x) = x^2 + 4$. Find each of the following.

5. $g(3)$
6. $f(-2)$
7. $f(x-h)$
8. $g(x+h)$

[5.2] *Find the distance between each of the given pairs of points, and find the slope of the line segment joining them.*

9. $(2, 0)$ and $(-3, 4)$
10. $(-6, 1)$ and $(-8, 2)$

[5.3] *Find the equation, in standard form, of the line passing through each of the given points and having the given slope.*

11. $(2, -7)$, $m = 4$
12. $(-6, 3)$, $m = \dfrac{1}{2}$

Write each equation in slope-intercept form; specify the slope and the y-intercept of the line.

13. $4x + y = 6$
14. $3x - 2y = 16$

Find the equation, in standard form, of the line with the given intercepts.

15. $x = -3; y = 2$
16. $x = \dfrac{1}{3}; y = -4$

Find an equation of the line which passes through the given point and is parallel to the graph of the given equation.

17. $4x - 3y = 12$; $(-2, 1)$
18. $x + 6y = 0$; $(3, -4)$

[5.4] *Find the x-intercepts, the axis of symmetry, and the maximum or minimum point of the graph of each function by analytic methods.*

19. $f(x) = x^2 - x - 6$
20. $f(x) = -x^2 + 7x - 10$

21. Graph the function of Exercise 19. Draw a line segment from $(4, 0)$ to $(4, f(4))$.
22. Graph the function of Exercise 20. Draw a line segment from $(3, 0)$ to $(3, f(3))$.

[5.5]
23. Use synthetic division to find $P(2)$, where $P(x) = x^3 - 3x^2 + 2x + 1$.
24. Use synthetic division to find $P(-3)$, where $P(x) = x^4 - 3x^2 - 1$.
25. Use synthetic division to find coordinates for points on the graph of
$$P(x) = x^3 + 4x^2 + x - 6$$
for the following replacements for x: $-4, -3, -2, -1, 0, 1, 2, 3, 4$.
26. Graph the function in Exercise 25.

[5.6] *Determine all asymptotes and graph each function.*

27. $y = \dfrac{3}{x + 2}$
28. $y = \dfrac{2x}{(x + 3)(x - 4)}$
29. $y = \dfrac{x + 2}{x - 3}$
30. $y = \dfrac{2x}{x^2 - 3x + 2}$

6 Special Relations and Graphs

In the preceding chapter we studied polynomial functions, rational functions, and their graphs. In this chapter we shall consider the graphs of some functions that are neither polynomial nor rational, and also the graphs of some special relations that are not functions.

6.1 Special Functions

Absolute-value functions

Functions with equations involving the absolute value of one or both of the variables not only are useful in more advanced courses in mathematics; they also offer interesting properties in their own right. We recall that

$$|x| = \begin{cases} x, & \text{if } x \geq 0, \\ -x, & \text{if } x < 0. \end{cases}$$

Now consider the function defined by

$$y = |x|. \tag{1}$$

From the definition of $|x|$, for nonnegative x we have

$$y = x, \tag{2}$$

and for negative x,

$$y = -x. \tag{3}$$

If we graph (2) and (3) on the same set of axes, we have the graph of $y = |x|$ shown in Figure 6.1. In the case of any equation involving $|x|$ or $|f(x)|$, we can always plot individual points to deduce the graph. For instance, if

$$y = |x| + 1, \tag{4}$$

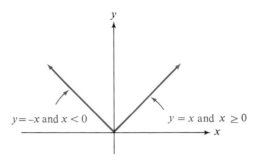

Figure 6.1

we can find solutions by assigning values to x and computing values for y. Some solutions of (4) are

$$(-2, 3), \quad (-1, 2), \quad (0, 1), \quad (1, 2), \quad \text{and} \quad (2, 3),$$

which can be graphed as in Figure 6.2-a. The graph of $y = |x| + 1$, where $x \in R$, appears in Figure 6.2-b. As an alternative approach, the definition of $|x|$ implies that

$$y = |x| + 1$$

is equivalent to

$$y = x + 1 \quad \text{for } x \geq 0,$$
$$y = -x + 1 \quad \text{for } x < 0,$$

and these can be graphed separately over the specified domains.

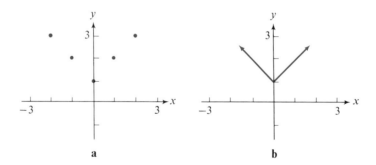

Figure 6.2

Related graphs

It is usually advisable, where possible, to avoid graphing large numbers of points. For instance, comparing Equations,

$$y = |x| \tag{1}$$

and

$$y = |x| + 1, \tag{4}$$

we observe that for each x the ordinate in (4) is one unit greater than that in (1); consequently, the graph of (4) is simply the graph of (1) with each ordinate increased by

6.1 Special Functions

one unit. Whenever we can, we should use such considerations to help us graph equations.

Bracket function

Another interesting function (sometimes called the **bracket function**) is defined by the equation

$$f(x) = [x], \qquad (5)$$

where the brackets denote "the greatest integer contained in," or "the greatest integer not greater than." Thus

$$[2] = 2, \quad \left[\frac{7}{4}\right] = 1, \quad [-2] = -2,$$

$$\left[-\frac{3}{2}\right] = -2, \quad \text{and} \quad \left[-\frac{5}{2}\right] = -3.$$

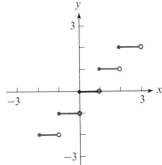

To graph (5), we consider unit intervals along the x-axis. If $0 \leq x < 1$, $[x]$ is 0, since the greatest integer contained in any number between 0 and 1 is 0. Similarly, if $1 \leq x < 2$, $[x]$ is 1; for $-2 \leq x < -1$, $[x]$ is -2; etc. The graph of (5), therefore, is as shown in Figure 6.3. The heavy dot on the left-hand endpoint of each line segment indicates that the endpoint is a part of the graph. The function defined by (5) is sometimes called a "step function," for an obvious reason.

Figure 6.3

Functions defined piecewise

The graph of a function such as

$$y = f(x) = \begin{cases} x^2, & \text{if } x \leq 1, \\ -x^2 + 2, & \text{if } x > 1 \end{cases} \qquad (6)$$

is usually obtained by graphing it separately over each subset of the domain. Thus to graph (6) we first graph $y = f(x) = x^2$ over the set $\{x \mid x \leq 1\}$ to obtain Figure 6.4-a. We then adjoin the graph of $y = f(x) = -x^2 + 2$ over the set $\{x \mid x > 1\}$ to obtain Figure 6.4-b, which is the complete graph of (6).

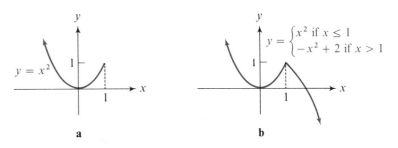

Figure 6.4

Exercise 6.1

Graph the function defined by the given equation over the domain $\{x \mid -5 \leq x \leq 5\}$.

1. $y = |x| + 2$
2. $y = -|x| + 3$
3. $f(x) = |x + 1|$
4. $F(x) = |x - 2|$
5. $y = -|2x - 1|$
6. $y = |3x + 2|$
7. $g(x) = |2x| - 3$
8. $y = |3x| + 2$
9. $y = |3x| - |x|$
10. $f(x) = |2x| + |x|$
11. $y = 3|x| - x$
12. $y = -2|x| + x$
13. $H(x) = |x^2|$
14. $y = |x|^2$
15. $y = |x + 1| - |x|$
16. $g(x) = |x + 1| - x$
17. $y = [x]$
18. $f(x) = [x] - 1$
19. $y = [x + 1]$
20. $y = [2x]$
21. $F(x) = [x] + x$
22. $y = \left[\dfrac{1}{2}x\right] + x$
23. $y = [x] - x$
24. $y = |[x]|$

25. $f(x) = \begin{cases} -x, & \text{if } x \leq -1 \\ 3x + 4, & \text{if } x > -1 \end{cases}$

26. $f(x) = \begin{cases} -2x + 1, & \text{if } x \leq 2 \\ -3 & \text{if } x > 2 \end{cases}$

27. $f(x) = \begin{cases} x^2 - 2, & \text{if } x \leq 1 \\ -x^3, & \text{if } x > 1 \end{cases}$

28. $f(x) = \begin{cases} -x^2 + 1, & \text{if } x \leq 0 \\ x^2, & \text{if } x > 0 \end{cases}$

29. The postage on a letter sent by first-class mail is c cents per ounce or fraction thereof. Write an equation relating the cost (C) of mailing a letter and the weight of the letter in ounces (x).

6.2 Inverse Functions

Inverse relations

If the two components of each ordered pair in a relation r are interchanged, then the resulting relation is called the **inverse relation** of r and is denoted by r^{-1}. For example, each of the relations

$$\{(1, 2), (3, 4), (5, 5)\}$$

and

$$\{(2, 1), (4, 3), (5, 5)\}$$

is the inverse of the other.

Inverse functions

By Definition 5.2, a function is a set of ordered pairs (x, y) such that no two have the same first components and different second components. When the components of every ordered pair in a function f are interchanged, the resulting relation may or

6.2 Inverse Functions

may not be a function. For example, if

$$(1, 5), \quad (2, 5), \quad \text{and} \quad (3, 6)$$

are ordered pairs in f, then

$$(5, 1), \quad (5, 2), \quad \text{and} \quad (6, 3)$$

are members of the relation formed by interchanging the components of these ordered pairs. Clearly these latter pairs cannot be members of a function since two of them, (5, 1) and (5, 2), have the same first components and different second components. If, however, a function f is **one-to-one**, that is, if no two different ordered pairs in f have the same second component (same value for y), then the relation obtained by interchanging the first and second components of every pair in the function will also be a function. This function is called the **inverse function** of f. The notation f^{-1} (read "f inverse") is frequently used to denote the inverse function of f.

Definition 6.1 *If the function f is such that no two of its ordered pairs with different first components have the same second component, then the **inverse function** f^{-1} is the set of ordered pairs obtained from f by interchanging the first and second components of each ordered pair in f.*

It is evident from this definition that the domain and range of f^{-1} are just the range and domain, respectively, of f. If $y = f(x)$ defines a function f, and if f is one-to-one, then $x = f(y)$ defines the inverse of f. For example, the inverse of the function defined by

$$y = 3x + 2 \tag{1}$$

is defined by

$$x = 3y + 2, \tag{2}$$

where the variables have been interchanged. When y is expressed in terms of x, we have

$$y = \frac{1}{3}x - \frac{2}{3}. \tag{2'}$$

Equations (2) and (2') are equivalent. In general, the equation $x = Q(y)$ is equivalent to $y = Q^{-1}(x)$.

Inverse of $y = x^n$

For any positive integer n, the inverse of the function

$$y = x^n \tag{3}$$

is defined by

$$x = y^n. \tag{4}$$

If n is an odd positive integer, then we can express Equation (4) equivalently as

$$y = \sqrt[n]{x}.$$

Thus if n is an odd positive integer and $f(x) = x^n$, then $f^{-1}(x) = \sqrt[n]{x}$.

If n is an even positive integer, then for positive values of x Equation (4) has two solutions, $\sqrt[n]{x}$ and $-\sqrt[n]{x}$. This reflects the fact that, for n an even positive integer, Equation (3) does not represent a one-to-one function. However, since Equation (4) was obtained from Equation (3) by interchanging the roles of x and y, restricting x to positive values in Equation (3) is equivalent to restricting y to positive values in Equation (4). Under those restrictions Equation (4) has only one solution, $y = \sqrt[n]{x}$. Thus, if n is an even positive integer and $f(x) = x^n$, $x \geq 0$, then $f^{-1}(x) = \sqrt[n]{x}$.

Example Find the inverse of the function $y = x^2 - 1$, $x \leq 0$.

Solution Interchanging the variables in the given equation, we obtain

$$x = y^2 - 1, \quad y \leq 0.$$

Solving for y in terms of x, we obtain

$$y = -\sqrt{x + 1},$$

where the minus sign is chosen to guarantee that $y \leq 0$.

The graphs of inverse relations are related in an interesting way. To see this, we first, in Figure 6.5, observe that the graphs of the ordered pairs (a, b) and (b, a) are always reflections of each other about the graph of $y = x$. Therefore, because for every ordered pair (a, b) in Q the ordered pair (b, a) is in Q^{-1}, the graphs of $y = Q^{-1}(x)$ and $y = Q(x)$ are reflections of each other about the graph of $y = x$.

Figure 6.6 shows the graphs of the linear function

$$y = 4x - 3 \tag{5}$$

and its inverse

$$y = \frac{1}{4}(x + 3),$$

together with the graph of $y = x$.

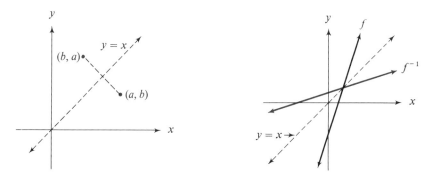

Figure 6.5   **Figure 6.6**

6.2 Inverse Functions

Since an element in the domain of the inverse Q^{-1} of a function Q is the range element in the corresponding ordered pair of Q, and vice versa, it follows that for every x in the domain of Q,

$$Q^{-1}(Q(x)) = x$$

(read "Q inverse of Q of x is equal to x"), and, for every x in the domain of Q^{-1},

$$Q(Q^{-1}(x)) = x$$

(read "Q of Q inverse of x is equal to x"). Using (3) above as an example, we note that if f is the linear function defined by

$$f(x) = 4x - 3,$$

then f is a one-to-one function. Now f^{-1} is defined by

$$f^{-1}(x) = \frac{1}{4}(x + 3),$$

and we see that

$$f^{-1}(f(x)) = \frac{1}{4}((4x - 3) + 3) = x$$

and

$$f(f^{-1}(x)) = 4\left(\frac{1}{4}(x + 3)\right) - 3 = x.$$

Exercise 6.2

Graph each given function f and its inverse function f^{-1}, using the same set of axes.

Example $f: x + 3y = 6$

Solution   Interchange the variables in the defining equation.

$f^{-1}: y + 3x = 6$

Graph f and f^{-1}.

1. $f = \{(-2, 3), (4, 7), (5, 9)\}$
2. $f = \{(-3, 1), (2, -1), (3, 4)\}$
3. $f = \{(-1, -1), (2, 2), (3, 3)\}$
4. $f = \{(-4, 4), (0, 0), (4, -4)\}$

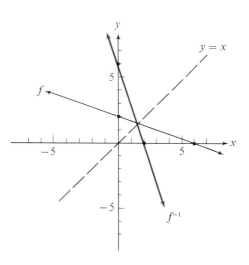

5. $f: y = 2x + 6$
6. $f: y = 3x - 6$
7. $f: y = 4 - 2x$
8. $f: y = 6 + 3x$
9. $f: 3x - 4y = 12$
10. $f: x - 6y = 6$
11. $f: 4x + y = 4$
12. $f: 2x - 3y = 12$
13. $f: y - x^2 = 0, x \leq 0$
14. $f: x^3 - y = 0$

In Exercises 15–22, each equation defines a one-to-one function F. Find an equation defining F^{-1} and show that $F(F^{-1}(x)) = F^{-1}(F(x)) = x$. Hint: Solve explicitly for y and let $y = F(x)$.

15. $y = x$
16. $y = -x$
17. $2x + y = 4$
18. $x - 2y = 4$
19. $3x - 4y = 12$
20. $3x + 4y = 12$
21. $y - x^2 = 2, x \geq 0$
22. $y + x^3 = -4$

6.3 Variation as a Functional Relationship

Direct variation

There are two types of functional relationships, widely used in the sciences, to which custom has assigned special names. First, any function defined by an equation of the form

$$y = kx \quad (k \text{ a positive constant}) \tag{1}$$

furnishes an example of **direct variation**. The variable y is said to **vary directly** as the variable x. Another example of direct variation is

$$y = kx^2 \quad (k \text{ a positive constant}), \tag{1a}$$

which indicates that y varies directly as the square of x. In general,

$$y = kx^n \quad (k \text{ a positive constant and } n > 0) \tag{1b}$$

is the assertion that y varies directly as the nth power of x.

We find examples of such variation in the relationships existing between the radius of a circle and the circumference and area. Thus

$$c = 2\pi r \tag{2}$$

asserts that the circumference of a circle varies directly as the radius, while

$$A = \pi r^2 \tag{3}$$

expresses the fact that the area of a circle varies directly as the square of the radius. Since for each r, (2) and (3) associate only one value of c or A, both of these equations define functions; (2) is a linear function and (3) is a quadratic function.

6.3 Variation as a Functional Relationship

Inverse variation

The second important type of variation arises from the equation

$$xy = k \quad (k \text{ a positive constant}), \tag{4}$$

in which x and y are said to **vary inversely**. When (4) is written in the form

$$y = \frac{k}{x}, \tag{5}$$

y is said to vary inversely as x. Similarly, if

$$y = \frac{k}{x^2}, \tag{5a}$$

y is said to vary inversely as the square of x, and so forth. As an example of inverse variation, consider the set of rectangles with area 24 square units. Since the area of a rectangle is given by

$$lw = A,$$

we have

$$lw = 24,$$

and the length and width can be seen to vary inversely.

Since (5) or (5a) associates only one y with each x ($x \neq 0$), an inverse variation defines a function, with domain $\{x \mid x \neq 0\}$.

The names "direct" and "inverse," as applied to variation, arise from the fact that in direct variation an assignment of increasing absolute values of x results in an association with increasing absolute values of y, whereas in inverse variation an assignment of increasing absolute values of x results in an association with decreasing absolute values of y.

Constant of variation

The constant involved in equations defining direct or inverse variation is called the **constant of variation**. If we know that one variable varies directly or inversely as another, and if we have one set of associated values for the variables, we can find the constant of variation involved. For example, suppose we know that y varies directly as x^2, and that $y = 4$ when $x = 7$. We express the fact that y varies directly as x^2 by writing

$$y = kx^2, \tag{6}$$

and then substitute 7 for x and 4 for y in (6) to obtain

$$4 = k \cdot 7^2 = k \cdot 49,$$

from which

$$k = \frac{4}{49}.$$

The equation specifically expressing the direct variation is

$$y = \frac{4}{49} x^2.$$

6 Special Relations and Graphs

Joint variation

In the event that one variable varies as the product of two or more other variables, we refer to the relationship as **joint variation**. Thus, if y varies jointly as u, v, and w, we have

$$y = kuvw. \tag{7}$$

Also, direct and inverse variation may take place concurrently. That is, y may vary directly as x and inversely as z, giving rise to the equation

$$y = k\frac{x}{z}.$$

It should be pointed out that the way in which the word "variation" is used herein is a technical one, and when the ideas of direct, inverse, or joint variation are encountered, you should always think of equations of the form (1), (5), or (7). For instance, the equations

$$y = 2x + 1 \quad \text{and} \quad y = \frac{1}{2}x - 2$$

do not describe examples of variation within our meaning of the word.

Proportionality and variation

An alternative term frequently used to describe the variation relationship discussed in this section is the word "proportional." Thus, to say that "y is directly proportional to x" or "y is inversely proportional to x" is another way of describing direct and inverse variation.

The use of the word "proportion" arises from the fact that any two solutions of an equation expressing a direct variation satisfy a fractional equation of the form

$$\frac{a}{b} = \frac{c}{d},$$

which is commonly called a **proportion**. For example, consider the situation in which the volume of a gas varies directly as the absolute temperature and inversely as the pressure, and can accordingly be represented by a relation of the form

$$V = \frac{kT}{P}. \tag{8}$$

For any set of values T_1, P_1, and V_1,

$$k = \frac{V_1 P_1}{T_1}, \tag{8a}$$

and for any other set of values T_2, P_2, and V_2,

$$k = \frac{V_2 P_2}{T_2}. \tag{8b}$$

6.3 Variation as a Functional Relationship

Equating the right-hand members of (8a) and (8b), we get

$$\frac{V_1 P_1}{T_1} = \frac{V_2 P_2}{T_2}, \qquad (8c)$$

from which any one of the six values can be determined if the other five values are known.

Exercise 6.3

Solve. *In Exercises 1–8, first find the constant of variation.*

Example If V varies directly as T and inversely as P, and $V = 40$ when $T = 300$ and $P = 30$, find V when $T = 324$ and $P = 24$.

Solution Write an equation expressing the relationship between the variables.

$$V = \frac{kT}{P} \qquad (1)$$

Substitute the initially known values for V, T, and P. Solve for k.

$$40 = \frac{k(300)}{30}$$

$$4 = k$$

Rewrite Equation (1) with k replaced by 4.

$$V = \frac{4T}{P}$$

Substitute the second set of values for T and P and solve for V.

$$V = \frac{4(324)}{24} = 54$$

1. If y varies directly as x^2, and $y = 9$ when $x = 3$, find y when $x = 4$.
2. If r varies directly as s and inversely as t, and $r = 12$ when $s = 8$ and $t = 2$, find r when $s = 3$ and $t = 6$.
3. The distance a particle falls in a certain medium is directly proportional to the square of the length of time it falls. If the particle falls 16 feet in two seconds, how far will it fall in 10 seconds?
4. In Exercise 3, how far will the particle fall *during* the seventh second?

5. The pressure exerted by a liquid at a given point varies directly as the depth of the point beneath the surface of the liquid. If a certain liquid exerts a pressure of 40 pounds per square foot at a depth of 10 feet, what is the pressure at 40 feet?

6. The volume (V) of a gas varies directly as its temperature (T) and inversely as its pressure (P). A gas occupies 20 cubic feet at a temperature of 300° A (absolute) and a pressure of 30 pounds per square inch. What will the volume be if the temperature is raised to 360° A and the pressure decreased to 20 pounds per square inch?

7. The maximum-safe uniformly distributed load (L) for a horizontal beam varies jointly as its breadth (b) and the square of the depth (d), and inversely as the length (l). An 8-foot beam with $b = 2$ feet and $d = 4$ feet will safely support a uniformly distributed load of up to 750 pounds. How many uniformly distributed pounds will an 8-foot beam support safely if $b = 2$ and $d = 6$?

8. The resistance (R) of a wire varies directly as its length (l) and inversely as the square of its diameter (d). Fifty feet of wire of diameter 0.012 inch has a resistance of 10 ohms. What is the resistance of 50 feet of the same type of wire if the diameter is increased to 0.015 inch?

Represent the relationship as a proportion by eliminating the constant of variation and then solving for the required variable.

Example

If V varies directly as T and varies inversely as P, and $V = 40$ when $T = 300$ and $P = 30$, find V when $T = 324$ and $P = 24$.

Solution

Write an equation expressing the relationship between the variables.

$$V = \frac{kT}{P}$$

Solve for k.

$$k = \frac{VP}{T}$$

Write a proportion relating the variables for two different sets of conditions.

$$\frac{V_1 P_1}{T_1} = \frac{V_2 P_2}{T_2}$$

Substitute the known values of the variables.

$$\frac{(40)(30)}{300} = \frac{V_2(24)}{324}$$

Solve for V_2.

$$V_2 = \frac{(324)(40)(30)}{300(24)} = 54$$

9. Exercise 1 of this set.
10. Exercise 2 of this set.
11. Exercise 5 of this set.
12. Exercise 6 of this set.
13. Exercise 7 of this set.
14. Exercise 8 of this set.
15. From the formula for the circumference of a circle, $c = \pi d$, show that the ratio of the circumference of two circles equals the ratio of their respective diameters.
16. From the formula for the area of a circle, $A = \pi r^2$, show that the ratio of the areas of two circles equals the ratio of the squares of their respective radii.
17. Graph on the same set of axes the linear functions defined by $y = kx$, $x \geq 0$, when $k = 1$, 2, and 3, respectively. Note that the constant of variation and the slope of the graph of the equation are the same.
18. Graph on the same set of axes the quadratic functions defined by $y = kx^2$, $x \geq 0$, when $k = 1$, 2, and 3, respectively. What effect does a change in k have on the graph of $y = kx^2$?
19. Graph on the same set of axes the equations $y = kx$, $y = kx^2$, and $y = kx^3$, when $k = 2$ and $x \geq 0$. What effect does increasing n have on the graph of $y = kx^n$?
20. Graph on the same set of axes the functions defined by $xy = k$, for $k = 1$ and 2, and $x > 0$. What effect does a change in k have on the graph of $xy = k$?
21. Graph on the same set of axes the functions defined by $xy = k$, $x^2 y = k$, and $x^3 y = k$, for $k = 2$ and $x > 0$. What effect does increasing n have on the graph of $x^n y = k$?
22. Show that if y varies directly as x, and z varies directly as x, then $y + z$ varies directly as x.

6.4 Conic Sections

In addition to equations typified by $y = ax^2 + bx + c$, which were discussed in Section 5.4, there are three other types of second-degree equations in two variables with graphs that are of particular interest. We shall discuss each of them separately.

First, consider the relation

$$x^2 + y^2 = 25. \tag{1}$$

Solving this equation explicitly for y, we have

$$y = \pm\sqrt{25 - x^2}. \tag{2}$$

Assigning values to x, we find the following ordered pairs in the relation:

$(-5, 0)$, $(-4, 3)$, $(-3, 4)$, $(0, 5)$, $(3, 4)$, $(4, 3)$, $(5, 0)$,

$(-4, -3)$, $(-3, -4)$, $(0, -5)$, $(3, -4)$, $(4, -3)$.

Plotting these points on the plane, we have the graph shown in Figure 6.7-a.

Connecting these points with a smooth curve, we obtain the graph in Figure 6.7-b, a **circle** with radius 5 and center at the origin.

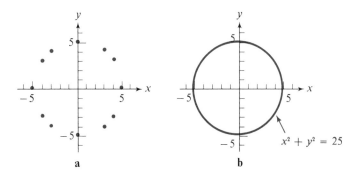

Figure 6.7

Equation of a circle

Since the number 25 in the right-hand member of the defining equation of (1) clearly determines the length of the radius of the circle, we can generalize and observe that any relation defined by an equation of the form

$$x^2 + y^2 = r^2, \quad r > 0,$$

has as its graph a circle with radius r and with center at the origin. (See Exercises 30–32 for a more general approach.)

Note that in the preceding example it is not necessary to assign any values to x for which $|x| > 5$, because then y^2 would have to be negative. Since, except for -5 and $+5$, each permissible value for x is associated with two values for y—one positive and one negative—the relation (1) is not a function. We could represent the relationship specified by (2) by using two equations,

$$y = f(x) = \sqrt{25 - x^2} \tag{3a}$$

and

$$y = g(x) = -\sqrt{25 - x^2}, \tag{3b}$$

whose graphs would appear as in Figure 6.8-a and 6.8-b, respectively. Each of

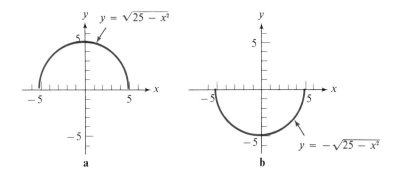

Figure 6.8

these equations does define a function. In both cases, the domain of the function is
$$\{x \mid |x| \leq 5\},$$
while the ranges differ. The range of the function defined by (3a) is
$$\{y \mid 0 \leq y \leq 5\},$$
and that of the function defined by (3b) is
$$\{y \mid -5 \leq y \leq 0\}.$$

The second kind of quadratic (second-degree) equation of interest, which actually has the first as a limiting case, is that typified by the equation
$$4x^2 + 9y^2 = 36. \qquad (4)$$
We obtain solutions of (4) by first solving the equation explicitly for y,
$$y = \pm \frac{2}{3}\sqrt{9 - x^2},$$
and then assigning values to x satisfying $-3 \leq x \leq 3$. (Why these values only?) We obtain, for example,

$$(-3, 0), \ \left(-2, \tfrac{2}{3}\sqrt{5}\right), \ \left(-1, \tfrac{4}{3}\sqrt{2}\right), \ (0, 2), \ \left(1, \tfrac{4}{3}\sqrt{2}\right), \ \left(2, \tfrac{2}{3}\sqrt{5}\right), \ (3, 0),$$

$$\left(-2, -\tfrac{2}{3}\sqrt{5}\right), \ \left(-1, -\tfrac{4}{3}\sqrt{2}\right), \ (0, -2), \ \left(1, -\tfrac{4}{3}\sqrt{2}\right), \ \left(2, -\tfrac{2}{3}\sqrt{5}\right).$$

Equation of an ellipse

Locating the corresponding points on the plane and connecting them with a smooth curve, we have the graph shown in Figure 6.9. This curve is called an **ellipse**. In general, the graph of the equation
$$Ax^2 + By^2 = C; \quad A, B, C > 0; \quad A \neq B$$
is an ellipse with center at the origin.

The third kind of quadratic equation with which we are presently concerned is typified by
$$x^2 - y^2 = 9.$$

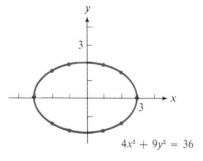

Figure 6.9

Solving this equation for y, we obtain
$$y = \pm \sqrt{x^2 - 9},$$
which has as a part of its solution set the ordered pairs
$$(-5, 4), \ (-4, \sqrt{7}), \ (-3, 0), \ (3, 0), \ (4, \sqrt{7}), \ (5, 4),$$
$$(-5, -4), \ (-4, -\sqrt{7}), \ (4, -\sqrt{7}), \ (5, -4).$$

Equation of a hyperbola

Plotting the corresponding points and connecting them with a smooth curve, we obtain the curve shown in Figure 6.10. This curve is called a **hyperbola.** In general, the equation

$$Ax^2 - By^2 = C; \quad A, B, C > 0$$

graphs into a hyperbola with center at the origin, (opening horizontally) and the equation

$$By^2 - Ax^2 = C; \quad A, B, C > 0$$

graphs into a hyperbola with center at the origin (opening vertically).

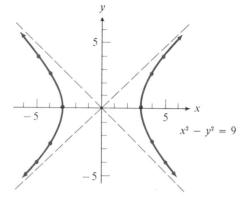

Figure 6.10

Asymptotes of a hyperbola

The dashed lines shown in Figure 6.10 are called **asymptotes** of the graph. They comprise the graph of the equation $x^2 - y^2 = 0$. While we shall not discuss the notion in detail here, it is true that, in general, the graph of $Ax^2 - By^2 = C$ "approaches" the two straight lines in the graph of $Ax^2 - By^2 = 0$ for each $A, B, C > 0$ (see Exercise 35).

The graphs of the foregoing relations, together with the parabola of the preceding section, are called **conic sections**, or **conics**, because such curves result from the intersection of a plane and a right circular cone, as shown in Figure 6.11.

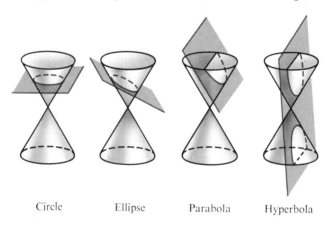

Figure 6.11

Use of the form of an equation in graphing

You should make use of the form of the defining equation as an aid in graphing quadratic relations. The ideas developed in this section and Section 5.4 may be summarized as follows:

1. A quadratic equation of the form

$$y = ax^2 + bx + c, \quad a \neq 0,$$

has a graph that is a parabola, opening upward if $a > 0$ and downward if $a < 0$.

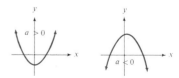

6.4 Conic Sections

Similarly, an equation of the form

$$x = ay^2 + by + c, \quad a \neq 0,$$

has a graph that is a parabola, opening to the right if $a > 0$ and to the left if $a < 0$.

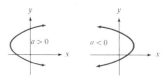

2. A quadratic equation of the form

$$Ax^2 + By^2 = C, \quad A^2 + B^2 \neq 0,$$

has a graph that is

(a) a **circle** if $A = B$ and A, B, and C have like signs;

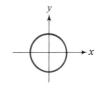

(b) an **ellipse** if $A \neq B$ and A, B, and C have like signs;

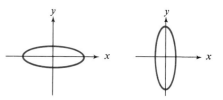

(c) a **hyperbola** if A and B are opposite in sign and $C \neq 0$; A & B are Not the same

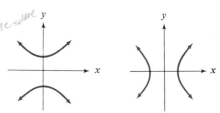

(d) **two distinct lines** through the origin if A and B are opposite in sign and $C = 0$ (see Exercise 25);

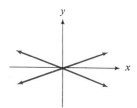

(e) **two distinct parallel lines** if one of A and $B = 0$ and the other has the same sign as C (see Exercise 26);

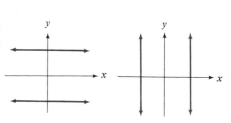

170   6 Special Relations and Graphs

(f) **two coincident parallel lines** (one line) through the origin if one of A and $B = 0$ and also $C = 0$ (see Exercise 27);

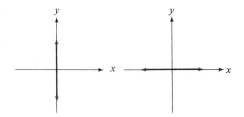

(g) a **point** if A and B are both >0 or both <0 and $C = 0$ (see Exercise 28);

(h) the **null set**, $\emptyset$, if A and B are both ≥ 0 and $C < 0$, or if A and B are both ≤ 0 and $C > 0$ (see Exercise 29).

Sketch of the graph of a conic section

After you recognize the general form of the curve, the graph of a few points should suffice to sketch the complete graph. The intercepts, for instance, are always easy to locate. Consider the equation

$$x^2 + 4y^2 = 8. \quad (5)$$

By comparing Equation (5) with the above Definition 2(b), we note immediately that its graph is an ellipse. If $y = 0$, then $x = \pm\sqrt{8}$; and if $x = 0$, then $y = \pm\sqrt{2}$. We can accordingly sketch the graph of (5) as in Figure 6.12

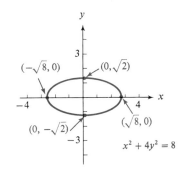

Figure 6.12

As another example, consider the equation

$$x^2 - y^2 = 3. \quad (6)$$

By comparing Equation (6) with 2(c), we see that its graph is a hyperbola. If $y = 0$, then $x = \pm\sqrt{3}$; and if $x = 0$, then y^2 would have to be negative, an impossibility in the field of real numbers. Thus the graph does not cross the y-axis. By assigning a few other arbitrary values to one of the variables, say x, e.g., $(4, \quad)$ and $(-4, \quad)$, we can find additional ordered pairs

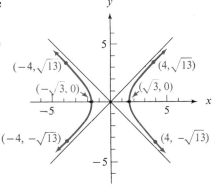

Figure 6.13

6.4 Conic Sections

$$(4, \sqrt{13}), (4, -\sqrt{13}), (-4, \sqrt{13}), (-4, -\sqrt{13})$$

which satisfy (6). The graph can then be sketched as shown in Figure 6.13. Observe that the asymptotes, which are the graphs of $x^2 - y^2 = 0$, are given by $y = x$ and $y = -x$. They can be helpful in sketching the hyperbola.

Exercise 6.4

(a) Rewrite each of the equations equivalently with y as the left-hand member.
(b) Write each equation equivalently as two separate equations, each of which defines a function.
(c) State the common domain of each of the functions defined in (b).

Example $4x^2 + y^2 = 36$

Solution
(a) $y^2 = 36 - 4x^2$
$y = \pm 2\sqrt{9 - x^2}$

(b) $y = 2\sqrt{9 - x^2}$
$y = -2\sqrt{9 - x^2}$

(c) The domain for each function is $\{x \mid -3 \leq x \leq 3\}$.

1. $x^2 + y^2 = 4$
2. $x^2 + y^2 = 9$
3. $9x^2 + y^2 = 36$
4. $4x^2 + y^2 = 4$
5. $x^2 + 4y^2 = 16$
6. $x^2 + 9y^2 = 4$
7. $2x^2 + 3y^2 = 24$
8. $4x^2 + 3y^2 = 12$
9. $x^2 - y^2 = 1$
10. $4x^2 - y^2 = 1$
11. $y^2 - x^2 = 9$
12. $4y^2 - 9x^2 = 36$

Name and sketch the graph of each of the following relations. Specify the intercepts, and for the hyperbolas give equations of the asymptotes.

Example $4x^2 - 36 = -9y^2$

Solution Rewrite the defining equation equivalently in standard form.

$$4x^2 + 9y^2 = 36$$

By inspection, the graph is an ellipse; x-intercepts are -3 and 3; y-intercepts are -2 and 2.

Sketch the graph.

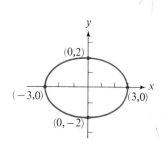

13. $x^2 + y^2 = 49$
14. $x^2 + y^2 = 64$
15. $4x^2 + 25y^2 = 100$
16. $x^2 + 2y^2 = 8$
17. $4x^2 = 4y^2$
18. $x^2 - 9y^2 = 0$
19. $x^2 = 9 + y^2$
20. $x^2 = 2y^2 + 8$
21. $4x^2 + 4y^2 = 1$
22. $9x^2 + 9y^2 = 2$
23. $3x^2 - 12 = -4y^2$
24. $12 - 3y^2 = 4x^2$

25. Graph $4x^2 - y^2 = 0$. Generalize from the result and discuss the graph of any equation of the form $Ax^2 - By^2 = 0$, $A, B > 0$.

26. Graph $x^2 = 4$. Generalize from the result and discuss the graph of any equation of the form $Ax^2 = C$, $A, C > 0$.

27. Graph $x^2 = 0$. Generalize from the result and discuss the graph of any equation of the form $Ax^2 = 0$, $A \neq 0$.

28. Graph $4x^2 + y^2 = 0$. Generalize from the result and discuss the graph of any equation of the form $Ax^2 + By^2 = 0$, $A, B > 0$.

29. Explain why the graph of $x^2 + y^2 = -1$ is the null set. Generalize from the result and discuss the graph of any equation of the form

$$Ax^2 + By^2 = C, \quad A^2 + B^2 \neq 0, \quad A, B \geq 0, C < 0.$$

30. Use the distance formula to show that the graph of the equation given by $x^2 + y^2 = r^2$, $r \geq 0$, is the set of all points located a distance r from the origin.

31. Show that the graph of $(x - 2)^2 + (y + 3)^2 = 16$ is a circle with center at $(2, -3)$ and radius 4. *Hint:* Use the distance formula.

32. Show that the graph of $(x - h)^2 + (y - k)^2 = r^2$, $r > 0$ is a circle with center at (h, k) and radius r.

33. Show that the general form of an equation for an ellipse, $Ax^2 + By^2 = C$, can be written in the intercept form $\dfrac{x^2}{a^2} + \dfrac{y^2}{b^2} = 1$, where a and b are x- and y-intercepts, respectively.

34. Show that the general form of an equation for a hyperbola, $Ax^2 - By^2 = C$, can be written in the intercept form $\dfrac{x^2}{a^2} - \dfrac{y^2}{b^2} = 1$, where a is an x-intercept.

35. By solving $Ax^2 - By^2 = C$ $(A, B, C > 0)$ for y, obtain the expression

$$y = \pm \sqrt{\dfrac{A}{B}} |x| \left(\sqrt{1 - \dfrac{C}{Ax^2}} \right),$$

and argue that the graph of $Ax^2 - By^2 = C$ approaches the straight-line graphs of $y = \pm \sqrt{A/B}\, |x|$ as $|x|$ increases.

36. Graph $x^2 + y^2 = 25$ and $4x^2 + y^2 = 36$ on the same set of axes. What is the significance of the coordinates of the points of intersection?

6.5 Linear and Quadratic Inequalities

In previous sections we discussed the graphs of relations and functions defined by equations. In this section we shall discuss the graphs of relations defined by inequalities.

Associated equation of an inequality

An inequality of the form

$$Ax + By + C \leq 0 \quad \text{or} \quad Ax + By + C < 0,$$

A and B not both 0, can be graphed on the plane, but the graph will be a region of the plane rather than a straight line. For example, consider the inequality

$$2x + y - 3 < 0. \tag{1}$$

When the inequality is rewritten in the equivalent form

$$y < -2x + 3, \tag{2}$$

we see that solutions (x, y) are such that for each x, y is less than $-2x + 3$. The graph of the equation

$$y = -2x + 3 \tag{3}$$

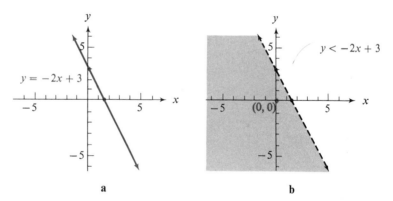

Figure 6.14

is simply a straight line, as illustrated in Figure 6.14-a. To graph the inequality, we need only observe that any point below this line has a y-coordinate that satisfies (2), and consequently the solution set of inequality (2) corresponds to the entire region below the line. The region is indicated on the graph with shading. That the line itself is not in the graph is shown by means of a broken line, as in Figure 6.14-b. Had the inequality been

$$2x + y - 3 \leq 0,$$

the line would be a part of the graph and would be shown as a solid line. In general, the graphs of solution sets of

$$Ax + By + C < 0 \quad \text{or} \quad Ax + By + C > 0$$

are the points in a half-plane on one side of the graph of the **associated equation**

$$Ax + By + C = 0,$$

depending on the constants and inequality symbols involved.

Determination of graph of an inequality

To determine which half-plane to shade in constructing graphs of first-degree inequalities, one can select any point in either half-plane and test its coordinates in the defining sentence to see whether or not the selected point lies in the graph. If the coordinates satisfy the sentence, then the half-plane containing the selected point is shaded; if not, the opposite half-plane is shaded. A very convenient point to use in this process is the origin, provided the origin is not contained in the graph of the associated equation. Thus, in the foregoing example, the replacement of x and y by 0 in

$$2x + y - 3 \leq 0$$

results in

$$2(0) + (0) - 3 \leq 0,$$

which is true, and hence the half-plane containing the origin is shaded.

Inequalities do not ordinarily define functions, according to our definition in Section 5.1, because it usually is not true that each element of the domain is associated with a unique element in the range.

Quadratic inequalities

Inequalities of the form

$$y < ax^2 + bx + c \tag{4}$$

or

$$y > ax^2 + bx + c \tag{5}$$

can be graphed in the same manner in which we graphed *linear* inequalities in two variables. We first graph the associated equation and then shade an appropriate region as required. For instance, to graph

$$y < x^2 + 2, \tag{6}$$

we first graph

$$y = x^2 + 2. \tag{7}$$

Then, as in the graphing of linear inequalities, the coordinates of a selected point not on the graph of the associated equation can be used to determine which of the two

6.5 Linear and Quadratic Inequalities

resulting regions of the plane is the graph of the inequality. Upon substitution of 0 for x and 0 for y in $y < x^2 + 2$, we have $(0) < (0)^2 + 2$, a true statement. Hence the region containing the origin is shaded. Since the graph of (7) is not part of the graph of (6), a broken curve is used (Figure 6.15).

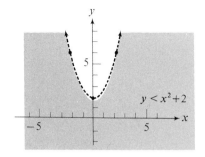

Figure 6.15

Exercise 6.5

Graph each inequality in R^2.

Example $2x + y \geq 4$

Solution Solve the inequality explicitly for y.

$$y \geq 4 - 2x$$

Graph the equality $y = 4 - 2x$.

Observe that the origin is not part of the graph of the inequality, because $(0, 0)$ is not a member of the given relation; that is, $0 \not\geq 4 - 2(0)$. Hence the region above the graph of the equation $y = 4 - 2x$ is shaded.

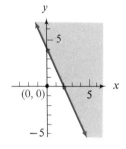

The line is included in the graph.

1. $y < x$
2. $y > x$
3. $y \leq x + 2$
4. $y \geq x - 2$
5. $x + y < 5$
6. $2x + y < 2$
7. $x - y < 3$
8. $x - 2y < 5$
9. $x \leq 2y - 4$
10. $2x \leq y + 1$
11. $3 \geq 2x - 2y$
12. $0 \geq x + y$

Example $x > 2$

Solution Graph $x = 2$.

Shade region to the right of the graph of $x = 2$, that is, the set of all points such that $x > 2$.

The line is excluded from the graph.

13. $x > 0$
14. $y < 0$
15. $x < 0$
16. $x < -2$

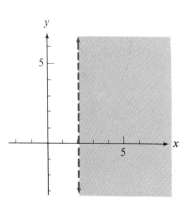

17. $-1 < x < 5$
18. $0 \le y \le 1$
19. $|x| < 3$
20. $|y| > 1$
21. $|x| + |y| \le 1$
22. $|x| + |y| \ge 1$

Hint: Consider the graphs in each quadrant separately:

$$x, y \ge 0; \quad x \le 0, y \ge 0; \quad x, y \le 0; \quad \text{and} \quad x \ge 0, y \le 0.$$

Graph.

Example $y \ge x^2 + 2x$

Solution First graph $y = x^2 + 2x$.

Use $(-1, 0)$ as a test point.
Is $0 \ge (-1)^2 + 2(-1)$? Yes.
Then $(-1, 0)$ is in the graph.
Shade the portion of the plane above the curve.

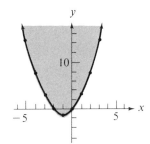

23. $y > x^2$
24. $y < x^2$
25. $y \ge x^2 + 3$
26. $y \le x^2 + 3$
27. $y < 3x^2 + 2x$
28. $y > 3x^2 + 2x$
29. $y \le x^2 + 3x + 2$
30. $y \ge x^2 + 3x + 2$
31. $y \ge 2x^2 - 5x + 1$
32. $y \le 2x^2 - 5x + 1$

33. Graph the set of points whose coordinates satisfy both
$$y \le 4 - x^2 \quad \text{and} \quad y \ge x^2 - 4.$$

34. Graph the set of points whose coordinates satisfy both
$$y \le 1 - x^2 \quad \text{and} \quad y \ge -1.$$

Chapter Review

[6.1] Graph the function defined by the given equation over the domain $\{x \mid -5 \le x \le 5\}$.

1. $f(x) = |x| + 4$
2. $g(x) = |x - 4|$
3. $g(x) = |x| - 1$
4. $h(x) = |x - 1|$
5. $f(x) = \begin{cases} x - 1, & \text{if } x \le 2 \\ \frac{1}{2}x, & \text{if } x \ge 2 \end{cases}$
6. $f(x) = \begin{cases} x + 4, & \text{if } x \le -1 \\ -3x, & \text{if } x > -1 \end{cases}$

Chapter Review

[6.2] *Graph each function f and its inverse f^{-1}, using the same set of axes.*

7. $f = \{(-3, 1), (-1, 3), (2, 4)\}$
8. $f: 2x - y = 6$
9. $f: y - x^2 = 0, \quad x \geq 0$
10. $f: x^4 - y = 0, \quad x \leq 0$

[6.3]
11. If y varies directly as x^2, and $y = 20$ when $x = 2$, find y when $x = 5$.
12. If r varies directly as x^2 and inversely as z^3, and $r = 4$ when $x = 3$ and $z = 2$, find r when $x = 2$ and $z = 3$.
13. The number of posts needed to string a telephone line over a given distance varies inversely as the distance between posts. If it takes 80 posts separated by 120 feet to string a wire between two points, how many posts would be required if the posts were 150 feet apart?
14. The speed of a gear varies directly as the number of teeth it contains. If a gear with 10 teeth rotates at 240 revolutions per minute (RPM), with what speed would a gear with 18 teeth revolve under the same conditions?

[6.4] *Name and graph the relation defined by the given equation.*

15. $x^2 - 16 = 4y^2$
16. $2y^2 = 8 - 2x^2$
17. $x^2 = 36 - 4y^2$
18. $x^2 - y - 9 = 0$
19. $x^2 - 9y^2 = 0$
20. $x^2 + 9y^2 = 0$

[6.5] *Graph the given inequality in R^2.*

21. $2x - y < 6$
22. $2y + x > 0$
23. $-2 \leq y < 3$
24. $3 < x \leq 5$
25. $y < x^2 - 9$
26. $y \geq x^2 + 6x + 5$

7 Exponential and Logarithmic Functions

In Chapters 1 through 4 we studied power expressions, such as x^2, that involve a variable base and a fixed exponent. In this chapter we shall consider the base fixed and the exponent variable, as in 2^x. We shall study properties of functions such as $y = 2^x$ and properties of their inverse functions.

7.1 Exponential Functions

We begin by discussing the graphs and some algebraic properties of exponential functions. Powers of the form b^x, where $b > 0$ and x denotes a rational number, were discussed in Section 3.2. The base b is restricted here to positive real numbers to ensure that b^x be real for all rational numbers x.

We shall assume that b^x ($b > 0$) exists for all real values of x and that the laws of exponents given in Theorem 3.1 on page 64, appropriately reworded for real-number exponents, are valid for such powers.

Since for any given $b > 0$ and for each real number x there is only one value for b^x, the equation

$$f(x) = b^x \quad (b > 0) \qquad (1)$$

defines a function. Because $1^x = 1$ for all real numbers x, (1) defines a constant function if $b = 1$. If $b \neq 1$, we say that (1) defines an **exponential function**. Exponential functions can perhaps be visualized most clearly by considering their graphs. We will illustrate two typical examples, in which $0 < b < 1$ and $b > 1$, respectively. Assigning values to x in the equations

$$f(x) = \left(\frac{1}{2}\right)^x \quad \text{and} \quad f(x) = 2^x,$$

7.1 Exponential Functions

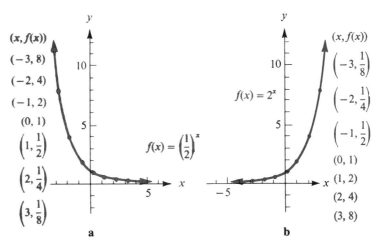

Figure 7.1

we find some ordered pairs in each solution set and sketch the graphs shown in Figures 7.1-a, and 7.1-b, respectively.

Increasing and decreasing functions

Notice that the graph of the function determined by $f(x) = (1/2)^x$ goes down to the right, and the graph of the function determined by $f(x) = 2^x$ goes up to the right. Thus the former function is a **decreasing function** and the latter is an **increasing function**. In either case, the domain is the set of real numbers, and the range is the set of positive real numbers.

Base e

One particularly useful exponential function is that defined by

$$y = e^x, \qquad (2)$$

where e is an irrational number with decimal approximation 2.71828. Tables for values of e^x have been prepared by advanced mathematical procedures. You can obtain values for e^x from Table II (page 468) or by using a scientific calculator.

Exercise 7.1

Find the second component of each of the ordered pairs that makes the pair a solution of the corresponding equation.

1. $y = 3^x$; (0,), (1,), (2,)
2. $y = -2^x$; (0,), (1,), (2,)
3. $y = -5^x$; (0,), (1,), (2,)
4. $y = 4^x$; (0,), (1,), (2,)
5. $y = \left(\dfrac{1}{2}\right)^x$; (−3,), (0,), (3,)
6. $y = \left(\dfrac{1}{3}\right)^x$; (−3,), (0,), (3,)
7. $y = 10^x$; (−2,), (1,), (0,)
8. $y = 10^{-x}$; (0,), (1,), (2,)

Graph the specified function.

9. $y = 4^x$ 10. $y = 5^x$ 11. $y = 10^x$ 12. $y = 10^{-x}$

13. $y = 2^{-x}$ 14. $y = 3^{-x}$ 15. $y = \left(\dfrac{1}{3}\right)^x$ 16. $y = \left(\dfrac{1}{4}\right)^x$

17. $y = e^x$ 18. $y = e^{-x}$

19. Graph $f: y = 10^x$ and $f^{-1}: x = 10^y$ on the same set of axes.
20. Graph $f: y = 2^x$ and $f^{-1}: x = 2^y$, on the same set of axes.

Solve for x by inspection.

21. $10^x = \dfrac{1}{100}$ 22. $\left(\dfrac{1}{2}\right)^x = 16$ 23. $16^x = 8$ 24. $8^x = 16$

Determine an integer n such that $n < x < n + 1$.

25. $3^x = 16.2$ 26. $4^x = 87.1$ 27. $10^x = 0.016$ 28. $2^x = 6$

29. The growth of bacteria in a certain culture is given by the formula
$$N = N_0 e^{0.04t},$$
where N_0 is the number (in thousands) of bacteria present initially and N is the number present at the end of t hours. Graph the equation for the condition that $N_0 = 10$.

30. Use the information in Problem 29 and either the table for e^x or a calculator to determine the number of bacteria present:
    a. after 8 hours   b. after 16 hours   c. after 24 hours

7.2 Logarithmic Functions

The inverse of an exponential function has an important role in many areas of mathematics.

Inverse of an exponential function

In the exponential function
$$y = b^x \quad (b > 0, b \neq 1), \qquad (1)$$
illustrated in Figure 7.1 for $b = 1/2$ and $b = 2$, there is only one x associated with each y. Thus by Definition 6.1 we have the inverse function
$$x = b^y \quad (b > 0, b \neq 1). \qquad (2)$$
Observe that the relation $x > 0$ is implied by Equation (2), because there is no real number y for which b^y is not positive.

7.2 Logarithmic Functions

The graphs of functions of this form can be illustrated by the example

$$x = 10^y.$$

We consider x the variable denoting an element in the domain. In the defining equation we assign particular values for x, say

$$(0.01, \ ), (0.1, \ ), (1, \ ), (10, \ ), (100, \ ),$$

to obtain the ordered pairs

$$(0.01, -2), (0.1, -1), (1, 0), (10, 1), (100, 2).$$

These can be graphed and connected with a smooth curve as shown in Figure 7.2.

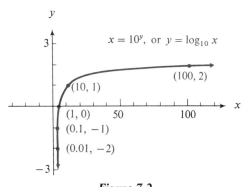

Figure 7.2

It is always useful to be able to express the variable denoting an element in the range explicitly in terms of the variable denoting an element in the domain. To do this in Equation (2), we use the notation

$$y = \log_b x \quad (x > 0, b > 0, b \neq 1). \tag{3}$$

Thus we have

$$y = \log_b x \quad \text{if and only if} \quad x = b^y.$$

Here, $\log_b x$ is read "the logarithm to the base b of x," or "the logarithm of x to the base b." The function defined by Equation (3) is called a **logarithmic function**.

From the graph in Figure 7.2, we generalize from $\log_{10} x$ to $\log_b x$, and observe that for $b > 1$, a logarithmic function has the following properties:

1. The domain is the set of positive real numbers, and the range is the set of all real numbers.

2. $\log_b x < 0$ for $0 < x < 1$, $\log_b x = 0$ for $x = 1$, and $\log_b x > 0$ for $x > 1$.

Logarithmic and exponential statements

It should be recognized that Equations (2) and (3) are two equations determining the same function, in the same way that $x = y + 4$ and $y = x - 4$ determine the same function, and we may use whichever equation suits our purpose. Thus, exponential statements may be written in logarithmic form, and logarithmic statements may be written in exponential form.

Examples

a. $5^2 = 25$ can be written as $\log_5 25 = 2$.

b. $8^{1/3} = 2$ can be written as $\log_8 2 = \frac{1}{3}$.

c. $3^{-3} = \frac{1}{27}$ can be written as $\log_3 \frac{1}{27} = -3$.

d. $\log_{10} 100 = 2$ can be written as $10^2 = 100$.

e. $\log_3 81 = 4$ can be written as $3^4 = 81$.

f. $\log_2 \frac{1}{2} = -1$ can be written as $2^{-1} = \frac{1}{2}$.

The logarithmic function associates with each positive number x the exponent y such that the power b^y is equal to x. In other words, we can think of $\log_b x$. as an exponent on b. Thus

$$b^{\log_b x} = x.$$

Laws of Logarithms

Since a logarithm is an exponent, the following theorem follows directly from the properties of powers with real-number exponents.

Theorem 7.1   If $x_1 > 0$, $x_2 > 0$, $b > 0$, $b \neq 1$, and m is any real number, then

I $\log_b (x_1 x_2) = \log_b x_1 + \log_b x_2$,

II $\log_b \frac{x_2}{x_1} = \log_b x_2 - \log_b x_1$,

III $\log_b (x_1)^m = m \log_b x_1$.

The validity of I can be shown as follows: Since

$$x_1 = b^{\log_b x_1} \quad \text{and} \quad x_2 = b^{\log_b x_2},$$

it follows that

$$x_1 x_2 = b^{\log_b x_1} \cdot b^{\log_b x_2} = b^{\log_b x_1 + \log_b x_2},$$

and, by the definition of a logarithm,

$$\log_b (x_1 x_2) = \log_b x_1 + \log_b x_2.$$

The validity of II and III can be established in a similar way.

Examples

a. $\log_{10} 100 = \log_{10} (10 \cdot 10)$
    $= \log_{10} 10 + \log_{10} 10$
    $= 1 + 1 = 2$

b. $\log_2 \frac{1}{4} = \log_2 1 - \log_2 4$
    $= 0 - 2$
    $= -2$

7.2 Logarithmic Functions

c. $\log_3 81 = \log_3 3^4$
$= 4 \log_3 3$
$= 4 \cdot 1 = 4$

d. $\log_2 2^{-1} = -1 \log_2 2$
$= -1 \cdot 1$
$= -1$

Exercise 7.2

Express in logarithmic notation.

1. $4^2 = 16$
2. $5^3 = 125$
3. $3^3 = 27$
4. $8^2 = 64$
5. $\left(\dfrac{1}{2}\right)^2 = \dfrac{1}{4}$
6. $\left(\dfrac{1}{3}\right)^2 = \dfrac{1}{9}$
7. $8^{-1/3} = \dfrac{1}{2}$
8. $64^{-1/6} = \dfrac{1}{2}$
9. $10^2 = 100$
10. $10^0 = 1$
11. $10^{-1} = 0.1$
12. $10^{-2} = 0.01$

Express in exponential notation.

13. $\log_2 64 = 6$
14. $\log_5 25 = 2$
15. $\log_3 9 = 2$
16. $\log_{16} 256 = 2$
17. $\log_{1/3} 9 = -2$
18. $\log_{1/2} 8 = -3$
19. $\log_{10} 1{,}000 = 3$
20. $\log_{10} 1 = 0$

Find the value of each expression.

21. $\log_7 49$
22. $\log_2 32$
23. $\log_4 64$
24. $\log_3 \dfrac{1}{3}$
25. $\log_5 \dfrac{1}{5}$
26. $\log_3 3$
27. $\log_2 2$
28. $\log_{10} 10$
29. $\log_{10} 100$
30. $\log_{10} 1$
31. $\log_{10} 0.1$
32. $\log_{10} 0.01$

Solve for x, y, or b.

Examples

a. $\log_2 x = 3$
b. $\log_b 2 = \dfrac{1}{2}$
c. $\log_{1/4} 16 = y$

Solutions

Determine the solution by inspection or by first writing the equation equivalently in exponential form.

a. $2^3 = x$
$x = 8$

b. $b^{1/2} = 2$
$(b^{1/2})^2 = (2)^2$
$b = 4$

c. $\left(\dfrac{1}{4}\right)^y = 16$
$(4^{-1})^y = 16$
$4^{-y} = 16$
$y = -2$

33. $\log_3 9 = y$
34. $\log_5 125 = y$
35. $\log_b 8 = 3$
36. $\log_b 625 = 4$
37. $\log_4 x = 3$
38. $\log_{1/2} x = -5$

39. $\log_2 \frac{1}{8} = y$ 40. $\log_5 5 = y$ 41. $\log_b 10 = \frac{1}{2}$

42. $\log_b 0.1 = -1$ 43. $\log_2 x = 2$ 44. $\log_{10} x = -3$

Express as the sum or difference of simpler logarithmic quantities.

Example $\log_b \left(\frac{xy}{z}\right)^{1/2}$

Solution   By Theorem 7.1-III,

$$\log_b \left(\frac{xy}{z}\right)^{1/2} = \frac{1}{2} \log_b \left(\frac{xy}{z}\right).$$

By Theorem 7.1-I and 7.1-II,

$$\frac{1}{2} \log_b \left(\frac{xy}{z}\right) = \frac{1}{2} (\log_b x + \log_b y - \log_b z).$$

Thus,

$$\log_b \left(\frac{xy}{z}\right)^{1/2} = \frac{1}{2} (\log_b x + \log_b y - \log_b z)$$

45. $\log_b (xy)$ 46. $\log_b (xyz)$ 47. $\log_b \left(\frac{x}{y}\right)$

48. $\log_b \left(\frac{xy}{z}\right)$ 49. $\log_b x^5$ 50. $\log_b x^{1/2}$

51. $\log_b \sqrt[3]{x}$ 52. $\log_b \sqrt[3]{x^2}$ 53. $\log_b \sqrt{\frac{x}{z}}$

54. $\log_b \sqrt{xy}$ 55. $\log_{10} \sqrt[3]{\frac{xy^2}{z}}$ 56. $\log_{10} \sqrt[5]{\frac{x^2 y}{z^3}}$

Express as a single logarithm with coefficient 1.

Example $\frac{1}{2} (\log_b x - \log_b y)$

Solution   By Theorems 7.1-II and 7.1-III,

$$\frac{1}{2} (\log_b x - \log_b y) = \frac{1}{2} \log_b \left(\frac{x}{y}\right) = \log_b \left(\frac{x}{y}\right)^{1/2}.$$

57. $\log_b 2x + 3 \log_b y$ 58. $3 \log_b x - \log_b 2y$

59. $\frac{1}{2} \log_b x + \frac{2}{3} \log_b y$ 60. $\frac{1}{4} \log_b x - \frac{3}{4} \log_b y$

7.3 Logarithms to the Base 10 and the Base e

61. $3 \log_b x + \log_b y - 2 \log_b z$ **62.** $\frac{1}{3}(\log_b x + \log_b y - 2 \log_b z)$

63. $\log_{10}(x - 2) + \log_{10} x - 2 \log_{10} z$

64. $\frac{1}{2}(\log_{10} x - 3 \log_{10} y - 5 \log_{10} z)$

65. Show that $\log_b 1 = 0$ for $b > 0$, $b \neq 1$. *Hint*: Express in exponential form.

66. Show that $\log_b b = 1$ for $b > 0$, $b \neq 1$.

67. Prove Theorem 7.1-II.

68. Prove Theorem 7.1-III.

7.3 Logarithms to the Base 10 and the Base e

There are two logarithmic functions of special interest; one is defined by

$$y = \log_{10} x, \qquad (1)$$

and the other by

$$y = \log_e x. \qquad (2)$$

These functions possess similar properties; but because we are more familiar with the number 10, we shall first consider $\log_{10} x$.

Values of $\log_{10} x$

Values for $\log_{10} x$ are called **logarithms to the base 10**, or **common logarithms**. From the definition of $\log_{10} x$,

$$10^{\log_{10} x} = x \quad (x > 0); \qquad (3)$$

that is, $\log_{10} x$ is the exponent that must be placed on 10 so that the resulting power is x. Our concern in this section is that of finding $\log_{10} x$ for each positive x. First, $\log_{10} x$ can easily be determined for all values of x that are *integral* powers of 10 by inspection, directly from the definition of a logarithm:

$$\log_{10} 10 = \log_{10} 10^1 = 1,$$
$$\log_{10} 100 = \log_{10} 10^2 = 2;$$

similarly,

$$\log_{10} 1 = \log_{10} 10^0 = 0,$$
$$\log_{10} 0.1 = \log_{10} 10^{-1} = -1,$$
$$\log_{10} 0.01 = \log_{10} 10^{-2} = -2,$$
$$\log_{10} 0.001 = \log_{10} 10^{-3} = -3.$$

Values for $\log_{10} x$ in case x is not an integral power of 10 can be obtained by using a table of logarithms (Table I on pages 466 and 467) and Theorem 7.1-I or by using a calculator. We shall consider how these values are obtained by using the table, and the examples and the answers to the exercises will reflect this procedure. If you use a calculator, you might obtain a slightly different approximation for these values.

x	0	1	2	3	4	5	6	7	8	9
3.8	.5798	.5809	.5821	.5832	.5843	.5855	.5866	.5877	.5888	.5899
3.9	.5911	.5922	.5933	.5944	.5955	.5966	.5977	.5988	.5999	.6010
4.0	.6021	.6031	.6042	.6053	.6064	.6075	.6085	.6096	.6107	.6117
4.1	.6128	.6138	.6149	.6160	.6170	.6180	.6191	.6201	.6212	.6222
4.2	.6232	.6243	.6253	.6263	.6274	.6284	.6294	.6304	.6314	.6325
4.3	.6335	.6345	.6355	.6365	.6375	.6385	.6395	.6405	.6415	.6425
4.4	.6435	.6444	.6454	.6464	.6474	.6484	.6493	.6503	.6513	.6522
4.5	.6532	.6542	.6551	.6561	.6571	.6580	.6590	.6599	.6609	.6618
4.6	.6628	.6637	.6646	.6656	.6665	.6675	.6684	.6693	.6702	.6712

Figure 7.3

For $1 < x < 10$, the table on pages 466 and 467 provides values of $\log_{10} x$, directly. Consider the excerpt from this table shown in Figure 7.3. Each number in the column headed by x represents the first two digits of x, while each number in the row opposite x contains the third digit of x. The digits located at the intersection of a row and a column form the logarithm of x. For example, to find $\log_{10} 4.25$, we look at the intersection of the row opposite 4.2 under x and the column headed by 5. Thus

$$\log_{10} 4.25 = 0.6284.$$

The equality sign is used here in an inexact sense; $\log_{10} 4.25 \approx 0.6284$ is more proper, because $\log_{10} 4.25$ is irrational and cannot be precisely represented by a rational number. We shall follow customary usage, however, writing $=$ instead of $\approx$, and leave the intent to the context.

Characteristic and mantissa

Now suppose we wish to find $\log_{10} x$ for values of x outside the range of the table—that is, for $0 < x < 1$ or $x > 10$, as shown in Figure 7.4. This can be done quite readily by first representing the number in scientific notation—that is, as the product of a number between 1 and 10 and a power of 10—and then applying Theorem 7.1-I. For example,

$$\log_{10} 42.5 = \log_{10} (4.25 \times 10^1) = \log_{10} 4.25 + \log_{10} 10^1$$
$$= 0.6284 + 1 = 1.6284,$$
$$\log_{10} 425 = \log_{10} (4.25 \times 10^2) = \log_{10} 4.25 + \log_{10} 10^2$$
$$= 0.6284 + 2 = 2.6284.$$

Observe that the decimal portion of the logarithm is always 0.6284, and the integral portion is just the exponent on 10 when the number is written in scientific notation.

7.3 Logarithms to the Base 10 and the Base e

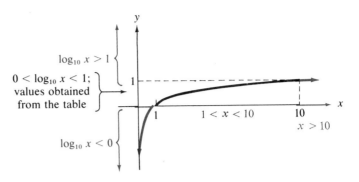

Figure 7.4

This process can be reduced to a mechanical one by considering $\log_{10} x$ to consist of two parts, an integral part (called the **characteristic**) and a nonnegative decimal fraction part (called the **mantissa**). Thus the table of values for $\log_{10} x$ for $1 < x < 10$ can be looked upon as a table of mantissas for $\log_{10} x$ for all $x > 0$.

For example, to find $\log_{10} 43,700$, we first write

$$\log_{10} 43,700 = \log_{10}(4.37 \times 10^4).$$

Upon examining the table of logarithms, we find that $\log_{10} 4.37 = 0.6405$, so that

$$\log_{10} 43,700 = 0.6405 + 4 = 4.6405.$$

Now consider an example of the form $\log_{10} x$ for $0 < x < 1$. To find $\log_{10} 0.00402$, we first write

$$\log_{10} 0.00402 = \log_{10}(4.02 \times 10^{-3}).$$

Examining the table, we find that $\log_{10} 4.02$ is 0.6042. Upon adding 0.6042 to the characteristic -3, we obtain

$$\log_{10} 0.00402 = 0.6042 - 3 = -2.3958.$$

In the latter expression, the decimal portion of the logarithm is no longer 0.6042 as it is for all numbers $x > 1$ whose first three significant digits are 402. However, for hand calculation it is customary and usually more convenient to leave the logarithm in the form $0.6042 - 3$, in which the decimal part is positive. (You may own or have seen electronic calculators that do otherwise.) Alternatively, we can write

$$\log_{10} 0.00402 = 0.6042 - 3 \tag{4}$$
$$= 0.6042 + (7 - 10)$$
$$= 7.6042 - 10, \tag{5}$$

and again the decimal part is positive. The expressions

$$6.6042 - 9, \quad 1.6042 - 4, \quad 4.6042 - 7$$

are equally valid representations, and any one of these forms may sometimes be more useful than forms (4) and (5) above.

Values of
antilog₁₀ x

It is possible to reverse the process described above. In other words, given $\log_{10} x$ we can find x. In this event, x is referred to as the **antilogarithm** (antilog$_{10}$) of $\log_{10} x$. Thus for any real number y,

$$x = \text{antilog}_{10} y$$

is equivalent to

$$\log_{10} x = y.$$

Example

Find the value of antilog$_{10}$ 1.6395.

Solution

We are looking for a value of x so that $\log_{10} x = 1.6395$. We observe that the mantissa of $\log_{10} x$ is 0.6395 and that the characteristic is 1. We locate 0.6395 in the table of mantissas and see that the associated value is 4.36. Thus

$$x = \text{antilog}_{10} 1.6395 = \text{antilog}_{10} (0.6395 + 1)$$
$$= 4.36 \times 10^1 = 43.6.$$

antilog₁₀ x
equals 10ˣ

Note that $\log_{10} y = x$ and $10^x = y$ are equivalent; furthermore, note from the equation $\log_{10} y = x$ that $y = \text{antilog}_{10} x$. Hence,

$$\text{antilog}_{10} x = 10^x.$$

Thus, values for antilog$_{10}$ x can be obtained from a table of powers of 10^x or by using the 10^x key on a calculator.

Examples

Find the value of each expression.

a. antilog$_{10}$ 2

b. antilog$_{10}$ 1.6395

Solutions

a. antilog$_{10}$ 2 = 10^2
   = 100

b. antilog$_{10}$ 1.6395 = $10^{1.6395}$; using a calculator, we obtain the value 43.6, the same value that we obtained in the preceding example.

If we seek the common logarithm of a number that is not an entry in the table (for example, $\log_{10}$ 3,712), or if we seek x when $\log_{10} x$ is not an entry in the table, it is possible to obtain a good approximation by using the value in the table closest to the given value.

Values of
log_e x

Values for $\log_e x$ (sometimes written as ln x) are called **logarithms to the base e**, or **natural logarithms**. From the definition of $\log_b x$,

$$e^{\log_e x} = x \quad (x > 0);$$

that is, $\log_e x$ is the exponent that must be placed on e so that the resulting power is x.

Table III on page 469 includes values of $\log_e x$, $0.1 \leq x \leq 190$. Values for $\log_e x$ outside of this interval can be obtained in the same way that we obtained values for $\log_{10} x$ except that we no longer have integer characteristics.

7.3 Logarithms to the Base 10 and the Base e

Example   Find $\log_e 278$.

Solution   We first observe that $278 = 2.78 \times 10^2$. Using Table III, we have

$$\begin{aligned}\log_e 278 = \log_e(2.78 \times 10^2) &= \log_e 2.78 + 2\log_e 10 \\ &= 1.0296 + 2(2.3026) \\ &= 1.0296 + 4.6052 \\ &= 5.6348.\end{aligned}$$

Example   Find $\log_e 0.278$.

Solution   This time, we observe that $0.278 = 2.78 \times 10^{-1}$. Hence

$$\begin{aligned}\log_e 0.278 = \log_e(2.78 \times 10^{-1}) &= \log_e 2.78 + \log_e 10^{-1} \\ &= \log_e 2.78 - \log_e 10 \\ &= 1.0296 - 2.3026 \\ &= -1.2730.\end{aligned}$$

As one would expect, the logarithm is negative.

The values in the above examples were obtained using Table III. If a calculator is used, slightly more accurate answers will be obtained.

antilog$_e$ x equals e^x   The process used in the preceding examples can be reversed to find antilog$_e$ x. If, however, a table for the exponential function $y = e^x$ is available (see Table II on page 468), the antilog$_e$ x can be read directly from this table, since

$$\text{antilog}_e x = e^x.$$

Alternatively, a calculator can be used to obtain such values.

Examples   a. Find antilog$_e$ 3.2.   b. Find antilog$_e$ (-3.2).

Solutions   The values can be read directly from Table II or by using a calculator.

a. antilog$_e$ $3.2 = e^{3.2} = 24.533$ b. antilog$_e$ $(-3.2) = e^{-3.2} = 0.0408$

Exercise 7.3

Find the characteristic of the given logarithm.

Examples   a. $\log_{10} 248$ b. $\log_{10} 0.0057$

Solutions   Represent the number in scientific notation.

a. $\log_{10}(2.48 \times 10^2)$ b. $\log_{10}(5.7 \times 10^{-3})$

Solutions Continued on Overleaf

The exponent on the base 10 is the characteristic.

2   -3, or $7 - 10$

1. $\log_{10} 312$
2. $\log_{10} 0.02$
3. $\log_{10} 0.00851$
4. $\log_{10} 8.01.$
5. $\log_{10} 0.00031$
6. $\log_{10} 0.0004$
7. $\log_{10} (15 \times 10^3)$
8. $\log_{10} (820 \times 10^4)$

Find the logarithm

Examples

a. $\log_{10} 16.8$
b. $\log_{10} 0.043$

Solutions

Represent the number in scientific notation.

a. $\log_{10} (1.68 \times 10^1)$
b. $\log_{10} (4.3 \times 10^{-2})$

Determine the mantissa from the table.

0.2253   0.6335

Add the characteristic as determined by the exponent on the base 10.

1.2253 $8.6335 - 10$

9. $\log_{10} 6.73$
10. $\log_{10} 891$
11. $\log_{10} 0.813$
12. $\log_{10} 0.00214$
13. $\log_{10} 0.08$
14. $\log_{10} 0.000413$
15. $\log_{10} (2.48 \times 10^2)$
16. $\log_{10} (5.39 \times 10^{-3})$

Find the antilogarithm

Example

$\text{antilog}_{10} 2.7364$

Solution

Locate the mantissa in the body of the table of mantissas and determine the associated antilog_{10} (a number between 1 and 10); write the characteristic 2 as an exponent on the base 10.

$$\text{antilog}_{10} 2.7364 = \text{antilog}_{10} (0.7364 + 2)$$
$$= 5.45 \times 10^2 = 545$$

Alternative solution

$\text{antilog}_{10} 2.7364 = 10^{2.7364}$; use a calculator to obtain the value 545 (to three places).

17. $\text{antilog}_{10} 0.6128$
18. $\text{antilog}_{10} 0.2504$
19. $\text{antilog}_{10} 0.5647$
20. $\text{antilog}_{10} 3.9258$
21. $\text{antilog}_{10} (8.8075 - 10)$
22. $\text{antilog}_{10} (3.9722 - 5)$
23. $\text{antilog}_{10} 3.7388$
24. $\text{antilog}_{10} 2.0086$
25. $\text{antilog}_{10} (6.8561 - 10)$
26. $\text{antilog}_{10} (1.8156 - 4)$

Find the logarithm directly from Table III or by using a calculator.

7.4 Applications of Logarithms

Examples a. $\log_e 12$ b. $\log_e 198$

Solutions a. $\log_e 12 = 2.4849$
b. $\log_e 198 = \log_e(1.98 \times 10^2)$
$= \log_e 1.98 + 2\log_e 10$
$= 0.6931 + 2(2.3026)$
$= 5.2983$

27. $\log_e 3$ 28. $\log_e 8$ 29. $\log_e 17$ 30. $\log_e 98$
31. $\log_e 327$ 32. $\log_e 107$ 33. $\log_e 450$ 34. $\log_e 605$

Find the antilogarithm directly from Table II or by using a calculator.

Examples a. $\text{antilog}_e 0.85$ b. $\text{antilog}_e(-3.4)$

Solutions a. $\text{antilog}_e 0.85 = e^{0.85}$ b. $\text{antilog}_e(-3.4) = e^{-3.4}$
$\qquad = 2.3396$ $\qquad = 0.0334$

35. $\text{antilog}_e 0.50$ 36. $\text{antilog}_e 1.5$ 37. $\text{antilog}_e 3.4$
38. $\text{antilog}_e 4.5$ 39. $\text{antilog}_e 0.231$ 40. $\text{antilog}_e 1.43$
41. $\text{antilog}_e(-0.15)$ 42. $\text{antilog}_e(-0.95)$ 43. $\text{antilog}_e(-2.5)$
44. $\text{antilog}_e(-4.2)$ 45. $\text{antilog}_e(-0.255)$ 46. $\text{antilog}_e(-3.65)$

47. By definition, $\log_e x = N$ can be written in exponential form as $e^N = x$. Show that $N = \log_{10} x / \log_{10} e$, and hence that

$$\log_e x = \frac{\log_{10} x}{\log_{10} e}.$$

48. Use the results of Exercise 47 to show that

$$\log_e x = 2.3026 \log_{10} x.$$

49. Find a value for $\log_3 18$ by using a table of logarithms to the base 10. *Hint:* Let $y = \log_3 18$ and write in exponential form.

50. Find a value for $\log_2 3$ by using a table of logarithms to the base 10. *Hint:* Let $y = \log_2 3$ and write in exponential form.

7.4 Applications of Logarithms

Logarithms were first used to perform numerical computations that involve products, quotients, and powers. Calculators and high-speed computing devices have made this application obsolete. However, logarithms have wide applicability in other areas. In particular, they are useful for solving exponential equations in which the variable occurs in an exponent.

Example Find the solution set of $5^x = 7$.

Solution Since $5^x > 0$ for all x, we have
$$\log_{10} 5^x = \log_{10} 7;$$
from Theorem 7.1-III,
$$x \log_{10} 5 = \log_{10} 7.$$
Dividing each member by $\log_{10} 5$, we obtain
$$x = \frac{\log_{10} 7}{\log_{10} 5} = \frac{0.8451}{0.6990} \approx 1.209.$$

Note that in seeking the decimal approximation, the logarithms are *divided*, not subtracted.

Example The amount A that results from a deposit of P dollars after t years when interest is compounded continuously at an annual rate of $r \%$ is given by the formula
$$A = Pe^{rt/100}.$$
If a savings and loan company compounds interest continuously, and if the annual rate is $5\frac{3}{4}\%$, how long will it take for a deposit of \$5,000 to grow to \$6,000?

Solution Substituting appropriate values in the given formula, we obtain the equation
$$6{,}000 = 5{,}000 e^{0.0575t}.$$
This is equivalent to
$$\frac{6}{5} = e^{0.0575t}.$$
Taking the logarithm to the base e of each member of this equation, we have
$$\log_e \frac{6}{5} = 0.0575t \log_e e = 0.0575t.$$
We can compute $\log_e(6/5)$ using the tables or a hand calculator and then solve for t, obtaining $t = 3.170$. Hence it takes slightly more than three years for the value of the account to reach \$6,000.

Example If a particular component of a certain manufactured system shows no aging effect (in other words, it does not wear out with use), then the probability that the component will not fail within t years is
$$P = e^{-kt},$$
where $k > 0$ is a constant that depends on the component. If $k = 2$ for a given component, find t so that the probability that the component will not fail within t years is $1/2$.

7.4 Applications of Logarithms

Solution Since we want $P = 1/2$ and we know $k = 2$, we have

$$\frac{1}{2} = e^{-2t}.$$

Taking the logarithm to the base e of each member of this equation, we have

$$\log_e \left(\frac{1}{2}\right) = -2t \log_e e,$$

$$-0.6931 = -2t,$$

$$t = \frac{-0.6931}{-2} = 0.3466.$$

Hence the probability is $1/2$ that the component will fail after 0.3466 years.

Exercise 7.4

Solve. Give the solution in logarithmic form using the base 10 and also as an approximation in decimal form.

Example $3^{x-2} = 16$

Solution Equating the $\log_{10}$ of both sides, we have

$$\log_{10} 3^{x-2} = \log_{10} 16.$$

By Theorem 7.1-III

$$(x - 2) \log_{10} 3 = \log_{10} 16,$$

from which

$$x - 2 = \frac{\log_{10} 16}{\log_{10} 3},$$

$$x = \frac{\log_{10} 16}{\log_{10} 3} + 2.$$

The solution is $\dfrac{\log_{10} 16}{\log_{10} 3} + 2 \approx 4.524.$

1. $2^x = 7$
2. $3^x = 4$
3. $3^{x+1} = 8$
4. $2^{x-1} = 9$
5. $7^{2x-1} = 3$
6. $3^{x+2} = 10$

Solve. Leave the result in the form of an equation equivalent to the given equation.

7. $y = x^n$, for n
8. $y = Cx^{-n}$, for n
9. $y = e^{kt}$, for t
10. $y = Ce^{-kt}$, for t
11. $y = 3 + 2 \log_5 (x - 1)$, for x
12. $y = a + b \log_c x^3$, for x

P dollars invested at an interest rate *r* compounded yearly yields an amount *A* after *n* years, given by $A = P(1 + r)^n$. If the interest is compounded *t* times yearly, the amount is given by

$$A = P\left(1 + \frac{r}{t}\right)^{tn}.$$

Example   One dollar compounded annually for 12 years yields $1.127. What is the rate of interest to the nearest 1/2 %?

Solution   Using the given equation, we obtain $(1 + r)^{12} = 1.127$. Equate $\log_{10}$ of each member and apply Theorem 7.1-III.

$$12 \log_{10}(1 + r) = \log_{10} 1.127 = 0.0531$$

Multiply each member by 1/12.

$$\log_{10}(1 + r) = \frac{1}{12}(0.0531) = 0.0044$$

Determine antilog$_{10}$ 0.0044 and solve for *r*.

$$\text{antilog}_{10} 0.0044 = 1 + r = 1.01$$
$$r = 0.01, \quad \text{or} \quad r = 1\%$$

13. One dollar compounded annually for 10 years yields $1.48. What is the rate of interest to the nearest 1/2 %?
14. How many years (nearest year) would it take for $1.00 to yield $2.19 if compounded annually at 4 %?
15. Find the compounded amount of $5,000 invested at 4 % for 10 years when compounded annually; when compounded semiannually.
16. Two men, *A* and *B*, each invested $10,000 at 4 % for 20 years with a bank that computed interest quarterly. *A* withdrew his interest at the end of each 3-month period, but *B* let his investment be compounded. How much more than *A* did *B* earn over the period of 20 years?

The present value of *A* dollars in *t* years when compounded continuously at *r* % annually is the amount that must be deposited in an account at the present so that it is worth *A* dollars in *t* years. The formula for present value (*PV*) is

$$PV = Ae^{-rt/100}.$$

17. What is the annual rate of interest of an account if when compounded continuously the present value of $10,000 in 10 years is $5,000?
18. At an annual interest rate of 7.25 % when compounded continuously, how long must $1,000 be kept in an account to attain a value of $2,000?

7.4 Applications of Logarithms

The chemist defines the pH (hydrogen potential) of a solution by

$$\text{pH} = \log_{10} \frac{1}{[\text{H}^+]} = \log_{10} [\text{H}^+]^{-1},$$

$$\text{pH} = -\log_{10} [\text{H}^+],$$

where $[\text{H}^+]$ is a numerical value for the concentration of hydrogen ions in aqueous solution in moles per liter.

Example Calculate the pH of a solution whose hydrogen ion concentration is 3.7×10^{-6}.

Solution Substitute 3.7×10^{-6} for $[\text{H}^+]$ in the relationship $\text{pH} = \log_{10} \frac{1}{[\text{H}^+]}$.

$$\text{pH} = \log_{10} \frac{1}{3.7 \times 10^{-6}}$$

$$= \log_{10} \left(\frac{1}{3.7} \times 10^6\right)$$

$$= \log_{10} 1 - \log_{10} 3.7 + \log_{10} 10^6$$

$$= 0 - 0.5682 + 6 = 5.432$$

19. Calculate the pH of a solution whose hydrogen ion concentration $[\text{H}^+]$ is 2.0×10^{-8}.
20. Calculate the pH of a solution whose hydrogen ion concentration is 6.3×10^{-7}.
21. Calculate the hydrogen ion concentration of a solution whose pH is 5.6.
22. Calculate the hydrogen ion concentration of a solution whose pH is 7.2.
23. The atmospheric pressure p, in inches of mercury, is given approximately by $p = 30.0(10)^{-0.09a}$, where a is the altitude in miles above sea level. What is the atmospheric pressure at sea level; at 3 miles above sea level?
24. Using the information in Exercise 23, find the atmospheric pressure 6 miles above sea level.
25. The amount of a radioactive element remaining at any time t is given by $y = y_0 e^{-0.4t}$, where t is in seconds and y_0 is the amount present initially. How much of the element would remain after 3 seconds if 40 grams were present initially?
26. The number of bacteria present in a culture is related to time by the formula $N = N_0 e^{0.04t}$, where N_0 is the number of bacteria present at time $t = 0$, and t is time in hours. If 10,000 bacteria are present 10 hours after the beginning of an experiment, how many were present when $t = 0$?
27. If a component in a system shows no aging effect and the probability of failure after 10 units of time is $1/10$, using the formula on page 192 find the value k associated with this component. The value $1/k$ is the average time until a failure occurs.

28. If a component in a system shows no aging effect and the probability of failure after 3 units of time is 0.001, find the average time until a failure occurs (see Exercise 27).

Chapter Review

[7.1]   1. Graph $y = 3^x$.   2. Graph $A = e^{0.05t}$.

[7.2]   *Express in logarithmic notation.*

3. $16^{-1/2} = \dfrac{1}{4}$ 4. $7^3 = 343$

Express in exponential notation.

5. $\log_2 8 = 3$ 6. $\log_{10} 0.0001 = -4$

Solve.

7. $\log_2 16 = y$ 8. $\log_{10} x = 3$

Express as the sum or difference of simpler logarithmic quantities.

9. $\log_{10} \sqrt[3]{xy^2}$ 10. $\log_{10} \dfrac{2R^3}{\sqrt{PQ}}$

Express as a single logarithm with coefficient 1.

11. $2 \log_b x - \dfrac{1}{3} \log_b y$ 12. $\dfrac{1}{3}(2 \log_{10} x + \log_{10} y) - 3 \log_{10} z$

[7.3]   *Find the logarithm.*

13. $\log_{10} 42$ 14. $\log_{10} 0.00314$
15. $\log_{10} 682$ 16. $\log_{10} 0.0414$

Find the antilogarithm.

17. antilog$_{10}$ 1.8287   18. antilog$_{10}$ (8.6684 − 10)
19. antilog$_{10}$ 0.4240   20. antilog$_{10}$ (9.8224 − 10)

Find the logarithm.

21. $\log_e 7$ 22. $\log_e 23$ 23. $\log_e 241$ 24. $\log_e 510$

Chapter Review

Find the antilogarithm.

25. $\text{antilog}_e\ 6.0$
26. $\text{antilog}_e\ 1.8$
27. $\text{antilog}_e(-3.5)$
28. $\text{antilog}_e(-0.42)$

[7.4] *Solve. Give the solution in logarithmic form using the base 10 and also as an approximation in decimal form.*

29. $3^x = 2$
30. $3^{x+1} = 80$

Solve. Leave the result in the form of an equation equivalent to the given equation.

31. $y = A + ke^{-t}$, for t
32. $y = N + N_0 \log_e x$, for x

33. Given $y = y_0 e^{-0.4t}$, find y if $y_0 = 30$ and $t = 5$.
34. Given $y = y_0 e^{-1.2t}$, find t if $y = 100$ and $y_0 = 10$.

Solve. Use the appropriate formula from pages 192, 194, and 195.

35. One dollar compounded quarterly for 5 years yields $1.25. What is the annual rate of interest to the nearest 1/4 %?
36. Calculate the hydrogen ion concentration of a solution if its pH is 6.2.
37. Find the interest rate at which $10,000 has a present value of $7,000 in 5 years with continuously compounded interest.
38. Find the average time until failure of a component that shows no aging effect if the probability of failure occurring after 3 units of time is 0.01.

8 Circular Functions

In Chapter 7 we discussed the exponential and logarithmic functions, which are useful in describing growth. Neither these functions, however, nor in fact the polynomial or rational functions which we have also considered, would be useful in describing cyclic phenomena. In this chapter we shall discuss the *circular functions*; because of their periodic nature, these are particularly appropriate in studying such phenomena.

8.1 The Functions Sine and Cosine

In this section we define the two basic circular functions.

Arc on a unit circle

Consider, intuitively, a point moving steadily in a counterclockwise direction around the circle with equation $x^2 + y^2 = 1$. At any given time, the moving point occupies a position on the circle; the point, in turn, is associated with an ordered pair (x, y). If the distance along the circle from the point $(1, 0)$ to the point (x, y) is designated by s, then we can associate the real number s with the ordered pair (x, y) (Figure 8.1-a). We shall assume that every arc of a circle has a length and that there is a one-to-one correspondence between the members of the set of nonnegative real numbers and the lengths of all arcs of the unit circle measured in a *counterclockwise* direction from the point $(1, 0)$ to points (x, y) on the circle. Let us agree that values of $s < 0$ denote lengths of arc measured from $(1, 0)$ in a *clockwise* direction to points (x, y) on the circle (Figure 8.1-b). This extends the one-to-one correspondence of the set of lengths of arcs on the unit circle to the entire set R of real numbers.

8.1 The Functions Sine and Cosine

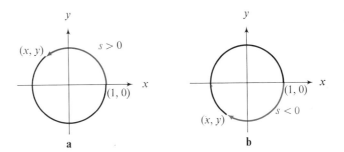

Figure 8.1

Since the circumference of a circle is $2\pi r$ and the radius of the unit circle is 1, the distance once around the unit circle is 2π, twice around is 4π, and so on. Furthermore, the distance halfway around is π, one-fourth of the way around is $\pi/2$, and so forth. The endpoints of selected arcs from the point (1, 0) and their corresponding arc lengths are shown in Figure 8.2: the arcs in (a) are in the *counterclockwise*

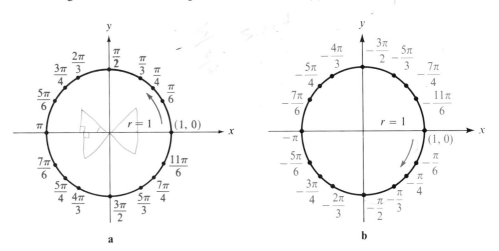

Figure 8.2

direction, and the arcs in (b) are in the *clockwise direction*. Especially useful are the two functions which associate the members of R with the members of the set of first components x and with the members of the set of second components y of the ordered pairs (x, y) corresponding to points on the unit circle. These functions are called the **cosine** and **sine**, respectively, and are defined as follows:

Definition 8.1 If (x, y) is the point at arc length s from $(1, 0)$ on the unit circle with equation $x^2 + y^2 = 1$, then the **cosine of** s is x, and the **sine of** s is y. We denote this by writing

$$\cos s = x \quad \text{and} \quad \sin s = y.$$

In other words, the first component of the point (x, y) located at arc length s from $(1, 0)$ on the unit circle is called the cosine of s, and the second component is called the sine of s; and we denote these by cos s and sin s, respectively. Thus, $(x, y) = (\cos s, \sin s)$.

Domain and range of cosine and sine

The domain of each of these functions is the set R. The range of the cosine function is the set of all first components of the ordered pairs corresponding to points on the unit circle, and hence is the set $\{x \mid -1 \leq x \leq 1\}$. Similarly, the range of the sine function is $\{y \mid -1 \leq y \leq 1\}$.

Signs of cosine and sine

Because cos s and sin s are simply the coordinates of points on the unit circle, they are positive or negative in accord with the values of x and y in the various quadrants. Table 8.1 summarizes in a convenient way the sign associated with $x = \cos s$ and $y = \sin s$ in each quadrant.

Table 8.1

Quadrant II	Quadrant I
x or cos $s < 0$	x or cos $s > 0$
y or sin $s > 0$	y or sin $s > 0$
Quadrant III	Quadrant IV
x or cos $s < 0$	x or cos $s > 0$
y or sin $s < 0$	y or sin $s < 0$

cos s and sin s in terms of each other

Because cos $s = x$ and sin $s = y$ are subject to the condition that $x^2 + y^2 = 1$, we have the following basic identity relating cos s and sin s.

Theorem 8.1 For every $s \in R$,

$$\cos^2 s + \sin^2 s = 1. \tag{1}$$

Note that for convenience we write $\cos^2 s$ for $(\cos s)^2$ and $\sin^2 s$ for $(\sin s)^2$. Now Theorem 8.1 and Table 8.1 can be used to write

$$\sin s = \begin{cases} \sqrt{1 - \cos^2 s} & \text{in Quadrants I and II,} \tag{2a} \\ -\sqrt{1 - \cos^2 s} & \text{in Quadrants III and IV,} \tag{2b} \end{cases}$$

and

$$\cos s = \begin{cases} \sqrt{1 - \sin^2 s} & \text{in Quadrants I and IV,} \tag{3a} \\ -\sqrt{1 - \sin^2 s} & \text{in Quadrants II and III.} \tag{3b} \end{cases}$$

Therefore, if either sin s or cos s is known and the quadrant in which the terminal point of the arc of length s lies can be determined, then we can find the value for the other function.

8.1 The Functions Sine and Cosine

Example Given that $\cos s = -3/5$ and $\pi < s < 3\pi/2$, find $\sin s$.

Solution Using (2b), we have
$$\sin s = -\sqrt{1 - \cos^2 s}$$
$$= -\sqrt{1 - \left(-\frac{3}{5}\right)^2} = -\sqrt{\frac{16}{25}} = -\frac{4}{5}.$$

Exercise 8.1

Determine the point on the figure which corresponds to the given arc length when measured from the point (1, 0). Do not refer to Figure 8.2.

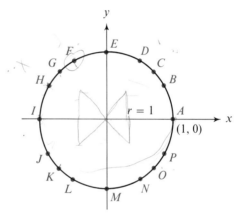

Example $\dfrac{3\pi}{2}$

Solution Since the circumference of the circle is 2π, and $3\pi/2$ is three-fourths of 2π, the point corresponding to the length $3\pi/2$ is M.

1. π
2. $\dfrac{\pi}{2}$
3. $\dfrac{\pi}{4}$
4. $\dfrac{\pi}{3}$
5. $\dfrac{5\pi}{6}$
6. $\dfrac{2\pi}{3}$
7. $\dfrac{5\pi}{4}$
8. $\dfrac{11\pi}{6}$
9. 3π
10. $\dfrac{7\pi}{2}$
11. $\dfrac{13\pi}{4}$
12. $\dfrac{13\pi}{6}$
13. $-\dfrac{\pi}{2}$
14. $-\dfrac{3\pi}{2}$
15. $-\dfrac{3\pi}{4}$
16. -3π

Use Table 8.1 to state whether the given function value is positive or negative.

Examples a. $\sin \dfrac{11\pi}{4}$ b. $\cos \dfrac{17\pi}{6}$

Solutions on Overleaf

Solutions   a. Since $11\pi/4$ terminates in the second quadrant, we find from Table 8.1 that $\sin 11\pi/4$ is positive.

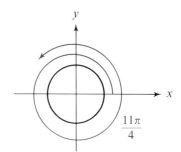

b. Since $17\pi/6$ terminates in the second quadrant, we find from Table 8.1 that $\cos 17\pi/6$ is negative.

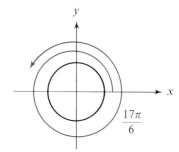

17. $\cos \dfrac{13\pi}{3}$

18. $\cos \dfrac{27\pi}{5}$

19. $\sin \dfrac{17\pi}{7}$

20. $\sin \dfrac{29\pi}{4}$

21. $\cos \dfrac{41\pi}{5}$

22. $\sin \dfrac{36\pi}{7}$

23. $\sin \left(-\dfrac{4\pi}{3}\right)$

24. $\cos \left(-\dfrac{7\pi}{5}\right)$

25. $\cos \left(-\dfrac{31\pi}{4}\right)$

26. $\sin \left(-\dfrac{17\pi}{3}\right)$

In Exercises 27–36, find the required function value and state the quadrant in which s terminates.

Example $\sin s$, given that $\cos s = \sqrt{7}/4$ and $\sin s < 0$.

Solution   Since $\cos s = \sqrt{7}/4$ and $\sin s < 0$, s terminates in Quadrant IV; and we see from (2b) on page 200 that

$$\sin s = -\sqrt{1 - \cos^2 s} = -\sqrt{1 - \left(\dfrac{\sqrt{7}}{4}\right)^2}$$

$$= -\sqrt{1 - \dfrac{7}{16}} = -\sqrt{\dfrac{9}{16}} = -\dfrac{3}{4}.$$

27. $\cos s$, given that $\sin s = 3/5$ and $\cos s > 0$.

28. sin s, given that cos s = 4/5 and sin s < 0.
29. sin s, given that cos s = 5/13 and sin s < 0.
30. cos s, given that sin s = 12/13 and cos s < 0.
31. cos s, given that sin s = −2/3 and cos s > 0.
32. sin s, given that cos s = −1/4 and sin s > 0.
33. sin s, given that cos s = $-\sqrt{3}/2$ and sin s < 0.
34. cos s, given that sin s = $-\sqrt{3}/2$ and cos s > 0.
35. cos s, given that sin s = −3/5 and cos s > 0.
36. sin s, given that cos s = −4/5 and sin s > 0.

8.2 Special Function Values of Cosine and Sine

Quadrantal function values

In general, finding cos s and sin s for a given real number s is a difficult matter. However, some values of cos s and sin s, corresponding to special values of s, are readily available.

Figure 8.3 shows the coordinates of four points on the unit circle with which we can associate specific values of s. For $s = 0$,

$$\cos 0 = 1 \quad \text{and} \quad \sin 0 = 0. \quad (1)$$

Furthermore, since the arc of a circle included in a quadrant is one-fourth of the circumference of the circle, that is, $2\pi/4 = \pi/2$, we know immediately that

$$\cos \frac{\pi}{2} = 0 \quad \text{and} \quad \sin \frac{\pi}{2} = 1, \quad (2)$$

$$\cos \pi = -1 \quad \text{and} \quad \sin \pi = 0, \quad (3)$$

$$\cos \frac{3\pi}{2} = 0 \quad \text{and} \quad \sin \frac{3\pi}{2} = -1. \quad (4)$$

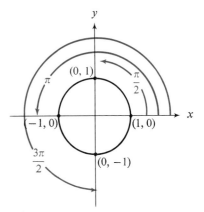

Figure 8.3

Notice that in finding values for cos s and sin s we are finding *coordinates of points* on the unit circle. The function values obtained in (1), (2), (3), and (4) are sometimes called **quadrantal values**, because the terminal point of each arc lies on an axis.

We can find other special values for cos s and sin s by using the geometry of the unit circle and the distance formula,

$$d^2 = (x_2 - x_1)^2 + (y_2 - y_1)^2.$$

Values for
$s = \pi/4$

First consider $s = \pi/4$. Figure 8.4 shows the unit circle and the designated value for s. Since (x, y) bisects the arc from $(1, 0)$ to $(0, 1)$, it follows from geometric considerations that $x = y$. Because $x^2 + y^2 = 1$, we have

$$x^2 + x^2 = 1,$$
$$2x^2 = 1,$$
$$x^2 = \frac{1}{2},$$

and

$$x = \frac{1}{\sqrt{2}} = \frac{\sqrt{2}}{2} \quad \text{or} \quad x = -\frac{1}{\sqrt{2}} = -\frac{\sqrt{2}}{2}.$$

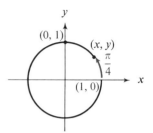

Figure 8.4

Now, both x and y are positive in the first quadrant, so the desired value for x is $1/\sqrt{2}$. Since $x = y$, y is also equal to $1/\sqrt{2}$, and hence we have

$$\cos\frac{\pi}{4} = \frac{1}{\sqrt{2}} = \frac{\sqrt{2}}{2} \quad \text{and} \quad \sin\frac{\pi}{4} = \frac{1}{\sqrt{2}} = \frac{\sqrt{2}}{2}.$$

Values for
$s = \pi/6$

Next, consider $s = \pi/6$, as pictured in Figure 8.5. If the ordered pair corresponding to $s = \pi/6$ is (x, y), then the ordered pair corresponding to $s = -\pi/6$ is $(x, -y)$. Now the arc from $(x, -y)$ to (x, y) is of length

$$\frac{\pi}{6} + \frac{\pi}{6} = \frac{\pi}{3}$$

and so is the length of the arc from (x, y) to $(0, 1)$. Because equal arcs of a circle subtend equal chords, the distance from $(0, 1)$ to (x, y) is the same as the distance from (x, y) to $(x, -y)$. Using the distance formula, we therefore have

$$(x - 0)^2 + (y - 1)^2 = (x - x)^2 + (-y - y)^2,$$

or

$$x^2 + y^2 - 2y + 1 = 4y^2.$$

Since $x^2 + y^2 = 1$, we substitute 1 for $x^2 + y^2$ in this equation and obtain

$$1 - 2y + 1 = 4y^2,$$
$$4y^2 + 2y - 2 = 0,$$
$$2(2y - 1)(y + 1) = 0,$$

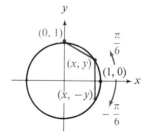

Figure 8.5

so that

$$y = \frac{1}{2} \quad \text{or} \quad y = -1.$$

8.2 Special Function Values of Cosine and Sine

Because (x, y) is in the first quadrant, we must select the value $1/2$ as the y-coordinate. Next, we use $x^2 + y^2 = 1$ to find a value for x. Thus,

$$x^2 + \left(\frac{1}{2}\right)^2 = 1,$$

from which

$$x = \pm \frac{\sqrt{3}}{2}.$$

Because (x, y) is in the first quadrant, we have $x = \sqrt{3}/2$, and it follows that

$$\cos \frac{\pi}{6} = \frac{\sqrt{3}}{2} \quad \text{and} \quad \sin \frac{\pi}{6} = \frac{1}{2}. \tag{5}$$

Values for $s = \pi/3$

From Equation (5), we can quickly find values for $\cos(\pi/3)$ and $\sin(\pi/3)$ by symmetry. Figure 8.6 shows the point (x, y) associated with s equal to $\pi/3$; clearly, the abscissa of (x, y) is the ordinate of $(\sqrt{3}/2, 1/2)$, and the ordinate of (x, y) is the abscissa of $(\sqrt{3}/2, 1/2)$. Hence

$$\cos \frac{\pi}{3} = \frac{1}{2} \quad \text{and} \quad \sin \frac{\pi}{3} = \frac{\sqrt{3}}{2}.$$

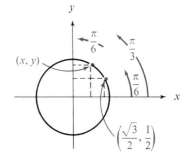

Figure 8.6

For ease of reference, the values of $\cos s$ and $\sin s$ for the special values of s discussed above are given in Table 8.2

Table 8.2

s	0	$\dfrac{\pi}{6}$	$\dfrac{\pi}{4}$	$\dfrac{\pi}{3}$	$\dfrac{\pi}{2}$	π	$\dfrac{3\pi}{2}$	2π
$\cos s$	1	$\dfrac{\sqrt{3}}{2}$	$\dfrac{\sqrt{2}}{2}$	$\dfrac{1}{2}$	0	-1	0	1
$\sin s$	0	$\dfrac{1}{2}$	$\dfrac{\sqrt{2}}{2}$	$\dfrac{\sqrt{3}}{2}$	1	0	-1	0

Values for $0 < s < \dfrac{\pi}{2}$

The values in Table 8.2 were obtained using geometric considerations. These techniques are not adequate for most values of s, but all values can be approximated using more sophisticated methods.

In Table V on page 471, decimal approximations are given for $\sin s$ and $\cos s$ for selected values of s between 0 and $\pi/2$ (≈ 1.57). The table is graduated in intervals of length 0.01. The table is read from top to bottom in the left-hand margin, with the desired function at the top of its particular column.

Examples   a. $\cos 0.73 \approx 0.7452$ b. $\sin 1.29 \approx 0.9608$

Notice that in Table V the letter x is used to represent an element in the domain. This is customary in many tables. In this usage, of course, x can be thought of as representing an arc length along the unit circle, just as s did. In fact, because x is ordinarily the variable used to represent an element in the domain of a function, we shall hereafter use x, t or other symbols where we heretofore used s.

Using a calculator

A scientific calculator can be used to obtain the function values listed in Table V. Furthermore, it enables us to find function values for numbers with more than three digits. Because the operating procedures for different types of calculators vary, you should refer to the instruction booklet to obtain this information for your particular calculator.

The circular-function values are obtained on a calculator by using the radian (RAD or "R") mode. (Radians will be discussed in detail in Chapter 9.) The number of digits that are shown in the display of a calculator vary. To be consistent with values obtained from Table V, all values obtained using a calculator will be rounded off to four places.

Examples   Find:

a. $\cos 0.73$ b. $\sin 1.43$

Solutions   Set the calculator in radian mode.

a. $\cos 0.73 \approx 0.74517 \approx 0.7452$ b. $\sin 1.43 \approx 0.99010 \approx 0.9901$

Notice that whether we use the tables or a hand calculator, the values which we obtain for $\cos x$ and $\sin x$ are, in most cases, approximations. However, for convenience, we shall generally use the symbol "$=$" instead of "$\approx$" in relationships that involve such approximations.

Exercise 8.2

Use Table 8.2 to find the value of the given expression.

Examples   a. $\cos \dfrac{\pi}{2} + \sin \pi$ b. $\left(\sin \dfrac{\pi}{2}\right)(\cos \pi)$

Solutions   a. From Table 8.2,

$\cos \dfrac{\pi}{2} = 0$ and $\sin \pi = 0$.

Thus,

$\cos \dfrac{\pi}{2} + \sin \pi = 0$.

b. From Table 8.2,

$\sin \dfrac{\pi}{2} = 1$ and $\cos \pi = -1$

Thus,

$\left(\sin \dfrac{\pi}{2}\right)(\cos \pi) = -1$.

8.2 Special Function Values of Cosine and Sine

1. $\cos \dfrac{\pi}{3}$
2. $\cos \dfrac{\pi}{6}$
3. $\sin \dfrac{\pi}{4}$
4. $\sin \dfrac{\pi}{6}$
5. $\cos 0$
6. $\cos \dfrac{3\pi}{2}$
7. $\sin \dfrac{\pi}{2}$
8. $\sin \pi$
9. $\cos \dfrac{\pi}{4} + \sin \dfrac{\pi}{2}$
10. $\sin \dfrac{\pi}{3} + \cos \dfrac{\pi}{6}$
11. $\left(\sin \dfrac{\pi}{4}\right)\left(\cos \dfrac{\pi}{4}\right)$
12. $(\cos 0)\left(\sin \dfrac{3\pi}{2}\right)$

Use a hand calculator or Table V to find the approximate value of the given expression.

Example $\quad \cos 1.43 + \sin 0.02$

Solution $\quad$ From Table V or a calculator, we find that
$$\cos 1.43 = 0.1403 \quad \text{and} \quad \sin 0.02 = 0.0200.$$
Thus,
$$\cos 1.43 + \sin 0.02 = 0.1603.$$

13. $\cos 0.33$
14. $\cos 0.67$
15. $\sin 0.75$
16. $\sin 0.25$
17. $\cos 1.41$
18. $\cos 1.17$
19. $\sin 1.34$
20. $\sin 1.05$
21. $\cos 0.11$
22. $\cos 0.64$
23. $\sin 1.25$
24. $\sin 1.48$
25. $\cos 1.17 + \sin 0.11$
26. $\cos 1.08 + \cos 1.00$
27. $(\sin 0.01)(\cos 1.10)$
28. $(\sin 1.50)(\cos 1.50)$
29. $\dfrac{\sin 0.15}{\cos 1.51}$
30. $\dfrac{\cos 1.11}{\sin 1.11}$

Assume that $d = K \sin 2\pi t$. Find the indicated value under the given conditions.

Example $\quad K$, given $d = 15$ and $t = 1/8$.

Solution $\quad$ We have
$$15 = K \sin 2\pi \left(\dfrac{1}{8}\right) = K \sin \dfrac{\pi}{4}.$$
Thus, from Table 8.2, we obtain
$$15 = \dfrac{\sqrt{2}}{2} K.$$
Therefore, $K = \dfrac{30}{\sqrt{2}} = 15\sqrt{2}.$

31. K, given d = 25 and t = 1/4.
32. K, given d = 10 and t = 3/4.
33. d, given K = 100 and t = 0.5.
34. d, given K = 75 and t = 0.25.

Assume that $E = 1.2 \cos 10\pi t$. Find the value of E for the given value of t.

35. $t = 0$
36. $t = \dfrac{1}{10}$
37. $t = \dfrac{1}{30}$
38. $t = \dfrac{1}{60}$
39. $t = 0.05$
40. $t = 0.15$

8.3 Function Values of cos x and sin x, for x ∈ R

In the preceding section we obtained function values of the sine and the cosine for certain arc lengths between 0 and $\pi/2$, on a unit circle. In this section we shall see how to find function values of the sine and cosine for any arc length.

Reference arcs

In order to find the values $\sin x$ and $\cos x$ where x is not between 0 and $\pi/2$, we shall use the notion of a reference arc.

Definition 8.2 *For any arc A on the unit circle, with initial point (1, 0) and with measure x, the **reference arc** is the arc on the unit circle with least nonnegative measure between the terminal point of A and the horizontal axis. The length of the reference arc is denoted $\bar{x}$.*

The reference arcs for positive arcs terminating in each of the four quadrants are illustrated in Figure 8.7.

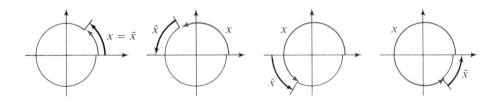

Figure 8.7

Examples Find the length of the reference arcs for the arc with the given length.

a. $\dfrac{8\pi}{9}$

b. $\dfrac{7\pi}{6}$

8.3 *Function Values of* cos *x and* sin *x, for x* ∈ *R*   209

Solutions

a. Observe that the arc with length $x = 8\pi/9$ terminates in the second quadrant, as shown in the figure. Thus,

$$\bar{x} = \pi - \frac{8\pi}{9} = \frac{\pi}{9}.$$

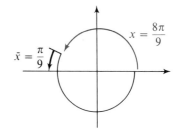

b. Observe that the arc with length $x = 7\pi/6$ terminates in the third quadrant, as shown in the figure. Thus,

$$\bar{x} = \frac{7\pi}{6} - \pi = \frac{\pi}{6}.$$

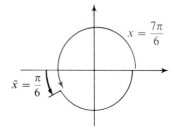

Note that negative arcs also have reference arcs, as illustrated in Figure 8.8.

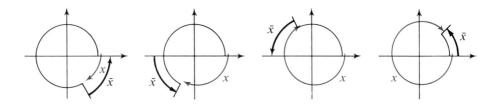

Figure 8.8

Example   Find the reference arc for the arc with length $x = -2.30$.

Solution   Note that the arc terminates in Quadrant III, as shown in the figure. We use $\pi \approx 3.14$ to obtain

$$\bar{x} = -2.30 - (-1)\pi = \pi - 2.30$$
$$\approx 3.14 - 2.30 = 0.84.$$

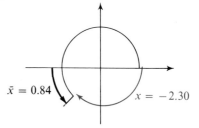

Consider an arc of length x, starting from the point $(1, 0)$, which terminates in the second quadrant (see Figure 8.9-a on page 210).

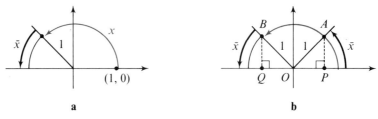

Figure 8.9

|cos x|
= cos x̄,
|sin x|
= sin x̄

Construct an arc of length x̄ (the reference arc length) from the point (1, 0) (see Figure 8.9-b). It can be shown using elementary geometry that triangles POA and QOB are congruent. This implies that $PA = QB$ and $OP = OQ$. But note that cos x and sin x are the abscissa and ordinate, respectively, of the point B, and that cos x̄ and sin x̄ are the abscissa and ordinate, respectively, of the point A. Thus, from Figure 8.9-b, we have

$$|\cos x| = QB = PA = \cos \bar{x}$$

and

$$|\sin x| = OQ = OP = \sin \bar{x}.$$

Although we assumed that x terminates in the second quadrant, the argument we used is quite general and can be used to prove the following theorem.

Theorem 8.2   For $x \in R$,

$$|\cos x| = \cos \bar{x} \quad \text{and} \quad |\sin x| = \sin \bar{x},$$

where x̄ is the length of the reference arc corresponding to x.

Calculating cos x, sin x

The fact that every reference arc length is between 0 and $\pi/2$, along with Theorem 8.2, enables us to compute cos x and sin x for any arc length x. We simply perform the following steps.

1. Find the reference arc length x̄.
2. Use Theorem 8.2 to find $|\cos x|$ and $|\sin x|$.
3. Use Table 8.1 to determine the correct algebraic sign for cos x and sin x.

Example

Compute cos $(3\pi/4)$ and sin $(3\pi/4)$ using Table 8.2.

Solution

The length of the reference arc is

$$\bar{x} = \pi - \frac{3\pi}{4} = \frac{\pi}{4},$$

as shown in the figure.

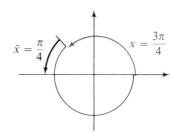

8.3 Function Values of cos x and sin x, for x ∈ R

From Theorem 8.2 and Table 8.2,

$$\left|\cos\frac{3\pi}{4}\right| = \cos\frac{\pi}{4} = \frac{\sqrt{2}}{2} \quad \text{and} \quad \left|\sin\frac{3\pi}{4}\right| = \sin\frac{\pi}{4} = \frac{\sqrt{2}}{2}.$$

Since the arc of length $3\pi/4$ terminates in the second quadrant, we know from Table 8.1 that $\cos(3\pi/4) < 0$ and $\sin(3\pi/4) > 0$. Thus, we have

$$\cos\frac{3\pi}{4} = -\frac{\sqrt{2}}{2} \quad \text{and} \quad \sin\frac{3\pi}{4} = \frac{\sqrt{2}}{2}.$$

Using a calculator

Most calculators will determine values for $\cos x$ and $\sin x$ directly for all real numbers x. Furthermore, because of the fact that calculators use a more accurate approximation to π than is used to obtain reference arcs ($\pi \approx 3.14$), the values obtained on a calculator will be more accurate than those that are obtained by using reference arcs. To be consistent with answers obtained by using reference arcs with the tables, *we shall use reference arcs in conjunction with a calculator to compute all values of the circular functions in all examples and exercises.*

Example

Compute $\cos 5.31$ and $\sin 5.31$ using a hand calculator or Table V.

Solution

The length of the reference arc is

$$\bar{x} = 2\pi - 5.31 \approx 6.28 - 5.31$$
$$= 0.97,$$

as shown in the figure.
From Theorem 8.2 and a calculator or Table V,

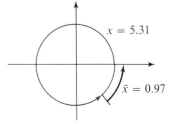

$$|\cos 5.31| = \cos 0.97 = 0.5653 \quad \text{and} \quad |\sin 5.31| = \sin 0.97 = 0.8249.$$

Since the arc of length 5.31 terminates in the fourth quadrant, we know from Table 8.1 that $\cos 5.31 > 0$ and $\sin 5.31 < 0$. Thus, we have

$$\cos 5.31 = 0.5653 \quad \text{and} \quad \sin 5.31 = -0.8249.$$

Example

Compute $\cos(-2.31)$ and $\sin(-2.31)$ using a hand calculator or Table V.

Solution

The length of the reference arc is

$$\bar{x} = -2.31 - (-1)\pi = \pi - 2.31$$
$$\approx 3.14 - 2.31 = 0.83,$$

as shown in the figure.

From a calculator or Table V and Theorem 8.2, we have

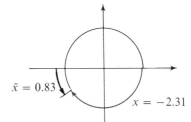

$$|\cos(-2.31)| = \cos 0.83 = 0.6749 \quad \text{and} \quad |\sin(-2.31)| = \sin 0.83 = 0.7379.$$

Solution Continued on Overleaf

Since the arc of length -2.31 terminates in the third quadrant, we know from Table 8.1 that $\cos x < 0$ and $\sin x < 0$. Thus, we have

$$\cos(-2.31) = -0.6749 \quad \text{and} \quad \sin(-2.31) = -0.7379.$$

Exercise 8.3

Use Table 8.2, a hand calculator, or Table V to compute the indicated value.

Example

$$\sin \frac{11\pi}{6}$$

Solution

The length of the reference arc is

$$\bar{x} = 2\pi - \frac{11\pi}{6} = \frac{\pi}{6}.$$

Now, from Table 8.2 and Theorem 8.2, we have

$$\left| \sin \frac{11\pi}{6} \right| = \sin \frac{\pi}{6} = \frac{1}{2}.$$

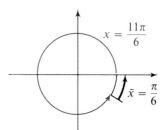

Since $x = 11\pi/6$ terminates in the fourth quadrant, we find from Table 8.1 that

$$\sin \frac{11\pi}{6} = -\frac{1}{2}.$$

1. $\cos \dfrac{5\pi}{6}$
2. $\sin \dfrac{2\pi}{3}$
3. $\cos \dfrac{4\pi}{3}$
4. $\sin \dfrac{7\pi}{6}$
5. $\cos \dfrac{7\pi}{4}$
6. $\sin \dfrac{5\pi}{3}$
7. $\sin \dfrac{4\pi}{3}$
8. $\cos \dfrac{2\pi}{3}$
9. $\sin \dfrac{7\pi}{4}$
10. $\cos \dfrac{7\pi}{6}$
11. $\sin \dfrac{3\pi}{4}$
12. $\cos \dfrac{5\pi}{3}$

Example $\cos 1.73$

Solution The length of the reference arc is

$$\bar{x} = \pi - 1.73 \approx 3.14 - 1.73$$
$$= 1.41.$$

Now, from a calculator or Table V, by Theorem 8.2 we have

$$|\cos 1.73| = \cos 1.41 = 0.1601.$$

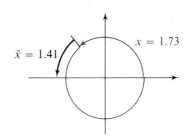

8.3 *Function Values of* $\cos x$ *and* $\sin x$, *for* $x \in R$

Since $x = 1.73$ terminates in the second quadrant, we find from Table 8.1 that
$$\cos 1.73 = -0.1601.$$

13. $\cos 2.11$
14. $\sin 1.69$
15. $\cos 3.51$
16. $\sin 4.00$
17. $\cos 6.15$
18. $\sin 5.15$
19. $\sin 1.58$
20. $\cos 1.59$
21. $\sin 3.16$
22. $\cos 3.18$
23. $\sin 4.75$
24. $\cos 4.72$

Example $\cos(-1.68)$

Solution The length of the reference arc is
$$\bar{x} = \pi - 1.68 \approx 3.14 - 1.68$$
$$= 1.46.$$
Now, from a calculator or Table V, by Theorem 8.2 we have

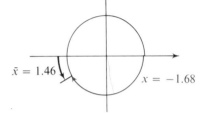

$$|\cos(-1.68)| = \cos 1.46 = 0.1106.$$
Since $x = -1.68$ terminates in the third quadrant, we find from Table 8.1 that
$$\cos(-1.68) = -0.1106.$$

25. $\cos\left(-\dfrac{\pi}{3}\right)$
26. $\sin\left(-\dfrac{\pi}{6}\right)$
27. $\cos\left(-\dfrac{3\pi}{4}\right)$
28. $\sin\left(-\dfrac{5\pi}{6}\right)$
29. $\cos\left(-\dfrac{4\pi}{3}\right)$
30. $\sin\left(-\dfrac{7\pi}{6}\right)$
31. $\sin(-1.85)$
32. $\cos(-3.00)$
33. $\sin(-3.57)$
34. $\cos(-4.11)$
35. $\sin(-5.11)$
36. $\cos(-6.20)$

Example $\cos \dfrac{7\pi}{3}$

Solution Notice that the arc with length $x = 7\pi/3$ makes more than one full revolution. The length of the reference arc is

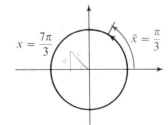

$$\bar{x} = \dfrac{7\pi}{3} - 2\pi = \dfrac{\pi}{3}.$$

Now, from Table 8.2 and Theorem 8.2, we obtain
$$\left|\cos \dfrac{7\pi}{3}\right| = \cos \dfrac{\pi}{3} = \dfrac{1}{2}.$$

Solution Continued on Overleaf

Since $x = 7\pi/3$ terminates in the fourth quadrant, we therefore have

$$\cos \frac{7\pi}{3} = \frac{1}{2}.$$

37. $\cos \dfrac{13\pi}{6}$ 38. $\sin \dfrac{9\pi}{4}$ 39. $\cos \dfrac{17\pi}{4}$

40. $\sin \dfrac{14\pi}{3}$ 41. $\cos \left(-\dfrac{7\pi}{3}\right)$ 42. $\sin \left(-\dfrac{11\pi}{4}\right)$

43. $\sin 6.40$ 44. $\cos 7.04$ 45. $\sin 14.25$

46. $\cos 16.00$ 47. $\sin(-6.85)$ 48. $\cos(-15.00)$

8.4 Graphs of $y = \cos x$ and $y = \sin x$

In the preceding three sections we discussed the definition and evaluation of the cosine function and the sine function. In this section we shall define the concept of a periodic function and use it to graph the functions $y = \cos x$ and $y = \sin x$.

Periodic functions

The graphs of some functions display interesting cyclical characteristics. For example, the x-axis shown in Figure 8.10 can clearly be divided into equal successive

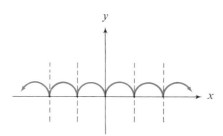

Figure 8.10

subintervals in such a way that the graph over each subinterval is a repetition of the graph over every other equal subinterval. Functions that display this property are said to be "periodic," and the length of each of the equal subintervals is called a **period** of the function.

Definition 8.3   If f is a function, such that for some $p \in R$, $p \neq 0$,

$$f(x + p) = f(x - p) = f(x),$$

then f is **periodic** with period p.

If there is a least positive number p for which the function is periodic, then p is called the **fundamental period** of the function. Thus, for example, Figure 8.11 shows part of the graph of a periodic function with fundamental period 5.

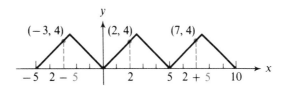

Figure 8.11

Period of the sine and cosine functions

Both sine and cosine are periodic functions. In particular, for any real number x we have (see Exercises 39 and 40)

$$\sin(x + 2\pi) = \sin x$$

and

$$\cos(x + 2\pi) = \cos x.$$

Hence, 2π is a period for both functions. It can be shown (although we shall not do so here) that 2π is the fundamental period of both.

Graph of $y = \sin x$

The graph of $y = \sin x$ on the interval $0 \leq x \leq 2\pi$ can be obtained by plotting several ordered pairs over this interval, as shown in Figure 8.12. The graphs of the ordered pairs are then joined with a smooth curve to obtain the graph shown in Figure 8.13.

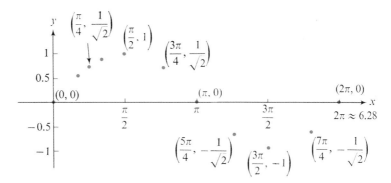

Figure 8.12

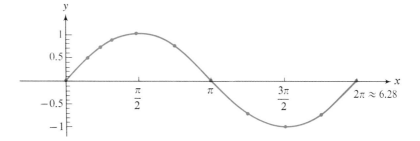

Figure 8.13

Since the sine function has fundamental period 2π, we can obtain the graph of $y = \sin x$ over an extended interval by repeating the pattern in Figure 8.13 in both directions along the x-axis. For example, the graph of $y = \sin x$ over the interval $-\pi \leq x \leq 4\pi$ is shown in Figure 8.14.

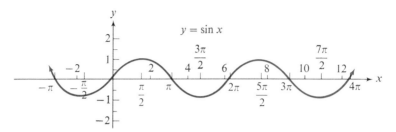

Figure 8.14

The zeros of the function, $k\pi$, where $k = 0, \pm 1, \pm 2, \pm 3, \ldots$—namely the values of x associated with the points at which the curve crosses the x-axis—appear on the graph. Graphs with this characteristic form are called **sine waves** or **sinusoids**. The portion of the graph over any fundamental period of the function is called a **cycle** of the sine wave. Half the difference of the maximum and minimum ordinates on such a curve is called the **amplitude** of the wave. Thus, for the graph of $y = \sin x$, the amplitude of the wave is

$$\frac{1}{2}[1 - (-1)] = 1.$$

Graph of cosine

The graph of the cosine function can be obtained in the same manner as the graph of the sine function. We first obtain the coordinates of several points on the graph, then by connecting these points with a smooth curve, as shown in Figure 8.15, we obtain one cycle of the graph of $y = \cos x$. Duplicating this pattern in both directions along the x-axis, we have a representative portion of the entire graph of the cosine function, as shown in Figure 8.16.

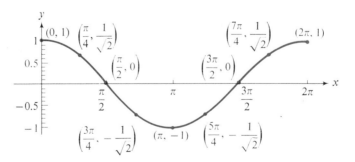

Figure 8.15

Notice that the graph of the cosine function is also a sinusoid, with fundamental period 2π and amplitude 1. Furthermore, the zeros of the function, $\pi/2 + k\pi$, where $k = 0, \pm 1, \pm 2, \pm 3, \ldots$, are evident.

8.4 Graphs of $y = \cos x$ and $y = \sin x$

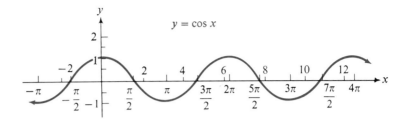

Figure 8.16

Functions defined by equations of the form

$$y = A \sin (Bx + C) \quad \text{and} \quad y = A \cos (Bx + C),$$

where A, B, and C are constants and A and B are not 0, always have sine waves for graphs. With variations in the numbers A, B, and C, the graphs are variously situated with respect to the origin and have a variety of amplitudes and periods. To analyze such graphs, let us consider the effect of A on the graphs of the functions defined by $y = A \sin x$ and $y = A \cos x$.

Graphs of
$y = A \sin x$
and
$y = A \cos x$

For each value of x, each ordinate to the graph of $y = A \sin x$ is A times the ordinate to the graph of $y = \sin x$. Therefore, the amplitude of the graph of $y = A \sin x$ is $|A|$ times the amplitude of the graph of $y = \sin x$; that is, the amplitude of $y = A \sin x$ is $|A|$. Of course, the graph of $y = A \cos x$ is a similar modification of the graph of $y = \cos x$.

Example

Graph $y = 3 \sin x$, $-\pi \leq x \leq 4\pi$.

Solution

It may be helpful first to sketch $y = \sin x$, $0 \leq x \leq 2\pi$, as a reference. Then, since the amplitude of $y = 3 \sin x$ is 3, we can sketch the desired graph on the same coordinate system over the interval $0 \leq x \leq 2\pi$ by making each ordinate 3 times the corresponding ordinate of the graph of $y = \sin x$.

We can then extend this cycle to include the entire interval $-\pi \leq x \leq 4\pi$, as shown in the figure. In this figure, one cycle is sketched with a colored line for emphasis. Note also that, in this and some of the succeeding figures, different unit lengths are used on the x- and y-axes. For $y = 3 \sin x$, the fundamental period p is the same as for $y = \sin x$, namely $p = 2\pi$.

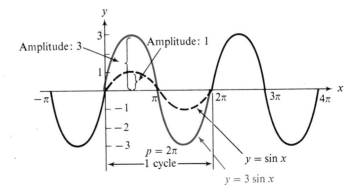

Graphs of
y = sin Bx
and
y = cos Bx

Next let us examine the way the graph of

$$y = \sin Bx, \quad B \neq 0,$$

differs from the graph of $y = \sin x$. Note first that sin Bx has values between -1 and 1 inclusive, as has sin x. Also,

$$\sin(Bx + 2\pi) = \sin Bx,$$

just as $\sin(x + 2\pi) = \sin x$. If we factor B from $Bx + 2\pi$, however, we have $B(x + 2\pi/B)$, and hence the function defined by $y = \sin Bx$ has period

$$p = \frac{2\pi}{|B|}.$$

We use $|B|$ instead of B to ensure a positive number for the period. It can be shown, although it is not done here, that $2\pi/|B|$ is the fundamental period of the function. Hence the graph of $y = \sin Bx$ is a sine wave with amplitude 1; it completes one cycle over the interval $0 \leq x \leq 2\pi/|B|$.

Example Graph $y = \cos 2x$, $\dfrac{-3\pi}{2} \leq x \leq 2\pi$.

Solution Let us first sketch a cycle of $y = \cos x$, $0 \leq x \leq 2\pi$, as a reference. Since

$$p = \frac{2\pi}{|B|} = \frac{2\pi}{2} = \pi,$$

we next sketch a cycle of the graph of $y = \cos 2x$ over the interval $0 \leq x \leq \pi$ (shown in color) on the same coordinate system, and extend the cycle obtained over the interval $-3\pi/2 \leq x \leq 2\pi$.

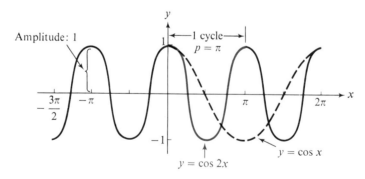

Example Graph $y = -4 \sin \dfrac{1}{2} x$, $-2\pi \leq x \leq 4\pi$.

Solution We first sketch the graph of $y = \sin x$ as a reference. Since $A = -4$, each ordinate of the graph of $y = -4 \sin(x/2)$ is the negative of the ordinate of the graph of $y = 4 \sin(x/2)$. Since

8.4 *Graphs of* $y = \cos x$ *and* $y = \sin x$

$$p = \frac{2\pi}{|B|} = \frac{2\pi}{\frac{1}{2}} = 4\pi,$$

there is one cycle in the interval $0 \leq x \leq 4\pi$. With this information, we sketch one cycle (shown in color) and then extend the graph over the interval $-2\pi \leq x \leq 4\pi$.

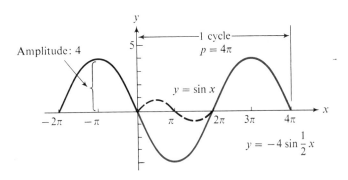

Graphs of
$y = \sin(x + C)$
and
$y = \cos(x + C)$

Finally, let us compare the graphs of

$$y = \sin(x + C) \quad \text{and} \quad y = \sin x,$$

where $C > 0$. For $x = -C$, we have

$$\sin(x + C) = \sin(-C + C) = \sin 0 = 0.$$

Similarly, for any real number x_1, the ordinate of $\sin(x + C)$ at $x_1 - C$ will be the same as the ordinate of $\sin x$ at x_1. Thus if x is thought of as measuring time, each value y occurs C units of time earlier on $y = \sin(x + C)$ than on $y = \sin x$. Hence the graph of $y = \sin(x + C)$ is said to **lead** the graph of $y = \sin x$ by C. The number C, itself, is called the **phase shift** of the wave. If $C < 0$, then the graph of $y = \sin(x + C)$ is obtained by shifting the graph of $y = \sin x$ $|C|$ units to the right and is said to **lag** the graph of $y = \sin x$.

We use all of the foregoing information about the effect of A, B, and C on the graph of $y = A \sin(Bx + C)$, or of $y = A \cos(Bx + C)$, to help sketch the graph of such an equation.

Example

Sketch the graph of $y = 3 \sin\left(2x + \frac{\pi}{3}\right)$.

Solution

First, let us rewrite the equation by factoring 2 from the expression in parentheses:

$$y = 3 \sin 2\left(x + \frac{\pi}{6}\right).$$

By inspecting this equation, we note the following things about the graph:

1. It is a sine wave.
2. It has amplitude 3.

Solution Continued on Overleaf

3. It has period $2\pi/2 = \pi$.
4. It leads the graph of $y = 3 \sin 2x$ by $\pi/6$.

With these facts, we can quickly sketch the graph shown.

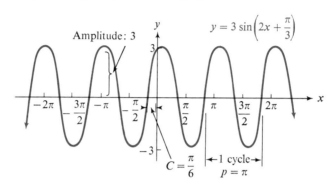

Example

Sketch the graph of $y = \dfrac{1}{2} \sin \dfrac{\pi x}{2}$.

Solution

By inspection, we note the following things concerning the graph:

1. It is a sine wave.
2. It has amplitude $1/2$.
3. It has period $\dfrac{2\pi}{\pi/2} = 4$.

Since the period is 4, we use integers as elements of the domain. Scaling the x-axis in integral units facilitates sketching the graph as shown.

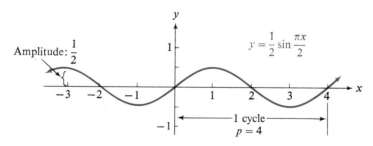

Exercise 8.4

Sketch the graph of the given equation over the interval $-2\pi \le x \le 2\pi$. Give the amplitude, period, and phase shift.

1. $y = 2 \sin x$
2. $y = 3 \cos x$
3. $y = \dfrac{1}{2} \cos x$

8.4 Graphs of $y = \cos x$ and $y = \sin x$

4. $y = \dfrac{1}{3} \cos x$
5. $y = -4 \sin x$
6. $y = -\dfrac{1}{2} \cos x$

7. $y = \sin 2x$
8. $y = \cos 3x$
9. $y = \cos \dfrac{1}{3} x$

10. $y = \cos \dfrac{1}{2} x$
11. $y = -3 \sin 2x$
12. $y = -\dfrac{1}{2} \sin 3x$

13. $y = \sin (x + \pi)$
14. $y = \cos \left(x - \dfrac{\pi}{2} \right)$

15. $y = 2 \cos \left(x - \dfrac{\pi}{4} \right)$
16. $y = 3 \sin \left(x + \dfrac{\pi}{6} \right)$

17. $y = 3 \sin 2 \left(x - \dfrac{\pi}{3} \right)$
18. $y = 2 \cos 3 \left(x + \dfrac{\pi}{4} \right)$

19. $y = 2 \sin \pi x$
20. $y = -3 \cos \dfrac{\pi}{2} x$

21. $y = -\dfrac{1}{2} \cos \dfrac{\pi}{3} x$
22. $y = \dfrac{1}{4} \sin \dfrac{\pi}{4} x$

From the respective graph, determine the zeros (over the specified domain) of the function defined by the equation in the given exercise.

23. Exercise 7
24. Exercise 8
25. Exercise 9
26. Exercise 10
27. Exercise 19
28. Exercise 20

Example   Sketch the graph of $y = \sin x + 2 \cos x$ over the interval $0 \leq x \leq 2\pi$.

Solution   First, sketch the graphs of $y = \sin x$ and $y = 2 \cos x$ on the same coordinate system over the given interval. The ordinate of the graph of $y = \sin x + 2 \cos x$ at each point x on the x-axis is the *algebraic* sum of the corresponding ordinates of $y = \sin x$ and $y = 2 \cos x$.
Thus for $x = \pi/6$,

$$\sin x = 0.5,$$

$$2 \cos x = \dfrac{2\sqrt{3}}{2} \approx 1.7,$$

and

$$\sin x + 2 \cos x \approx 2.2.$$

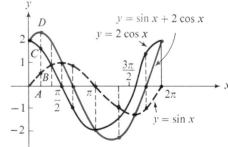

Now the ordinate can be approximated graphically by adding the directed line segments from the x-axis to the curves at $x = \pi/6$:

$$\overline{AB} + \overline{AC} = \overline{AB} + \overline{BD} = \overline{AD}.$$

If this is done for a few selected values of x, we obtain a good approximation for the curve.

Sketch the graph of the given equation over the interval $-2\pi \le x \le 2\pi$.

29. $y = \sin x + \cos x$
30. $y = 3 \sin x + \cos x$
31. $y = \sin 2x + \frac{1}{2} \cos x$
32. $y = \sin 3x + 2 \cos \frac{1}{2} x$
33. $y = \sin x - 2 \cos x$
34. $y = 3 \cos x - \sin 2x$
35. $y = 3 \sin x + 1$
36. $y = 4 \sin 2x - 3$
37. $y = x + \cos x$
38. $y = 2x - \cos x$

39. Show for any real number x, $\sin (x + 2\pi) = \sin x$.
40. Show for any real number x, $\cos (x + 2\pi) = \cos x$.

8.5 Tangent Function

The functions sine and cosine can be used to define other periodic functions. One of these is the tangent function which we shall study in this section.

Definition 8.4   For $x \in R$, $\cos x \ne 0$, the **tangent of x** is denoted by **tan x**, and

$$\tan x = \frac{\sin x}{\cos x}.$$

Domain and range of tangent

Because $\tan x = \sin x / \cos x$, the domain of the tangent function is R, with the exception of the real numbers x for which $\cos x = 0$. In other words, we except real numbers of the form $\pi/2 + k\pi$, $k \in J$. The range of the tangent function is R. We can use Definition 8.4 together with Table 8.2, page 205, to find tan x for all values of x included in the table. For example,

$$\tan 0 = \frac{\sin 0}{\cos 0} = \frac{0}{1} = 0,$$

$$\tan \frac{\pi}{6} = \frac{\sin (\pi/6)}{\cos (\pi/6)}$$

$$= \frac{1/2}{\sqrt{3}/2} = \frac{1}{\sqrt{3}},$$

and so forth. The values of tan x obtained in this way are displayed in Table 8.3. By using Equations (2) or (3) on page 200, we can compute tan x given only the value of cos x or sin x, respectively, and the quadrant in which x lies.

8.5 Tangent Function

Table 8.3

x	0	$\dfrac{\pi}{6}$	$\dfrac{\pi}{4}$	$\dfrac{\pi}{3}$	$\dfrac{\pi}{2}$	π	$\dfrac{3\pi}{2}$
$\tan x$	0	$\dfrac{1}{\sqrt{3}}$	1	$\sqrt{3}$	not def.	0	not def.

Example If $\sin x = 3/5$ and $\pi/2 \leq x \leq \pi$, find $\cos x$ and $\tan x$.

Solution By Equation (3b) on page 200, for $\pi/2 \leq x \leq \pi$,
$$\cos x = -\sqrt{1 - \sin^2 x}.$$
Then, since $\sin x = 3/5$,
$$\cos x = -\sqrt{1 - \left(\frac{3}{5}\right)^2} = -\frac{4}{5}.$$

By Definition 8.4,
$$\tan x = \frac{\sin x}{\cos x} = \frac{3/5}{-4/5} = -\frac{3}{4}.$$

Period of $y = \tan x$

Observe that since the sine and cosine functions have period 2π, so has the tangent function. Note, however, that the points at arc length x and $x + \pi$ from $(1, 0)$ on the unit circle lie on the same line through the origin (see Figure 8.17). Thus, if $x \neq \pi/2 + k\pi$, $k \in J$, the slope of segment OA is the same as the slope of segment OB, and we have

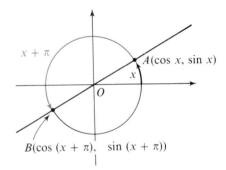

Figure 8.17

$$\frac{\sin(x + \pi) - 0}{\cos(x + \pi) - 0} = \text{slope of } OB = \text{slope of } OA = \frac{\sin x - 0}{\cos x - 0}.$$

Thus,
$$\tan(x + \pi) = \frac{\sin(x + \pi)}{\cos(x + \pi)} = \frac{\sin x}{\cos x} = \tan x,$$

and π is also a period of the tangent function. It can be shown (although we shall not do so) that π is the fundamental period of the tangent function.

Graph of
$y = \tan x$

We can obtain the graph of $y = \tan x$ by first plotting several points over an interval of length π (the fundamental period of the tangent function). Then we join these points with a smooth curve as shown in Figure 8.18.

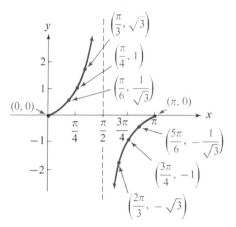

Figure 8.18

Observe that the line $x = \pi/2$ is a vertical asymptote of the graph of $y = \tan x$ since $\tan x = \sin x/\cos x$ and $\sin \pi/2 \neq 0$ but $\cos \pi/2 = 0$.

We can now duplicate the pattern in Figure 8.18 in both directions along the x-axis to obtain the graph of $y = \tan x$ over an extended interval. Several cycles of the graph are shown in Figure 8.19.

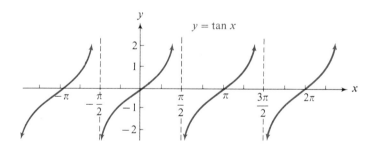

Figure 8.19

It is evident from the graph that the range of the tangent function is the set of all real numbers. Furthermore, it is evident that any integral multiple of π is a zero of $y = \tan x$, and that the curve has vertical asymptotes at $x = \pi/2 + k\pi$ for every integer k.

Graph of
$y = \tan Bx$

Because for any real number x,

$$\tan(Bx + \pi) = \tan Bx, \quad B \neq 0,$$

and $Bx + \pi = B(x + \pi/B)$, we have

$$\tan B(x + \pi/B) = \tan Bx.$$

8.5 Tangent Function

Thus, $\pi/|B|$ is a period of $y = \tan Bx$, and is in fact the fundamental period since π is the fundamental period of $\tan x$. Thus, the graph of $y = \tan Bx$ has the same basic shape as the graph of $y = \tan x$, but it repeats over intervals of length $\pi/|B|$ instead of π.

Example

Graph $y = \tan 2x$ over the interval $-\pi \le x \le \pi$.

Solution

It is helpful first to sketch the graph of $y = \tan x$ over the interval $-\pi/2 \le x \le \pi/2$ as a reference (see the dashed curve in the figure). Since the fundamental period of $y = \tan 2x$ is

$$p = \frac{\pi}{|B|} = \frac{\pi}{2},$$

we next sketch a cycle of $y = \tan 2x$ over the interval $0 \le x \le \pi/2$ on the same coordinate system and extend the cycle obtained over the interval $-\pi \le x \le \pi$.

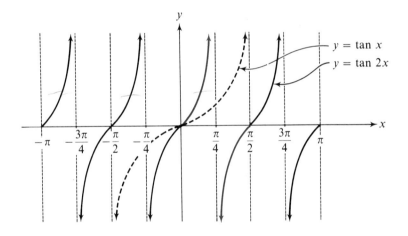

We can sketch the graph of the function $y = A \tan (Bx + C)$ using the same procedure that was used in Section 8.4 to graph $y = A \sin (Bx + C)$ and $y = A \cos (Bx + C)$.

Exercise 8.5

Use Table 8.2 and Definition 8.4 to compute the indicated value if it exists; if not, then so indicate.

1. $\tan 0$
2. $\tan \dfrac{\pi}{6}$
3. $\tan \dfrac{\pi}{4}$
4. $\tan \dfrac{\pi}{3}$
5. $\tan \dfrac{\pi}{2}$
6. $\tan \pi$
7. $\tan \dfrac{3\pi}{2}$
8. $\tan 2\pi$.

Use Definition 8.4 to compute the indicated value if it exists; if not, then so indicate.

Example $\tan \dfrac{5\pi}{3}$

Solution We first use the reference arc to compute $\sin 5\pi/3$ and $\cos 5\pi/3$, obtaining

$$\sin \frac{5\pi}{3} = -\frac{\sqrt{3}}{2} \quad \text{and} \quad \cos \frac{5\pi}{3} = \frac{1}{2}.$$

Thus, from Definition 8.4 we have

$$\tan \frac{5\pi}{3} = \frac{\sin \dfrac{5\pi}{3}}{\cos \dfrac{5\pi}{3}} = \frac{-\dfrac{\sqrt{3}}{2}}{\dfrac{1}{2}}$$

$$= -\frac{\sqrt{3}}{1} = -\sqrt{3}.$$

9. $\tan \dfrac{2\pi}{3}$ 10. $\tan \dfrac{3\pi}{4}$ 11. $\tan \dfrac{7\pi}{6}$ 12. $\tan \dfrac{7\pi}{4}$

13. $\tan \left(-\dfrac{\pi}{6}\right)$ 14. $\tan \left(-\dfrac{3\pi}{4}\right)$ 15. $\tan \dfrac{5\pi}{2}$ 16. $\tan \dfrac{13\pi}{6}$

Use the definitions and periodicity of the circular functions, together with Table V to obtain an approximation for the given function value.

Example $\tan 3.31$

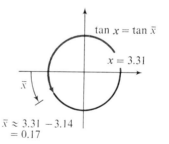

Solution $\tan 3.31 \approx \tan (3.31 - 3.14)$

$= \tan 0.17$

≈ 0.1717

17. $\tan 4.52$ 18. $\tan 5.61$ 19. $\tan 7.45$ 20. $\tan 8.30$
21. $\tan 12.50$ 22. $\tan 27.30$ 23. $\tan (-9.32)$ 24. $\tan (-6.41)$

Sketch the graph of each function over the interval $-2\pi \leq x \leq 2\pi$, and list the zeros and vertical asymptotes of each.

25. $y = \tan \dfrac{x}{2}$ 26. $y = \tan \dfrac{x}{3}$

27. $y = \tan\left(x + \dfrac{\pi}{2}\right)$

28. $y = \tan\left(2x + \dfrac{\pi}{2}\right)$

8.6 Other Circular Functions

The functions sine and cosine can be used to define other periodic functions. In this section we shall discuss three such functions that are commonly used in mathematics and science.

Cotangent

The first of these periodic functions is the cotangent function.

Definition 8.5 For $x \in R$, $\sin x \neq 0$, *the **cotangent** of x is denoted by* $\cot x$, *and*

$$\cot x = \frac{\cos x}{\sin x}.$$

Since $\cot x = \cos x / \sin x$, the domain of the cotangent function is R, excepting those real numbers for which $\sin x = 0$, that is, all real numbers except those of the form $k\pi$, where $k \in J$. The range of cotangent is R, and, like the tangent function, cotangent has fundamental period π. Note that because $\cot x = \cos x / \sin x$ and $\tan x = \sin x / \cos x$,

$$\cot x = \frac{1}{\tan x} \quad \text{and} \quad \tan x = \frac{1}{\cot x} \tag{1}$$

for those values of x for which each expression is defined.

Since $\cot x$ is the reciprocal of $\tan x$, $x \neq k\pi$, $k \in J$, we can find the function values $\cot x$ by first finding $\tan x$ and then computing its reciprocal. Since most calculators will not compute $\cot x$ with a single keystroke, this method is usually used to compute $\cot x$.

Examples Compute each of the following values.

a. $\cot \dfrac{\pi}{4}$

b. $\cot 1.32$

Solutions

a. By using Equation (1) and Table 8.3, we have

$$\cot \frac{\pi}{4} = \frac{1}{\tan \dfrac{\pi}{4}} = \frac{1}{1} = 1.$$

b. By using a calculator and Equation (1), we obtain

$$\cot 1.32 = \frac{1}{\tan 1.32} = 0.2562.$$

Graph of
y = cot *x*

We can sketch the graph of the cotangent function by plotting some points over $0 < x < \pi$ and joining the points with a curve. The graph is shown in Figure 8.20. Note that the graph has vertical asymptotes at $x = k\pi$, $k \in J$ and x-intercepts at $x = (\pi/2) + k\pi$, $k \in J$.

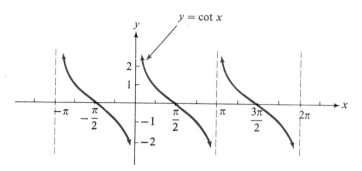

Figure 8.20

Secant

Another periodic function is the secant.

Definition 8.6 For $x \in R$, $\cos x \neq 0$, the **secant** of *x* is denoted by **sec** *x*, and

$$\sec x = \frac{1}{\cos x}.$$

Since sec *x* is the reciprocal of cos *x*, the domain of the secant function is *R*, except for those values of *x* for which $\cos x = 0$, namely, $x = \pi/2 + k\pi$, $k \in J$. Because $|\cos x| \leq 1$ for all $x \in R$, sec *x* is never less than 1 in absolute value; the range of the secant function is $\{y \mid |y| \geq 1\}$. Since the cosine function has fundamental period 2π, the secant function also has fundamental period 2π.

As in the case of the cotangent the values sec *x* usually cannot be obtained with a single keystroke on a calculator, and thus the fact that sec *x* is the reciprocal of cos *x*, $x \neq \pi/2 + k\pi$, $k \in J$, is used to compute sec *x*.

Examples

Compute each of the following values.

a. $\sec \dfrac{\pi}{6}$
b. sec 1.41

Solutions

a. By using Definition 8.6 and Table 8.2, we have

$$\sec \frac{\pi}{6} = \frac{1}{\cos \dfrac{\pi}{6}} = \frac{1}{\sqrt{3}/2}$$

$$= \frac{2}{\sqrt{3}} = \frac{2\sqrt{3}}{3}.$$

8.6 Other Circular Functions

b. By using a calculator, we obtain

$$\sec 1.41 = \frac{1}{\cos 1.41} = 6.2459.$$

Graph of $y = \sec x$

We can sketch the graph of the secant function by plotting some points over $0 < x < 2\pi$ and joining the points with a curve. The graph is shown in Figure 8.21. Note that the graph has vertical asymptotes at $x = (\pi/2) + k\pi$, $k \in J$.

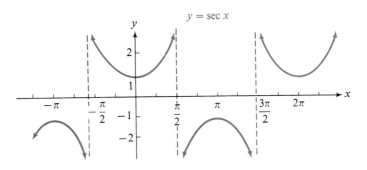

Figure 8.21

Cosecant

The last function we shall discuss in this section is the cosecant.

Definition 8.7   For $x \in R$, $\sin x \neq 0$, the **cosecant of x** is denoted by **csc x**, and

$$\csc x = \frac{1}{\sin x}.$$

Because csc x is the reciprocal of sin x, the domain of the cosecant function contains all real numbers except those for which $\sin x = 0$; that is, the domain is all of R except the numbers $k\pi$, $k \in J$. The cosecant function has range $\{y \mid |y| \geq 1\}$ and fundamental period 2π.

As in the case of the secant function the values csc x usually cannot be obtained with a single keystroke on a calculator, and thus the fact that csc x is the reciprocal of sin x, $x \neq k\pi$, $k \in J$, is used to compute csc x.

Examples

Compute each of the following values.

a. $\csc \dfrac{\pi}{6}$

b. csc 1.32

Solutions

a. By using Definition 8.7 and Table 8.2, we have

$$\csc \frac{\pi}{6} = \frac{1}{\sin \dfrac{\pi}{6}} = \frac{1}{1/2} = 2.$$

Solution Continued on Overleaf

b. By using a calculator, we obtain

$$\csc 1.32 = \frac{1}{\sin 1.32} = 1.0323.$$

Graph of $y = \csc x$

We can sketch the graph of the cosecant function by plotting some points over $0 < x < 2\pi$ and joining the points with a curve. The graph is shown in Figure 8.22. Note that the graph has vertical asymptotes at $x = k\pi$, $k \in J$.

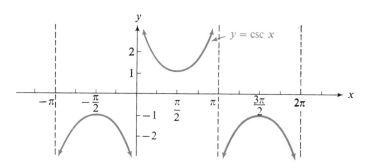

Figure 8.22

Functions of the form $y = A \cot(Bx + C)$, $y = A \sec(Bx + C)$, and $y = A \csc(Bx + C)$ can be graphed using the same methods that were used for $y = A \sin(Bx + C)$, $y = A \cos(Bx + C)$, and $y = A \tan(Bx + C)$ in Sections 8.4 and 8.5.

Exercise 8.6

Example

Use the appropriate definition and Table 8.2 to compute the indicated value if it, exists; if not, then so indicate.

$\cot \pi$

Solution

Using Table 8.2 and Definition 8.5, we have

$$\cot \pi = \frac{\cos \pi}{\sin \pi} = \frac{-1}{0}.$$

Thus, $\cot \pi$ is undefined.

1. $\cot \dfrac{\pi}{2}$
2. $\cot \dfrac{\pi}{3}$
3. $\sec \dfrac{\pi}{4}$
4. $\sec \dfrac{\pi}{3}$
5. $\csc \dfrac{\pi}{3}$
6. $\csc \dfrac{\pi}{2}$
7. $\csc \pi$
8. $\sec \dfrac{\pi}{2}$

Example

$\sec \dfrac{4\pi}{3}$

8.7 Inverse Circular Functions

Solution Using reference arcs, we obtain

$$\cos \frac{4\pi}{3} = -\frac{1}{2}.$$

Thus, from Definition 8.6, we have

$$\sec \frac{4\pi}{3} = \frac{1}{\cos \frac{4\pi}{3}} = \frac{1}{-1/2}$$

$$= \frac{-2}{1} = -2$$

9. $\cot \dfrac{7\pi}{6}$ 10. $\cot \dfrac{5\pi}{4}$ 11. $\sec \dfrac{7\pi}{4}$

12. $\sec \dfrac{3\pi}{4}$ 13. $\csc \dfrac{5\pi}{6}$ 14. $\csc \dfrac{5\pi}{3}$

15. $\cot(-\pi)$ 16. $\cot\left(\dfrac{13\pi}{4}\right)$ 17. $\sec\left(-\dfrac{\pi}{6}\right)$

18. $\sec \dfrac{13\pi}{4}$ 19. $\csc\left(-\dfrac{5\pi}{2}\right)$ 20. $\csc \dfrac{11\pi}{4}$

Use the appropriate definition and a hand calculator or Table V to compute the indicated value.

21. $\cot 4.31$ 22. $\cot 3.51$ 23. $\sec 1.05$
24. $\sec 2.41$ 25. $\csc 1.65$ 26. $\csc 5.32$
27. $\cot(-2.10)$ 28. $\cot 8.31$ 29. $\sec 3.64$
30. $\sec 12.51$ 31. $\csc(-3.00)$ 32. $\csc 8.65$

8.7 Inverse Circular Functions

In many applications it is necessary to find a value x which yields a given value of y for a function $y = f(x)$. For example, we may want to find a number x such that $\sin x = 0.5$. We encountered a similar problem with the exponential function in Section 7.2, and solved it by considering its inverse, the logarithm function. In this section we shall consider the inverses of the circular functions.

Inverse relations Recall from Section 6.2 that the inverse of a function is the relation obtained by interchanging the components x and y of each ordered pair in the function and that the graph of the inverse of a function can be obtained by reflecting the graph of the function in the line with equation $y = x$. Furthermore, the resulting relation is a function if and only if the original function is one-to-one. Now consider the sine function,

$$y = \sin x, \qquad (1)$$

and the relation,

$$x = \sin y. \tag{2}$$

Both graphs are shown in Figure 8.23 where it is evident that each is the reflection of the other in the line $y = x$. Equation (2) does not define a function, because for each element x in its domain, $\{x \mid -1 \leq x \leq 1\}$, there are an unlimited number of elements in its range.

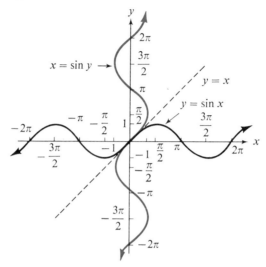

Figure 8.23

The graphs of the relations

$$x = \cos y$$

and

$$x = \tan y$$

are shown in Figure 8.24. Note that these equations also do not define functions.

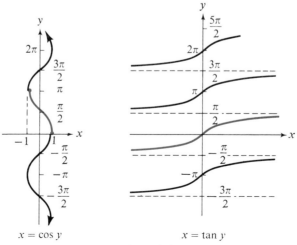

Figure 8.24

8.7 Inverse Circular Functions

Inverse functions

By suitably restricting the domains of the circular functions—that is, by suitably restricting the ranges of the relations obtained by interchanging the variables—we can define an inverse function for each circular function. For the sine function, we restrict the range to the interval $-\pi/2 \leq y \leq \pi/2$, and we have the inverse function

$$x = \sin y, \quad -\frac{\pi}{2} \leq y \leq \frac{\pi}{2}. \qquad (3)$$

The graph of this inverse is shown in Figure 8.25.

To write (3) in a form in which y is expressed explicitly in terms of x, we use the special notation

$$y = \text{Sin}^{-1} x$$

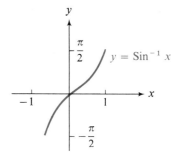

Figure 8.25

(read "y is equal to the inverse sine of x"), with the understanding, of course, that $\text{Sin}^{-1} x$ does not denote $1/\text{Sin } x$. Note that we use a capital S with this inverse notation. Similar notation is used for the other inverse circular functions, where the ranges have been suitably restricted in accordance with the following definition.

Definition 8.8 The six inverse circular functions are

$$y = \text{Sin}^{-1} x \leftrightarrow x = \sin y, \quad -\frac{\pi}{2} \leq y \leq \frac{\pi}{2}, *$$

$$y = \text{Cos}^{-1} x \leftrightarrow x = \cos y, \quad 0 \leq y \leq \pi,$$

$$y = \text{Tan}^{-1} x \leftrightarrow x = \tan y, \quad -\frac{\pi}{2} < y < \frac{\pi}{2},$$

$$y = \text{Csc}^{-1} x \leftrightarrow x = \csc y, \quad -\frac{\pi}{2} \leq y \leq \frac{\pi}{2} \ (y \neq 0),$$

$$y = \text{Sec}^{-1} x \leftrightarrow x = \sec y, \quad 0 \leq y \leq \pi \ \left(y \neq \frac{\pi}{2}\right),$$

$$y = \text{Cot}^{-1} x \leftrightarrow x = \cot y, \quad -\frac{\pi}{2} \leq y \leq \frac{\pi}{2} \ (y \neq 0).$$

Although the particular ranges in Definition 8.8 are the ones customarily chosen, the choices actually are quite arbitrary. They have generally been selected to involve small values of y, to have relatively simple graphs, and of course to yield a one-to-one correspondence between domain and range. They are also values that can be obtained with most scientific calculators. The graphs of $y = \text{Cos}^{-1} x$ and

---

* The symbol "$\leftrightarrow$" should be read "is equivalent to."

$y = \text{Tan}^{-1} x$ are shown in Figures 8.26-a and b, respectively. The remaining inverse circular functions are less frequently used, and their graphs are not shown.

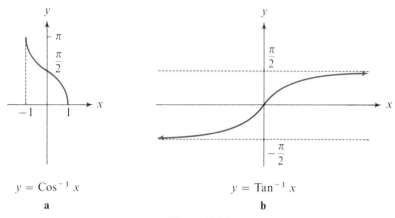

Figure 8.26

The inverse circular function values are sometimes named by using the prefix "Arc" with the function name. For example, $\text{Sin}^{-1} x$ is sometimes represented by Arcsin x (read "arcsin x").

It is sometimes helpful to interpret an expression involving inverse notation by first rewriting the expression equivalently without using this notation.

Examples Find the value of each of the following in terms of rational multiples of π.

  a. $\text{Sin}^{-1} \dfrac{1}{2}$

  b. $\text{Arcsec} \dfrac{2}{\sqrt{3}}$

Solutions a. Let $y = \text{Sin}^{-1}(1/2)$. Then by Definition 8.8,

$$y = \text{Sin}^{-1} \frac{1}{2} \leftrightarrow \sin y = \frac{1}{2}, \quad -\frac{\pi}{2} \le y \le \frac{\pi}{2}.$$

From Table 8.2, $y = \pi/6$. Hence,

$$y = \text{Sin}^{-1} \frac{1}{2} = \frac{\pi}{6}.$$

  b. Let $y = \text{Arcsec} \dfrac{2}{\sqrt{3}}$. Then by Definition 8.8,

$$y = \text{Arcsec} \frac{2}{\sqrt{3}} \leftrightarrow \sec y = \frac{2}{\sqrt{3}}, \quad 0 \le y \le \pi,$$

which by Definition 8.6 is equivalent to

$$\cos y = \frac{\sqrt{3}}{2}, \quad 0 \le y \le \pi.$$

8.7 Inverse Circular Functions

From Table 8.2, $y = \pi/6$. Hence,

$$y = \text{Arcsec } \frac{2}{\sqrt{3}} = \frac{\pi}{6}.$$

With a little practice, you will be able to find an inverse function value directly. The following examples show the process.

Examples Find the value of each of the following in terms of rational multiples of π.

    a. $\text{Arctan } \dfrac{1}{\sqrt{3}}$ b. $\text{Cos}^{-1} \dfrac{1}{\sqrt{2}}$

Solutions a. The number y between 0 and $\pi/2$ such that $\tan y = 1/\sqrt{3}$ is $\pi/6$. Hence,

$$\text{Arctan } \frac{1}{\sqrt{3}} = \frac{\pi}{6}.$$

     b. The number between 0 and π such that $\cos y = 1/\sqrt{2}$ is $\pi/4$. Hence,

$$\text{Cos}^{-1} \frac{1}{\sqrt{2}} = \frac{\pi}{4}.$$

We must give special attention to the ranges specified in Definition 8.8 when finding values of functions that have negative values in their domains.

Examples Find the value of the following expressions in terms of rational multiples of π.

    a. $\text{Arctan}\left(-\dfrac{1}{\sqrt{3}}\right)$ b. $\text{Cos}^{-1}\left(-\dfrac{1}{\sqrt{2}}\right)$

Solutions a. The number y between $-\pi/2$ and $\pi/2$ such that $\tan y = -1/\sqrt{3}$ is $-\pi/6$. Hence,

$$\text{Arctan}\left(-\frac{1}{\sqrt{3}}\right) = -\frac{\pi}{6}.$$

     b. The number y between 0 and π such that $\cos y = -1/\sqrt{2}$ is $3\pi/4$. Hence,

$$\text{Cos}^{-1}\left(-\frac{1}{\sqrt{2}}\right) = \frac{3\pi}{4}.$$

Using a table or a calculator Table V or a hand calculator, in radian mode, with inverse-circular-function capability can be used to find approximations for function values that are not listed in Table 8.2. It sometimes happens that the exact value being sought is not in Table V. If this occurs, simply choose the closest value in the table. A calculator will be used to compute inverse-circular-function values in all examples and exercises. Answers obtained in this way may vary slightly from those obtained using Table V if the exact value sought is not in the table.

Some calculators have a key marked INV or ARC which is used to calculate the values of the inverse circular function.

Example Compute $\text{Tan}^{-1} 0.2236$.

Solution It may be helpful first to let $y = \text{Tan}^{-1} 0.2236$ and then rewrite the equation equivalently as

$$\tan y = 0.2236, \quad -\frac{\pi}{2} < y < \frac{\pi}{2}.$$

From Table V or by using a calculator in radian mode, we obtain $y = 0.22$. Since $-\pi/2 < 0.22 < \pi/2$, we have

$$\text{Tan}^{-1} 0.2236 = 0.22.$$

Exercise 8.7

Find the value of the given expression. Give the result in terms of rational multiples of π.

Examples
a. $\text{Arccos}\dfrac{1}{2}$
b. $\text{Tan}^{-1}(-1)$

Solutions

a. We seek a number y such that
$$\cos y = \frac{1}{2}, \quad 0 \le y \le \pi.$$
The number is $\dfrac{\pi}{3}$.

b. We seek a number y such that
$$\tan y = -1, \quad -\frac{\pi}{2} < y < \frac{\pi}{2}.$$
The number is $-\dfrac{\pi}{4}$.

1. $\text{Arcsin}\dfrac{1}{2}$
2. $\text{Arctan}\sqrt{3}$
3. $\text{Cot}^{-1} 1$
4. $\text{Cos}^{-1}\dfrac{1}{\sqrt{2}}$
5. $\text{Cos}^{-1}\dfrac{1}{2}$
6. $\text{Arcsin}\dfrac{\sqrt{3}}{2}$
7. $\text{Arctan}\left(-\dfrac{1}{\sqrt{3}}\right)$
8. $\text{Cot}^{-1}(-1)$

Use a calculator or Table V in the Appendix to find the value of the given expression.

Example $\text{Cos}^{-1} 0.9940$

Solution It may be helpful to let $y = \text{Cos}^{-1} 0.9940$ and then rewrite this equivalently as
$$\cos y = 0.9940, \quad 0 \le y \le \pi.$$

8.7 Inverse Circular Functions

By using a calculator in radian mode, or from Table V, we obtain $y = 0.11$. Since $0 \leq 0.11 \leq \pi$, we have
$$\text{Cos}^{-1} 0.9940 = 0.11.$$

9. $\text{Tan}^{-1} 0.1003$
10. $\text{Arctan } 3.467$
11. $\text{Sin}^{-1} 0.3802$
12. $\text{Cos}^{-1} 0.6675$
13. $\text{Arcsin } (-0.8624)$
14. $\text{Arccos } (-0.3902)$
15. $\text{Arcsec } 1.053$
16. $\text{Csc}^{-1} 1.422$

Find the value for the given expression if the value exists.

Examples

a. $\text{Cos}^{-1} (\tan \pi)$
b. $\sin \left(\text{Cos}^{-1} \dfrac{1}{2} \right)$

Solutions

a. Since $\tan \pi = 0$,
$$\text{Cos}^{-1} (\tan \pi) = \text{Cos}^{-1} 0 = \dfrac{\pi}{2}.$$

b. Since $\text{Cos}^{-1} \dfrac{1}{2} = \dfrac{\pi}{3}$,
$$\sin \left(\text{Cos}^{-1} \dfrac{1}{2} \right) = \sin \left(\dfrac{\pi}{3} \right) = \dfrac{\sqrt{3}}{2}.$$

17. $\text{Sin}^{-1} \left(\cos \dfrac{\pi}{4} \right)$
18. $\text{Cos}^{-1} \left(\sin \dfrac{\pi}{2} \right)$
19. $\text{Tan}^{-1} \left(\tan \dfrac{\pi}{3} \right)$
20. $\text{Sin}^{-1} \left(\sin \dfrac{3\pi}{2} \right)$
21. $\sin \left(\text{Arccos } \dfrac{1}{2} \right)$
22. $\tan \left(\text{Arcsin } \dfrac{\sqrt{3}}{2} \right)$
23. $\cos (\text{Cot}^{-1} (-\sqrt{3}))$
24. $\sin (\text{Tan}^{-1} (-1))$
25. $\sin \left(\text{Arcsin } \dfrac{1}{2} \right)$
26. $\cos (\text{Tan}^{-1} 0)$
27. $\tan \left(\text{Arcsin } \dfrac{12}{13} \right)$
28. $\sin \left(\text{Cos}^{-1} \dfrac{3}{5} \right)$

Find the value for the given expression if the value exists.

29. $\text{Arccos } (\sin (\text{Arctan } (-1)))$
30. $\sin (\text{Cos}^{-1} (\tan 0))$
31. $\sin \left(\text{Sin}^{-1} \dfrac{1}{2} + \text{Cos}^{-1} \dfrac{1}{2} \right)$
32. $\cos \left(\text{Sin}^{-1} \dfrac{1}{\sqrt{2}} + \text{Cos}^{-1} \dfrac{1}{\sqrt{2}} \right)$
33. $\sin (\text{Tan}^{-1} 1 + \text{Sin}^{-1} 1)$
34. $\cos (\text{Tan}^{-1} 0 + \text{Sin}^{-1} 0)$

Solve the given equation for x in terms of y.

Example

$y = 2 \text{ Arcsin } 3x$

Solution on Overleaf

Solution We have
$$\frac{y}{2} = \text{Arcsin } 3x,$$
and from the definition of Arcsine, we write the expression as
$$3x = \sin\frac{y}{2}, \quad \text{and then as} \quad x = \frac{1}{3}\sin\frac{y}{2},$$
with $-\frac{\pi}{2} \le \frac{y}{2} \le \frac{\pi}{2}$, or $-\pi \le y \le \pi$.

35. $y = 3 \text{ Arccos } 2x$

36. $y = 2 \text{ Sin}^{-1}\frac{x}{5}$

37. $y = \frac{1}{2}\text{Tan}^{-1}(x + \pi)$

38. $y = \frac{2}{3}\text{Cot}^{-1}(\pi x)$

Chapter Review

[8.1] State whether the given function value is positive or negative.

1. $\cos\frac{2\pi}{3}$ **2.** $\cos\frac{16\pi}{3}$ **3.** $\sin\left(\frac{-8\pi}{7}\right)$ **4.** $\sin\left(\frac{-\pi}{6}\right)$

Find the required function value and state the quadrant in which s terminates.

5. $\sin s$, given that $\cos s = 2/3$ and $\sin s < 0$.
6. $\cos s$, given that $\sin s = 3/8$ and $\cos s > 0$.

[8.2] Use Table 8.2 to find the value of the given expression.

7. $\cos\frac{\pi}{6} + \sin \pi$

8. $\sin\frac{\pi}{4} - \cos\frac{\pi}{4}$

9. $(\sin \pi)\left(\cos\frac{\pi}{3}\right)$

10. $\dfrac{\cos \pi/3}{\sin \pi/3}$

Use Table V or a hand calculator to find the approximate value of the given expression.

11. $\cos 1.48 + \sin 1.05$

12. $\cos 1.35 - \sin 0.45$

13. $(\sin 0.31)(\cos 1.25)$

14. $\dfrac{\sin 1.01}{\cos 1.31}$

Assume that $d = K \cos 2\pi t$. Find the indicated value under the given conditions. Use $\pi \approx 3.14$.

15. K, given $d = 25$ and $t = 1/2$.
16. d, given $K = 75$ and $t = 0.05$.

[8.3] Use Table 8.2 or a hand calculator or Table V to compute the indicated value.

17. $\cos \dfrac{3\pi}{4}$ **18.** $\cos \dfrac{11\pi}{6}$ **19.** $\sin \dfrac{5\pi}{4}$

20. $\sin \dfrac{5\pi}{6}$ **21.** $\cos 3.28$ **22.** $\cos 6.11$

23. $\sin 2.51$ **24.** $\sin 3.35$ **25.** $\cos(-1.53)$

26. $\cos 6.56$ **27.** $\sin(-1.65)$ **28.** $\sin 7.21$

[8.4] Find the amplitude, period, and phase shift for the indicated function.

29. $y = 2 \sin 3x$ **30.** $y = \dfrac{3}{2} \cos\left(x - \dfrac{\pi}{4}\right)$ **31.** $y = 4 \sin(2x - \pi)$

Sketch the graph of the indicated function over the interval $-\pi \leq x \leq 2\pi$.

32. $y = \dfrac{1}{3} \cos 2x$ **33.** $y = \sin\left(x + \dfrac{\pi}{3}\right)$ **34.** $y = \cos(3x + \pi/2)$

[8.5] Use Definition 8.4 and Table 8.2, Table V, or a hand calculator to compute the indicated value.

35. $\tan \dfrac{5\pi}{4}$ **36.** $\tan \dfrac{4\pi}{3}$ **37.** $\tan 1.95$

38. $\tan 3.65$ **39.** $\tan(-2.18)$ **40.** $\tan 7.25$

Sketch the graph of each function over the interval $-\pi \leq x \leq 2\pi$.

41. $y = \tan \dfrac{x}{4}$ **42.** $y = 2 \tan 2x$.

[8.6] Use the appropriate definition and Table 8.2 to compute the indicated value if it exists; if not, then so indicate.

43. $\cot \dfrac{\pi}{6}$ **44.** $\sec \dfrac{\pi}{6}$ **45.** $\csc \dfrac{\pi}{2}$ **46.** $\sec \dfrac{\pi}{2}$

Use the appropriate definition to approximate the indicated value if it exists; if not, then so indicate.

47. $\cot \dfrac{3\pi}{4}$
48. $\sec \dfrac{9\pi}{4}$
49. $\csc \left(-\dfrac{\pi}{6}\right)$
50. $\cot 3.85$
51. $\sec 9.56$
52. $\csc(-1.65)$

[8.7] Find the value of the given expression if the value exists. Express the result in terms of rational multiples of π.

53. $\text{Sin}^{-1} \dfrac{1}{\sqrt{2}}$
54. $\text{Cos}^{-1} 1$
55. $\text{Tan}^{-1} \dfrac{1}{\sqrt{3}}$
56. $\text{Arccot}\, \sqrt{3}$
57. $\text{Arcsec}(-2)$
58. $\text{Csc}^{-1}(-2)$

Use Table V or a hand calculator to find the value of the given expression.

59. $\text{Arctan}\, 1.398$
60. $\text{Arcsin}\, 0.9320$
61. $\text{Cos}^{-1} 0.9550$
62. $\text{Sec}^{-1} 1.609$
63. $\text{Arccsc}\, 1.771$
64. $\text{Cot}^{-1} 0.9712$

Find the value of the given expression if the value exists.

65. $\sin \left[\text{Sin}^{-1} 0.25\right]$
66. $\text{Cos}^{-1}(\cos 4\pi)$
67. $\tan \left(\text{Sin}^{-1} \dfrac{\sqrt{2}}{2}\right)$
68. $\sin(\text{Tan}^{-1} \sqrt{3})$

9 Trigonometric Functions; Vectors

9.1 Angles and Their Measure

Standard position

In studying geometry, we learn that an angle is the union of two rays with a common endpoint (Figure 9.1-a) and that an angle can be designated by naming a point on each ray together with the common endpoint of the rays, or else by simply assigning a single symbol, say α, to the angle. The common endpoint of the rays is called the **vertex** of the angle, and the rays are called the **sides** of the angle.

Now each angle in the plane is congruent ($\cong$) to an angle with one side along the positive x-axis and vertex at the origin (Figure 9.1-b). Such an angle is said to be in **standard position.** The side $\overrightarrow{OC'}$ of $\angle A'OC'$ in Figure 9.1-b is called the **initial side** of the angle, the side $\overrightarrow{OA'}$ is called the **terminal side** and the angle can be visualized as being formed by a rotation from the initial side $\overrightarrow{OC'}$ into the terminal side $\overrightarrow{OA'}$. In trigonometry, the *amount of rotation* is considered along with the angle itself. If the terminal side of an angle in standard position lies in a given quadrant, we say that the angle is *in* that quadrant.

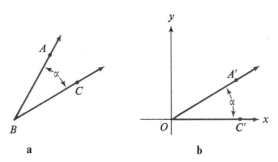

Figure 9.1

Angle measurement

A measure is assigned to an angle by means of a circle with center at the vertex of the angle. For convenience, we shall restrict this discussion to the set of angles in standard position, although the process described is perfectly general. If the circumference of a circle of radius $r > 0$ is divided into p arcs of equal length, then each of the arcs will have length $2\pi r/p$. Starting at the point $(r, 0)$, we can scale the circumference of the circle in both the counterclockwise and clockwise directions from this point, in terms of this arc length as a unit. In doing this, it is customary to assign *negative* numbers in the *clockwise* direction, and *positive* numbers in the *counterclockwise* direction. The terminal side of any angle α in standard position intercepts the circumference at a point, and one of the scale numbers (arc lengths) assigned to that point becomes the measure of the angle, depending on the amount of rotation involved.

The two most commonly encountered units of angle measure are the **degree** and the **radian**. In degree measure, the circumference of a circle is divided into 360 arcs of equal length, and hence the unit arc is of length

$$\frac{2\pi r}{360} = \frac{\pi}{180} r,$$

where r is the radius of the circle (see Figure 9.2-a). For this measure, we use the notation $^\circ$. Thus, the angle shown in Figure 9.2-a has measure 1°.

In radian measure, the circumference is divided into 2π arcs of equal length, and hence the length of the unit arc (see Figure 9.2-b) is

$$\frac{2\pi r}{2\pi} = r.$$

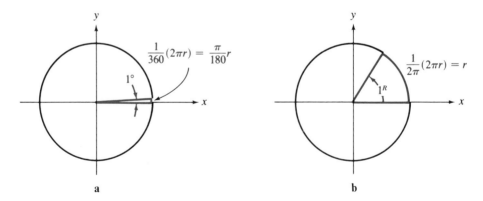

Figure 9.2

For this measure we use the notation ^{R}. Thus, in radian measure, the length of the unit arc is the length of the radius of the circle, and the circumference contains just 2π of these units. Thus, the angle in Figure 9.2-b has measure 1^R. Recall from geometry that the lengths of arcs intercepted by a given central angle on concentric circles are proportional to the circumferences of the circles. *It follows that the measure assigned to an angle in terms of a given unit is independent of the radius of the measuring circle.* In particular, if the circle that is used as the basis for radian

9.1 Angles and Their Measure

measure has radius one, then the length of the unit arc is 1. For this reason, the **unit circle**, that is, the circle of radius one in standard position, is customarily used as a reference for the measure of an angle in radians.

Conversion formulas

Since one full rotation of the terminal side of an angle in standard position yields an angle with measures of 360° or $2\pi^R$, we write

$$360° = 2\pi^R.$$

Thus,

$$1^R = \left(\frac{180}{\pi}\right)°, \tag{1}$$

and

$$1° = \left(\frac{\pi}{180}\right)^R. \tag{2}$$

Equations (1) and (2) are **conversion formulas** for expressing the relationships between degrees and radians.

Examples

Find the degree measure, to the nearest tenth of a degree, of the angle whose radian measure is given.

a. $\dfrac{\pi^R}{3}$
b. 0.62^R

Solutions

a. $\dfrac{\pi^R}{3} = \left(\dfrac{180}{\pi} \cdot \dfrac{\pi}{3}\right)° = 60°$

b. $0.62^R = \left(\dfrac{180}{\pi} \cdot 0.62\right)°$

$= \dfrac{111.6°}{\pi} \approx 35.5°$

In Example (b) above, we use the calculator value for π, 3.1415926. We shall use this value for π in our computations in examples and in the answers for the exercises. Slightly different values may be expected if a different approximation for π is used.

Examples

Find the radian measure, to the nearest hundredth of a radian, of the angle whose degree measure is given.

a. 30°
b. 300°

Solutions

a. $30° = \left(\dfrac{\pi}{180} \cdot 30\right)^R = \dfrac{\pi^R}{6} \approx 0.52^R$ b. $300° = \left(\dfrac{\pi}{180} \cdot 300\right)^R = \dfrac{5\pi^R}{3} \approx 5.24^R$

Note that, in these examples, the number of units in one measure does not equal the number of units in another measure; for example, $30 \neq \pi/6$. But an angle whose measure is 30° is congruent to an angle whose measure is $(\pi/6)^R$, and we express this fact by writing $30° = (\pi/6)^R$. This is analogous to 36 inches = 3 feet, 6 feet = 2 yards, and so forth, when indicating lengths of line segments in different units.

If the measure of an angle α is given in radians, then the length s of the intercepted arc of a circle with given radius r can be found directly. Thus since

$$\frac{s}{2\pi r} = \frac{\alpha^R}{2\pi^R},$$

we have

$$s = r \cdot \alpha^R. \qquad (3)$$

If the measure of an angle is given in degrees, it can first be changed to radian measure and then the length of the intercepted arc can be found directly from (3).

Example Find the length of the arc intercepted by a central angle of 135° in a circle with a radius of 5 centimeters.

Solution We first change 135° to radian measure.

$$135° = \left(\frac{\pi}{180} \cdot 135\right)^R = \frac{3\pi^R}{4}.$$

Then

$$s = r \cdot \alpha^R = 5 \cdot \frac{3\pi}{4} = \frac{15\pi}{4} \approx 11.8.$$

Therefore, the arc length to the nearest tenth is 11.8 centimeters.

Note that if the intercepted arc is part of a unit circle, then Equation (3) guarantees that the length of the arc is precisely the radian measure of the angle subtending the arc.

Coterminal angles Note that although angles such as α_1, α_2, and α_3 in Figure 9.3 have the same initial side and the same terminal side, their measures are different because the amounts of rotation involved are different. Such angles are called **coterminal**. The measures of α_1 and α_3 are positive, and the measure of α_2 is negative.

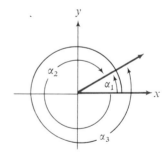

Figure 9.3

Exercise 9.1

Find the degree measure of the angle whose radian measure is as given.

1. a. 0^R b. $\dfrac{\pi^R}{2}$ c. π^R d. $\dfrac{3\pi^R}{2}$ e. $2\pi^R$

9.1 Angles and Their Measure

2. By filling in the blank spaces, complete the following table, which compares the radian measure and degree measure of angles that are of frequent occurrence.

°	30°	45°	60°	120°	135°	150°
R	.523	.7853	$\frac{\pi}{3}^R$	2.09	2.356	$\frac{5\pi}{6}^R$

°	210°	225°	240°	300°	315°	330°
R	3.665	$\frac{5\pi}{4}^R$	$\frac{4\pi}{3}^R$	$\frac{5\pi}{3}^R$	5.49	5.75

Find the degree measure, to the nearest tenth of a degree, of the angle whose radian measure is as given.

Example 1.65^R

Solution   By Equation (1),

$$1^R = \left(\frac{180}{\pi}\right)^\circ.$$

Thus, we have

$$1.65^R = 1.65\left(\frac{180}{\pi}\right)^\circ \approx 94.5^\circ.$$

3. $\dfrac{2\pi^R}{9}$ **4.** $\dfrac{3\pi^R}{5}$ **5.** $\dfrac{7\pi^R}{5}$ **6.** $\dfrac{5\pi^R}{8}$

7. 0.30^R **8.** 1.25^R **9.** 3.62^R **10.** 9.14^R

Find the radian measure, to the nearest hundredth of a radian, of the angle whose degree measure is as given.

Example $85°$

Solution   By Equation (2),

$$1° = \left(\frac{\pi}{180}\right)^R.$$

Solution Continued on Overleaf

Thus, we have

$$85° = 85\left(\frac{\pi}{180}\right)^R = 1.48^R.$$

11. 20°   **12.** 50°   **13.** 130°   **14.** 310°

15. 420°   **16.** 580°   **17.** 750°   **18.** 800°

19. What is the degree measure, to the nearest hundredth of a degree, of an angle whose radian measure is 1^R?

20. What is the radian measure, to the nearest thousandth of a radian, of an angle whose degree measure is 1°?

On a circle with given radius, find the length of the arc intercepted by the angle whose measure is as given.

Example $r = 3$ in.; 120°

Solution The measure of the angle is first changed to radian measure. Thus

$$120° = \left(\frac{\pi}{180} \cdot 120\right)^R = \frac{2\pi^R}{3}.$$

Then, by Equation (3) on page 244,

$$s = r \cdot \alpha^R = 3 \cdot \frac{2\pi}{3} = 2\pi \approx 6.28.$$

Therefore, the desired arc length is approximately 6.28 in.

21. $r = 4$ cm; $(\pi/6)^R$ **22.** $r = 5.2$ in.; π^R **23.** $r = 1.2$ m; 0.60^R

24. $r = 3.6$ ft; 1^R **25.** $r = 2$ m; 135°   **26.** $r = 4.3$ yd; 180°

Find the degree measure of the angle subtending an arc of the given length on a unit circle.

Example $\frac{\pi}{6}$

Solution Since the arc in question lies on a unit circle, its length is precisely the radian measure of the angle subtending the arc. Thus, we need to convert $(\pi/6)^R$ to degrees. By Equation (1) on page 243,

$$1^R = \left(\frac{180}{\pi}\right)°,$$

so

$$\frac{\pi}{6}^R = \left(\frac{\pi}{6} \cdot \frac{180}{\pi}\right)° = 30°.$$

9.2 Functions of Angles

27. $\dfrac{\pi}{2}$
28. $\dfrac{3\pi}{2}$
29. π
30. 2π
31. $\dfrac{\pi}{4}$
32. $\dfrac{\pi}{3}$

Find all angles α satisfying $-360° \leq \alpha \leq 720°$ that are coterminal with the angle whose measure is given. Write a representation for the measures of all angles coterminal with the angle.

Examples a. $24°$ b. $-184°$

Solutions

a. $24° + 360° = 384°$
$24° - 360° = -336°$
$24° + 360°k,\ k \in J$

b. $-184° + 360° = 176°$
$-184° + 720° = 536°$
$-184° + 360°k,\ k \in J$

33. $30°$
34. $-150°$
35. $-240°$
36. $190°$
37. $420°$
38. $683°$
39. $-330°$
40. $-271°$

9.2 Functions of Angles

Historically, interest in the circular functions arose from the study of angles and triangles. In defining the circular functions, we used the notion of arc length on a unit circle to associate a real number (the arc length from the point (1, 0) to a point on the circle) with another real number. Thus we defined functions with real numbers for *both* domain and range.

Geometric ratios in the coordinate plane

Now, consider any angle α in standard position, let (x_1, y_1) and (x_2, y_2) be any two distinct points (except the origin) in its terminal side or ray, and then let $r_1 = \sqrt{x_1^2 + y_1^2}$ and $r_2 = \sqrt{x_2^2 + y_2^2}$ (Figure 9.4). We can show that the following equalities of ratios hold:

$$\dfrac{x_1}{y_1} = \dfrac{x_2}{y_2} \quad (y_1, y_2 \neq 0),$$

$$\dfrac{x_1}{r_1} = \dfrac{x_2}{r_2},$$

and

$$\dfrac{y_1}{r_1} = \dfrac{y_2}{r_2}.$$

The proofs of these are left as exercises.

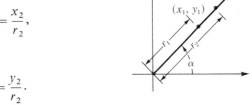

Figure 9.4

Although the figure shows the terminal side of α in the first quadrant, the ratios are equal for coordinates of points on the terminal side of an angle in any quadrant. Because these ratios do not depend on the choice of a point in the terminal side of α, we can use them to define some new functions, each with *the set of angles in standard position* as domain, and with sets of real numbers as ranges. Traditionally, six such assignments are made, and the names given the resulting functions are those we have used to name the circular functions.

Trigonometric ratios

Definition 9.1 *Let α be an angle in standard position, let $(x, y) \neq (0, 0)$ be any point on the terminal side of α, and let $r = \sqrt{x^2 + y^2}$. Then*

$$\cos \alpha = \frac{x}{r} \qquad\qquad \cot \alpha = \frac{x}{y}, \quad y \neq 0$$

$$\sin \alpha = \frac{y}{r} \qquad\qquad \sec \alpha = \frac{r}{x}, \quad x \neq 0$$

$$\tan \alpha = \frac{y}{x}, \quad x \neq 0 \qquad\qquad \csc \alpha = \frac{r}{y}, \quad y \neq 0.$$

These ratios define the six **trigonometric functions**.

Example Find the value of each of the six trigonometric functions of α, if the terminal side of α contains the point $(-3, 5)$.

Solution First we compute r as the *positive* square root of $x^2 + y^2$.

$$r = \sqrt{x^2 + y^2} = \sqrt{(-3)^2 + 5^2} = \sqrt{9 + 25} = \sqrt{34}.$$

Then, by Definition 9.1 we have

$$\cos \alpha = \frac{x}{r} = -\frac{3}{\sqrt{34}}, \qquad \sec \alpha = \frac{r}{x} = -\frac{\sqrt{34}}{3},$$

$$\sin \alpha = \frac{y}{r} = \frac{5}{\sqrt{34}}, \qquad \csc \alpha = \frac{r}{y} = \frac{\sqrt{34}}{5},$$

$$\tan \alpha = \frac{y}{x} = \frac{5}{-3} = -\frac{5}{3}, \qquad \cot \alpha = \frac{x}{y} = \frac{-3}{5} = -\frac{3}{5}.$$

Because every angle in the plane is congruent to an angle in standard position, the definitions of the trigonometric functions can be extended to assign the same numbers to every angle congruent to a given angle α. Thus, while these functions are defined in terms of angles in standard position, they can be viewed as applying to the set of all angles in the plane. Moreover, since congruent angles have the same measure, we can identify angles in the domain of each function with a particular unit of measure. Thus, we write

$$\sin 30° \quad \text{and} \quad \sin \frac{\pi^R}{6}$$

9.2 Functions of Angles

as abbreviations for "the sine of an angle whose measure is 30 degrees" and "the sine of an angle whose measure is $\pi/6$ radians," respectively.

Relationship between circular and trigonometric functions

The fact that the circular functions are defined using the unit circle, together with the fact that the unit circle can be used to assign measures to angles, makes it reasonable to expect a very close relationship to exist between the circular and trigonometric functions. Such is indeed the case.

Since the trigonometric functions of an angle α have been defined in terms of the coordinates of *any* point other than the origin on the terminal side of α, we can arbitrarily choose the point where the terminal side of the angle α intersects the unit circle. This is the point $P(\cos x, \sin x)$, where x is the radian measure of α (Figure 9.5). Thus, the trigonometric functions cosine and sine are related to the circular functions cosine and sine by

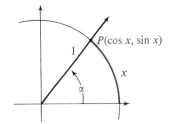

Figure 9.5

$$\cos \alpha = \frac{\cos x}{1} = \cos x, \qquad (1)$$

and, similarly,

$$\sin \alpha = \frac{\sin x}{1} = \sin x. \qquad (2)$$

From this, we see that the remaining four trigonometric functions are related to the corresponding circular functions in a similar way.

In Section 9.1 we agreed to name an angle by its measure. In particular, if the radian measure of α is x, then we write $\alpha = x^R$; and if the degree measure of α is t, then we write $\alpha = t°$. Thus, by replacing α by x^R and $t°$ in (1), we obtain

$$\cos x^R = \cos x \quad \text{and} \quad \cos t° = \cos x.$$

For example,

$$\cos \frac{\pi}{3}^R = \cos \frac{\pi}{3} = \frac{1}{2} \quad \text{and} \quad \cos 60° = \cos \frac{\pi}{3} = \frac{1}{2}. \qquad (3)$$

The values of the trigonometric functions for some special angles are given in Table 9.1 on page 250. These values can be obtained by the same method used to obtain Equation (3). The entries are left for you to verify.

Examples

a. $\cos 45° = \cos \dfrac{\pi}{4} = \dfrac{1}{\sqrt{2}}$

b. $\tan \pi^R = \tan \pi = 0$

Table 9.1

°	R	sin α	csc α	cos α	sec α	tan α	cot α
0°	0^R	0	not defined	1	1	0	not defined
30°	$\frac{\pi}{6}^R$	$\frac{1}{2}$	2	$\frac{\sqrt{3}}{2}$	$\frac{2}{\sqrt{3}}$	$\frac{1}{\sqrt{3}}$	$\sqrt{3}$
45°	$\frac{\pi}{4}^R$	$\frac{1}{\sqrt{2}}$	$\sqrt{2}$	$\frac{1}{\sqrt{2}}$	$\sqrt{2}$	1	1
60°	$\frac{\pi}{3}^R$	$\frac{\sqrt{3}}{2}$	$\frac{2}{\sqrt{3}}$	$\frac{1}{2}$	2	$\sqrt{3}$	$\frac{1}{\sqrt{3}}$
90°	$\frac{\pi}{2}^R$	1	1	0	not defined	not defined	0
180°	π	0	not defined	-1	-1	0	not defined
270°	$\frac{3\pi}{2}^R$	-1	-1	0	not defined	not defined	0

Signs of the trigonometric functions

The quadrant in which an angle lies determines whether a given trigonometric function of this angle is positive or negative. Figure 9.6 presents this in convenient tabular form.

QUADRANT	I. $x > 0, y > 0$ 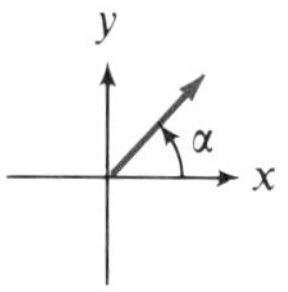	II. $x < 0, y > 0$ 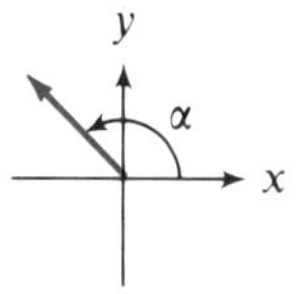	III. $x < 0, y < 0$	IV. $x > 0, y < 0$
$\sin \alpha = y/r$ $\csc \alpha = r/y$	+	+	−	−
$\cos \alpha = x/r$ $\sec \alpha = r/x$	+	−	−	+
$\tan \alpha = y/x$ $\cot \alpha = x/y$	+	−	+	−

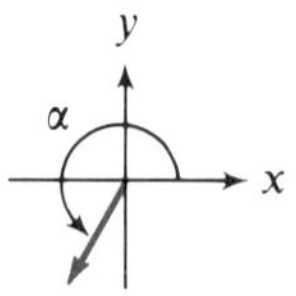

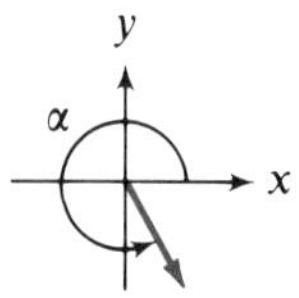

Figure 9.6

9.2 Functions of Angles

Tables for trigonometric functions

Trigonometric function values that are not listed in Table 9.1 can be obtained from Table V in the Appendix if the radian measure of α is known. Table VI in the Appendix gives trigonometric function values of angles with given degree measure. Table VI is graduated in intervals of 6 minutes (6'), one minute being equal to one-sixtieth of a degree. Observe that the table reads from top to bottom for $0° \leq \alpha \leq 45°$, where the function values are identified at the top of the page, and from bottom to top for $45° \leq \alpha \leq 90°$, where the function values are identified at the bottom of the page.

Reference angles

Note that if the radian measure of an angle is not between 0^R and 1.57^R, then Table V does not contain the trigonometric function values for that angle. Similarly, if the degree measure of an angle is not between 0° and 90°, then Table VI does not contain the trigonometric function values for that angle. For angles whose measures are not in these ranges, we use the concept of a reference angle in the same way that we used a reference arc with the circular functions.

Definition 9.2 *For any angle α in standard position, the **reference angle** $\bar{\alpha}$ is the least positive angle which the terminal ray of α forms with the x-axis.*

Figure 9.7 illustrates the reference angle for angles lying in each of the four quadrants.

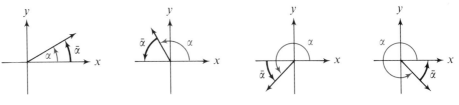

Figure 9.7

Examples

Find the measure of the reference angle for each angle. Use $\pi \approx 3.14$.

a. 135°   b. 3.50^R

Solutions

a. Since $\alpha = 135°$ terminates in the second quadrant, we have

$$\bar{\alpha} = 180° - 135°$$
$$= 45°.$$

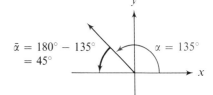

b. Since $\alpha = 3.50^R$ terminates in the third quadrant, we have

$$\bar{\alpha} = 3.50^R - \pi^R$$
$$\approx 3.50^R - 3.14^R$$
$$= 0.36^R.$$

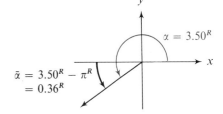

The following theorem is analogous to Theorem 8.2 of Section 8.3, and since the proof is similar, it is omitted.

Theorem 9.1 If T is a trigonometric function and $\bar{\alpha}$ is the reference angle for the angle α, then

$$|T(\alpha)| = T(\bar{\alpha}).$$

We can now use reference angles to compute the values of the trigonometric functions in the same way that reference arcs were used to compute the values of the circular functions.

Example tan 225°

Solution The reference angle for 225° is

$$\bar{\alpha} = 225° - 180°$$
$$= 45°.$$

Thus, from Table 9.1 and Theorem 9.1, we have

$$|\tan 225°| = \tan 45° = 1.$$

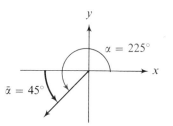

Hence, either $\tan 225° = 1$ or $\tan 225° = -1$. Since the terminal side of $\alpha = 225°$ lies in the third quadrant, we must have

$$\tan 225° = 1.$$

Hand calculators Hand calculators can be used to obtain values for the trigonometric functions in the same way they were used to compute values for the circular functions when the measure of the angle is given in either radians or degrees. Care must be taken, however, to ensure that the calculator is in degree mode when the measure of the angle is given in degrees, and in radian mode when the measure of the angle is given in radians.

Example tan 149°

Solution The reference angle for 149° is

$$\bar{\alpha} = 180° - 149°$$
$$= 31°.$$

Thus, we have

$$|\tan 149°| = \tan 31°$$
$$= 0.6009.$$

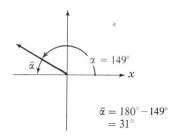

Hence, either $\tan 149° = 0.6009$ or $\tan 149° = -0.6009$. Since the terminal side of $\alpha = 149°$ lies in the second quadrant, we must have

$$\tan 149° = -0.6009.$$

9.2 Functions of Angles

Inverse trigonometric functions

Inverse functions for each of the six trigonometric functions can be defined by considering the domains of these functions to contain angles in standard position with measures suitably restricted. The restrictions ordinarily used are the same as those used in Definition 8.8 (page 233) with y representing the radian measure of an angle. The same notation is used for both the inverse trigonometric functions and the inverse circular functions.

Note that the elements of the ranges of the inverse trigonometric functions are angles in standard position. However, we are frequently interested in the *measure* of the angle rather than the angle itself. If the radian measure of the angle is desired, a hand calculator in radian mode or Table V should be used; if the degree measure of the angle is desired, a hand calculator in degree mode or Table VI should be used. As in the case of the inverse circular functions, it sometimes happens that the exact value desired is not in the table. If this occurs, then we choose the value in the table closest to the desired value. In such a case the value obtained from the table may vary slightly from the value obtained on a hand calculator. In all examples and exercises a hand calculator will be used to compute values of the inverse trigonometric functions.

Example Compute $\text{Sin}^{-1}\left(\dfrac{1}{\sqrt{2}}\right)$. Express your answer in (a) radians and (b) degrees.

Solution We want an angle y with $\sin y = \dfrac{1}{\sqrt{2}}$ and $(-\pi/2)^R \leq y \leq (\pi/2)^R$. From Table 9.1, we find $y = (\pi/4)^R$, and converting to degrees, we have (a) $(\pi/4)^R$ and (b) $45°$.

Exercise 9.2

Find the value of each of the six trigonometric functions of α for the terminal side of α containing the given point.

1. $(3, 4)$
2. $(5, -12)$
3. $(\sqrt{2}, -\sqrt{2})$
4. $(-1, \sqrt{3})$
5. $(-6, -8)$
6. $(-4, -3)$
7. $(0, 3)$
8. $(-7, 0)$

Determine in which quadrant the terminal side of the angle α lies.

Example $\sin \alpha > 0$, $\cos \alpha < 0$

Solution Since $\sin \alpha > 0$, the terminal side of α lies in Quadrant I or II; since $\cos \alpha < 0$, the terminal side of α lies in Quadrant II or III. Therefore, the terminal side of α lies in Quadrant II.

9. $\sin \alpha < 0$, $\cos \alpha > 0$
10. $\cos \alpha > 0$, $\tan \alpha < 0$
11. $\sec \alpha > 0$, $\sin \alpha > 0$
12. $\tan \alpha < 0$, $\sin \alpha > 0$
13. $\sec \alpha > 0$, $\csc \alpha < 0$
14. $\cot \alpha > 0$, $\sin \alpha < 0$

254   9 *Trigonometric Functions; Vectors*

Use Table 9.1 and reference angles to find the value of the given expression.

Example   cos 150°

Solution   The reference angle for 150° is

$$\bar{\alpha} = 180° - 150°$$
$$= 30°.$$

Thus, we have

$$|\cos 150°| = \cos 30° = \frac{\sqrt{3}}{2}.$$

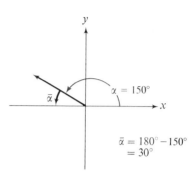

Hence, either $\cos 150° = \frac{\sqrt{3}}{2}$ or $\cos 150° = -\frac{\sqrt{3}}{2}$. Since the terminal side of the angle lies in the second quadrant, we must have

$$\cos 150° = -\frac{\sqrt{3}}{2}.$$

15. tan 330°   16. cot 120°   17. cos 225°
18. tan 150°   19. sin 330°   20. cos 240°
21. sin (−30°)   22. tan (−60°)   23. sin (−240°)
24. cos (−135°)   25. tan (−300°)   26. tan (−750°)

Use a hand calculator or Table V or Table VI, as appropriate, to find function values to four significant figures for the given expression.

27. sin 32°   28. cos 49°   29. tan 33°
30. cot 51°   31. sec 17°   32. csc 84°
33. cos 0.63^R 34. sin 0.42^R 35. cot 1.42^R
36. tan 1.03^R 37. csc $\frac{5\pi^R}{9}$ 38. sec $\frac{5\pi^R}{12}$
39. sin 132°   40. cos 153°   41. tan 320°
42. sin 312°   43. cos (−130°)   44. tan (−605°)
45. sin 2.07^R 46. cos 3.51^R 47. tan 4.21^R
48. tan 6.00^R 49. cos (−1.63^R)   50. sin (−12.32^R)

Use a hand calculator or Table V and Table VI to compute (a) the radian measure and (b) the degree measure of the indicated value.

9.3 Right Triangles

Example $\text{Cos}^{-1}\ 0.7071$

Solution   We are seeking the measure of an angle α such that $\cos \alpha = 0.7071$ and $0^R \leq \alpha \leq \pi^R$, or $0° \leq \alpha \leq 180°$. Using Tables V and VI, we have (a) 0.79^R and (b) $45°$.

51. $\text{Sin}^{-1}\ 1.000$ 52. $\text{Cos}^{-1}\ 0.5000$ 53. $\text{Tan}^{-1}\ 1.000$

54. Arcsin 0.0993   55. Arccos 0.9816   56. Arctan 0.3249

57. Show (geometrically) that if (x_1, y_1) and (x_2, y_2) are the coordinates of any two points (except the origin) on the terminal side (ray) of an angle in the first quadrant, and $r_1 = \sqrt{x_1^2 + y_1^2}$, $r_2 = \sqrt{x_2^2 + y_2^2}$, then the following equalities of ratios hold:

$$\frac{x_1}{y_1} = \frac{x_2}{y_2} \quad (y_1, y_2 \neq 0),$$

$$\frac{x_1}{r} = \frac{x_2}{r}, \quad \text{and} \quad \frac{y_1}{r} = \frac{y_2}{r}.$$

58. Under the same conditions as in Exercise 57, explain how the equalities of these ratios hold for the coordinates of two points on the terminal side of an angle whose terminal side is in the second, third, and fourth quadrants.

9.3 Right Triangles

An examination of Figure 9.8-a makes it evident that in the first quadrant, any point (x, y) on the terminal side of an angle α determines a right triangle with sides

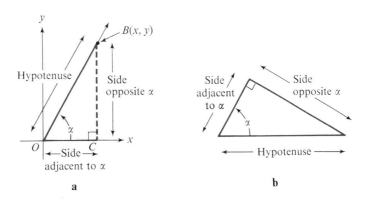

Figure 9.8

measuring x and y and with hypotenuse of length $r = \sqrt{x^2 + y^2}$. Therefore, as special cases of trigonometric function values, we can consider the following **trigonometric ratios** of the sides of a right triangle:

$$\sin \alpha = \frac{\text{length of side opposite } \alpha}{\text{length of hypotenuse}},$$

$$\cos \alpha = \frac{\text{length of side adjacent to } \alpha}{\text{length of hypotenuse}},$$

$$\tan \alpha = \frac{\text{length of side opposite } \alpha}{\text{length of side adjacent to } \alpha}.$$

Similar statements can be made about $\csc \alpha$, $\sec \alpha$, and $\cot \alpha$.

It is not necessary that the right triangle be oriented in such a way that α is in standard position. The foregoing relations involving ratios of the lengths of the sides of a right triangle are equally applicable to a right triangle in any position, as suggested in Figure 9.8-b.

Solution of a triangle

If we are given some parts of a triangle, that is, the measures of some angles and the lengths of some sides, and asked to find the remaining measures, we are asked to **solve** the triangle. In general, we shall give solutions to the nearest tenth of a unit for lengths and to the nearest 6' for the degree measures of angles.

Example

If one angle of a right triangle measures 47° and the side adjacent to this angle has length 24, solve the triangle.

Solution

We first make a sketch illustrating the situation. Generally the hypotenuse is labeled c, the legs a and b, and the angles opposite a, b, and c are labeled α, β, and γ, respectively, or A, B, and C, respectively. In this case, we note first that

$$\beta = 90° - \alpha = 90° - 47° = 43°.$$

Next we have

$$\tan 47° = \frac{a}{b} = \frac{a}{24}, \quad \text{or} \quad a = 24 \tan 47°,$$

and

$$\cos 47° = \frac{b}{c} = \frac{24}{c}, \quad \text{or} \quad c = \frac{24}{\cos 47°}.$$

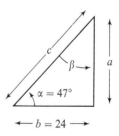

From Table VI or by using a calculator, we obtain

$$\tan 47° = 1.072 \quad \text{and} \quad \cos 47° = 0.6820,$$

so that

$$a = 24(1.072) \approx 25.7,$$

and

$$c = \frac{24}{0.6820} \approx 35.2.$$

We selected the trigonometric ratios for tan 47° and cos 47° in the above example instead of cot 47° and sec 47°, because the values for tangent and cosine could more easily be obtained on a calculator.

Right triangles play a useful role in finding one trigonometric function value, given another.

Example   If $\tan \alpha = 2/3$ and α is in Quadrant I, find $\sin \alpha$.

Solution   Sketch a right triangle and label the sides so that $\tan \alpha$ is as given. By the Pythagorean theorem, the hypotenuse of the triangle has length

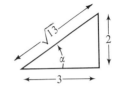

$$\sqrt{2^2 + 3^2} = \sqrt{13}, \quad \text{so that} \quad \sin \alpha = \frac{2}{\sqrt{13}}.$$

Note in this example that once the length of the hypotenuse is determined, *all* of the values for trigonometric functions of α are available by inspection.

Right triangles are helpful also in quadrants other than the first if care is taken to assign *directed* (*positive or negative*) *distances* as lengths of the sides, while viewing the length of the hypotenuse as always positive. This is simply a variation on our procedures for finding function values using reference angles.

Example   If $\cot \alpha = 3/7$ and $\sin \alpha < 0$, find $\cos \alpha$.

Solution   Since $\cot \alpha > 0$ while $\sin \alpha < 0$, α can lie only in Quadrant III. Sketch a right triangle in Quadrant III and label the sides with directed lengths so that $\cot \bar{\alpha} = \frac{3}{7}$.

By the Pythagorean theorem,

$$OA = \sqrt{(-3)^2 + (-7)^2}$$
$$= \sqrt{9 + 49} = \sqrt{58}.$$

By inspection, then, $\cos \alpha = \dfrac{-3}{\sqrt{58}}.$

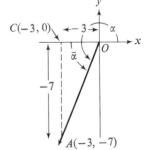

Triangle interpretation of inverse notation   It is sometimes useful to interpret an expression including inverse notation by using right triangles.

Example Evaluate $\cos\left(\text{Arcsin}\frac{\sqrt{3}}{4}\right)$.

Solution We construct a right triangle as shown, with hypotenuse of length 4 and one leg of length $\sqrt{3}$, such that

$$\alpha = \text{Arcsin}\frac{\sqrt{3}}{4} \quad \left(\sin\alpha = \frac{\sqrt{3}}{4}\right).$$

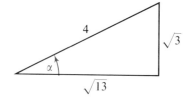

From the triangle, we see that

$$\cos\left(\text{Arcsin}\frac{\sqrt{3}}{4}\right) = \cos\alpha = \frac{\sqrt{13}}{4}.$$

Right triangles can also be used if the element in the domain is positive but not specified.

Example Write $\sin(\text{Cos}^{-1} x)$ as an equivalent expression without inverse notation.

Solution We construct a right triangle as shown, with hypotenuse of length 1 and one leg of length x, such that

$$\alpha = \text{Cos}^{-1} x \quad (\cos\alpha = x).$$

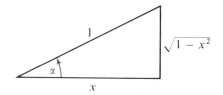

From the triangle, we see that

$$\sin(\text{Cos}^{-1} x) = \sin\alpha = \sqrt{1 - x^2}.$$

Exercise 9.3

Solve the right triangle ABC. Consider $C = 90°$. Give lengths to the nearest tenth of a unit and angle measures to the nearest 6'.

1. $a = 6, b = 8$
2. $a = 5.7, c = 6.2$
3. $b = 120, B = 54°$
4. $a = 3.5, B = 48°$
5. $c = 16, A = 22° 42'$
6. $b = 12.5, A = 14° 24'$

7. In the right triangle ABC, find a if $\sin A = 4/5$ and $c = 20$.
8. In the right triangle ABC, find b if $\tan A = 3/4$ and $a = 12$.
9. Find the other trigonometric function values of θ if $\sin\theta = -1/2$ and the terminal side of θ is in the third quadrant.
10. Find the other trigonometric function values of θ if $\tan\theta = -7/24$ and the terminal side of θ is in the fourth quadrant.
11. Find the other trigonometric function values of θ if $\cos\theta = -3/7$ and the terminal side of θ is in the second quadrant.

9.3 Right Triangles

12. Find the other trigonometric function values of θ if $\sin \theta = 2/3$ and the terminal side of θ is in the second quadrant.

Use right triangles to compute the indicated value.

Example $\sin\left(\text{Tan}^{-1} \dfrac{4}{3}\right)$

Solution We draw a right triangle as shown, with legs of length 4 and 3, such that

$$\alpha = \text{Tan}^{-1} \dfrac{4}{3} \quad \left(\tan \alpha = \dfrac{4}{3}\right).$$

From the triangle, we have

$$\sin\left(\text{Tan}^{-1} \dfrac{4}{3}\right) = \sin \alpha = \dfrac{4}{5}.$$

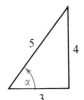

13. $\tan\left(\text{Arcsin} \dfrac{1}{3}\right)$

14. $\sin\left(\text{Cos}^{-1} \dfrac{2}{3}\right)$

15. $\cos\left(\text{Tan}^{-1} \dfrac{2}{3}\right)$

16. $\cot\left(\text{Tan}^{-1} 5\right)$

Write the given expression without inverse notation. Assume x and y are positive.

17. $\cos(\text{Arcsin } x)$
18. $\cot(\text{Arctan } x)$
19. $\tan(\text{Sin}^{-1} y)$
20. $\sin(\text{Tan}^{-1} y)$

21. Show that $\text{Arcsin} \dfrac{2}{5} = \text{Arctan} \dfrac{2}{\sqrt{21}}$. *Hint*: Sketch a triangle.

22. Show that $\text{Arccos} \dfrac{1}{3} = \text{Arccot} \dfrac{1}{2\sqrt{2}}$.

23. Find the value of $\dfrac{\sin \theta + 2 \cos \theta - \tan \theta}{1 - \cot \theta + \sec \theta}$ if $\tan \theta = 1$ and the terminal side of θ is in the third quadrant.

24. Find the value of $\dfrac{\csc^2 \theta + \sin \theta - 1}{2 - \tan^2 \theta + \cot^2 \theta}$ if $\sin \theta = \dfrac{1}{2}$ and the terminal side of θ is in the second quadrant.

In Exercises 25 and 26, given the information pertaining to the figure, find the length of the line segment denoted by x.

25. Given: $BA = 25,$
$\alpha = 16°,$
$\beta = 12°,$

Line BC is a straight line.

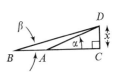

26. Given: $AB = 350$,
$\alpha = 21°$,
$\beta = 8°$,

Line BD is a straight line.

27. A rectangle is 80 meters long and 64 meters wide. Find the degree measures of the angles between a diagonal and the sides.

28. The sides of an isosceles triangle have lengths 6, 6, and 8 meters. Find the degree measure of each angle in the triangle.

29. The angle of elevation (the angle between the line of sight and the horizontal) from a point 58.3 meters from the base of a cliff to the base of a fortress wall on top of the cliff is 60°. The angle of elevation from the same spot to the top of the wall is 65°. How tall is the wall?

30. A pilot is preparing to fly from city A to city B. City A is 100 miles due south of city C, which is 60 miles due west of city B. Determine the heading the pilot should fly (assume that there is no wind and that due North is a 0° heading and due East is a 90° heading).

31. Government surveyors are measuring the height of a mountain. They measure the angle of elevation from point A to the top of the mountain as 68°, and from point B to the top of the mountain as 34°. Assume point B is at the same elevation as point A, and is 1 kilometer (1000 meters) farther from the mountain than point A. How high above A and B is the top of the mountain in meters? (Cf. Exercise 25.)

32. A man standing on the top of a building 35 meters high measures the angle of elevation of the top of the building across the street to be 32°. He measures the angle of depression of the base of the same building to be 55°. How far away is the building across the street and how tall is it?

33. A radar antenna is atop a 30-meter building and is elevated to an angle of 28° with the horizontal. Assume an airplane is flying toward the radar station at an altitude of 5000 meters AGL (above ground level). How far away from the radar station will the airplane be when it first appears on the air traffic controller's radar screen?

34. If the radar antenna in the preceding problem were on the ground, how would the answer be affected? How much is the answer actually changed?

9.4 Law of Sines

In Section 9.3 we used trigonometric ratios to solve right triangles. In this section we shall discuss one technique that can sometimes be used in solving triangles that are not right triangles.

The fact that the area $\mathscr{A}$ of a triangle is equal to one-half the product of the length of its base and the length of its altitude gives us an immediate expression for its area

9.4 Law of Sines

in terms of the lengths of two sides of the triangle and the measure of the included angle. Thus, from Figure 9.9, we have

$$\mathscr{A} = \frac{1}{2} c(b \sin \alpha) = \frac{1}{2} bc \sin \alpha$$

and

$$\mathscr{A} = \frac{1}{2} c(a \sin \beta) = \frac{1}{2} ac \sin \beta.$$

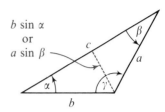

Figure 9.9

It can also be shown that

$$\mathscr{A} = \frac{1}{2} ab \sin \gamma.$$

Because each triangle has only one area, we can equate the right-hand members of these equations to obtain

$$\frac{1}{2} bc \sin \alpha = \frac{1}{2} ac \sin \beta = \frac{1}{2} ab \sin \gamma.$$

Multiplying each member here by $2/(abc)$, we obtain the following result, called the **law of sines.**

Theorem 9.2 *If α, β, and γ are the angles of a triangle and a is the length of the side opposite α, b is the length of the side opposite β, and c is the length of the side opposite γ, then*

$$\frac{\sin \alpha}{a} = \frac{\sin \beta}{b} = \frac{\sin \gamma}{c}.$$

Solving triangles

The law of sines can be used to solve certain triangles. Let us first consider the case in which two angles and one side of a triangle are given.

Example

Solve the triangle for which

$$\alpha = 45°, \quad \beta = 60°, \quad \text{and} \quad a = 10.$$

Solution

It is helpful first to make a sketch. Then, from Theorem 9.2, we have

$$\frac{\sin 45°}{10} = \frac{\sin 60°}{b},$$

Solution Continued on Overleaf

from which

$$b = \frac{10 \sin 60°}{\sin 45°} = 10\left(\frac{0.8660}{0.7071}\right)$$

$$\approx 12.24 \approx 12.2.$$

Next, we observe that

$$\gamma = 180° - \alpha - \beta = 180° - 45° - 60° = 75°.$$

From Theorem 9.2, we have

$$\frac{\sin 45°}{10} = \frac{\sin 75°}{c}.$$

Then, from Table VI or a calculator,

$$c = \frac{10 \sin 75°}{\sin 45°} = 10\left(\frac{0.9659}{0.7071}\right) \approx 13.66 \approx 13.7.$$

Thus, we have

$$b \approx 12.2, \quad c \approx 13.7, \quad \text{and} \quad \gamma = 75°.$$

Ambiguous case

If the lengths of two sides of a triangle, say a and b, and the measure of an angle opposite one of them, say α, are given, we may encounter ambiguity, depending on the value of a in relation to those of b and α. In Figure 9.10, where α is acute, we hold b and α constant and observe the possible situations as a assumes different values.

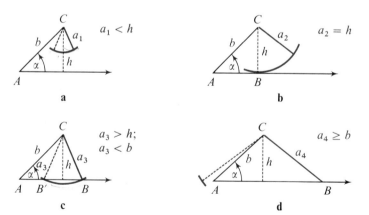

Figure 9.10

First we determine the length h of the altitude of a triangle with the given measures for the acute angle α and the adjacent side b. Since $\sin \alpha = h/b$,

$$h = b \sin \alpha.$$

Now consider the four cases illustrated in Figure 9.10.

9.4 Law of Sines

Case I (Figure a). The length a_1 of the given side satisfies the inequality $a_1 \leq b \sin \alpha$. Since a is less than h, the given values are such that no triangle is possible.

Case II (Figure b). The length a_2 of the given side satisfies the equation

$$a_2 = b \sin \alpha.$$

We have a unique solution—a right triangle.

Case III (Figure c). The length a_3 of the given side is such that

$$a_3 > b \sin \alpha \quad \text{and} \quad a_3 < b.$$

We have two triangles possible, $\triangle ABC$ and $\triangle AB'C$.

Case IV (Figure d). The length a_4 of the given side is such that

$$a_4 \geq b.$$

Here we also have a unique solution.

Of course, if the given angle α is obtuse (Figure 9.11), then any specified lengths of the sides and measures of the angles permit only two possibilities:

1. $a \leq b$, no triangle (as shown in Figure 9.11-a);
2. $a > b$, one oblique triangle (as shown in Figure 9.11-b).

Figure 9.11

Example Solve the triangle for which $a = 4$, $b = 3$, and $\beta = 45°$.

Solution We first sketch a figure and observe that this gives rise to the ambiguous case in which we are given the lengths of two sides and the measure of an angle opposite one of them. We then check for the number of possible solutions and observe that $b > a \sin \beta$ and $b < a$. Therefore, there are two triangles with the given measurements. Using Theorem 9.2, we next determine a value for $\sin \alpha$. We have

$$\frac{\sin \alpha}{4} = \frac{\sin 45°}{3},$$

$$\sin \alpha = \frac{4}{3}(0.7071) = 0.9428.$$

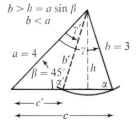

Because $\sin \alpha > 0$ in Quadrants I and II, from Table VI or by using a calculator, we have either

$$\alpha \approx 70.5° \quad \text{or} \quad \alpha' \approx 180° - 70.5° = 109.5°.$$

Solution Continued on Overleaf

If $\alpha \approx 70.5°$, then $\gamma \approx 180° - 45° - 70.5° = 64.5°$, and we therefore have

$$\frac{\sin 45°}{3} \approx \frac{\sin 64.5°}{c},$$

so that

$$c \approx 3\left(\frac{0.9026}{0.7071}\right) \approx 3.8.$$

If $\alpha' \approx 109.5°$, then $\gamma' \approx 180° - 45° - 109.5° = 25.5°$, and we have

$$\frac{\sin 45°}{3} \approx \frac{\sin 25.5°}{c'},$$

so that

$$c' \approx 3\left(\frac{0.4305}{0.7071}\right) \approx 1.8.$$

Thus, the two solutions for the triangle are

$$\alpha \approx 70.5°, \quad \gamma \approx 64.5°, \quad c \approx 3.8$$

and

$$\alpha' \approx 109.5°, \quad \gamma' \approx 25.5°, \quad c' \approx 1.8.$$

Exercise 9.4

In all exercises, give lengths to the nearest tenth and angle measures to the nearest 6' or tenth of a degree.

In Exercises 1–6, solve the triangle.

1. $b = 10, \quad \beta = 30°, \quad \alpha = 80°$
2. $a = 64, \quad \beta = 36°, \quad \gamma = 82°$
3. $c = 78.1, \quad \alpha = 58.1°, \quad \gamma = 63.2°$
4. $b = 1.02, \quad B = 41.6°, \quad C = 80.6°$
5. $a = 84.2, \quad A = 110°, \quad C = 22° 6'$
6. $c = 0.94, \quad A = 41° 18', \quad B = 96° 54'$

Determine the number of triangles that satisfy the given conditions.

Example

$b = 34, \quad c = 12, \quad C = 30°$

Solution

Sketch a figure and determine

$h = 34 \sin 30° = 34(0.5000) = 17.$

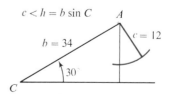

Since $12 < 17$, we have $c < b \sin C$, and no triangle exists for the given data.

9.4 Law of Sines

7. $a = 5.4$, $b = 7.0$, $B = 30°$
8. $b = 4.9$, $c = 3.2$, $C = 30°$
9. $a = 31.1$, $c = 41.3$, $C = 30°$
10. $a = 42.3$, $b = 20.7$, $B = 30°$
11. $b = 16.2$, $c = 14.3$, $C = 20°$
12. $a = 141$, $b = 182$, $B = 20°$
13. $a = 4.6$, $b = 2.3$, $B = 30°$
14. $b = 68.1$, $c = 41.3$, $C = 30°$

Solve the triangle.

15. $a = 4.8$, $c = 3.9$, $\alpha = 113°$
16. $a = 3.2$, $b = 2.6$, $\beta = 54°$
17. $b = 6.21$, $c = 4.39$, $\beta = 42° 42'$
18. $a = 179$, $b = 212$, $\beta = 114° 12'$
19. $b = 1.8$, $c = 1.5$, $\gamma = 32.5°$
20. $a = 9.4$, $b = 8.6$, $\beta = 54.4°$
21. $a = 8.4$, $b = 6.9$, $B = 62° 12'$
22. $b = 0.42$, $c = 0.21$, $B = 31° 54'$
23. $a = 13.84$, $b = 6.92$, $A = 60°$
24. $a = 4.72$, $c = 9.44$, $A = 30°$
25. $b = 420$, $c = 610$, $B = 33° 24'$
26. $a = 5.42$, $b = 6.82$, $A = 43° 30'$

27. From a window in a tower, 85 feet above the ground, the angle of elevation to the top of a nearby building measures 34° 30'. From a point on the ground directly below the window, the angle of elevation to the top of the same building measures 50° 24'. Find the height of the building.

28. A pilot is flying away from city A at a heading of 138° (0° is due north and 90° is due east). When he is 10 kilometers away from A, he reports to the weather station in city A that there is a large storm centered due east of the city and 15° north of east from his position. How far is the storm centered from the city?

29. Engineers are planning to tunnel straight through a mountain. They take three measurements. Point A is at the base of the western side of the mountain, and the angle of elevation from point A to the top of the mountain is 48°. Point B is 1 kilometer west of point A, and the angle of elevation to the top of the mountain is 25°. Point C is at the base of the mountain due east of point A, and the angle of elevation from point C to the top of the mountain is 28°. How far must they tunnel; that is, how wide is the mountain at the base?

30. A woman standing on a river bank is trying to decide if she can swim across the river. On the opposite bank there are two towers which she knows to be about 1 kilometer from one another. She approximates the angles formed by her lines of sight and the river bank to be 30° for tower A and 45° for tower B. Approximately how wide is the river?

31. If an equilateral triangle is inscribed in a circle of radius 4, what is the length of each side of the triangle?

32. A regular pentagon is inscribed in a circle of radius 25. Find the area and the circumference of the pentagon. *Hint:* Divide the pentagon into five congruent triangles, compute the area of each, and multiply by 5.

9.5 Law of Cosines

The law of sines introduced in Section 9.4 cannot be used to solve a triangle when we are given only the length of each side of the triangle, or when we are given two sides and the included angle. In this section we shall consider a relationship which will help us solve such triangles.

Let us denote the angles of a triangle by α, β, and γ and the sides opposite these angles by a, b, and c, respectively (Figure 9.12-a). Assume all three angles are acute (the case in which the triangle contains an obtuse angle is left as an exercise). Assume that the line segment through the vertex of γ and perpendicular to side c is of length h and that side c of the triangle lies along the positive x-axis of a coordinate system, as shown in Figure 9.12-b.

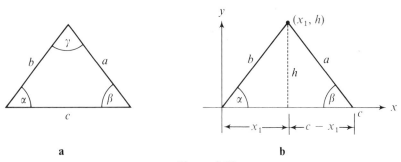

Figure 9.12

We observe from this figure that

$$x_1^2 = b^2 - h^2 \tag{1}$$

and

$$x_1 = b \cos \alpha. \tag{2}$$

Using the distance formula and Equations (1) and (2), we have

$$\begin{aligned}a^2 &= (c - x_1)^2 + h^2 \\ &= c^2 - 2cx_1 + x_1^2 + h^2 \\ &= c^2 - 2cb \cos \alpha + b^2 - h^2 + h^2 \\ &= b^2 + c^2 - 2bc \cos \alpha.\end{aligned}$$

This argument can be repeated using the other sides of the triangle to establish the following theorem, which is known as the **law of cosines.**

9.5 Law of Cosines

Theorem 9.3 *If α, β, and γ are the angles of a triangle, and a, b, and c are the lengths of the sides opposite α, β, and γ, respectively, then*

$$c^2 = a^2 + b^2 - 2ab \cos \gamma, \tag{3}$$

$$b^2 = a^2 + c^2 - 2ac \cos \beta, \tag{4}$$

$$a^2 = b^2 + c^2 - 2bc \cos \alpha. \tag{5}$$

Solving triangles

This theorem has many applications, among them the solution of certain triangles. If we are given the measure of an angle and the lengths of the adjacent sides, we can use the law of cosines to help us find the remaining parts of the triangle.

Example

Solve the triangle for which

$$\alpha = 100°, \quad b = 10, \quad \text{and} \quad c = 12.$$

Solution

Make a sketch of the triangle and label the sides and angles. We can first find a by using the law of cosines in the form

$$a^2 = b^2 + c^2 - 2bc \cos \alpha.$$

Thus,

$$a^2 = 10^2 + 12^2 - 2(10)(12) \cos 100°,$$
$$= 100 + 144 - 240(-0.1736)$$
$$= 244 + 41.68 \approx 285.7.$$

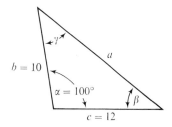

Using a calculator, or Table I, we find $\sqrt{285.7} \approx 16.9$. Next, to find β, we again use the law of cosines (we could also use the law of sines because now we know the length of a side opposite a given angle) in the form

$$b^2 = a^2 + c^2 - 2ac \cos \beta,$$

or

$$\cos \beta = \frac{b^2 - a^2 - c^2}{-2ac}.$$

With $a^2 = 285.7$, $b^2 = 100$, and $c^2 = 144$, we obtain

$$\cos \beta = \frac{100 - 285.7 - 144}{-2(16.9)(12)} \approx 0.8129.$$

Hence, $\beta \approx 35.6°$. Since the sum of the angles of a triangle is 180°, we can find an approximation for γ by noting that

$$\gamma = 180° - \alpha - \beta$$
$$\approx 180° - 100° - 35.6° = 44.4°.$$

For the remaining parts of the triangle, we accordingly have

$$a \approx 17, \quad \beta \approx 35.6° \quad \text{and} \quad \gamma \approx 44.4°.$$

Since Equations (3), (4), and (5) in Theorem 9.3 are relationships between three sides and one angle of a triangle, we can, as we did in the preceding example, use any of these forms to solve for the measure of an angle, given the lengths of the three sides.

Exercise 9.5

Find the remaining parts of the triangle. In all exercises in this set, give lengths and areas to the nearest tenth and angle measurements to the nearest 6'.

1. $a = 10$, $b = 4$, $\gamma = 30.6°$
2. $b = 14.2$, $c = 7.9$, $\alpha = 64.2°$
3. $a = 4.9$, $c = 6.8$, $\beta = 122° \, 24'$
4. $a = 241$, $c = 104$, $\beta = 148° \, 12'$
5. $b = 14.6$, $c = 6.21$, $A = 80° \, 42'$
6. $b = 9.4$, $c = 10.2$, $A = 100° \, 54'$
7. $a = 5$, $b = 8$, $c = 7$
8. $a = 4.5$, $b = 5.3$, $c = 2.8$

9. Find the greatest angle of the triangle whose sides are 5.1, 4.2, and 4.5.
10. Find the least angle of the triangle whose sides are 29.5, 33.2, and 41.4.
11. Two points A and B are on opposite sides of a pond. A surveyor at a point C determines that A is 92 feet from C, B is 128 feet from C, and the measure of the angle between CA and CB is 37°. Determine the distance between A and B.
12. A ship navigator determines his position C to be 48 miles from Port A and 80 miles from Port B. The distance between Ports A and B is 57 miles. Find the degree measure of the angle between CA and CB.
13. A professional golfer has hit his drive to the right. He estimates the angle formed by the line from the tee to the hole and the line from the tee to his ball is 30°. He paces the distance from the tee to his ball and finds that distance to be 240 yards. Assume the distance from the tee to the hole is 460 yards. How far is his ball from the hole?
14. A weather balloon is hovering between cities A and B at an altitude of 7000 meters AGL (above ground level). The angle of elevation from city B to the balloon is 38°. Assume that the distance from city A to city B is 20 kilometers. Find the distance between the balloon and city A, and the angle of elevation from city A to the balloon.
15. A hiker starts out from ranger station A on a heading of 20° east of north. He walks to a campground and stops to rest. His pedometer reads 5.75 miles. Assume ranger station B is 10 miles due north of station A. The hiker now wants to hike from his resting place to station B. What compass heading should he take?

16. The distance between two U.S. Forest Service observation platforms is 4 miles. A fire is spotted by observers on both platforms, and the angles formed by the lines joining the fire and the stations and the line joining the stations are 12° and 20°. How far is each platform from the fire?

17. Show that $a^2 = b^2 + c^2 - 2bc \cos \alpha$, where $a, b, c,$ and α are as shown in the figure.

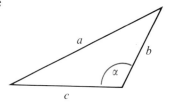

18. Use the law of cosines to show that if $a, b,$ and c are the lengths of the sides of a triangle, and α is the angle opposite side a, then

   a. $a^2 < b^2 + c^2$ if α is acute,
   b. $a^2 = b^2 + c^2$ if α is a right angle,
   c. $a^2 > b^2 + c^2$ if α is obtuse.

   Part b is the Pythagorean Theorem.

9.6 Geometric Vectors

A single real number is adequate to measure a physical quantity such as length, mass, or temperature which can be described by a magnitude only. Furthermore, the real number can be associated with a point or a line segment on the real number line. However, a complete description of a physical quantity such as velocity, acceleration, and force requires that direction as well as magnitude be specified. Such quantities can be represented by a pair of real numbers (a *vector*) and associated with a line segment in the plane pointing in an assigned direction (a *geometric vector*).

Two-dimensional geometric vectors

In this section we shall only consider the notion of a geometric vector in a plane and several applications in which these vectors can be used to set up models in applied problems.

Definition 9.3 A *two-dimensional geometric vector* is a line segment with a specified direction.

A geometric vector can be denoted by a letter such as **v**. It can also be named by using the names of the endpoints. For example, in Figure 9.13 on page 270 the vector can be denoted by **AB**, where A is the initial point and B is the terminal point.

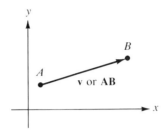

Figure 9.13

The following concepts are useful when working with vectors (see Figure 9.14).

Definition 9.4 *For each geometric vector* **v**, *the* **norm** *or* **magnitude** *of* **v** *is the length of* **v** *and is denoted by* $\|\mathbf{v}\|$.

Definition 9.5 *The direction angle of a geometric* **v** *is the angle* α, *where* $-180° < \alpha \leq 180°$, *such that the initial side of* α *is the ray from the initial point of* **v** *parallel to the x-axis and directed in the positive x-direction, and the terminal side of* α *is the ray from the initial point of* **v** *and containing* **v**.

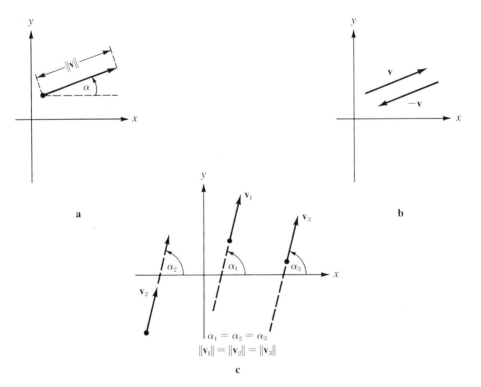

Figure 9.14

9.6 · Geometric Vectors

Definition 9.6 For each geometric vector **v**, the geometric vector $-\mathbf{v}$ has the same norm as **v** and has a direction angle whose measure differs from the direction angle of **v** by 180°.

Definition 9.7 Geometric vectors $\mathbf{v}_1$ and $\mathbf{v}_2$ are equivalent if they have the same norm and same direction angle.

The symbol $=$ is used to denote the relationship of equivalent geometric vectors. Note that we can imagine "sliding" a geometric vector onto any vector equivalent to it, without at any time altering the direction angle of the vector (see Figure 9.14-c).

Operations on geometric vectors The following two operations pair two geometric vectors to form a third geometric vector.

Definition 9.8 For all geometric vectors $\mathbf{v}_1$ and $\mathbf{v}_2$, their **sum** or **resultant** $\mathbf{v}_1 + \mathbf{v}_2$ is a geometric vector having as its initial point the initial point of $\mathbf{v}_1$ (or an equivalent geometric vector) and as its terminal point the terminal point of $\mathbf{v}_2$ (or an equivalent geometric vector), where the terminal point of $\mathbf{v}_1$ is the initial point of $\mathbf{v}_2$.

From Definitions 9.7 and 9.8, the sum $\mathbf{v}_1 + \mathbf{v}_2$ in Figure 9.15-a equals $\mathbf{v}_1 + \mathbf{v}_2'$, where $\mathbf{v}_2'$ is equivalent to $\mathbf{v}_2$. Note that a triangle is formed by $\mathbf{v}_1$, $\mathbf{v}_2$, and $\mathbf{v}_3$. Hence, we say that geometric vectors can be added according to the "the triangle law." Also note in Figure 9.15-b, that if $\mathbf{v}_1$ and $\mathbf{v}_2$ have the same initial point, the sum $\mathbf{v}_1 + \mathbf{v}_2$ is a diagonal of the parallelogram with adjacent sides $\mathbf{v}_1$ and $\mathbf{v}_2$. Hence, we also say that geometric vectors can be added according to "the parallelogram law." Figure 9.15-c illustrates $\mathbf{v}_1 + \mathbf{v}_2$ in the case where $\mathbf{v}_1$ and $\mathbf{v}_2$ are collinear and have the same direction.

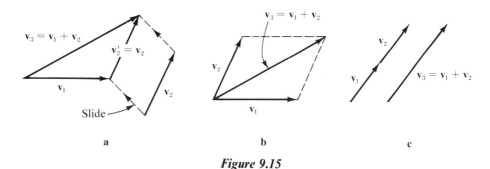

Figure 9.15

From a vector viewpoint, real numbers are called **scalars**.

Definition 9.9 For any geometric vector **v** and any scalar c, the **product of the vector by a scalar**, $c\mathbf{v}$, is a geometric vector with magnitude $\|c\mathbf{v}\| = |c| \cdot \|\mathbf{v}\|$, having the same direction angle as **v** if $c > 0$ and the same direction angle as $-\mathbf{v}$ if $c < 0$.

In Figure 9.16, $2\mathbf{v}$ and $\frac{1}{2}\mathbf{v}$ are in the same direction as $\mathbf{v}$, and $\|2\mathbf{v}\| = 2\|\mathbf{v}\|$ and $\|\frac{1}{2}\mathbf{v}\| = \frac{1}{2}\|\mathbf{v}\|$. Also, $-2\mathbf{v}$ is in the opposite direction to $\mathbf{v}$, and $\|-2\mathbf{v}\| = 2\|\mathbf{v}\|$.

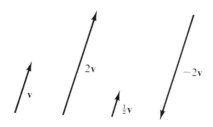

Figure 9.16

In Definition 9.9, if $c = 0$, then $0\mathbf{v}$ is the **zero geometric vector** with magnitude 0. It is considered to have any direction angle.

Applications As we noted, many physical quantities are specified by magnitude and direction. Therefore, geometric vectors are sometimes helpful in setting up models which can be solved by trigonometric methods. In doing this, we sometimes make use of *projections* of vectors on the coordinate axes, as illustrated in the following examples.

Example Find the magnitudes of the vector projections on the x- and y-axes of the vector $\mathbf{v}$ with magnitude $\|\mathbf{v}\| = 8$ and direction angle $\alpha = 52°$.

Solution Sketching the vector $\mathbf{v}$, with initial point at the origin, we want to find $\|\mathbf{v}_x\|$ and $\|\mathbf{v}_y\|$. The projections $\mathbf{v}_x$ and $\mathbf{v}_y$ are the geometric vectors that are obtained by "constructing" a perpendicular line from the terminal point of $\mathbf{v}$ to the x- and y-axis, respectively. Hence,

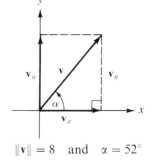

$\|\mathbf{v}\| = 8$ and $\alpha = 52°$

$$\cos 52° = \frac{\|\mathbf{v}_x\|}{8} \quad \text{and} \quad \sin 52° = \frac{\|\mathbf{v}_y\|}{8},$$

from which

$$\|\mathbf{v}_x\| = 8 \cos 52° \quad \text{and} \quad \|\mathbf{v}_y\| = 8 \sin 52°$$
$$\approx 8(0.6157) \quad\quad\quad\quad \approx 8(0.7880)$$
$$= 4.93 \quad\quad\quad\quad\quad\quad = 6.30$$

Hence, the magnitude of the x-projection is 4.9256 and the magnitude of the y-projection is 6.304.

Example Two forces of 3 and 7 pounds act on an object at an angle of 30° with respect to each other. Find the magnitude of the resultant force and the angle it forms with the 3-pound force.

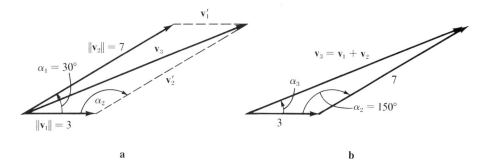

Figure 9.17

Solution   The conditions can be described by geometric vectors as suggested in Figure 9.17-a. Since v_2' is parallel to v_2, we have from geometry that $\alpha_2 = 150°$ (the interior angles on the same side of a transversal are supplementary).

Now, using the law of cosines for the triangle in Figure 9.17-b, we obtain

$$\|v_3\| = \sqrt{3^2 + 7^2 - 2(3)(7)\cos 150°}$$
$$\approx 9.715.$$

Hence, the magnitude of the resultant force is about 9.715. Then, using the law of sines,

$$\frac{7}{\sin \alpha_3} = \frac{9.715}{\sin 150°};$$

$$\sin \alpha_3 = \frac{7 \sin 150°}{9.715}$$

$$\approx \frac{7(0.5000)}{9.715}$$

$$\approx 0.3603.$$

From Table VI or a calculator, to the nearest tenth of a degree, $\alpha_3 = 21.1°$.

Example   An airplane flies 340 miles per hour in a direction 30° counter-clockwise from the east for two hours and then flies due north at the same speed for one hour. How far from the starting point is the airplane at the end of this time?

Solution   The conditions can be described by geometric vectors as suggested in Figure 9.18-a, where we wish to find $\|v_3\| = \|v_1 + v_2\|$. Since α_2 in Figure 9.18-b is 60° and since v_2 is parallel to the y-axis (due north), we have from geometry that $\alpha_3 = 120°$. Now, using the law of cosines, we obtain

$$\|v_3\| = \sqrt{(680)^2 + (340)^2 - 2(680)(340)\cos 120°}$$
$$\approx \sqrt{809{,}200}.$$

Hence, the airplane is approximately $\sqrt{809{,}200}$, or about 900 miles from its starting point.

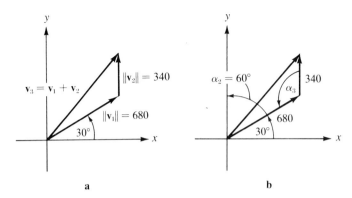

Figure 9.18

Exercise 9.6

In Exercises 1–10, use the following geometric vectors.

If $\|v_1\| = 1$ and v_1 is in the direction of the positive y-axis, estimate the magnitude and the measure of the direction angle for each geometric vector in Exercises 1–4.

1. v_1 2. v_2 3. v_3 4. v_4

Using free-hand methods, represent each of the following by a single geometric vector.

5. $v_1 + v_2$ 6. $v_2 + v_3$ 7. $2v_3$
8. $\frac{1}{2}v_4$ 9. $2v_1 + v_4$ 10. $v_2 + 2v_3$

Find the magnitudes of the vector projections on the x- and y-axis of the vector **v** with the given magnitude and direction angle.

11. $\|v\| = 3\sqrt{2}$ and $\alpha = 45°$ 12. $\|v\| = 4$ and $\alpha = 60°$
13. $\|v\| = 7$ and $\alpha = 150°$ 14. $\|v\| = 2$ and $\alpha = 90°$

In Exercises 15–18, two forces act on a point in a plane at an angle with the given measures. Find the magnitude of the resultant force and the angle this force makes with the smaller given force.

Chapter Review

15. Forces of 5 and 7 pounds, acting at an angle of 90° with respect to each other.
16. Forces of 6 and 12 pounds, acting at an angle of 45° with respect to each other.
17. Forces of 2 and 8 pounds, acting at an angle of 120° with respect to each other.
18. Forces of 3 and 9 pounds, acting at an angle of 148° with respect to each other.
19. An airplane flies 800 miles due south and then flies 600 miles 60° clockwise from north. How far from the starting point is the airplane at the end of this time?
20. An airplane is flying at 20,000 feet on a heading of 300° clockwise from north at 450 miles per hour. The wind velocity at this level is 90 miles per hour from 240° clockwise from north. Find the ground speed and the direction of the flight over the ground.
21. A weight of 160 pounds is placed on a smooth plane inclined at an angle of 30° with the horizontal (see figure). What is the minimum force that must be exerted against the weight and parallel to the plane to prevent the weight from slipping? *Hint:* Resolve the vertical force into components in the direction of the plane and perpendicular to it.

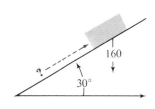

22. A force of 40 pounds is directed to the right. What vertical force must be applied so that the magnitude of the resultant force is 80 pounds? What angle does the resultant force make with the horizontal?

Chapter Review

[9.1] Find the degree measure, to the nearest tenth of a degree, of the angle whose radian measure is as given.

1. $\dfrac{3\pi^R}{8}$
2. 8.72^R

Find the radian measure, to the nearest hundredth of a radian, of the angle whose degree measure is as given.

3. $40°$
4. $610°$

On a circle with given radius, find the length of the arc intercepted by the angle whose measure is as given.

5. $r = 4.8"$; $\dfrac{\pi^R}{3}$
6. $r = 1.6'$; $150°$

Find all angles α, where $-360° \leq \alpha \leq 720°$, *that are coterminal with the angle whose measure is given.*

7. $-330°$ **8.** $518°$

[9.2] *Determine in which quadrant the terminal side of the angle α lies.*

9. $\tan \alpha > 0$ and $\cos \alpha < 0$ **10.** $\cos \alpha < 0$ and $\sin \alpha < 0$

Find the value of each expression.

11. $\cos 210°$ **12.** $\sin 330°$ **13.** $\tan 225°$
14. $\cos 120°$ **15.** $\sin(-210°)$ **16.** $\cos(-315°)$

Find approximations for the given expression. Use an appropriate table or a hand calculator.

17. $\cos 23°$ **18.** $\tan 1.24^R$ **19.** $\sin 120°$ **20.** $\cos(-135°)$
21. $\sin 192°$ **22.** $\tan 422°$ **23.** $\cos 3.68^R$ **24.** $\sin(-5.12^R)$

Use the appropriate table or a hand calculator to compute (a) the radian measure and (b) the degree measure of the indicated value.

25. Arcsin 0.5000   **26.** $\text{Cos}^{-1}\, 0.9963$

In all exercises in this review, give lengths to the nearest tenth and angle measure to the nearest 6', or tenth of a degree.

[9.3] *In Exercises 27–30, solve the right triangle ABC. Assume* $C = 90°$.

27. $a = 9$, $b = 12$ **28.** $c = 11$, $\alpha = 41°$
29. $a = 5$, $A = 12°$ **30.** $a = 5$, $c = 11$.

Find the other trigonometric function values of θ where θ is one acute angle of a right triangle.

31. $\sin \theta = 3/5$ **32.** $\tan \theta = 5/12$

33. Each of two surveyors is located 62.7 meters from a flagpole. If the angle formed by the lines of sight from the first surveyor to the second, and from the first to the flagpole, is 36°, how far apart are the two surveyors?

34. A T.V. antenna installer needs to run a guy wire from the top of an antenna 2 meters tall to the edge of a building. If the angle of depression (the angle between the line of sight and the horizontal) from the top of the antenna to the edge of the building is 28°, how much wire is needed?

Chapter Review

[9.4] Solve the triangle.

35. $b = 20$, $\beta = 25°$, $\alpha = 75°$
36. $a = 12$, $c = 15$, $\alpha = 30°$
37. $b = 11.4$, $B = 35°$, $C = 95°$
38. $c = 12.5$, $A = 100°$, $C = 20°$

39. Two men, 500 feet apart, observe a balloon between them that is in the same vertical plane with them. The respective angles of elevation of the balloon are observed by the men to measure 80° 12′ and 52° 54′. Find the height of the balloon above the ground.

40. An airplane flies straight and level between two cities. A person in city A measures the angle of elevation to the airplane as 58°. At the same time a person in city B measures the angle of elevation to the airplane as 30°. If the distance between the two cities is 10 kilometers, and if both cities are at sea level, what is the altitude of the airplane?

[9.5] Solve the triangle.

41. $a = 12$, $b = 3$, $\gamma = 25°$
42. $b = 12.8$, $c = 8.35$, $\alpha = 80° 42′$
43. $a = 8$, $b = 10$, $c = 9$
44. $a = 5$, $b = 4$, $c = 3$

45. The straight-line distance between city A and city B is approximately 1566 miles, between city A and city C is approximately 608 miles, and between city B and city C is approximately 1169 miles. Find the measures of the angles of the triangle formed by the three cities.

46. The captain of a rescue ship receives messages in the form of light signals from two ships which have collided. The captain estimates that his ship is 1800 meters from one of the damaged ships and 2100 meters from the other, and from the light signals he finds that the angle formed by the lines joining his ship with the two damaged ships is 4°. How far apart have the damaged ships drifted?

[9.6] 47. Find the magnitude of the projections on the x- and y-axis of the vector **v** with a direction angle $\alpha = 30°$ and $\|\mathbf{v}\| = 12$.

48. Find the magnitude of the projections on the x- and y-axis of the vector **v** with a direction angle $\alpha = 135°$ and $\|\mathbf{v}\| = 16.8$.

49. Two forces with magnitudes of 6 and 10 pounds act at an angle of 72° with respect to each other. Find the magnitude of the resultant force and the angle this force makes with the larger of the given forces.

50. A boat has a maximum speed of 18.2 miles per hour in still water. If the boat crossed a river at right angles to the current and has a "drift angle" of 8°, find the speed of the current.

10 Identities and Conditional Equations

In this chapter we shall consider identities and conditional equations which involve trigonometric function values. We shall use lower case Greek letters such as α, β, γ, and θ to represent angles only, and we shall use lower case English letters such as x, y, and z to represent either angles or real numbers.

10.1 Basic Identities

The fact that $\sin x$ and $\cos x$ are related by the equation

$$\sin^2 x + \cos^2 x = 1, \qquad (1)$$

and the facts that

$$\tan x = \frac{\sin x}{\cos x}, \qquad \cot x = \frac{\cos x}{\sin x},$$

$$\sec x = \frac{1}{\cos x}, \quad \text{and} \quad \csc x = \frac{1}{\sin x},$$

suggest that the trigonometric functions are related in a variety of ways. We use equations to display such relationships. For example, since $\sin^2 x + \cos^2 x = 1$, and because $\cos^2 x \neq 0$ for any x for which $\tan x$ and $\sec x$ are defined, we can multiply each member by $1/\cos^2 x$ to obtain the equation

$$\frac{\sin^2 x}{\cos^2 x} + \frac{\cos^2 x}{\cos^2 x} = \frac{1}{\cos^2 x},$$

from which we get

10.1 Basic Identities

$1 + \tan^2 x$

$$\tan^2 x + 1 = \sec^2 x. \tag{2}$$

We call the above relationships **identities** because they are true for all values of x for which all of the function values involved are defined. Hereafter, in the absence of specific statements of restrictions on variables in identities, we shall assume that the replacement sets of the variables contain no values for which the expressions are undefined.

$1 + \cot^2 x$

We can obtain another useful identity by starting with Equation (1). Since $\sin^2 x \neq 0$ for any x for which $\cot x$ and $\csc x$ are defined, we can multiply each member of Equation (1) by $1/\sin^2 x$ to obtain the equation

$$\frac{\sin^2 x}{\sin^2 x} + \frac{\cos^2 x}{\sin^2 x} = \frac{1}{\sin^2 x},$$

from which we get

$$1 + \cot^2 x = \csc^2 x. \tag{3}$$

$\sin(-x)$
$\cos(-x)$
$\tan(-x)$

The relationships

$$\sin(-x) = -\sin x, \tag{4}$$

$$\cos(-x) = \cos x, \tag{5}$$

and

$$\tan(-x) = -\tan x \tag{6}$$

are also important identities. To see why they are valid, note that for the arcs measuring x and $-x$ in Figure 10.1 triangles AOP and BOP are congruent.

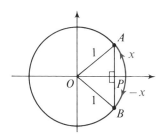

Figure 10.1

Thus, $OA = OB$ and $AP = BP$. Since the ordinate of the point B is negative, we have

$$\sin(-x) = -\frac{BP}{OB} = -\frac{AP}{OA}$$

$$= -\sin x$$

and

$$\cos(-x) = \frac{OP}{OB} = \frac{OP}{OA}$$
$$= \cos x,$$

from which we obtain

$$\tan(-x) = \frac{\sin(-x)}{\cos(-x)} = \frac{-\sin x}{\cos x}$$
$$= -\tan x.$$

The geometric arguments to demonstrate the validity of identities (4), (5), and (6) for arcs terminating in quadrants other than the first quadrant are similar.

Methods to prove identities

There is no general method for proving identities. The following suggestions may help you to determine a feasible method to use:

1. It is often helpful if the more complicated member of an equation is simplified so that it is the same as the simpler member.
2. It is sometimes helpful first to change all terms of an equation to expressions involving only sines and/or cosines, and then to simplify the members.
3. Always look for a possible application of an algebraic procedure such as factoring or simplifying fractions.

Example

Verify that $\dfrac{\sin \alpha}{1 - \cos \alpha} - \cot \alpha = \dfrac{1}{\sin \alpha}$ is an identity.

Solution

In this case, it is helpful to rewrite the equation in terms of $\sin \alpha$ and $\cos \alpha$ only. Writing $\cos \alpha / \sin \alpha$ for $\cot \alpha$, we have

$$\frac{\sin \alpha}{1 - \cos \alpha} - \frac{\cos \alpha}{\sin \alpha} = \frac{1}{\sin \alpha}.$$

Writing the left-hand member as a single fraction, we obtain

$$\frac{\sin^2 \alpha - (1 - \cos \alpha) \cos \alpha}{(1 - \cos \alpha) \sin \alpha} = \frac{1}{\sin \alpha},$$

from which

$$\frac{\sin^2 \alpha - \cos \alpha + \cos^2 \alpha}{(1 - \cos \alpha) \sin \alpha} = \frac{1}{\sin \alpha},$$

$$\frac{(\sin^2 \alpha + \cos^2 \alpha) - \cos \alpha}{(1 - \cos \alpha) \sin \alpha} = \frac{1}{\sin \alpha},$$

$$\frac{1 - \cos \alpha}{(1 - \cos \alpha) \sin \alpha} = \frac{1}{\sin \alpha}.$$

Since $1 - \cos \alpha$ is restricted from 0 in the original equation, we can rewrite the left-hand member here and arrive at the equivalent equation

10.1 Basic Identities

$$\frac{1}{\sin \alpha} = \frac{1}{\sin \alpha},$$

which is clearly an identity.

Example Verify that $\dfrac{\sec \alpha}{\sin \alpha} = \tan \alpha + \cot \alpha$ is an identity.

Solution It is again helpful to rewrite the equation in terms of $\sin \alpha$ and $\cos \alpha$ only. Writing $\sin \alpha/\cos \alpha$, $\cos \alpha/\sin \alpha$, and $1/\cos \alpha$ for $\tan \alpha$, $\cot \alpha$, and $\sec \alpha$, respectively, we have

$$\frac{1/\cos \alpha}{\sin \alpha} = \frac{\sin \alpha}{\cos \alpha} + \frac{\cos \alpha}{\sin \alpha}.$$

Writing each member of this equation as a single fraction, we get

$$\frac{1}{\cos \alpha \sin \alpha} = \frac{\sin^2 \alpha + \cos^2 \alpha}{\cos \alpha \sin \alpha}.$$

Hence, we can rewrite the right-hand member to obtain

$$\frac{1}{\cos \alpha \sin \alpha} = \frac{1}{\cos \alpha \sin \alpha},$$

which is clearly an identity.

Example Verify that $\dfrac{1 - \sin x}{\cos x} = \dfrac{\cos x}{1 + \sin x}$ is an identity.

Solution In this case, we shall use algebraic procedures to show that the left-hand member is equivalent to the right-hand member.

Since $\cos x \neq 0$, we can multiply the left-hand member by $\cos x/\cos x$ to obtain

$$\frac{\cos x(1 - \sin x)}{\cos^2 x} = \frac{\cos x}{1 + \sin x}. \tag{7}$$

The equation $\sin^2 x + \cos^2 x = 1$ can be written equivalently as

$$\cos^2 x = 1 - \sin^2 x = (1 - \sin x)(1 + \sin x).$$

Thus, we can rewrite the left-hand member of (7) to obtain

$$\frac{\cos x(1 - \sin x)}{(1 + \sin x)(1 - \sin x)} = \frac{\cos x}{1 + \sin x},$$

from which

$$\frac{\cos x}{1 + \sin x} = \frac{\cos x}{1 + \sin x},$$

which is clearly an identity.

Proving equations are not identities

If an equation is not an identity, this fact can be established by simply finding a counterexample. For example, substituting $\pi/6$ for x in the equation

$$\sin x \cdot \cos x = \tan x \qquad (8)$$

yields

$$\frac{1}{2} \cdot \frac{\sqrt{3}}{2} = \frac{1}{\sqrt{3}},$$

which is a false statement. Thus, by finding a counterexample, we have shown that Equation (8) is not an identity.

In using a counterexample, you must, of course, select a value for the variable for which each member of the equation is defined.

Exercise 10.1

Transform the first expression so that it is identical to the second expression. Assume suitable restrictions on the variable.

1. $\cos x \tan x$; $\sin x$
2. $\sin x \sec x$; $\tan x$
3. $\sin^2 x \cot^2 x$; $\cos^2 x$
4. $\dfrac{\cos^2 x}{\cot^2 x}$; $\sin^2 x$
5. $\cos^2 \alpha (1 + \tan^2 \alpha)$; 1
6. $(\csc^2 \alpha - 1)$; $\cot^2 \alpha$
7. $\sec \alpha \csc \alpha - \cot \alpha$; $\tan \alpha$
8. $(1 - \cos^2 \alpha)(1 + \cot^2 \alpha)$; 1
9. $\dfrac{\sin \theta \sec \theta}{\tan \theta}$; 1
10. $\cos \theta \tan \theta \csc \theta$; 1
11. $(\sec^2 \theta - 1)(\csc^2 \theta - 1)$; 1
12. $\dfrac{\cos \theta - \sin \theta}{\cos \theta}$; $1 - \tan \theta$
13. $\dfrac{1}{1 + \sin \theta} + \dfrac{1}{1 - \sin \theta}$; $2 \sec^2 \theta$
14. $\dfrac{1 + \tan^2 \theta}{\csc^2 \theta}$; $\tan^2 \theta$

Verify that the equation is an identity.

15. $\sin x \cot x = \cos x$
16. $\tan x \csc x = \sec x$
17. $\dfrac{\sin \theta}{\tan \theta} = \cos(-\theta)$
18. $\dfrac{\sin(-\theta) \sec \theta}{\tan(-\theta)} = 1$
19. $\tan^2 x \cos^2 x = \sin^2 x$
20. $(\csc^2 x - 1)(\sin^2 x) = \cos^2 x$
21. $\dfrac{(1 + \sin \theta)(1 - \sin \theta)}{\cos \theta} = \cos \theta$
22. $\dfrac{(1 + \cos \theta)(1 - \cos \theta)}{\sin \theta} = \sin \theta$
23. $\sec x - \cos x = \sin x \tan x$
24. $\sec x - \sin x \tan x = \cos x$
25. $\dfrac{1 + \tan^2 x}{\tan^2 x} = \csc^2 x$
26. $\dfrac{\sin^2 x}{1 - \cos x} = \dfrac{1 + \sec x}{\sec x}$

10.2 Special Formulas for the Cosine Function

27. $\tan^2 \alpha - \sin^2 \alpha = \sin^2 \alpha \tan^2 \alpha$

28. $\cot^2 \alpha + \sec^2 \alpha = \tan^2 \alpha + \csc^2 \alpha$

29. $\tan \alpha + \sec \alpha = \dfrac{1}{\sec \alpha - \tan \alpha}$

30. $\dfrac{\sec \alpha + \csc \alpha}{1 + \tan \alpha} = \csc \alpha$

31. $\dfrac{\cos x + \sin x}{\sin x} = 1 + \cot x$

32. $\dfrac{1 - \tan^2 x}{\tan x} = \cot x - \tan x$

33. $\dfrac{1}{1 + \sin x} + \dfrac{1}{1 - \sin x} = 2 \sec^2 x$

34. $\tan x + \cot x = \sec x \csc x$

35. $\dfrac{1 + \sec \theta}{\sec \theta} = \dfrac{\sin^2 \theta}{1 - \cos \theta}$

36. $\dfrac{1 - \cos \alpha}{\sin \alpha} = \dfrac{\sin \alpha}{1 + \cos \alpha}$

37. $\dfrac{1 + \sin \alpha}{\cos \alpha} = \dfrac{\cos \alpha}{1 - \sin \alpha}$

38. $\dfrac{1}{\sec x - \tan x} = \sec x + \tan x$

39. $\dfrac{1 - \cos \alpha}{1 + \cos \alpha} = \dfrac{\sec \alpha - 1}{\sec \alpha + 1}$

40. $\dfrac{\cos x}{\sec x - \tan x} = \dfrac{\cos^2 x}{1 - \sin x}$

41. $\tan^4 \theta + \tan^2 \theta = \sec^4 \theta - \sec^2 \theta$

42. $\cos^4 \theta - \sin^4 \theta = \cos^2 \theta - \sin^2 \theta$

Use a counterexample to show that the equation is not an identity.

43. $\tan^2 \theta + \sec^2 \theta = 1$

44. $\sin^2 \theta - \cos^2 \theta = 1$

45. $\sin \theta + \tan \theta = \cos \theta$

46. $\tan^2 \theta = 2 \tan \theta$

Verify that the equation is an identity.

47. $(a \sin \alpha + b \cos \alpha)^2 + (a \cos \alpha - b \sin \alpha)^2 = a^2 + b^2$

48. $\dfrac{\cot \alpha - 1}{\cot \alpha + 1} = \dfrac{1 - \tan \alpha}{1 + \tan \alpha}$

49. $\left(\dfrac{\sec \theta - 1}{\sec \theta + 1}\right) \sin^2 \theta + \sin^2 \theta - 2 = -2 \cos \theta$

50. $\dfrac{\tan^2 \theta + 2 \tan \theta + 1}{\cot^2 \theta + 2 \cot \theta + 1} + 1 = \dfrac{1}{1 - \sin^2 \theta}$

10.2 Special Formulas for the Cosine Function

In this section we shall derive several important formulas involving the cosine which are used in calculus and other advanced mathematics courses.

$\cos(x_1 + x_2)$

Figure 10.2 shows selected points on the unit circle with their coordinates. Since the length of arc from $(1, 0)$ to $(\cos(x_1 + x_2), \sin(x_1 + x_2))$ is $x_1 + x_2$, and from $(\cos x_1, -\sin x_1)$ to $(\cos x_2, \sin x_2)$ the arc length is also $x_1 + x_2$, the respective chords have equal lengths. Using the distance formula to express this fact, we have

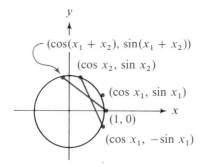

Figure 10.2

$$\sqrt{[\cos(x_1 + x_2) - 1]^2 + [\sin(x_1 + x_2) - 0]^2}$$
$$= \sqrt{(\cos x_2 - \cos x_1)^2 + [\sin x_2 - (-\sin x_1)]^2},$$

or, squaring both members,

$$[\cos(x_1 + x_2) - 1]^2 + [\sin(x_1 + x_2)]^2$$
$$= (\cos x_2 - \cos x_1)^2 + (\sin x_2 + \sin x_1)^2.$$

If we now perform the operations indicated, we have

$$\cos^2(x_1 + x_2) - 2\cos(x_1 + x_2) + 1 + \sin^2(x_1 + x_2)$$
$$= \cos^2 x_2 - 2\cos x_1 \cos x_2 + \cos^2 x_1 + \sin^2 x_2 + 2\sin x_1 \sin x_2 + \sin^2 x_1.$$

Regrouping terms, we obtain

$$[\cos^2(x_1 + x_2) + \sin^2(x_1 + x_2)] - 2\cos(x_1 + x_2) + 1$$
$$= (\cos^2 x_2 + \sin^2 x_2) + (\cos^2 x_1 + \sin^2 x_1) - 2\cos x_1 \cos x_2 + 2\sin x_1 \sin x_2,$$

and since $\cos^2 x + \sin^2 x = 1$, it follows that

$$1 - 2\cos(x_1 + x_2) + 1 = 1 + 1 - 2\cos x_1 \cos x_2 + 2\sin x_1 \sin x_2.$$

This simplifies to the identity

$$\mathbf{\cos(x_1 + x_2) = \cos x_1 \cos x_2 - \sin x_1 \sin x_2.} \tag{1}$$

Example Find an exact expression for $\cos 105°$.

Solution Write $105°$ as $60° + 45°$. By Equation (1),

$$\cos 105° = \cos(60° + 45°) = \cos 60° \cos 45° - \sin 60° \sin 45°$$
$$= \left(\frac{1}{2}\right)\left(\frac{\sqrt{2}}{2}\right) - \left(\frac{\sqrt{3}}{2}\right)\left(\frac{\sqrt{2}}{2}\right)$$
$$= \frac{\sqrt{2}(1 - \sqrt{3})}{4}.$$

10.2 Special Formulas for the Cosine Function

$\cos(x_1 - x_2)$ If we replace x_2 in Equation (1) by $-x_2$, we obtain

$$\cos(x_1 - x_2) = \cos x_1 \cos(-x_2) - \sin x_1 \sin(-x_2),$$

and since $\cos(-x_2) = \cos x_2$ and $\sin(-x_2) = -\sin x_2$, we have the identity

$$\cos(x_1 - x_2) = \cos x_1 \cos x_2 + \sin x_1 \sin x_2. \tag{2}$$

Equations (1) and (2) are called the **sum formula** and **difference formula**, respectively, for the cosine function.

Example   Find an exact expression for $\cos 15°$.

Solution   Write $15°$ as $60° - 45°$. By Equation (2),

$$\cos 15° = \cos(60° - 45°)$$
$$= \cos 60° \cos 45° + \sin 60° \sin 45°$$
$$= \left(\frac{1}{2}\right)\left(\frac{\sqrt{2}}{2}\right) + \left(\frac{\sqrt{3}}{2}\right)\left(\frac{\sqrt{2}}{2}\right)$$
$$= \frac{\sqrt{2}(1+\sqrt{3})}{4}.$$

We can make use of Equations (1) and (2) to obtain formulas for $\cos 2x$ and $\cos x/2$. These formulas are known respectively as the **double-angle** and **half-angle formulas**, although they are equally valid when x is a real number.

$\cos 2x$ By letting $x_1 = x_2 = x$ in Equation (1) we get

$$\cos(x + x) = \cos x \cos x - \sin x \sin x,$$

or

$$\cos 2x = \cos^2 x - \sin^2 x. \tag{3}$$

Alternative forms of the equation for $\cos 2x$ can be obtained by using the substitutions

$$\cos^2 x = 1 - \sin^2 x \quad \text{and} \quad \sin^2 x = 1 - \cos^2 x$$

in the right-hand member of (3). The results are the formulas:

$$\cos 2x = 1 - 2\sin^2 x, \tag{4}$$
$$\cos 2x = 2\cos^2 x - 1. \tag{5}$$

Example   Given that $\cos x = -3/5$, find $\cos 2x$.

Solution   Using Equation (5), we have

$$\cos 2x = 2\left(-\frac{3}{5}\right)^2 - 1 = 2\left(\frac{9}{25}\right) - 1 = \frac{18}{25} - 1 = -\frac{7}{25}.$$

$\cos\left(\dfrac{x}{2}\right)$ By replacing x with $x/2$ in Equation (5), we get

$$\cos 2\left(\dfrac{x}{2}\right) = 2\cos^2 \dfrac{x}{2} - 1,$$

or

$$\cos x = 2\cos^2 \dfrac{x}{2} - 1.$$

Solving for $\cos x/2$, we obtain

$$\cos \dfrac{x}{2} = \pm\sqrt{\dfrac{1 + \cos x}{2}}.$$

Now, because $\cos(x/2) > 0$ for $x/2$ terminating in Quadrants I and IV, we take the positive square root for $x/2$ terminating in these quadrants. Because $\cos(x/2) < 0$ for $x/2$ terminating in Quadrants II and III, we take the negative square root in these quadrants. Thus, we have the identity

$$\cos \dfrac{x}{2} = \begin{cases} \sqrt{\dfrac{1 + \cos x}{2}} & \text{when } \dfrac{x}{2} \text{ terminates in Quadrant I or IV;} \quad (6a) \\ -\sqrt{\dfrac{1 + \cos x}{2}} & \text{when } \dfrac{x}{2} \text{ terminates in Quadrant II or III.} \quad (6b) \end{cases}$$

Example Given that $\cos x = 3/7$ and $0 \le x/2 \le \pi/2$, find $\cos x/2$.

Solution Using Equation (6a), we have

$$\cos \dfrac{x}{2} = \sqrt{\dfrac{1 + \cos x}{2}} = \sqrt{\dfrac{1 + \frac{3}{7}}{2}} = \sqrt{\dfrac{5}{7}}.$$

Exercise 10.2

Find an exact value for each expression.

1. $\cos 195°$
2. $\cos 345°$
3. $\cos 255°$
4. $\cos 165°$
5. $\cos 75°$
6. $\cos 22° 30'$

Example Given that $\cos x = -4/5$, find $\cos 2x$.

Solution By Equation (5), we have

$$\cos 2x = 2\cos^2 x - 1 = 2\left(-\dfrac{4}{5}\right)^2 - 1$$

$$= \dfrac{32}{25} - 1 = \dfrac{7}{25}.$$

10.3 Special Formulas for the Sine Function

7. Given that $\cos x = -3/5$, find:

    a. $\cos 2x$, $\dfrac{\pi}{2} \leq x \leq \pi$ **b.** $\cos \dfrac{x}{2}$, $\dfrac{\pi}{2} \leq x \leq \pi$

8. Given that $\cos x = 3/4$, find:

    a. $\cos 2x$, $\dfrac{3\pi}{2} \leq x \leq 2\pi$ **b.** $\cos \dfrac{x}{2}$, $\dfrac{3\pi}{2} \leq x \leq \pi$

Prove the identity.

9. $\cos(\pi + x) = -\cos x$ **10.** $\cos(\pi - x) = -\cos x$

11. $\cos\left(\dfrac{\pi}{2} + x\right) = -\sin x$ **12.** $\cos\left(\dfrac{\pi}{2} - x\right) = \sin x$

13. $\cos\left(\dfrac{3\pi}{2} + x\right) = \sin x$ **14.** $\cos\left(\dfrac{3\pi}{2} - x\right) = -\sin x$

15. $\sin^2 x = \dfrac{1 - \cos 2x}{2}$ **16.** $\cos^2 x = \dfrac{1 + \cos 2x}{2}$

17. $\cos 2\theta = \dfrac{1 - \tan^2 \theta}{1 + \tan^2 \theta}$ **18.** $\dfrac{2}{1 + \cos \theta} = \sec^2 \theta$

19. $\cos 3x = 4\cos^3 x - 3\cos x$, $0 \leq x \leq \dfrac{\pi}{2}$ *Hint:* $\cos 3x = \cos(2x + x)$

20. $\cos 4x = 1 - 8\sin^2 x + 8\sin^4 x$ *Hint:* $\cos 4x = \cos 2(2x)$

21. Show that for any real numbers x, and $h \neq 0$,

$$\dfrac{\cos(x + h) - \cos x}{h} = \cos x \left(\dfrac{\cos h - 1}{h}\right) - \sin x \left(\dfrac{\sin h}{h}\right).$$

10.3   Special Formulas for the Sine Function

In Section 10.2 we derived the sum, difference, double-angle, and half-angle formulas for the cosine function. In this section we obtain similar formulas for the sine function.

$\sin(x_1 + x_2)$ Note that from Equation (1) of Section 10.2 (page 284),

$$\cos\left(\dfrac{\pi}{2} - x\right) = \cos\dfrac{\pi}{2} \cos x + \sin\dfrac{\pi}{2} \sin x$$
$$= 0 \cdot \cos x + 1 \cdot \sin x,$$

from which

$$\cos\left(\dfrac{\pi}{2} - x\right) = \sin x. \qquad (1)$$

If we now set $x = x_1 + x_2$, we obtain

$$\cos\left(\frac{\pi}{2} - (x_1 + x_2)\right) = \sin(x_1 + x_2).$$

This can be rewritten as

$$\cos\left[\left(\frac{\pi}{2} - x_1\right) - x_2\right] = \sin(x_1 + x_2),$$

and if the left-hand member is expanded by means of Equation (2) of Section 10.2 (page 285), we find that

$$\cos\left(\frac{\pi}{2} - x_1\right)\cos x_2 + \sin\left(\frac{\pi}{2} - x_1\right)\sin x_2 = \sin(x_1 + x_2). \qquad (2)$$

From Equation (1) above with $x = \pi/2 - x_1$, we have

$$\sin\left(\frac{\pi}{2} - x_1\right) = \cos\left[\frac{\pi}{2} - \left(\frac{\pi}{2} - x_1\right)\right]$$
$$= \cos x_1.$$

Thus, we obtain

$$\cos\left(\frac{\pi}{2} - x_1\right)\cos x_2 + \sin\left(\frac{\pi}{2} - x_1\right)\sin x_2 = \sin x_1 \cos x_2 + \cos x_1 \sin x_2,$$

which in conjunction with Equation (2) above establishes the identity

$$\sin(x_1 + x_2) = \sin x_1 \cos x_2 + \cos x_1 \sin x_2. \qquad (3)$$

$\sin(x_1 + x_2)$ By replacing x_2 in Equation (3) with $-x_2$, we have

$$\sin(x_1 - x_2) = \sin x_1 \cos(-x_2) + \cos x_1 \sin(-x_2).$$

Substituting $\cos x_2$ for $\cos(-x_2)$ and $-\sin x_2$ for $\sin(-x_2)$ in the right-hand member, we obtain the identity

$$\sin(x_1 - x_2) = \sin x_1 \cos x_2 - \cos x_1 \sin x_2. \qquad (4)$$

Relationships (3) and (4), respectively, are called the **sum formula** and the **difference formula** for the sine function.

Example   Find an exact value for $\sin 75°$.

Solution   Write $75°$ as $30° + 45°$. Then by Equation (3),

$$\sin 75° = \sin(30° + 45°)$$
$$= \sin 30° \cos 45° + \cos 30° \sin 45°$$

10.3 Special Formulas for the Sine Function

$$= \left(\frac{1}{2}\right)\left(\frac{\sqrt{2}}{2}\right) + \left(\frac{\sqrt{3}}{2}\right)\left(\frac{\sqrt{2}}{2}\right)$$

$$= \frac{\sqrt{2}(1 + \sqrt{3})}{4}.$$

sin 2x The double-angle formula for the sine function can be obtained by letting $x_1 = x_2 = x$ in Equation (3):

$$\sin(x + x) = \sin x \cos x + \cos x \sin x,$$

or

$$\sin 2x = 2 \sin x \cos x. \tag{5}$$

sin $\frac{x}{2}$ The half-angle formula for the sine function can be obtained from the identity

$$\cos 2x = 1 - 2 \sin^2 x.$$

By replacing x with $x/2$, we have

$$\cos 2\left(\frac{x}{2}\right) = 1 - 2 \sin^2 \frac{x}{2},$$

from which

$$\cos x = 1 - 2 \sin^2 \frac{x}{2},$$

$$\sin^2 \frac{x}{2} = \frac{1 - \cos x}{2},$$

and

$$\sin \frac{x}{2} = \pm\sqrt{\frac{1 - \cos x}{2}}.$$

Now, because $\sin(x/2) > 0$ for $x/2$ terminating in Quadrants I and II, we take the positive square root for $x/2$ terminating in these quadrants, and because $\sin(x/2) < 0$ for $x/2$ terminating in Quadrants III and IV, we take the negative square root for $x/2$ terminating in these quadrants. Thus, we have the identity

$$\sin \frac{x}{2} = \begin{cases} \sqrt{\dfrac{1 - \cos x}{2}} & \text{if } \dfrac{x}{2} \text{ terminates in Quadrant I or II;} \quad (6a) \\ -\sqrt{\dfrac{1 - \cos x}{2}} & \text{if } \dfrac{x}{2} \text{ terminates in Quadrant III or IV.} \quad (6b) \end{cases}$$

Example Given that $\cos x = \dfrac{3}{7}$ and $0 \leq \dfrac{x}{2} \leq \dfrac{\pi}{2}$, find $\sin(x/2)$.

Solution on Overleaf

Solution Using Equation (6a), we have

$$\sin \frac{x}{2} = \sqrt{\frac{1 - \cos x}{2}} = \sqrt{\frac{1 - 3/7}{2}} = \sqrt{\frac{2}{7}}.$$

Exercise 10.3

Find an exact value for each expression.

1. $\sin 15°$
2. $\sin 105°$
3. $\sin 255°$
4. $\sin 165°$
5. $\sin 195°$
6. $\sin 22° 30'$

Example Given that $\sin x = 4/5$, find $\sin 2x$, $\frac{\pi}{2} \le x \le \pi$.

Solution Since $\sin x = 4/5$ and x terminates in Quadrant II,

$$\cos x = -\sqrt{1 - \sin^2 x} = -\sqrt{1 - \left(\frac{4}{5}\right)^2} = -\sqrt{\frac{9}{25}} = -\frac{3}{5}.$$

From Equation (5) on page 289, we have

$$\sin 2x = 2 \sin x \cos x = 2\left(\frac{4}{5}\right)\left(-\frac{3}{5}\right) = -\frac{24}{25}.$$

7. Given that $\cos x = -3/5$, find:

   a. $\sin \frac{x}{2}$, $3\pi \le x \le \frac{7\pi}{2}$

   b. $\sin 2x$, $\pi \le x \le \frac{3\pi}{2}$

8. Given that $\sin x = \frac{\sqrt{5}}{3}$, find:

   a. $\sin \frac{x}{2}$, $0 \le x \le \frac{\pi}{2}$

   b. $\sin 2x$, $\frac{\pi}{2} \le x \le \pi$

Prove the identity.

9. $\sin (\pi + x) = -\sin x$
10. $\sin (\pi - x) = \sin x$
11. $\sin \left(\frac{\pi}{2} + x\right) = \cos x$
12. $\sin \left(\frac{\pi}{2} - x\right) = \cos x$
13. $\sin \left(\frac{3\pi}{2} + x\right) = -\cos x$
14. $\sin \left(\frac{3\pi}{2} - x\right) = -\cos x$
15. $\sin 2\alpha = \frac{2 \tan \alpha}{1 + \tan^2 \alpha}$
16. $\frac{2}{\sin 2\alpha} = \tan \alpha + \cot \alpha$
17. $\frac{1 + \cos 2x}{\sin 2x} = \cot x$
18. $\sin 2x \sec x = 2 \sin x$

10.4 Special Formulas for the Tangent Function

19. $\sin 3x = 3 \sin x - 4 \sin^3 x$ *Hint:* $\sin 3x = \sin(2x + x)$.
20. $\sin 4x = 4 \sin x \cos x - 8 \sin^3 x \cos x$ *Hint:* $\sin 4x = \sin 2(2x)$.
21. Show that for any real numbers x and $h \neq 0$,
$$\frac{\sin(x+h) - \sin x}{h} = \sin x \left(\frac{\cos h - 1}{h}\right) + \cos x \left(\frac{\sin h}{h}\right).$$

10.4 Special Formulas for the Tangent Function

In this section we derive the sum, difference, double-angle, and half-angle formulas for the tangent function.

$\tan(x_1 + x_2)$ By Definition 8.4,
$$\tan(x_1 + x_2) = \frac{\sin(x_1 + x_2)}{\cos(x_1 + x_2)}.$$

So by Equation (1) of Section 10.2 and Equation (3) of Section 10.3, we have
$$\tan(x_1 + x_2) = \frac{\sin x_1 \cos x_2 + \cos x_1 \sin x_2}{\cos x_1 \cos x_2 - \sin x_1 \sin x_2}.$$

Dividing numerator and denominator of the right-hand member by $\cos x_1 \cos x_2$, we get
$$\tan(x_1 + x_2) = \frac{\dfrac{\sin x_1 \cos x_2}{\cos x_1 \cos x_2} + \dfrac{\cos x_1 \sin x_2}{\cos x_1 \cos x_2}}{\dfrac{\cos x_1 \cos x_2}{\cos x_1 \cos x_2} - \dfrac{\sin x_1 \sin x_2}{\cos x_1 \cos x_2}}.$$

After simplifying each term in the right-hand member,
$$\tan(x_1 + x_2) = \frac{\tan x_1 + \tan x_2}{1 - \tan x_1 \tan x_2}. \tag{1}$$

$\tan(x_1 - x_2)$ By first replacing x_2 in Equation (1) with $-x_2$, we have
$$\tan(x_1 - x_2) = \frac{\tan x_1 + \tan(-x_2)}{1 - \tan x_1 \tan(-x_2)}.$$

Substituting $-\tan x_2$ for $\tan(-x_2)$ in the right-hand member, we obtain
$$\tan(x_1 - x_2) = \frac{\tan x_1 - \tan x_2}{1 + \tan x_1 \tan x_2}. \tag{2}$$

10 Identities and Conditional Equations

Example   Find an exact value for tan 15°.

Solution   Write 15° as 60° − 45°. Then by Equation (2) we have

$$\tan 15° = \tan(60° - 45°)$$
$$= \frac{\tan 60° - \tan 45°}{1 + \tan 60° \tan 45°}$$
$$= \frac{\sqrt{3} - 1}{1 + \sqrt{3}}.$$

tan 2x   By replacing both x_1 and x_2 with x in Equation (1), we get

$$\tan(x + x) = \frac{\tan x + \tan x}{1 - \tan x \tan x},$$

from which we obtain the identity

$$\tan 2x = \frac{2 \tan x}{1 - \tan^2 x}. \tag{3}$$

$\tan \frac{x}{2}$ By Definition 8.4, we have

$$\tan \frac{x}{2} = \frac{\sin \frac{x}{2}}{\cos \frac{x}{2}}.$$

Then, by multiplying numerator and denominator by $2 \sin(x/2)$, we have

$$\tan \frac{x}{2} = \frac{2 \sin^2 \frac{x}{2}}{2 \sin \frac{x}{2} \cos \frac{x}{2}}.$$

Substituting $1 - \cos x$ for $2 \sin^2(x/2)$ and $\sin x$ for $2 \sin(x/2) \cos(x/2)$, we obtain

$$\tan \frac{x}{2} = \frac{1 - \cos x}{\sin x}. \tag{4}$$

Example   Given that $\cos x = -\frac{3}{5}, \frac{\pi}{2} < x < \pi$, find

     a. $\tan 2x$ b. $\tan \frac{x}{2}$

Solution   Since $\cos x = -\frac{3}{5}$ and x terminates in Quadrant II,

10.4 Special Formulas for the Tangent Function

$$\sin x = \sqrt{1 - \cos^2 x} = \sqrt{1 - \left(-\frac{3}{5}\right)^2} = \sqrt{\frac{16}{25}} = \frac{4}{5}.$$

Hence,

$$\tan x = \frac{\sin x}{\cos x} = \frac{4/5}{-3/5} = -\frac{4}{3}.$$

a. From Equation (3),

$$\tan 2x = \frac{2 \tan x}{1 - \tan^2 x} = \frac{2(-4/3)}{1 - (-4/3)^2}$$

$$= \frac{-8/3}{1 - 16/9} = \frac{-8/3}{-7/9} = \frac{24}{7}.$$

b. From Equation (4),

$$\tan \frac{x}{2} = \frac{1 - \cos x}{\sin x} = \frac{1 - (-3/5)}{4/5}$$

$$= \frac{8/5}{4/5} = 2.$$

Relationships similar to those developed in this and the preceding sections can be derived for cotangent, secant, and cosecant. For practical purposes, however, the formulas for cosine, sine, and tangent are the only ones necessary in view of the fact that

$$\sec x = \frac{1}{\cos x}, \quad \csc x = \frac{1}{\sin x}, \quad \text{and} \quad \cot x = \frac{1}{\tan x}.$$

For convenience, let us make a single list of some of the more important trigonometric identities. While no restrictions are given for the variables here, it is necessary to keep such restrictions in mind when using any identity. Furthermore, the variable x can be viewed as representing either an element in the set of real numbers or an element in the set of all angles.

Summary of Identities

1. $\sin^2 x + \cos^2 x = 1$
2. $\cos(-x) = \cos x$
3. $\sin(-x) = -\sin x$
4. $\cos(x_1 + x_2) = \cos x_1 \cos x_2 - \sin x_1 \sin x_2$
5. $\cos(x_1 - x_2) = \cos x_1 \cos x_2 + \sin x_1 \sin x_2$
6. $\sin(x_1 + x_2) = \sin x_1 \cos x_2 + \cos x_1 \sin x_2$
7. $\sin(x_1 - x_2) = \sin x_1 \cos x_2 - \cos x_1 \sin x_2$

8. a. $\cos 2x = \cos^2 x - \sin^2 x$
   b. $\cos 2x = 2\cos^2 x - 1$
   c. $\cos 2x = 1 - 2\sin^2 x$

9. $\sin 2x = 2 \sin x \cos x$

10. $\cos \dfrac{x}{2} = \pm \sqrt{\dfrac{1 + \cos x}{2}}$

11. $\sin \dfrac{x}{2} = \pm \sqrt{\dfrac{1 - \cos x}{2}}$

12. $\tan x = \dfrac{\sin x}{\cos x}$

13. $\tan(-x) = -\tan x$

14. $\tan^2 x + 1 = \sec^2 x$

15. $\cot^2 x + 1 = \csc^2 x$

16. $\sec x = \dfrac{1}{\cos x}$

17. $\csc x = \dfrac{1}{\sin x}$

18. $\cot x = \dfrac{\cos x}{\sin x}$

19. $\cot x = \dfrac{1}{\tan x}$

20. $\tan(x_1 + x_2) = \dfrac{\tan x_1 + \tan x_2}{1 - \tan x_1 \tan x_2}$

21. $\tan(x_1 - x_2) = \dfrac{\tan x_1 - \tan x_2}{1 + \tan x_1 \tan x_2}$

22. $\tan 2x = \dfrac{2 \tan x}{1 - \tan^2 x}$

23. $\tan \dfrac{x}{2} = \dfrac{1 - \cos x}{\sin x}$

24. $\tan \dfrac{x}{2} = \dfrac{\sin x}{1 + \cos x}$

Exercise 10.4

Find the exact value of each expression.

Example $\tan 75°$

Solution Write $75°$ as $75° = 30° + 45°$. Then

$$\tan 75° = \tan(30° + 45°) = \dfrac{\tan 30° + \tan 45°}{1 - \tan 30° \tan 45°}$$

$$= \dfrac{\dfrac{1}{\sqrt{3}} + 1}{1 - \left(\dfrac{1}{\sqrt{3}}\right)(1)} = \dfrac{\sqrt{3} + 3}{3 - \sqrt{3}}.$$

1. $\tan 105°$
2. $\tan 255°$
3. $\tan 22° 30'$
4. $\tan 112° 30'$
5. $\tan 67° 30'$
6. $\tan 195°$

7. Given that $\sin x = \dfrac{3}{5}$, $0 < x < \dfrac{\pi}{2}$, find:

   a. $\tan 2x$
   b. $\tan \dfrac{x}{2}$

8. Given that $\cos x = -\frac{4}{5}$, $\frac{\pi}{2} < x < \pi$, find:

   a. $\tan 2x$ b. $\tan \frac{x}{2}$

Prove each identity.

9. $\tan(\pi - x) = -\tan x$
10. $\tan(\pi + x) = \tan x$
11. $\tan(2\pi - x) = -\tan x$
12. $\tan(2\pi + x) = \tan x$
13. $\cot \theta - \cot 2\theta = \dfrac{\sec^2 \theta}{2 \tan \theta}$
14. $\tan \dfrac{x}{2} = \dfrac{\sin x}{1 + \cos x}$
15. $\tan 3x = \dfrac{3 \tan x - \tan^3 x}{1 - 3 \tan^2 x}$ *Hint:* $\tan 3x = \tan(2x + x)$
16. $\tan 4x = \dfrac{4 \tan x - 4 \tan^3 x}{1 - 6 \tan^2 x + \tan^4 x}$ *Hint:* $\tan 4x = \tan(2x + 2x)$.
17. $2 \tan 4x = \dfrac{4 \tan 2x}{1 - \tan^2 2x}$
18. $\tan 2x = \dfrac{1 - \cos 4x}{\sin 4x}$
19. $\tan\left(\dfrac{3x}{2}\right) = \dfrac{1 + 3 \cos x - 4 \cos^3 x}{3 \sin x - 4 \sin^3 x}$

    Hint: See Exercise 19 in Sections 10.2 and 10.3.
20. $\tan\left(\dfrac{3x}{2}\right) = \dfrac{\sin x + \sin 2x}{\cos x + \cos 2x}$ *Hint:* $\tan\left(\dfrac{3x}{2}\right) = \tan\left(x + \dfrac{x}{2}\right)$.
21. $\tan\left(\dfrac{\pi}{2} - x\right) = \cot x$
22. $\tan\left(\dfrac{\pi}{2} + x\right) = -\cot x$
23. $\tan\left(\dfrac{3\pi}{2} - x\right) = \cot x$
24. $\tan\left(\dfrac{3\pi}{2} + x\right) = -\cot x$

10.5 Conditional Equations

In Section 10.1 we observed that certain equations involving trigonometric function values are *identities*; that is, they are true for all permissible values of any variables involved. In this section we shall be concerned with *conditional equations* involving these functions, namely, equations that are not satisfied by all permissible values of the variables involved.

Various procedures, including the methods applicable to algebraic equations, exist for solving conditional trigonometric equations. We shall use Tables 8.2 and 8.3 when the results can be expressed in terms of the special values in those tables, and Table V or VI or a calculator otherwise. In the examples and in the exercises where

Table V or VI or a calculator are required, the solutions are expressed to the nearest tenth of a unit.

Because the trigonometric functions are periodic, we should expect equations involving these functions to have infinite solution sets.

Example Solve $\sin x = -1$ for:

a. x a real number,

b. x an angle with measure given in radians,

c. x an angle with measure given in degrees.

Solution From the entries in Table 8.2 (page 205) and the fact that the sine is periodic, we observe that the solutions are:

a. $\left\{ x \mid x = \dfrac{3\pi}{2} + k \cdot 2\pi,\ k \in J \right\}$,

b. $\left\{ x \mid x = \left(\dfrac{3\pi}{2} + k \cdot 2\pi\right)^R,\ k \in J \right\}$,

c. $\{x \mid x = (270 + k \cdot 360)°,\ k \in J\}$.

where J is the set of integers.

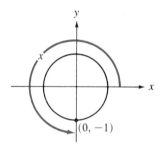

Next let us consider an example in which a reference angle proves helpful.

Example Solve $\sin \theta = \sqrt{3}/2$ for θ an angle measured in degrees.

Solution From the entries in Table 8.2 we observe that

$$\sin 60° = \sqrt{3}/2.$$

Hence, $\theta_1 = 60°$ is the positive acute angle that satisfies the equation. Using $60°$ as the measure of the reference angle and the fact that $\sin \theta$ is positive only in the first and second quadrants, we see that $\theta_2 = 120°$ is also a solution. Since the sine function has period $360°$, the solution set is the union of

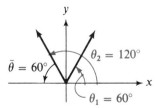

$\{\theta \mid \theta = (60 + k \cdot 360)°,\ k \in J\}$ and $\{\theta \mid \theta = (120 + k \cdot 360)°,\ k \in J\}$.

Using the symbol $\cup$ to denote the union of sets, we can write the solution set as

$\{\theta \mid \theta = (60 + k \cdot 360)°,\ k \in J\} \cup \{\theta \mid \theta = (120 + k \cdot 360)°,\ k \in J\}$.

10.5 Conditional Equations

We can use identities to rewrite equations involving more than one function value as equivalent equations involving values of only one function.

Example Solve $\sin \alpha = \cos \alpha$ over the interval $0^R \leq \alpha \leq (\pi/2)^R$.

Solution Since $\sin \alpha \neq 0$ when $\cos \alpha = 0$, the equation is not satisfied if $\cos \alpha = 0$. We can accordingly assume that $\cos \alpha \neq 0$ and hence can multiply each member by $1/\cos \alpha$ to obtain

$$\frac{\sin \alpha}{\cos \alpha} = 1,$$

or, equivalently,

$$\tan \alpha = 1.$$

Therefore, by Table 9.1, the solution set in $0^R \leq \alpha \leq \frac{\pi^R}{2}$ is $\left\{\frac{\pi^R}{4}\right\}$.

Recall that a product ab equals 0, if and only if $a = 0$ or $b = 0$ (or both). This property is sometimes useful in solving trigonometric equations in which one member is 0 and the other member is in factored form.

Example Solve $(2 \cos \theta - 1)(\sin \theta - 1) = 0$ over $0^R \leq \theta < 2\pi^R$.

Solution If $(2 \cos \theta - 1)(\sin \theta - 1) = 0$, then either $2 \cos \theta - 1 = 0$ or $\sin \theta - 1 = 0$. If $2 \cos \theta - 1 = 0$, then $\cos \theta = 1/2$; if $\sin \theta - 1 = 0$, then $\sin \theta = 1$. Thus, the values of θ satisfying these equations are of the form

$$\left(\frac{\pi}{3} + k \cdot 2\pi\right)^R, \quad \left(\frac{5\pi}{3} + k \cdot 2\pi\right)^R, \quad \text{or} \quad \left(\frac{\pi}{2} + k \cdot 2\pi\right)^R, \quad k \in J.$$

Over the interval $0^R \leq \theta < 2\pi^R$, the solution set is

$$\left\{\frac{\pi^R}{3}, \frac{5\pi^R}{3}, \frac{\pi^R}{2}\right\}.$$

Note that, in the foregoing example, if the equation had been presented in the form

$$2 \sin \theta \cos \theta - \sin \theta - 2 \cos \theta + 1 = 0,$$

it would first have been necessary to factor the equation into the given form before proceeding with the solution.

Exercise 10.5

In Exercises 1–12, find the solution set of the given equation in (**a**) degree measure and (**b**) radian measure.

1. $\cos \theta = \dfrac{1}{2}$
2. $\sin \theta = \dfrac{\sqrt{2}}{2}$
3. $\tan \theta = \sqrt{3}$

4. $\cot \theta = -1$
5. $\sec \theta - \sqrt{2} = 0$
6. $\csc \theta + 1 = 0$
7. $4 \sin \theta - 1 = 0$
8. $2 \tan \theta + 3 = 0$
9. $3 \cot \theta - 1 = 0$
10. $3 \cos \theta + 1 = 0$
11. $2 \sec \theta - 5 = 0$
12. $3 \csc \theta + 8 = 0$

In Exercises 13–18, find the solution set of the given equation, where x denotes a real number.

Example $\sqrt{3} \sin x + \cos x = 0$

Solution By first noting that $\cos x \neq 0$ in any solution, we can divide each member by $\cos x$ to produce

$$\sqrt{3} \frac{\sin x}{\cos x} + 1 = 0,$$

from which

$$\sqrt{3} \tan x + 1 = 0,$$

or

$$\tan x = -\frac{1}{\sqrt{3}},$$

an equation that involves only one function value. By inspection, the solution set is

$$\left\{ x \mid x = \frac{5\pi}{6} + k\pi, \quad k \in J \right\}.$$

13. $\sin x - \sqrt{3} \cos x = 0$
14. $3 \sin x + \cos x = 0$
15. $\tan^2 x - 1 = 0$
16. $2 \sin^2 x - 1 = 0$
17. $\sin^2 x - \cos^2 x = 1$
18. $\sin x + \cos x \tan x = 3$

In Exercises 19–22, find the solutions of the given equation for θ a member of the set of angles such that $0° \leq \theta < 360°$.

Example $2 \cos^2 \theta \tan \theta - \tan \theta = 0$

Solution Factoring $\tan \theta$ from the terms in the left-hand member, we have

$$\tan \theta (2 \cos^2 \theta - 1) = 0,$$

from which either

$$\tan \theta = 0 \quad \text{or} \quad 2 \cos^2 \theta - 1 = 0.$$

If $\tan \theta = 0$, either $\theta = 0°$ or $\theta = 180°$. If $2 \cos^2 \theta - 1 = 0$, then $\cos^2 \theta = 1/2$, $\cos \theta = \pm 1/\sqrt{2}$, and θ is an odd multiple of $45°$. Therefore, the solution set over $0° \leq \theta < 360°$ is

$$\{0°, 45°, 135°, 180°, 225°, 315°\}.$$

10.6 Conditional Equations for Multiples

19. $(2 \sin \theta - 1)(2 \sin^2 \theta - 1) = 0$ **20.** $(\tan \theta - 1)(2 \cos \theta + 1) = 0$
21. $2 \sin \theta \cos \theta + \sin \theta = 0$ **22.** $\tan \theta \sin \theta - \tan \theta = 0$

In Exercises 23–28, find the solutions of the given equation over the interval $0^R \leq \alpha < 2\pi^R$.

Example $\tan^2 \alpha + \sec \alpha - 1 = 0$

Solution   Since $\tan^2 \alpha = \sec^2 \alpha - 1$, the given equation can be written as

$$\sec^2 \alpha - 1 + \sec \alpha - 1 = 0,$$
$$\sec^2 \alpha + \sec \alpha - 2 = 0,$$
$$(\sec \alpha + 2)(\sec \alpha - 1) = 0.$$

If $\sec \alpha + 2 = 0$, then $\sec \alpha = -2$; if $\sec \alpha - 1 = 0$, then $\sec \alpha = 1$. Over the interval $0^R \leq \alpha < 2\pi^R$, the required solution set is

$$\left\{\frac{2\pi^R}{3}, \frac{4\pi^R}{3}\right\} \cup \{0^R\} = \left\{0^R, \frac{2\pi^R}{3}, \frac{4\pi^R}{3}\right\}.$$

23. $\tan^2 \alpha - 2 \tan \alpha + 1 = 0$ **24.** $4 \sin^2 \alpha - 4 \sin \alpha + 1 = 0$
25. $\cos^2 \alpha + \cos \alpha = 2$ **26.** $\cot^2 \alpha = 5 \cot \alpha - 4$
27. $\sec^2 \alpha + 3 \tan \alpha - 11 = 0$ **28.** $\tan^2 \alpha + 4 = 2 \sec^2 \alpha$

29. Solve $\sin^2 \alpha + \sin \alpha - 1 = 0$ *Hint:* Use the quadratic formula.
30. Solve $\tan^2 \alpha = \tan \alpha + 3$ *Hint:* Use the quadratic formula.
31. Approximate a solution of $x/2 - \sin x = 0$, where x denotes a real number, $x \neq 0$, by graphical methods. *Hint:* Graph $y_1 = x/2$ and $y_2 = \sin x$ on the same coordinate system, and determine a value of x for which $y_1 = y_2$.
32. Approximate a positive solution of $\cos x = x^2$, where x denotes a real number, by graphical methods.

10.6 Conditional Equations for Multiples

Equations that contain trigonometric function values such as $\sin 2\alpha$ and $\cos 3\alpha$ need further consideration.

Example   Solve $\sqrt{2} \cos 3\alpha = 1$ over the interval $0° \leq \alpha \leq 360°$.

Solution   The equation can be written equivalently as $\cos 3\alpha = 1/\sqrt{2}$. Therefore, we have

$$3\alpha = 45° + k \cdot 360° \quad \text{or} \quad 3\alpha = 315° + k \cdot 360°, \quad k \in J,$$

from which, by dividing each member by 3, we obtain

$$\alpha = 15° + k \cdot 120° \quad \text{or} \quad \alpha = 105° + k \cdot 120°, \quad k \in J.$$

Notice that 0, 1, and 2 are the only replacements for k that will yield values of α in the interval $0° \leq \alpha \leq 360°$. Therefore, the solution set over this interval is

$$\{15°, 135°, 255°, 105°, 225°, 345°\}.$$

A judicious selection of one or more of the identities encountered earlier can frequently help you solve certain kinds of equations. The following examples illustrate two such selections.

Example Solve $\sin \alpha \cos \alpha = 1/4$ over the interval $0^R \leq \alpha \leq 2\pi^R$.

Solution Since $2 \sin \alpha \cos \alpha = \sin 2\alpha$ for every value α, we multiply each member of the given equation by 2 to obtain the equivalent equation

$$2 \sin \alpha \cos \alpha = \frac{1}{2},$$

or, in terms of a single function value,

$$\sin 2\alpha = \frac{1}{2}.$$

Therefore, we have

$$2\alpha = \left(\frac{\pi}{6} + k \cdot 2\pi\right)^R \quad \text{or} \quad 2\alpha = \left(\frac{5\pi}{6} + k \cdot 2\pi\right)^R, \quad k \in J,$$

so that

$$\alpha = \left(\frac{\pi}{12} + k\pi\right)^R \quad \text{or} \quad \alpha = \left(\frac{5\pi}{12} + k\pi\right)^R, \quad k \in J.$$

Because we want the solutions over the interval $0^R \leq \alpha < 2\pi^R$, we consider 0 and 1 as replacements for k and obtain the solution set

$$\left\{\frac{\pi^R}{12}, \frac{5\pi^R}{12}, \frac{13\pi^R}{12}, \frac{17\pi^R}{12}\right\}.$$

Example Solve $\cos 2x = \sin x$ for $x \in R$.

Solution Since $\cos 2x = 1 - 2 \sin^2 x$, the equation $\cos 2x = \sin x$ can be written equivalently as

$$1 - 2 \sin^2 x = \sin x,$$
$$2 \sin^2 x + \sin x - 1 = 0,$$
$$(2 \sin x - 1)(\sin x + 1) = 0,$$

from which

$$\sin x = \frac{1}{2} \quad \text{or} \quad \sin x = -1.$$

Then as solution set we have

$$\left\{x \mid x = \frac{\pi}{6} + k \cdot 2\pi\right\} \cup \left\{x \mid x = \frac{5\pi}{6} + k \cdot 2\pi\right\} \cup \left\{x \mid x = \frac{3\pi}{2} + k \cdot 2\pi\right\}, \quad k \in J.$$

Exercise 10.6

In Exercises 1–12, solve the given equation for θ in the interval $0° \le \theta < 360°$.

1. $\cos 2\theta = \dfrac{\sqrt{2}}{2}$
2. $\tan 2\theta = \sqrt{3}$
3. $\sin \dfrac{1}{2}\theta = \dfrac{1}{2}$
4. $\cot \dfrac{1}{3}\theta = -1$
5. $\tan 3\theta = 0$
6. $\sin 4\theta = 1$
7. $\sin \theta \cos \theta = \dfrac{1}{2}$
8. $\cos^2 \theta - \sin^2 \theta = -1$
9. $\cos 2\theta + \sin 2\theta = 0$
10. $\sin \theta \cos \theta = \dfrac{\cos 2\theta}{2}$
11. $2 \cos^2 2\theta + \cos 2\theta - 1 = 0$
12. $\tan^2 2\theta + 2 \tan 2\theta + 1 = 0$

In Exercises 13–20, solve the given equation for real numbers x.

13. $\sin 2x - \cos x = 0$
14. $\cos 2x = \cos^2 x - 1$
15. $\cos 2x = \cos x - 1$
16. $\sin x = \sin 2x$
17. $\cos 2x \sin x + \sin x = 0$
18. $\sin 2x \cos x - \sin x = 0$
19. $\sin 4x - 2 \sin 2x = 0$ *Hint:* $\sin 4x = \sin 2(2x)$.
20. $\sin 3x + 4 \sin^2 x = 0$ *Hint:* $\sin 3x = \sin (x + 2x)$.

Chapter Review

[10.1] Verify that the given equation is an identity.

1. $\tan x = \sin x \cdot \sec x$
2. $\cos x - \sin x = (1 - \tan x) \cos x$
3. $\dfrac{1 - \tan^2 \alpha}{\tan \alpha} = \cot \alpha - \tan \alpha$
4. $\dfrac{\cos^2 \alpha}{1 - \sin \alpha} = \dfrac{\cos \alpha}{\sec \alpha - \tan \alpha}$
5. $\dfrac{1}{\cot^2 \theta \csc^2 \theta} = \dfrac{1}{\cot^2 \theta} - \dfrac{1}{\csc^2 \theta}$
6. $\dfrac{4}{\cos^2 \theta} = \dfrac{2}{1 - \sin \theta} + \dfrac{2}{1 + \sin \theta}$
7. $\dfrac{1 + \sin x}{\cos x} = \dfrac{1}{\sec x - \tan x}$
8. $\dfrac{1}{\cos^2 x} + \dfrac{1}{\sin^2 x} = \dfrac{1}{\cos^2 x \sin^2 x}$

[10.2] *Given that* $\cos \theta = -2/3$ *and* $90° \leq \theta \leq 180°$, *find an approximation for the given expression.*

9. $\cos(30° + \theta)$
10. $\cos(60° + \theta)$
11. $\cos(30° - \theta)$
12. $\cos(60° - \theta)$
13. $\cos 2\theta$
14. $\cos \theta/2$

[10.3] *Given that* $\sin \theta = 2/3$ *and* $90° \leq \theta \leq 180°$, *find an approximation for the given expression.*

15. $\sin(30° + \theta)$
16. $\sin(60° + \theta)$
17. $\sin(30° - \theta)$
18. $\sin(60° - \theta)$
19. $\sin 2\theta$
20. $\sin \theta/2$

[10.4]
21. $\tan(30° + \theta)$
22. $\tan(60° + \theta)$
23. $\tan(30° - \theta)$
24. $\tan(60° - \theta)$
25. $\tan 2\theta$
26. $\tan \theta/2$

[10.5] *Solve the given equation over the specified replacement set.*

27. $\cos^2 \theta - 1 = 0; \quad 0° \leq \theta \leq 360°$
28. $2\cos\theta \sin\theta - \cos\theta = 0; \quad 0° \leq \theta \leq 360°$
29. $\sin^2 \alpha - 3\sin\alpha + 2 = 0; \quad 0^R \leq \alpha \leq 2\pi^R$
30. $\sqrt{3} \tan x - 1 = 0$, where x is a real number.

[10.6] *Solve the given equation over the interval* $0° \leq \theta \leq 360°$.

31. $\sin 3\theta = \dfrac{\sqrt{3}}{2}$
32. $\cos^2 2\theta + 2\cos 2\theta + 1 = 0$

Solve the given equation for x a real number.

33. $\sin 2x - \cos 2x = 0$
34. $\tan^2 2x - 3 = 0$

11 Systems of Equations and Inequalities

In Chapters 5 and 6, we observed that the solution set of an equation or inequality in two variables, such as

$$ax + by + c = 0$$

or

$$ax + by + c \leq 0,$$

might contain infinitely many ordered pairs of numbers. It is often necessary to consider pairs of such sentences and to inquire whether or not the solution sets of the sentences contain ordered pairs in common. More specifically, we are interested in determining the members of the intersection of their solution sets.

11.1 Systems of Linear Equations in Two Variables

We shall begin by considering the system

$$a_1 x + b_1 y + c_1 = 0 \quad (a_1, b_1 \text{ not both } 0)$$
$$a_2 x + b_2 y + c_2 = 0 \quad (a_2, b_2 \text{ not both } 0)$$

and studying $A \cap B$, where

$$A = \{(x, y) \mid a_1 x + b_1 y + c_1 = 0\}$$

and

$$B = \{(x, y) \mid a_2 x + b_2 y + c_2 = 0\}.$$

In a geometric sense, because the graphs of both A and B are straight lines, we are confronted with three possibilities, as illustrated in Figure 11.1 on page 304.

   a. The graphs are the same line.
   b. The graphs are parallel but distinct lines.
   c. The graphs intersect in one and only one point.

These possibilities lead, correspondingly, to the conclusion that one and only one of the following is true for any given system of two such linear equations in x and y:

   a. The intersection of the two solution sets contains all those ordered pairs found in either one of the given solution sets; that is, the solution sets are equal.
   b. The intersection of the two solution sets is the null set.
   c. The intersection of the two solution sets contains exactly one ordered pair.

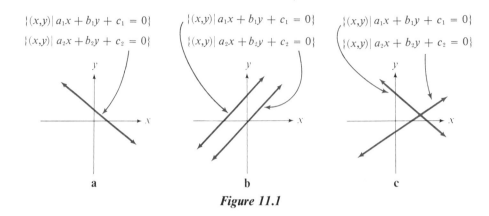

Figure 11.1

Dependency and consistency

In case (a), the left-hand members of the two equations in standard form are said to be **linearly dependent**, and the equations are **consistent**; in case (b), the left-hand members are **linearly independent**, and the equations are **inconsistent**; and in case (c), the left-hand members are linearly independent and the equations are consistent. If the two left-hand members are linearly dependent, then one of them can be obtained from the other through multiplication by a constant. For example,

$$2x + 4y - 8 \quad \text{and} \quad 6x + 12y - 24$$

are linearly dependent, but

$$2x + 4y - 8 \quad \text{and} \quad 6x + 12y - 23$$

are not.

You will recall that equivalent equations are defined to be equations that have the same solution set. Systems of equations may be said to be equivalent in a similar sense.

Definition 11.1 *If the solution set of one system of equations is equal to (the same as) the solution set of another system of equations, then the systems are **equivalent**.*

Generation of equivalent systems

In seeking the solution set of a system of equations, our procedure will be to generate equivalent systems until we arrive at a system of which the solution set is obvious. One way to obtain an equivalent system is to replace one equation by a certain *linear combination* of the equations. If $f(x, y)$ and $g(x, y)$ are polynomials, the polynomial $af(x, y) + bg(x, y)$, where a and b are not both zero, is said to be a linear combination of $f(x, y)$ and $g(x, y)$. We then refer to the equation

$$af(x, y) + bg(x, y) = 0$$

as a **linear combination** of the equations $f(x, y) = 0$ and $g(x, y) = 0$.

Theorem 11.1 *If either equation in the system*

$$f(x, y) = 0$$
$$g(x, y) = 0$$

is replaced by a linear combination of the two equations with nonzero coefficient for the replaced equation, the result is an equivalent system.

11.1 Systems of Linear Equations in Two Variables

Proof Compare the two systems:

$$f(x, y) = 0$$
$$g(x, y) = 0 \quad (1)$$

and

$$af(x, y) + bg(x, y) = 0$$
$$g(x, y) = 0, \quad (2)$$

where $a \neq 0$. If (x_1, y_1) is a solution to (1), then $f(x_1, y_1) = 0$ and $g(x_1, y_1) = 0$, so

$$af(x_1, y_1) + bg(x_1, y_1) = 0,$$

and (x_1, y_1) is a solution to (2). On the other hand, suppose (x_2, y_2) is a solution to (2). Then $g(x_2, y_2) = 0$ and

$$af(x_2, y_2) + bg(x_2, y_2) = 0.$$

Thus $af(x_2, y_2) = 0$ and since $a \neq 0$, $f(x_2, y_2) = 0$, so that (x_2, y_2) is a solution to (2). Therefore, (1) and (2) have the same solution set.

Theorem 11.1 is useful in solving linear systems of the form

$$a_1 x + b_1 y + c_1 = 0 \quad (a_1, b_1 \text{ not both 0}) \quad (3)$$
$$a_2 x + b_2 y + c_2 = 0 \quad (a_2, b_2 \text{ not both 0}). \quad (4)$$

By appropriate choice of multipliers a and b, the linear combination

$$a(a_1 x + b_1 y + c_1) + b(a_2 x + b_2 y + c_2) = 0 \quad (5)$$

will be free of one variable; that is, the coefficient of one variable will be 0. We can then form an equivalent system by substituting Equation (5) for either Equation (1) or Equation (4).

Example Solve

$$x - 3y + 5 = 0$$
$$2x + y - 4 = 0. \quad (6)$$

Solution We first form the linear combination

$$1(x - 3y + 5) + 3(2x + y - 4) = 0,$$

or

$$7x - 7 = 0,$$

from which

$$x - 1 = 0.$$

Replacing $x - 3y + 5 = 0$ with $x - 1 = 0$, we then have the equivalent system

$$x - 1 = 0$$
$$2x + y - 4 = 0. \quad (7)$$

We can next replace the second equation in (7) with the linear combination

$$-2(x - 1) + 1(2x + y - 4) = 0,$$

Solution Continued on Overleaf

or
$$y - 2 = 0,$$
to obtain the equivalent system
$$\begin{aligned} x - 1 &= 0 \\ y - 2 &= 0, \end{aligned} \tag{8}$$
from which, by inspection, we can obtain the unique solution (1, 2). Hence the solution set of (6) is $\{(1, 2)\}$. Graphs of the systems (6), (7), and (8) appear in Figure 11.2. The point of intersection in each case has coordinates (1, 2).

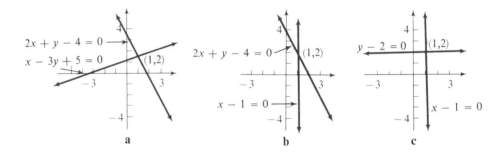

Figure 11.2

Solution of a system by substitution

Note that we could have proceeded somewhat differently from (7), as follows. Since any solution of the system (7) must be of the form $(1, y)$ (because any such ordered pair is a solution of $x - 1 = 0$), and when x is replaced by 1 in $2x + y - 4 = 0$, we obtain $y = 2$, the only ordered pair that satisfies both of the equations in (7) is (1, 2). Hence, again the solution set of (6) is $\{(1, 2)\}$. In this latter way of proceeding from (7) to the solution of (6), we have used what is known as the **method of substitution**.

Choice of multipliers for linear combinations

Observe that the multipliers used in the foregoing example were first 3 and 1, and later -2 and 1. These were chosen because they produced coefficients that were additive inverses, first for the terms in y and later for the terms in x. In general, the linear combination
$$b_2(a_1 x + b_1 y + c_1) - b_1(a_2 x + b_2 y + c_2) = 0$$
will always be free of y, and
$$a_2(a_1 x + b_1 y + c_1) - a_1(a_2 x + b_2 y + c_2) = 0$$

11.1 Systems of Linear Equations in Two Variables

will be free of x. For example, to form a useful linear combination of the equations in the system

$$2x + 3y + 7 = 0$$
$$5x + 4y + 3 = 0,$$

we might use 5 and -2, or -5 and 2, if we wish to obtain an equation that is free of x; and we might use 4 and -3, or -4 and 3, if we wish the result to obtain an equation that is free of y.

Criteria for dependent or inconsistent equations

If the coefficients of the variables in one equation in a system are proportional to the corresponding coefficients in the other equation, then the equations might be either dependent or inconsistent. The equations in the system

$$a_1 x + b_1 y + c_1 = 0$$
$$a_2 x + b_2 y + c_2 = 0 \quad (a_2, b_2, c_2 \neq 0)$$

are dependent if

$$\frac{a_1}{a_2} = \frac{b_1}{b_2} = \frac{c_1}{c_2},$$

and inconsistent if

$$\frac{a_1}{a_2} = \frac{b_1}{b_2} \neq \frac{c_1}{c_2}. \quad \text{(See Exercises 26 and 27.)}$$

Applications

Systems of linear equations are quite useful in expressing relationships in practical applications. By assigning separate variables to represent separate physical quantities, we can usually decrease the difficulty in symbolically representing these relationships.

Example

A company offers split-rail fence for sale in two prepackaged options. One option consists of 4 posts and 6 rails for \$31; the other consists of 3 posts and 4 rails for \$22. What are the individual values of posts and rails?

Solution

Let x represent the value of the posts in dollars and y represent the value of the rails in dollars. Then

$$4x + 6y = 31$$
$$3x + 4y = 22.$$

This system is solved in the first example in the exercises.

Exercise 11.1

Solve each system in Exercises 1–14.

Example

$$4x + 6y = 31 \quad (1)$$
$$3x + 4y = 22 \quad (2)$$

Solution

Replace (1) by 3 times itself and (2) by 4 times itself to obtain

$$12x + 18y = 93 \quad (1')$$
$$12x + 16y = 88. \quad (2')$$

Replace (1') by itself plus -1 times (2').

$$2y = 5 \quad (1'')$$
$$12x + 16y = 88 \quad (2'')$$

From (1''), $y = 2\tfrac{1}{2}$. Substituting this value into (2''), we obtain

$$12x + 40 = 88,$$

so $12x = 48$ and $x = 4$. The solution is therefore $(4, 2\tfrac{1}{2})$.

1. $x - y = 1$
   $x + y = 5$

2. $2x - 3y = 6$
   $x + 3y = 3$

3. $3x + y = 7$
   $2x - 5y = -1$

4. $2x - y = 7$
   $3x + 2y = 14$

5. $5x - y = -29$
   $2x + 3y = 2$

6. $6x + 4y = 12$
   $3x + 2y = 12$

7. $5x + 2y = 3$
   $x = 0$

8. $2x - y = 0$
   $x = -3$

9. $3x - 2y = 4$
   $y = -1$

10. $x + 2y = 6$
    $x = 2$

11. $\dfrac{1}{4}x - \dfrac{1}{3}y = -\dfrac{5}{12}$
    $\dfrac{1}{10}x + \dfrac{1}{5}y = \dfrac{1}{2}$

12. $\dfrac{2}{3}x - y = 4$
    $x - \dfrac{3}{4}y = 6$

13. $\dfrac{1}{7}x - \dfrac{3}{7}y = 1$
    $2x - y = -4$

14. $\dfrac{1}{3}x - \dfrac{2}{3}y = 2$
    $x - 2y = 6$

15. Find a and b so that the graph of $ax + by + 3 = 0$ passes through the points $(-1, 2)$ and $(-3, 0)$.

16. Find a and b so that the solution set of the system
$$ax + by = 4$$
$$bx - ay = -3$$
is $\{(1, 2)\}$.

11.1 Systems of Linear Equations in Two Variables

17. Recall that the slope-intercept form of the equation of a straight line is given by $y = mx + b$. Find an equation of the line that passes through the points $(0, 2)$ and $(3, -8)$.
18. Find an equation of the line that passes through the points $(-6, 2)$ and $(4, 1)$.
19. Find a linear relationship between centigrade temperature, C, and Fahrenheit temperature, F, given that $F = 32°$ when $C = 0°$, and $F = 212°$ when $C = 100°$.
20. A man has $1.80 in nickels and dimes, with three more dimes than nickels. How many dimes and nickels does he have?
21. How many pounds of an alloy containing 45% silver must be melted with an alloy containing 60% silver to obtain 40 pounds of a 48% silver alloy?
22. A man has three times as much money invested in 3% bonds as he has in stocks paying 5%. How much does he have invested in each if his yearly income from the investments is $1680?
23. A man has $1000 more invested at 5% than he has invested at 4%. If his annual income from the two investments together is $698, how much does he have invested at each rate?
24. An airplane travels 1260 miles in the same time that an automobile travels 420 miles. If the rate of the airplane is 120 miles per hour greater than the rate of the automobile, find the rate of each.
25. Two cars start together and travel in the same direction, one going twice as fast as the other. At the end of 3 hours, they are 96 miles apart. How fast is each traveling?
26. Show that the linear polynomials

$$a_1 x + b_1 y + c_1 \quad (a_1, b_1, c_1 \neq 0)$$
$$a_2 x + b_2 y + c_2 \quad (a_2, b_2, c_2 \neq 0)$$

are dependent if and only if

$$\frac{a_1}{a_2} = \frac{b_1}{b_2} = \frac{c_1}{c_2}.$$

Hint: Write the equations in slope-intercept form.

27. Show that the equations in the system

$$a_1 x + b_1 y + c_1 = 0$$
$$a_2 x + b_2 y + c_2 = 0 \quad (a_2, b_2, c_2 \neq 0)$$

are inconsistent if and only if

$$\frac{a_1}{a_2} = \frac{b_1}{b_2} \neq \frac{c_1}{c_2}.$$

Hint: Write the equations in slope-intercept form.

11.2 Systems of Linear Equations in Three Variables

Solutions of an equation in three variables

A solution of an equation in three variables, such as

$$x + 2y - 3z + 4 = 0, \qquad (1)$$

is an ordered triple of numbers (x, y, z), because all three of the variables must be replaced before we can decide whether or not the sentence is true. Thus, $(0, -2, 0)$ and $(-1, 0, 1)$ are solutions of Equation (1), whereas $(1, 1, 1)$ is not. There are, of course, infinitely many members in the solution set of such an equation.

Solution of a system

The solution set of a system of linear (first-degree) equations in three variables is the intersection of the solution sets of the separate equations in the system. We are primarily interested in systems involving three equations, such as

$$x + 2y - 3z + 4 = 0 \qquad (1)$$
$$2x - y + z - 3 = 0 \qquad (2)$$
$$3x + 2y + z - 10 = 0, \qquad (3)$$

where the solution set is

$$\{(x, y, z) \mid x + 2y - 3z + 4 = 0\} \cap \{(x, y, z) \mid 2x - y + z - 3 = 0\}$$
$$\cap \{(x, y, z) \mid 3x + 2y + z - 10 = 0\}.$$

We can find the members of this set by methods analogous to those used in the preceding section.

Theorem 11.2 *If any equation in the system*

$$f(x, y, z) = 0$$
$$g(x, y, z) = 0$$
$$h(x, y, z) = 0$$

is replaced by a linear combination, with nonzero coefficients, of itself and any one of the other equations in the system, then the result is an equivalent system.

The proof of this theorem is similar to that of Theorem 11.1 and is omitted here. Now, to see how this theorem applies, let us examine the foregoing system (1), (2), and (3). If we begin by replacing Equation (2) with the linear combination formed by multiplying Equation (1) by -2 and Equation (2) by 1, that is, with

$$-2(x + 2y - 3z + 4) + 1(2x - y + z - 3) = 0,$$

or

$$-5y + 7z - 11 = 0,$$

11.2 Systems of Linear Equations in Three Variables

we obtain the equivalent system

$$x + 2y - 3z + 4 = 0 \tag{1}$$
$$-5y + 7z - 11 = 0 \tag{2'}$$
$$3x + 2y + z - 10 = 0, \tag{3}$$

where (2') is free of x. Next, if we replace (3) by the sum of -3 times (1) and 1 times (3), we have

$$x + 2y - 3z + 4 = 0 \tag{1}$$
$$-5y + 7z - 11 = 0 \tag{2'}$$
$$-4y + 10z - 22 = 0, \tag{3'}$$

where both (2') and (3') are free of x. If now (3') is replaced by the sum of 4 times (2') and -5 times (3'), we have

$$x + 2y - 3z + 4 = 0 \tag{1}$$
$$-5y + 7z - 11 = 0 \tag{2'}$$
$$-22z + 66 = 0, \tag{3''}$$

which is equivalent to the original (1), (2), and (3). But, from (3''), we see that for any solution of (1), (2'), and (3''), $z = 3$; that is, the solution will be of the form $(x, y, 3)$. If 3 is substituted for z in (2'), we have

$$-5y + 7(3) - 11 = 0, \quad \text{or} \quad y = 2,$$

and any solution of the system must be of the form $(x, 2, 3)$. Substituting 2 for y and 3 for z in (1) gives

$$x + 2(2) - 3(3) + 4 = 0, \quad \text{or} \quad x = 1,$$

so that the single member of the solution set of (1), (2'), and (3'') is $(1, 2, 3)$. Therefore, the solution set of the original system is $\{(1, 2, 3)\}$.

The foregoing process of solving a system of linear equations can be reduced to a series of mechanical procedures.

Example   Solve.

$$x + 2y - z = -1 \tag{1}$$
$$x - 3y + z = 2 \tag{2}$$
$$2x + y + 2z = 6 \tag{3}$$

Solution   Multiply (1) by -1 and add the result to 1 times (2); multiply (1) by -2 and add the result to 1 times (3).

$$-5y + 2z = 3 \tag{4}$$
$$-3y + 4z = 8 \tag{5}$$

Multiply (4) by -3 and add the result to 5 times (5).

$$14z = 31 \tag{6}$$

Now (1), (4), and (6) constitute a set of equations equivalent to (1), (2), and (3).

$$x + 2y - z + 1 = 0 \tag{1}$$
$$-5y + 2z - 3 = 0 \tag{4}$$
$$14z - 31 = 0 \tag{6}$$

Solve for z in (6).

$$z = \frac{31}{14}$$

Substitute $31/14$ for z in (4).

$$-5y + 2\left(\frac{31}{14}\right) - 3 = 0$$

$$y = \frac{2}{7}$$

Substitute $2/7$ for y and $31/14$ for z in either (1), (2), or (3); say (1).

$$x + 2\left(\frac{2}{7}\right) - \frac{31}{14} + 1 = 0$$

$$x = \frac{9}{14}$$

The solution set is $\left\{\left(\frac{9}{14}, \frac{2}{7}, \frac{31}{14}\right)\right\}$.

Linear dependence and inconsistent equations

If at any step in the procedure used in the examples above, the resulting linear combination vanishes or yields a contradiction, the system contains linearly dependent left-hand members or else inconsistent equations, or both, and it either has an infinite number of members or else has no member in its solution set.

Graphs in three dimensions

By establishing a three-dimensional Cartesian coordinate system as shown in Figure 11.3, a one-to-one correspondence can be established between the points in a three-dimensional space and ordered triples of real numbers. If this is done, it can be shown that the graph of a linear equation in three variables is a plane. For example, the equation

$$x + y + z = 1$$

represents the plane through the points $(1, 0, 0)$, $(0, 1, 0)$, and $(0, 0, 1)$, as illustrated in Figure 11.3 Hence the solution set of a system of three linear

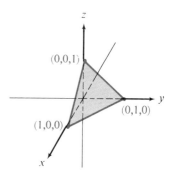

Figure 11.3

11.2 Systems of Linear Equations in Three Variables

equations in three variables consists of the coordinates of the common intersection of three planes. Figure 11.4 shows the possibilities for the relative positions of the plane graphs of three linear equations in three variables. In case (a), the common intersection consists of a single point, and hence the solution set of the corresponding system of three equations contains a single member. In cases (b), (c), and (d), the intersection is a line or a plane, and the solution of the corresponding system has infinitely many members. In cases (e), (f), (g) and (h), the three planes have no common intersection, and the solution set of the corresponding system is the null set. Linear equations corresponding to cases (a), (b), (c), and (d) are consistent, and the others inconsistent.

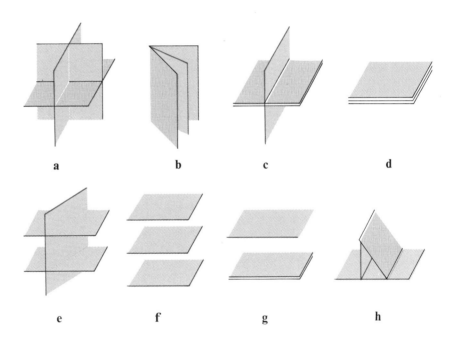

Figure 11.4

The use of linear combinations to solve systems of linear equations can be extended to cover cases of n equations in n variables. Clearly, as n grows larger, the time and effort necessary to find the solution set of a system increase correspondingly. Fortunately, modern high-speed computers can handle such computations in stride for fairly large values of n. There are analytic methods other than those exhibited here which become preferable for very large systems.

Exercise 11.2

Solve.

Example

$$x + 2y + z = 4 \quad (1)$$
$$2x + y - z = -1 \quad (2)$$
$$-x + y + z = 2 \quad (3)$$

Solution

Replace (2) by itself minus 2 times (1), and (3) by itself plus (1).

$$x + 2y + z = 4 \quad (1')$$
$$-3y - 3z = -9 \quad (2')$$
$$3y + 2z = 6 \quad (3')$$

Replace (3') by itself plus (2').

$$x + 2y + z = 4 \quad (1'')$$
$$-3y - 3z = -9 \quad (2'')$$
$$-z = -3 \quad (3'')$$

From (3''), obtain $z = 3$. Substitute $z = 3$ into (2'') to obtain $y = 0$. Substitute $z = 3$ and $y = 0$ into (1'') to obtain $x = 1$. The solution is therefore $(1, 0, 3)$.

1. $x + y + z = 2$
   $2x - y + z = -1$
   $x - y - z = 0$

2. $x + y + z = 1$
   $2x - y + 3z = 2$
   $2x - y - z = 2$

3. $x + y + 2z = 0$
   $2x - 2y + z = 8$
   $3x + 2y + z = 2$

4. $2x - 3y + z = 3$
   $x - y - 3z = -1$
   $-x + 2y - 3z = -4$

5. $x - 2y + z = -1$
   $2x + y - 3z = 3$
   $3x + 3y - 2z = 10$

6. $x - 2y + 4z = -3$
   $3x + y - 2z = 12$
   $2x + y - 3z = 11$

7. $4x - 2y + 3z = 4$
   $2x - y + z = 1$
   $3x - 3y + 4z = 5$

8. $x + 5y - z = 2$
   $3x - 9y + 3z = 6$
   $x - 3y + z = 4$

9. $x + z = 5$
   $y - z = -4$
   $x + y = 1$

10. $5y - 8z = -19$
    $5x - 8z = 6$
    $3x - 2y = 12$

11. $x - \frac{1}{2}y - \frac{1}{2}z = 4$
    $x - \frac{3}{2}y - 2z = 3$
    $\frac{1}{4}x + \frac{1}{4}y - \frac{1}{4}z = 0$

12. $x + 2y + \frac{1}{2}z = 0$
    $x + \frac{3}{5}y - \frac{2}{5}z = \frac{1}{5}$
    $4x - 7y - 7z = 6$

11.2 Systems of Linear Equations in Three Variables

Use three variables in the solution of each of the following problems.

Example The parabola $y = ax^2 + bx + c$ passes through the points $(-1, 9)$, $(1, 3)$, and $(3, 5)$. Find the coefficients a, b, and c.

Solution Since $(-1, 9)$ is on the parabola,
$$9 = a(-1)^2 + b(-1) + c$$
or
$$a - b + c = 9. \tag{1}$$

The other two points yield the equations
$$a + b + c = 3, \tag{2}$$
$$9a + 3b + c = 5. \tag{3}$$

Replace (2) by itself minus (1), and replace (3) by itself minus 9 times (1).
$$a - b + c = 9 \tag{1'}$$
$$2b = -6 \tag{2'}$$
$$12b - 8c = -76. \tag{3'}$$

From (2'), $b = -3$. Substitute $b = -3$ into (3') to obtain $c = 5$. Substitute $b = -3$ and $c = 5$ into (1') to obtain $a = 1$. Therefore the equation of the parabola is $y = x^2 - 3x + 5$.

13. The sum of three numbers is 15. The second equals two times the first and the third equals the second. Find the numbers.

14. The sum of three numbers is 2. The first number is equal to the sum of the other two, and the third number is the result of subtracting the first from the second. Find the numbers.

15. A box contains $6.25 in nickels, dimes, and quarters. There are 85 coins in all, with three times as many nickels as dimes. How many coins of each kind are there?

16. The perimeter of a triangle is 155 inches. The side x is 20 inches shorter than the side y, and the side y is 5 inches longer than the side z. Find the lengths of the sides of the triangle.

17. A man had $446 in ten-dollar, five-dollar, and one-dollar bills. There were 94 bills in all and 10 more five-dollar bills than ten-dollar bills. How many bills of each kind did he have?

18. Find values for a, b, and c so that the graph of $x^2 + y^2 + ax + by + c = 0$ will contain the points $(0, 0)$, $(6, 0)$, and $(0, 8)$.

19. The equation for a circle can be written $x^2 + y^2 + ax + by + c = 0$. Find the equation of the circle with graph containing the points $(2, 3)$, $(3, 2)$, and $(-4, -5)$.

20. Find values for a, b, and c so that the graph of $y = ax^2 + bx + c$ contains the points $(-1, 2)$, $(1, 6)$, and $(2, 11)$.

21. Three solutions of the equation $ax + by + cz = 1$ are $(0, 2, 1)$, $(6, -1, 2)$, and $(0, 2, 0)$. Find the coefficients a, b, and c.

22. Three solutions of the equation $ax + by + cz = 1$ are $(2, 1, 0)$, $(-1, 3, 2)$, and $(3, 0, 0)$. Find the coefficients a, b, and c.

23. Show that the system

$$x + y + 2z = 2$$
$$2x - y - z = 3$$

has an infinite number of members in its solution set. List two ordered triples that are solutions. *Hint:* Express x in terms of z alone, and express y in terms of z alone.

24. Give a geometric argument to show that any system of two consistent linear equations in three variables has an infinite number of solutions.

25. Show that the system

$$x + y + z = 3 \qquad (a)$$
$$x - 2y - z = -2 \qquad (b)$$
$$x + y + 2z = 4 \qquad (c)$$
$$2x - y + 2z = 4 \qquad (d)$$

has $\emptyset$ as its solution set. *Hint:* Does the solution set of the system (a), (b), and (c) have any member that satisfies (d)?

11.3 Systems of Nonlinear Equations

Substitution The substitution method is sometimes a convenient means of finding solution sets of systems in which nonlinear equations are present.

Example Solve.

$$x^2 + y^2 = 25 \qquad (1)$$
$$x + y = 7 \qquad (2)$$

Solution Equation (2) can be written equivalently in the form

$$y = 7 - x, \qquad (3)$$

and we can replace y in (1) by $(7 - x)$ from (3). This produces

$$x^2 + (7 - x)^2 = 25, \qquad (4)$$

11.3 Systems of Nonlinear Equations

which has as a solution set those values of x for which the ordered pair (x, y) is a common solution of (1) and (2). We can now find the solution set of (4):

$$x^2 + 49 - 14x + x^2 = 25,$$
$$x^2 - 7x + 12 = 0,$$
$$(x - 3)(x - 4) = 0,$$

which is satisfied if x is either 3 or 4. Now, by replacing x in the *first-degree equation* (3) by each of these numbers in turn, we obtain

$$y = 7 - 3 = 4$$

and

$$y = 7 - 4 = 3,$$

respectively, so that the solution set of the system (1) and (2) is $\{(3, 4), (4, 3)\}$.

If, in the foregoing example, we had substituted the values 3 and 4 that we obtained for x in the *second-degree equation* (1), we would have obtained some extraneous solutions that do not satisfy Equation (2). (See Exercise 31.)

Linear combination

If both of the equations in a system of two equations in two variables are of the second degree in both variables, the use of linear combinations of the equations often provides a simpler means of solution than does substitution.

Example

Solve.

$$3x^2 - 7y^2 + 15 = 0 \quad (5)$$
$$3x^2 - 4y^2 - 12 = 0 \quad (6)$$

Solution

By forming the linear combination of -1 times (5) and 1 times (6), we obtain

$$3y^2 - 27 = 0$$
$$y^2 = 9,$$

from which

$$y = 3 \quad \text{or} \quad y = -3,$$

and we have the y-components of the members of the solution set of the system (5) and (6). Substituting 3 for y in either (5) or (6), say (5), we have

$$3x^2 - 7(3)^2 + 15 = 0,$$
$$x^2 = 16,$$

from which

$$x = 4 \quad \text{or} \quad x = -4.$$

Thus, the ordered pairs $(4, 3)$ and $(-4, 3)$ are solutions of the system. Substituting -3 for y in (5) or (6) [this time we shall use (6)] gives us

$$3x^2 - 4(-3)^2 - 12 = 0,$$
$$x^2 = 16,$$

Solution Continued on Overleaf

so that

$$x = 4 \quad \text{or} \quad x = -4.$$

Thus the ordered pairs $(4, -3)$ and $(-4, -3)$ are solutions of the system, and accordingly the complete solution set is $\{(4, 3), (4, -3), (-4, 3), (-4, -3)\}$.

Linear combination and substitution

The solution of some systems requires the application of both linear combinations *and* substitution.

Example

Solve.

$$x^2 + y^2 = 5 \tag{7}$$
$$x^2 - 2xy + y^2 = 1 \tag{8}$$

Solution

By forming the linear combination of 1 times (7) and -1 times (8), we obtain

$$2xy = 4,$$
$$xy = 2. \tag{9}$$

The system (7) and (8) is equivalent to the system (7) and (9). This latter system can be solved by substitution. From (9), we have

$$y = \frac{2}{x}.$$

Replacing y in (7) by $2/x$, we obtain

$$x^2 + \left(\frac{2}{x}\right)^2 = 5,$$

$$x^2 + \frac{4}{x^2} = 5, \tag{10}$$

$$x^4 + 4 = 5x^2,$$

$$x^4 - 5x^2 + 4 = 0, \tag{11}$$

which is a quadratic in x^2. The left-hand member of (11), when factored, yields

$$(x^2 - 1)(x^2 - 4) = 0,$$

from which we obtain

$$x^2 - 1 = 0 \quad \text{or} \quad x^2 - 4 = 0,$$

so that

$$x = 1, \quad x = -1, \quad x = 2, \quad x = -2.$$

11.3 Systems of Nonlinear Equations

Since the step from (10) to (11) was not an elementary transformation, we are careful to note that these values of x all satisfy (10). Substituting 1, -1, 2, and -2 in turn for x in (9), we obtain the corresponding values for y, and thus the solution set of the system (7) and (9) or, equivalently, the solution set of the system (7) and (8), is $\{(1, 2), (-1, -2), (2, 1), (-2, -1)\}$.

There are other techniques involving substitution in conjunction with linear combinations that are useful in handling systems of higher-degree equations, but they all bear similarity to those illustrated.

Use of graphs as a check

In solving a set of equations, it is frequently helpful to sketch the graphs of the equations as a rough check on the algebraic solution. Any attempt to solve a set of two second-degree equations graphically on the real plane may produce only approximations to such real solutions as exist. We can expect ordinarily to find at most four and as few as no points of intersection, depending on the types of equations involved and on the coefficients and constants in the given equations.

Approximations to solutions

When the degree of any equation in a set of equations is greater than two, or when the left-hand member of one of the equations of the form $f(x, y) = 0$ is not a polynomial, it may be very difficult or impossible to find common solutions analytically. In this case, we can at least obtain approximations to any real solutions by graphical methods. Consider the system of equations

$$\begin{aligned} y &= 2^{-x} \\ y &= \sqrt{0.01x}, \end{aligned} \tag{12}$$

and the graphs of these equations in Figure 11.5.

Examining Figure 11.5-b, an enlargement of part of Figure 11.5-a, we observe that the curves intersect at approximately $(2.6, 0.16)$, and from geometric considerations we conclude that this ordered pair approximates the only member in the solution set of (12).

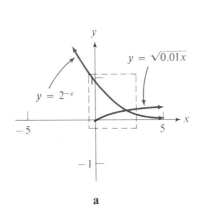
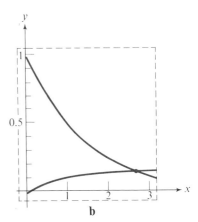

Figure 11.5

This result shows, in particular, that the equation

$$2^{-x} - \sqrt{0.01x} = 0$$

has just one root, approximately 2.6, and suggests a method of approximating solutions of similar equations.

Exercise 11.3

Solve by the method of substitution. Check the solutions by sketching the graphs of the equations and estimating the coordinates of any points of intersection.

Example

$$y = x^2 + 2x + 1 \quad (1)$$
$$y - x = 3 \quad (2)$$

Solution

Solve Equation (2) explicitly for y.

$$y = x + 3 \quad (2')$$

Substitute $x + 3$ for y in (1).

$$x + 3 = x^2 + 2x + 1 \quad (3)$$

Solve for x.

$$x^2 + x - 2 = 0$$
$$(x + 2)(x - 1) = 0$$
$$x = -2, \quad x = 1$$

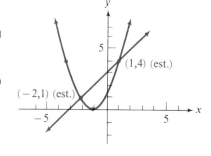

Substitute each of these values in turn in (2') to determine values for y.

If $x = -2$, then $y = 1$.
If $x = 1$, then $y = 4$.

The solution set is $\{(-2, 1), (1, 4)\}$.

1. $y = x^2 - 5$
   $y = 4x$

2. $y = x^2 - 2x + 1$
   $y + x = 3$

3. $x^2 + y^2 = 13$
   $x + y = 5$

4. $x^2 + 2y^2 = 12$
   $2x - y = 2$

5. $x + y = 1$
   $xy = -12$

6. $2x - y = 9$
   $xy = -4$

Solve using linear combinations.

Example

$$2x^2 + y^2 = 7 \quad (1)$$
$$x^2 - y^2 = 2 \quad (2)$$

11.3 Systems of Nonlinear Equations

Solution Replace (1) by itself minus 2 times (2).

$$3y^2 = 3 \qquad (1')$$
$$x^2 - y^2 = 2 \qquad (2')$$

From (1), $y^2 = 1$; substitution yields $x^2 = 3$. Therefore $x = \pm\sqrt{3}$ and $y = \pm 1$ so the solution set is $\{(\sqrt{3}, 1), (\sqrt{3}, -1), (-\sqrt{3}, 1), (-\sqrt{3}, -1)\}$.

7. $x^2 + y^2 = 10$
   $9x^2 + y^2 = 18$

8. $x^2 + 4y^2 = 52$
   $x^2 + y^2 = 25$

9. $x^2 - y^2 = 7$
   $2x^2 + 3y^2 = 24$

10. $x^2 + 4y^2 = 25$
    $4x^2 + y^2 = 25$

11. $4x^2 - 9y^2 + 132 = 0$
    $x^2 + 4y^2 - 67 = 0$

12. $16y^2 + 5x^2 - 26 = 0$
    $25y^2 - 4x^2 - 17 = 0$

13. $x^2 - xy + y^2 = 7$
    $x^2 + y^2 = 5$

14. $3x^2 - 2xy + 3y^2 = 34$
    $x^2 + y^2 = 17$

15. $3x^2 + 3xy - y^2 = 35$
    $x^2 - xy - 6y^2 = 0$

16. $x^2 - xy + y^2 = 21$
    $x^2 + 2xy - 8y^2 = 0$

Solve by graphing. Approximate components of solutions to the nearest half unit.

17. $y = 10^x$
    $x + y = 2$

18. $y = 2^x$
    $y - x = 2$

19. $y = \log_{10} x$
    $y = x^2 - 2x + 1$

20. $y = 10^x$
    $y = x^2$

21. $y = 10^{-x}$
    $y = \log_{10} x$

22. $y = 2^{x-3}$
    $y = \log_2 x$

23. $10^{-x} - \log_{10} x = 0$

24. $2^{x-3} - \log_2 x = 0$

25. How many real solutions are possible for simultaneous systems of linearly independent equations that consist of:
    a. two linear equations in two variables?
    b. one linear equation and one quadratic equation in two variables?
    c. two quadratic equations in two variables?
    Support each of your answers with sketches.

26. The sum of the squares of two positive numbers is 13. If twice the first number is added to the second, the sum is 7. Find the numbers.

27. The sum of two numbers is 6 and their product is 35/4. Find the numbers.

28. The annual income from an investment is $32. If the amount invested were $200 more and the rate 1/2% less, the annual income would be $35. What are the amount and rate of the investment?

29. At a constant temperature, the pressure P and volume V of a gas are related by the equation $PV = K$. The product of the pressure (in pounds per square inch) and the volume (in cubic inches) of a certain gas is 30 inch-pounds. If

the temperature remains constant as the pressure is increased 4 pounds per square inch, the volume is decreased by 2 cubic inches. Find the original pressure and volume of the gas.

30. What relationships must exist between the numbers a and b so that the solution set of the system

$$x^2 + y^2 = 25$$
$$y = ax + b$$

has two ordered pairs of real numbers? One ordered pair of real numbers? No ordered pairs of real numbers? *Hint:* Use substitution and consider the nature of the roots of the resulting quadratic equation.

31. Consider the system

$$x^2 + y^2 = 8 \qquad (1)$$
$$xy = 4. \qquad (2)$$

We can solve this system by substituting $4/x$ for y in (1) to obtain

$$x^2 + \frac{16}{x^2} = 8,$$

from which we obtain $x = 2$ or $x = -2$. Now if we obtain the y-components of the solution from (2), we find that for $x = 2$ we have $y = 2$, and that for $x = -2$ we have $y = -2$. But if we seek y-components from (1), for $x = 2$ we have $y = \pm 2$, and for $x = -2$, $y = \pm 2$. Discuss the fact that we seem to obtain two more solutions from (1) than from (2). What is the solution set of the system?

11.4 Systems of Inequalities

Graphs of systems of inequalities

In Chapter 6, we observed that the graph of the solution set of an inequality in two variables might be a region in the plane. The graph of the solution set of a system of inequalities in two variables, which consists of the intersection of the graphs of the inequalities in the system, might also be a plane region.

Example

Graph the solution set of the system

$$x + 2y \leq 6$$
$$2x - 3y \geq 12.$$

Solution

Graphing each inequality by the method discussed in Section 6.5, we obtain the figure shown, where the doubly shaded region is the graph of the solution set of the system. We have graphed

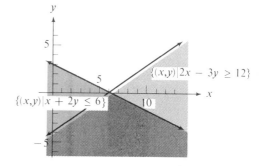

$\{(x, y) \mid x + 2y \leq 6\} \cap \{(x, y) \mid 2x - 3y \geq 12\}$.

11.4 Systems of Inequalities

Example Graph the solution set of the system
$$y \geq x + 2$$
$$y \geq x^2.$$

Solution The graph of each inequality is shaded as shown in the figure. The doubly shaded region constitutes the graph of the solution set of the system. We have graphed
$$\{(x, y) \mid y \geq x^2\} \cap \{(x, y) \mid y \leq x + 2\}.$$

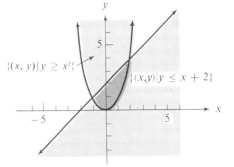

Example Graph the solution set of the system
$$y > 2$$
$$x > -2$$
$$x + y > 1.$$

Solution The triply shaded region in the figure constitutes the graph of the solution set of the system. We have graphed
$$\{(x, y) \mid y > 2\} \cap \{(x, y) \mid x > -2\}$$
$$\cap \{(x, y) \mid x + y > 1\}.$$

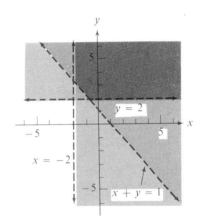

Exercise 11.4

By double or triple shading, indicate the region in the plane representing the solution set of each system.

1. $y \geq x + 1$
   $y \geq 5 - x$

2. $y \geq 4$
   $x \geq 2$

3. $y - x \geq 0$
   $y + x \geq 0$

4. $2y - x \geq 1$
   $x < -3$

5. $y + 3x < 6$
   $y > 2$

6. $x > 3$
   $x + y \geq 5$

7. $x = 3$
   $x + y < 4$

8. $y = 2$
   $2y + x < 3$

9. $x - 3y < 6$
   $y + x = 1$

10. $2x + y \geq 4$
    $x - y = -2$

11. $y > x^2 + 1$
    $x + y > 4$

12. $y < x^2 + 4$
    $x - y \leq 4$

13. $x^2 + y^2 \leq 25$
    $y > 3$

14. $x^2 + y^2 \geq 25$
    $y > x^2$

15. $9x^2 + 16y^2 \geq 144$
    $y < x^2$

16. $y \geq 2$
    $x \geq 2$
    $y \geq x$

17. $y \leq 3$
    $x \leq 2$
    $y < x$

18. $x^2 + y^2 \leq 36$
    $x \geq 3$
    $y \geq 3$

19. $x^2 + y^2 \leq 25$
    $y \geq x^2 - 4$
    $y \leq -x^2 + 4$

20. $y \leq \log_{10} x$
    $y \geq x - 1$

21. $y \leq 10^x$
    $y \leq 2 - x^2$

22. $9 \leq x^2 + y^2 \leq 16$
    $-1 \leq x - y \leq 1$

23. $9 \leq x^2 + y^2 \leq 16$
    $x^2 + 1 \leq y \leq x^2 + 3$

24. $x^2 - 2 \leq y \leq 2 - x^2$
    $|x| \geq 1$

11.5 Convex Sets—Polygonal Regions

From the examples in Section 11.4 it is apparent that the graph of the solution set of a system of linear inequalities in two variables is simply the common intersection of a number of open or closed half-planes. Any such intersection is an example of a *convex set*.

Definition 11.2 *A set $\mathscr{S}$ of points is a **convex set** if and only if, for each two points P and Q in $\mathscr{S}$, the line segment $\overline{PQ}$ lies entirely in $\mathscr{S}$.*

Figure 11.6 shows three sets of points in a plane. In Figures 11.6-a and 11.6-b are pictured examples of convex sets, while the set in Figure 11.6-c is not convex because part of the line segment $\overline{PQ}$ does not lie in the set. If the boundary of a convex set is a (closed) polygon, then the boundary is called a **convex polygon,** and the region enclosed by the polygon (including the boundary) is called a **closed convex polygonal set.**

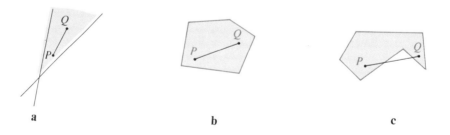

Figure 11.6

We have two immediate results from Definition 11.2 which are intuitively true and whose proofs are omitted.

Theorem 11.3 *Any half-plane is a convex set.*

11.5 Convex Sets—Polygonal Regions

Theorem 11.4 *The intersection of two convex sets is a convex set.*

As suggested by Figure 11.7, this latter result extends to any number of convex sets. If the intersection of the graphs of a system of linear inequalities constitutes a (closed) polygonal set $\mathcal{P}$, then we can locate the set $\mathcal{P}$ in the plane by graphing the equations associated with the given inequalities and identifying the vertices $\mathcal{P}$.

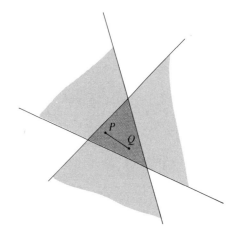

Figure 11.7

Example Locate and identify by shading the polygonal set specified by the system

$$x + y \leq 5$$
$$x - y \leq 2$$
$$x - y \geq -2$$
$$x \geq 0$$
$$y \geq 0.$$

Solution Graph the associated equations

$$x + y = 5$$
$$x - y = 2$$
$$x - y = -2$$
$$x = 0$$

and

$$y = 0,$$

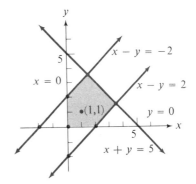

as shown. Note which half-plane is determined by each of the five inequalities, and shade the intersection of the five half-planes, The intersection is the pentagonal region shown. To check the result, note that (1, 1) is in the region and its coordinates satisfy each given inequality.

We should not conclude from the foregoing example that, simply because the five equations associated with the set of inequalities determine a convex pentagon, the pentagonal region is necessarily the graph of the system. Thus, if the inequality $y \geq 0$ in the example is replaced with $y \leq 0$, then the graph of the equations remains unchanged, but the graph of the system of inequalities is the triangular region shown in Figure 11.8-a. On the other hand, if the inequality $x + y \leq 5$ is replaced with $x + y \geq 5$, then the intersection of the graphs becomes the (infinite) rectangular region shown in Figure 11.8-b. Finally we observe that if both of the foregoing changes are made, that is, if $x + y \geq 5$ replaces $x + y \leq 5$ and $y \leq 0$ replaces $y \geq 0$, then the intersection of the graphs of the inequalities in the system is $\emptyset$.

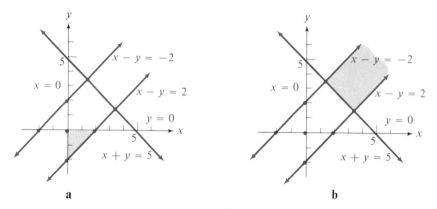

Figure 11.8

Exercise 11.5

Graph the convex polygonal set defined by the given system of inequalities.

1. $x - y \leq 2$
   $x - y \geq -2$
   $x + y \geq 2$
   $x + y \leq 6$

2. $0 \leq x \leq 5$
   $0 \leq y \leq 5$
   $x + y \leq 6$

3. $0 \leq x \leq 5$
   $0 \leq y \leq 4$
   $x + 2y \leq 10$
   $2x + y \leq 10$

4. $0 \leq x$
   $0 \leq y \leq 4$
   $x - 2y \leq 6$
   $2x + y \leq 12$
   $2x - y \leq 10$

5. $0 \leq x \leq 5$
   $y \geq x - 3$
   $x + y \leq 9$
   $3y \leq 2x + 12$

6. $0 \leq x$
   $0 \leq y \leq 6$
   $x + 2y \leq 13$
   $2x + y \leq 11$
   $3x + y \leq 15$

7. Repeat Exercise 1 with $x - y \leq 2$ replaced with $x - y \geq 2$. Is the graph a closed polygonal set?

8. Repeat Exercise 3 with $x + 2y \leq 10$ replaced with $x + 2y \geq 10$. Is the graph a closed polygonal set?

9. Repeat Exercise 4 with $0 \leq y \leq 4$ replaced with $y \leq 4$. Is the graph a closed polygonal set?

10. Repeat Exercise 6 without the condition $x \geq 0$. Is the graph a closed polygonal set?

11.6 Linear Programming

For each ordered pair $(x, y) \in R^2$, the linear expression $ax + by$ has a value. In particular, if $\mathcal{P} \subset R^2$, and $\mathcal{P}$ is a closed polygonal subset of R^2, then $ax + by$ has a value for each $(x, y) \in \mathcal{P}$. It can be shown that the following result holds for the values of $ax + by$ over $\mathcal{P}$.

Theorem 11.5 *If $\mathcal{P} \subset R^2$ is a closed convex polygonal set, then the expression $ax + by$ takes on its maximum and minimum values over $\mathcal{P}$ at one of the vertices of $\mathcal{P}$.*

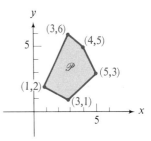

Thus if $\mathcal{P}$ is the closed convex polygonal set with vertices (1, 2), (3, 1), (5, 3), (4, 5), and (3, 6), as pictured in Figure 11.9, then the linear expression $3x - 2y$ has a value for every ordered pair (x, y) in, or on the boundary of, $\mathcal{P}$. In particular, at the vertices we have the values shown in the table below.

Figure 11.9

Vertex	(1, 2)	(3, 1)	(5, 3)	(4, 5)	(3, 6)
Value of $3x - 2y$	-1	7	9	2	-3

By inspection, the maximum of these values is 9 and the minimum is -3. By Theorem 11.5, then, 9 is the greatest value $3x - 2y$ has over the entire set $\mathcal{P}$, and its least value over $\mathcal{P}$ is -3.

Theorem 11.5 has many important applications.

Example A company manufactures two kinds of electric shavers, one using a cord and the other a cordless model. The company can make up to 500 cord models and 400 cordless models per day, but it can make a total of only 600 shavers per day. It takes 2 man-hours to manufacture the cord model and 3 man-hours to manufacture the cordless model, and the company has available at most 1400 man-hours per day. If there is a profit of \$2.50 on each cord shaver and \$3.50 on each cordless shaver, how many of each kind should the company make each day in order to realize the greatest profit?

Solution Let x = number of cord shavers made in one day;

y = number of cordless shavers made in one day.

We have the following restraints on x and y:

$$0 \leq x \leq 500, \quad 0 \leq y \leq 400,$$
$$x + y \leq 600, \quad 2x + 3y \leq 1400.$$

Solving the associated equations for these inequalities in pairs yields the vertices of the polygonal region shown in the figure. We wish to maximize

$$2.5x + 3.5y.$$

We find the values for this expression at the vertices to be:

Vertex	Profit
(0, 0)	$ 0
(0, 400)	$1400
(100, 400)	$1650
(400, 200)	$1700
(500, 100)	$1601
(500, 0)	$1250

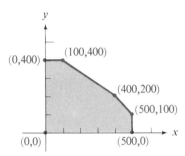

Clearly, the most profitable numbers of shavers are 400 cord models and 200 cordless models.

Exercise 11.6

Find the maximum and minimum values of the given expression over the given closed convex polygonal set for the specified set of Exercise 11.5, page 326.

1. $x + 3y$; the set in Exercise 1.
2. $3x + 5y$; the set in Exercise 2.
3. $3x + 5y$; the set in Exercise 3.
4. $x + 2y$; the set in Exercise 4.
5. $5x - 4y$; the set in Exercise 5.
6. $10x - 2y$; the set in Exercise 6.

For each Exercise 7–10, use Tables A and B below, which show the number of units of labor, machinery, materials, and overhead necessary for a manufacturer to produce one unit of each of two products x and y. The tables also show the total available units of each of these items.

Table A

	Units needed for 1 unit of product x	Units needed for 1 unit of product y	Total units available
Labor	1	6	120
Machinery	1	3	66
Material	3	2	86
Overhead	10	3	250

Table B

	Units needed for 1 unit of product x	Units needed for 1 unit of product y	Total units available
Labor	2	7	63
Machinery	5	1	50
Material	1	2	21
Overhead	1	1	14

Chapter Review

7. Using Table A, find the maximum profit if each unit of product x earns a profit of $50 and each unit of product y earns a profit of $60.

8. Using Table A find the maximum profit if each unit of product x earns a profit of $60 and each unit of product y earns a profit of $50.

9. Using Table B, find the maximum profit if each unit of product x earns a profit of $50 and each unit of product y earns a profit of $60.

10. Using Table B, find the maximum profit if each unit of product x earns a profit of $60 and each unit of product y earns a profit of $50.

11. An electronics company makes two kinds of electronic ranges, a standard model that earns a $50 profit and a deluxe model that earns a $60 profit. The company has machinery capable of producing any number of deluxe ranges up to 400 per month and any number of standard ranges up to 500 per month, but it has enough man-hours available to produce only 600 ranges of both kinds in a month. How many of each kind of range should the company produce to realize a maximum profit?

12. A pharmacy has 300 ounces of a drug it can use to make two different kinds of medicines. From each ounce of the drug, 15 bottles of medicine A or 25 bottles of medicine B can be produced. The pharmacy must keep all the bottles in its own storage room, which has a capacity of 6000 bottles in all. At least 600 bottles of medicine A and 1250 bottles of medicine B must be retained in inventory and the rest will be used. If medicine A sells for $2.75 per bottle and medicine B sells for $2.00 per bottle, how many ounces of the drug should be devoted to each medicine to maximize the total return?

Chapter Review

[11.1] Solve each system.

1. $x + 5y = 18$
   $x - y = -3$

2. $x + 5y = 11$
   $2x + 3y = 8$

3. $2x - 3y = 8$
   $3x + 2y = 7$

4. Find values for a and b so that the graph of $ax + by = 19$ passes through the points $(2, 3)$ and $(-3, 5)$.

[11.2] Solve each system.

5. $x + 3y - z = 3$
   $2x - y + 3z = 1$
   $3x + 2y + z = 5$

6. $x + y + z = 2$
   $3x - y + z = 4$
   $2x + y + 2z = 3$

7. $2x + 3y - z = -2$
   $x - y + z = 6$
   $3x - y + z = 10$

8. Find values for a, b, and c so that the graph of $y = ax^2 + bx + c$ contains the points $(-1, 9)$, $(0, 4)$, and $(1, 3)$.

[11.3] Solve each system.

9. $x^2 + y = 3$
    $5x + y = 7$

10. $x^2 + 3xy + x = -12$
    $2x - y = 7$

11. $2x^2 + 5y^2 - 53 = 0$
    $4x^2 + 3y^2 - 43 = 0$

12. Approximate the solution of the system

    $$y = 3^x$$
    $$y = 3 - x$$

    by graphical methods.

[11.4] By double shading, indicate the region representing the solution set of each system.

13. $y > x^2 - 4$
    $y < 2 - x$

14. $y + x^2 < 0$
    $y + 3 > 0$
    $y < x - 3$

[11.5] Graph the polygonal set defined by each system of inequalities.

15. $0 \leq x \leq 3$
    $0 \leq y \leq 4$
    $x + y \leq 5$

16. $0 \leq x \leq 3$
    $0 \leq y$
    $y \geq 2 + x$
    $x + y \leq 4$

[11.6] 17. Find the maximum and minimum values of $2x + y$ over the set in Exercise 15.

18. Find the maximum and minimum values of $3x + 2y$ over the set in Exercise 16.

19. Using Table A on page 328, find the maximum profit if each unit of product x earns a profit of $80 and each unit of product y earns a profit of $60.

20. Using Table B on page 328, find the maximum profit if each unit of product x earns a profit of $10 and each unit of product y earns a profit of $20.

12 Matrices and Determinants

The systems of equations studied in Chapter 11 are determined by the numerical coefficients. The rectangular array of coefficients is an example of a matrix. Matrices, as introduced in this chapter, are today much used in mathematics and engineering, and also in the physical, social, and life sciences.

12.1 Definitions; Matrix Addition

A **matrix** is a rectangular array of numbers (or other suitable entities), which are called the **entries** or **elements** of the matrix. In this book, we shall consider only real numbers as entries. A matrix is customarily displayed in a pair of brackets or parentheses (we shall use brackets). Thus

$$\begin{bmatrix} 1 & 2 & 3 \\ 4 & 5 & 6 \end{bmatrix} \quad \text{and} \quad \begin{bmatrix} 2 \\ 1 \end{bmatrix}$$

are matrices. The **order**, or **dimension**, of a matrix is the ordered pair having as first component the number of (horizontal) **rows** and as second component the number of (vertical) **columns** in the matrix. Thus,

$$\begin{bmatrix} 1 & 2 & 3 \\ 4 & 5 & 6 \end{bmatrix}, \quad \begin{bmatrix} 1 \\ 2 \\ 3 \end{bmatrix}, \quad \text{and} \quad \begin{bmatrix} a_1 & a_2 & a_3 & a_4 \\ b_1 & b_2 & b_3 & b_4 \\ c_1 & c_2 & c_3 & c_4 \\ d_1 & d_2 & d_3 & d_4 \end{bmatrix}$$

are 2×3 (read "two by three"), 3×1 (read "three by one"), and 4×4 (read "four by four") matrices, respectively. Note that the number of *rows* is given first, and then the number of *columns*. A matrix consisting of a single row is called a **row matrix** or a **row vector**, whereas a matrix consisting of a single column is called a **column matrix** or a **column vector**.

Matrices are frequently represented by capital letters. Thus, we might want to talk about the matrices A and B, where

$$A = \begin{bmatrix} a_1 & a_2 \\ b_1 & b_2 \end{bmatrix} \quad \text{and} \quad B = [b_1 \ b_2].$$

To show that A is a 2×2 matrix, we can write $A_{2 \times 2}$. Similarly, $B_{1 \times 2}$ is a matrix with one row and two columns.

To represent the entries of a matrix, either single or double subscript notation is employed. Consider any 3×3 matrix, A. We can represent A by

$$A = \begin{bmatrix} a_1 & a_2 & a_3 \\ b_1 & b_2 & b_3 \\ c_1 & c_2 & c_3 \end{bmatrix},$$

where a different letter is used for each row and a single subscript denotes the column in which each particular entry is located. Alternatively, we can use a different letter for each column, and let the subscript denote the row. Thus, we might write

$$A = \begin{bmatrix} a_1 & b_1 & c_1 \\ a_2 & b_2 & c_2 \\ a_3 & b_3 & c_3 \end{bmatrix}.$$

In either event, problems would clearly arise if we wanted to talk about a matrix containing a large number of rows or columns, because we would run out of letters.

A much more useful convention involves double subscripts, where a single letter, say a, is used to denote an entry in a matrix, and then *two* subscripts are appended, the first subscript telling in which *row* the entry occurs, and the second telling in which *column*. Thus, we write

$$A = \begin{bmatrix} a_{11} & a_{12} & a_{13} \\ a_{21} & a_{22} & a_{23} \\ a_{31} & a_{32} & a_{33} \end{bmatrix},$$

where a_{21} is the element in the second *row* and first *column*, a_{33} is the element in the third *row* and third *column*, and, if we wish to generalize, a_{ij} is the element in the ith *row* and jth *column*.

Definition 12.1   Two matrices, A and B, are **equal** if and only if both matrices are of the same order and $a_{ij} = b_{ij}$ for each i, j.

Thus,

$$\begin{bmatrix} 2 & 1 \\ 3 & 0 \end{bmatrix} = \begin{bmatrix} \frac{4}{2} & 2-1 \\ \sqrt{9} & 0 \end{bmatrix}, \quad \text{but} \quad \begin{bmatrix} 2 & 1 \\ 3 & 0 \end{bmatrix} \neq \begin{bmatrix} 2 & 3 \\ 1 & 0 \end{bmatrix}.$$

12.1 Definitions; Matrix Addition

Definition 12.2   The **transpose** of a matrix A, denoted by A^t, is the matrix in which the rows are the columns of A and the columns are the rows of A.

Thus,

$$\begin{bmatrix} 2 & 1 \\ 3 & 0 \end{bmatrix}^t = \begin{bmatrix} 2 & 3 \\ 1 & 0 \end{bmatrix} \text{ and } \begin{bmatrix} 1 & 2 & 3 \\ 4 & 5 & 6 \end{bmatrix}^t = \begin{bmatrix} 1 & 4 \\ 2 & 5 \\ 3 & 6 \end{bmatrix}.$$

Definition 12.3   The **sum** of two matrices of the same order, $A_{m \times n}$ and $B_{m \times n}$, is the matrix $(A + B)_{m \times n}$, in which the entry in the ith row and jth column is $a_{ij} + b_{ij}$, for $i = 1, 2, 3, \ldots, m$ and $j = 1, 2, 3, \ldots, n$.

For example,

$$\begin{bmatrix} 3 & 1 & 2 \\ 2 & 1 & 4 \end{bmatrix} + \begin{bmatrix} 1 & 0 & 2 \\ -1 & 3 & 0 \end{bmatrix} = \begin{bmatrix} 3+1 & 1+0 & 2+2 \\ 2+(-1) & 1+3 & 4+0 \end{bmatrix}$$

$$= \begin{bmatrix} 4 & 1 & 4 \\ 1 & 4 & 4 \end{bmatrix}.$$

Thus the sum of two matrices of the same order is obtained by adding the corresponding entries; the sum of two matrices of different orders is not defined.

Definition 12.4   A matrix with each entry equal to 0 is a **zero matrix**.

Zero matrices are generally denoted by the symbol **0**. This distinguishes the zero matrix from the real number 0. For example,

$$0_{2 \times 4} = \begin{bmatrix} 0 & 0 & 0 & 0 \\ 0 & 0 & 0 & 0 \end{bmatrix}$$

is the 2×4 zero matrix.

Definition 12.5   The **negative** of a matrix $A_{m \times n}$, denoted by $-A_{m \times n}$, is formed by replacing each entry in the matrix $A_{m \times n}$ with its additive inverse.

For example, if

$$A_{3 \times 2} = \begin{bmatrix} 3 & -1 \\ 2 & -2 \\ -4 & 5 \end{bmatrix}, \text{ then } -A_{3 \times 2} = \begin{bmatrix} -3 & 1 \\ -2 & 2 \\ 4 & -5 \end{bmatrix}.$$

The sum $B_{m \times n} + (-A_{m \times n})$ is called the **difference** of $B_{m \times n}$ and $A_{m \times n}$ and is denoted $B_{m \times n} - A_{m \times n}$.

Properties of sums

At this point, we are able to establish the following facts concerning sums of matrices with real-number entries.

Theorem 12.1 If A, B, and C are $m \times n$ matrices with real-number entries, then:

I $(A + B)_{m \times n}$ is a matrix with real-number entries.   *Closure law for addition.*

II $(A + B) + C = A + (B + C)$.   *Associative law for addition.*

III The matrix $\mathbf{0}_{m \times n}$ has the property that for every matrix $A_{m \times n}$,   *Additive-identity law.*

$$A + \mathbf{0} = A \quad \text{and} \quad \mathbf{0} + A = A.$$

IV For every matrix $A_{m \times n}$, the matrix $-A_{m \times n}$ has the property that   *Additive-inverse law.*

$$A + (-A) = \mathbf{0} \quad \text{and} \quad (-A) + A = \mathbf{0}.$$

V $A + B = B + A$.   *Commutative law for addition.*

Proof of 12.1-III Since each entry of the zero matrix is 0, it follows that the entries of $A_{m \times n} + \mathbf{0}_{m \times n}$ are $a_{ij} + 0 = a_{ij}$ and the entries of $\mathbf{0}_{m \times n} + A_{m \times n}$ are $0 + a_{ij} = a_{ij}$, and part III is proved.

For example,

$$\begin{bmatrix} a_{11} & a_{12} \\ a_{21} & a_{22} \end{bmatrix} + \begin{bmatrix} 0 & 0 \\ 0 & 0 \end{bmatrix} = \begin{bmatrix} a_{11} & a_{12} \\ a_{21} & a_{22} \end{bmatrix}.$$

Proof of 12.1-IV Let the entries of $A_{m \times n}$ and $-A_{m \times n}$ be a_{ij} and $-a_{ij}$, respectively. Since each entry of $A + (-A)$ is $a_{ij} - a_{ij}$, or 0, we have $A + (-A) = \mathbf{0}$. Similarly, $(-A) + A = \mathbf{0}$, and part IV is proved.

For example, if

$$A = \begin{bmatrix} 1 & -1 & 2 \\ 3 & -1 & 1 \end{bmatrix},$$

then

$$A + (-A) = \begin{bmatrix} 1 & -1 & 2 \\ 3 & -1 & 1 \end{bmatrix} + \begin{bmatrix} -1 & 1 & -2 \\ -3 & 1 & -1 \end{bmatrix} = \begin{bmatrix} 0 & 0 & 0 \\ 0 & 0 & 0 \end{bmatrix} = \mathbf{0}.$$

The proof of Parts I, II, and V of Theorem 12.1 are left as exercises.

12.1 Definitions; Matrix Addition

Exercise 12.1

State the order and find the transpose of each matrix.

Example

$$\begin{bmatrix} 2 & 4 \\ 1 & -3 \\ 6 & 0 \end{bmatrix}$$

Solution 3×2 matrix; $\begin{bmatrix} 2 & 4 \\ 1 & -3 \\ 6 & 0 \end{bmatrix}^t = \begin{bmatrix} 2 & 1 & 6 \\ 4 & -3 & 0 \end{bmatrix}$

1. $\begin{bmatrix} 6 & -1 \\ 2 & 3 \end{bmatrix}$
2. $\begin{bmatrix} 4 & 1 \\ 0 & -2 \end{bmatrix}$
3. $\begin{bmatrix} 2 & -7 & 3 \\ 1 & 4 & 0 \end{bmatrix}$

4. $\begin{bmatrix} -3 & 1 \\ 6 & 0 \\ 0 & 2 \end{bmatrix}$
5. $\begin{bmatrix} 2 & 3 & -1 \\ 4 & 0 & 1 \\ -2 & 3 & 1 \end{bmatrix}$
6. $\begin{bmatrix} 4 & -1 & -2 \\ 3 & 0 & 0 \\ 2 & 1 & 1 \end{bmatrix}$

7. $\begin{bmatrix} 4 & -3 & -1 & 0 \\ 2 & 1 & 1 & 6 \end{bmatrix}$
8. $\begin{bmatrix} -2 & 1 & 3 & 2 \\ 4 & 0 & 0 & -2 \\ -1 & 3 & 2 & 4 \end{bmatrix}$

Write each sum or difference as a single matrix.

Example

$$\begin{bmatrix} 2 & 1 & 4 \\ 3 & -1 & 0 \end{bmatrix} + \begin{bmatrix} 6 & 3 & 0 \\ -2 & 1 & 0 \end{bmatrix}$$

Solution

$$\begin{bmatrix} 2 & 1 & 4 \\ 3 & -1 & 0 \end{bmatrix} + \begin{bmatrix} 6 & 3 & 0 \\ -2 & 1 & 0 \end{bmatrix} = \begin{bmatrix} 2+6 & 1+3 & 4+0 \\ 3-2 & -1+1 & 0+0 \end{bmatrix} = \begin{bmatrix} 8 & 4 & 4 \\ 1 & 0 & 0 \end{bmatrix}$$

9. $\begin{bmatrix} 2 & 3 \\ 1 & 6 \end{bmatrix} + \begin{bmatrix} 1 & -2 \\ 2 & 3 \end{bmatrix}$
10. $\begin{bmatrix} 4 & -1 & 3 \\ 2 & 1 & 0 \end{bmatrix} + \begin{bmatrix} 3 & -1 & 0 \\ 4 & 0 & -2 \end{bmatrix}$

11. $\begin{bmatrix} 3 & 0 & -1 \\ 2 & 1 & 2 \end{bmatrix} + \begin{bmatrix} 6 & -1 & 0 \\ 0 & 2 & 4 \end{bmatrix}$
12. $[1 \quad 3 \quad 5 \quad 7] + [0 \quad -2 \quad 1 \quad 3]$

13. $\begin{bmatrix} 4 & -3 \\ 2 & 1 \end{bmatrix} - \begin{bmatrix} 6 & 0 \\ -2 & 1 \end{bmatrix}$
14. $\begin{bmatrix} 4 & -1 & 2 \\ 3 & 1 & -4 \end{bmatrix} - \begin{bmatrix} -1 & -1 & 2 \\ 3 & 1 & 4 \end{bmatrix}$

15. $\begin{bmatrix} 10 & 3 & 2 \\ 5 & 1 & 7 \\ 6 & 1 & 9 \end{bmatrix} - \begin{bmatrix} 8 & 12 & 15 \\ -2 & 5 & 6 \\ -3 & 1 & 9 \end{bmatrix}$
16. $\begin{bmatrix} 3 & -1 & 2 \\ 4 & -2 & 1 \\ 6 & 3 & 2 \end{bmatrix} - \begin{bmatrix} 2 & -1 & 2 \\ 4 & -1 & 1 \\ 6 & 3 & 1 \end{bmatrix}$

17. $\begin{bmatrix} 4 \\ 3 \\ -1 \end{bmatrix} + \begin{bmatrix} 6 \\ 0 \\ -2 \end{bmatrix}$

18. $\begin{bmatrix} 2 & 3 \\ 1 & 0 \\ -1 & 2 \end{bmatrix} + \begin{bmatrix} -2 & 0 \\ -3 & 0 \\ 4 & -1 \end{bmatrix}$

19. $\begin{bmatrix} 2 & 3 & 4 \\ -1 & 6 & 2 \\ 1 & 0 & 3 \end{bmatrix} + \begin{bmatrix} 0 & 0 & 0 \\ 0 & 0 & 0 \\ 0 & 0 & 0 \end{bmatrix}$

20. $\begin{bmatrix} 2 & -3 \\ 4 & -1 \\ -2 & 1 \end{bmatrix} + \begin{bmatrix} -2 & 3 \\ -4 & 1 \\ 2 & -1 \end{bmatrix}$

21. Use Theorem 12.1 to argue that $X + A = B$ and $X = B - A$ are equivalent matrix equations in the system of 2×2 matrices.

Solve each of the following matrix equations.

Example

$X + \begin{bmatrix} 2 & 3 \\ 1 & 7 \end{bmatrix} = \begin{bmatrix} 9 & -4 \\ 2 & 0 \end{bmatrix}$

Solution

$X = \begin{bmatrix} 9 & -4 \\ 2 & 0 \end{bmatrix} - \begin{bmatrix} 2 & 3 \\ 1 & 7 \end{bmatrix} = \begin{bmatrix} 7 & -7 \\ 1 & -7 \end{bmatrix}$

22. $X + \begin{bmatrix} 3 & -1 \\ 2 & 1 \end{bmatrix} = \begin{bmatrix} 5 & 1 \\ -3 & 5 \end{bmatrix}$

23. $X - \begin{bmatrix} -1 & 0 \\ 0 & 0 \end{bmatrix} = \begin{bmatrix} 3 & -1 \\ 2 & 1 \end{bmatrix}^t$

24. $X + \begin{bmatrix} 3 & 2 \\ -1 & 4 \end{bmatrix} = \begin{bmatrix} 2 & 6 \\ 1 & 5 \end{bmatrix} + \begin{bmatrix} -4 & -8 \\ -2 & 0 \end{bmatrix}$

25. $\begin{bmatrix} 1 & 3 \\ -1 & 0 \end{bmatrix}^t - \begin{bmatrix} 0 & 1 \\ 1 & 0 \end{bmatrix} = \begin{bmatrix} 2 & -2 \\ -1 & 3 \end{bmatrix}^t - X$

26. Show that $[A_{2 \times 2} + B_{2 \times 2}]^t = A^t_{2 \times 2} + B^t_{2 \times 2}$. Does an analogous result seem valid for $n \times n$ matrices?

27. Prove Theorem 12.1-I.   28. Prove Theorem 12.1-II.
29. Prove Theorem 12.1-V.

12.2 Matrix Multiplication

We shall be interested in two kinds of products involving matrices: (1) the product of a matrix and a real number, and (2) the product of two matrices.

Definition 12.6 The **product** of a real number c and an $m \times n$ matrix A with entries a_{ij} is the matrix cA with corresponding entries ca_{ij}, where $i = 1, 2, 3, \ldots, m$ and $j = 1, 2, 3, \ldots, n$.

For example,

$$3 \begin{bmatrix} 2 & 1 \\ 0 & 5 \end{bmatrix} = \begin{bmatrix} 3 \times 2 & 3 \times 1 \\ 3 \times 0 & 3 \times 5 \end{bmatrix} = \begin{bmatrix} 6 & 3 \\ 0 & 15 \end{bmatrix}.$$

12.2 Matrix Multiplication

Properties of products of matrices and real numbers

The following theorem states some simple algebraic laws for the multiplication of matrices by real numbers.

Theorem 12.2   If A and B are $m \times n$ matrices, and $c, d \in R$, then

    I cA is an $m \times n$ matrix,   V $1A = A$,

    II $c(dA) = (cd)A$,   VI $(-1)A = -A$,

    III $(c + d)A = cA + dA$,   VII $0A = \mathbf{0}$,

    IV $c(A + B) = cA + cB$,   VIII $c\mathbf{0} = \mathbf{0}$.

Proof We shall prove only Part IV, leaving the remaining parts as exercises. Since the elements of $A + B$ are of the form $a_{ij} + b_{ij}$, it follows, by definition, that the elements of $c(A + B)$ are of the form $c(a_{ij} + b_{ij})$. But, since a_{ij}, b_{ij}, and c denote real numbers, $c(a_{ij} + b_{ij}) = ca_{ij} + cb_{ij}$. Now, the elements of cA are of the form ca_{ij}, and those of cB are of the form cb_{ij}, so that the elements of $cA + cB$ are of the form $ca_{ij} + cb_{ij}$ and part IV is proved.

Products of matrices

Turning now to the *product of two matrices*, we have the following definitions.

Definition 12.7   The scalar product of the $1 \times p$ row matrix with entries $a_1, a_2, \ldots, a_p$, and the $p \times 1$ column matrix with entries $b_1, b_2, \ldots, b_p$ is

$$a_1 b_1 + a_2 b_2 + \cdots + a_p b_p.$$

Example   The product of $[1, 0, -1, 2]$ and $\begin{bmatrix} 3 \\ 1 \\ -2 \\ 4 \end{bmatrix}$ is

$$1 \cdot 3 + 0 \cdot 1 + (-1)(-2) + 2 \cdot 4 = 13.$$

Definition 12.8   The product of matrices $A_{m \times p}$ and $B_{p \times n}$ is the $m \times n$ matrix whose i, j entry is the scalar product of the ith row of A and the jth column of B.

Example   Multiply $\begin{bmatrix} 3 & 0 & 1 \\ 0 & 1 & 2 \end{bmatrix} \begin{bmatrix} 1 & -2 \\ -1 & 2 \\ 1 & 1 \end{bmatrix}$.

Solution   The product will be 2×2.

The 1, 1 entry is $[3 \ 0 \ 1] \begin{bmatrix} 1 \\ -1 \\ 1 \end{bmatrix} = 3 \cdot 1 + 0(-1) + 1 \cdot 1 = 4$;

the 1, 2 entry is $[3 \ 0 \ 1] \begin{bmatrix} -2 \\ 2 \\ 1 \end{bmatrix} = 3(-2) + 0(2) + 1 \cdot 1 = -5$;

Solution Continued on Overleaf

the 2, 1 entry is $[0 \ 1 \ 2] \begin{bmatrix} 1 \\ -1 \\ 1 \end{bmatrix} = 0 \cdot 1 + 1(-1) + (2 \cdot 1) = 1;$

the 2, 2 entry is $[0 \ 1 \ 2] \begin{bmatrix} -2 \\ 2 \\ 1 \end{bmatrix} = 0(-2) + 1 \cdot 2 + 2 \cdot 1 = 4.$

Therefore $\begin{bmatrix} 3 & 0 & 1 \\ 0 & 1 & 2 \end{bmatrix} \begin{bmatrix} 1 & -2 \\ -1 & 2 \\ 1 & 1 \end{bmatrix} = \begin{bmatrix} 4 & -5 \\ 1 & 4 \end{bmatrix}.$

Example If $A = \begin{bmatrix} 1 & 2 \\ -1 & 3 \end{bmatrix}$ and $B = \begin{bmatrix} 2 & 1 \\ 1 & 1 \end{bmatrix}$, find AB and BA.

Solution
$$AB = \begin{bmatrix} 1 & 2 \\ -1 & 3 \end{bmatrix} \begin{bmatrix} 2 & 1 \\ 1 & 1 \end{bmatrix} = \begin{bmatrix} 2+2 & 1+2 \\ -2+3 & -1+3 \end{bmatrix} = \begin{bmatrix} 4 & 3 \\ 1 & 2 \end{bmatrix}$$

$$BA = \begin{bmatrix} 2 & 1 \\ 1 & 1 \end{bmatrix} \begin{bmatrix} 1 & 2 \\ -1 & 3 \end{bmatrix} = \begin{bmatrix} 2-1 & 4+3 \\ 1-1 & 2+3 \end{bmatrix} = \begin{bmatrix} 1 & 7 \\ 0 & 5 \end{bmatrix}$$

Properties of products The foregoing example shows very clearly that the multiplication of matrices, in general, is *not commutative*. Thus, when discussing products of matrices, we must specify the *order* in which the matrices are to be considered as factors. For the product AB, we say that A is *right-multiplied* by B, and that B is *left-multiplied* by A.

Note that the definition of the product of two matrices, A and B, requires that the matrix A have the same number of *columns* as B has *rows*; the result, AB, then has the same number of rows as A and the same number of columns as B. Such matrices A and B are said to be **conformable** for multiplication. The fact that two matrices are conformable in the order AB, however, does not mean that they necessarily are conformable in the order BA.

Example If $A = \begin{bmatrix} 3 & 1 \\ 1 & 0 \\ 2 & 1 \end{bmatrix}$ and $B = \begin{bmatrix} 1 & -1 \\ 2 & 1 \end{bmatrix}$, find AB.

Solution Since A is a 3×2 matrix, and B is a 2×2 matrix, they are conformable for multiplication in the order AB. We have

$$AB = \begin{bmatrix} 3 & 1 \\ 1 & 0 \\ 2 & 1 \end{bmatrix} \begin{bmatrix} 1 & -1 \\ 2 & 1 \end{bmatrix} = \begin{bmatrix} 3+2 & -3+1 \\ 1+0 & -1+0 \\ 2+2 & -2+1 \end{bmatrix} = \begin{bmatrix} 5 & -2 \\ 1 & -1 \\ 4 & -1 \end{bmatrix}.$$

Note that the matrices A and B in the example above are not conformable in the order BA.

In much of the matrix work in this book, we shall focus our attention on matrices having the same number of rows as columns. For brevity, a matrix of order $n \times n$

12.2 Matrix Multiplication

is often called a **square matrix** of order n. Although many of the ideas we shall discuss are applicable to conformable matrices of any order, we shall apply the notions only to square matrices.

Theorem 12.3   If A, B, and C are $n \times n$ square matrices, then
$$(AB)C = A(BC).$$

If A is a square matrix, then A^2, A^3, etc. denote AA, $(AA)A$, etc.

Theorem 12.4   If A, B, and C are $n \times n$ square matrices, then
$$A(B + C) = AB + AC$$
and
$$(B + C)A = BA + CA.$$

The proofs of these theorems involve some complicated symbolism and are omitted here, but you will be asked to show their validity for the case of 2×2 matrixes in the exercises. Observe that, because matrix multiplication is not, in general, commutative, we must establish both the left-hand and the right-hand distributive property.

Definition 12.9   The **principal diagonal** of a square matrix is the ordered set of entries a_{jj}, extending from the upper left-hand corner to the lower right hand corner of the matrix. Thus, the principal diagonal contains a_{11}, a_{22}, a_{33}, etc.

For example, the principle diagonal of

$$\begin{bmatrix} 1 & 3 & -1 \\ 5 & 2 & 3 \\ 6 & 4 & 0 \end{bmatrix}$$

consists of 1, 2, and 0, in that order.

Definition 12.10   A **diagonal matrix** is a square matrix in which all entries not in the principal diagonal are 0.

Thus,

$$\begin{bmatrix} 4 & 0 \\ 0 & 2 \end{bmatrix} \quad \text{and} \quad \begin{bmatrix} 1 & 0 & 0 \\ 0 & 1 & 0 \\ 0 & 0 & 0 \end{bmatrix}$$

are diagonal matrices.

Definition 12.11 $I_{n \times n}$ denotes the diagonal matrix having 1's for entries on the principal diagonal.

For example,

$$I_{2 \times 2} = \begin{bmatrix} 1 & 0 \\ 0 & 1 \end{bmatrix} \text{ and } I_{4 \times 4} = \begin{bmatrix} 1 & 0 & 0 & 0 \\ 0 & 1 & 0 & 0 \\ 0 & 0 & 1 & 0 \\ 0 & 0 & 0 & 1 \end{bmatrix}.$$

The following properties are consequences of the definitions we have adopted.

Theorem 12.5 For each matrix $A_{n \times n}$,

$$A_{n \times n} I_{n \times n} = I_{n \times n} A_{n \times n} = A_{n \times n}.$$

Furthermore, $I_{n \times n}$ is the unique matrix having this property for all matrices $A_{n \times n}$.

Accordingly, $I_{n \times n}$ is the **identity element for multiplication** in the set of $n \times n$ square matrices. The proof of this theorem, for the illustrative case $n = 2$, is left as an exercise.

Order of multiplication The following result relates the order in which matrices can be multiplied by real numbers and by other matrices.

Theorem 12.6 If A and B are $n \times n$ square matrices, and a is a real number, then

$$a(AB) = (aA)B = A(aB).$$

The proof for the illustrative case $n = 2$ is left as an exercise.

Exercise 12.2

Write each product as a single matrix.

Examples a. $3 \begin{bmatrix} 2 & 1 \\ -1 & 3 \\ 2 & 0 \end{bmatrix}$ b. $\begin{bmatrix} 3 & 1 & -1 \\ 0 & -1 & 2 \end{bmatrix} \cdot \begin{bmatrix} 1 & -1 \\ 0 & 2 \\ 1 & 0 \end{bmatrix}$

Solutions a. $\begin{bmatrix} 6 & 3 \\ -3 & 9 \\ 6 & 0 \end{bmatrix}$ b. $\begin{bmatrix} 3+0-1 & -3+2+0 \\ 0+0+2 & 0-2+0 \end{bmatrix} = \begin{bmatrix} 2 & -1 \\ 2 & -2 \end{bmatrix}$

1. $-5 \begin{bmatrix} 0 & 1 & -1 \\ 3 & -1 & 2 \end{bmatrix}$ **2.** $2 \begin{bmatrix} 2 & 1 & 3 & -2 \\ 4 & 2 & 0 & -1 \\ 0 & 0 & -1 & 2 \end{bmatrix}$

12.2 Matrix Multiplication

3. $[1 \ -2] \cdot \begin{bmatrix} 3 \\ 2 \end{bmatrix}$

4. $[3 \ -2 \ 2] \cdot \begin{bmatrix} 1 \\ 0 \\ -2 \end{bmatrix}$

5. $\begin{bmatrix} 3 & -1 \\ 2 & 1 \end{bmatrix} \cdot \begin{bmatrix} 1 & -4 \\ 2 & 1 \end{bmatrix}$

6. $\begin{bmatrix} 1 & -5 \\ 0 & 2 \end{bmatrix} \cdot \begin{bmatrix} 3 & 1 \\ -1 & 2 \end{bmatrix}$

7. $\begin{bmatrix} 4 & -5 \\ 7 & 3 \end{bmatrix} \cdot \begin{bmatrix} 5 & -1 \\ -2 & 7 \end{bmatrix}$

8. $\begin{bmatrix} 1 & -2 \\ -3 & 1 \end{bmatrix} \cdot \begin{bmatrix} 5 & 1 \\ 0 & 2 \end{bmatrix}$

9. $\begin{bmatrix} -3 & 1 & 0 \\ 2 & 1 & 1 \end{bmatrix} \cdot \begin{bmatrix} 2 & 0 \\ 1 & -1 \\ 3 & 0 \end{bmatrix}$

10. $\begin{bmatrix} 1 & -1 & 0 \\ 2 & 1 & 3 \end{bmatrix} \cdot \begin{bmatrix} 4 & -1 \\ 2 & 0 \\ 1 & 1 \end{bmatrix}$

11. $\begin{bmatrix} -1 & 0 & 1 \\ 2 & 1 & 0 \\ 1 & 0 & 0 \end{bmatrix} \cdot \begin{bmatrix} 0 & 1 & 3 \\ 1 & 0 & 2 \\ -1 & 1 & 1 \end{bmatrix}$

12. $\begin{bmatrix} 2 & -3 & 1 \\ 0 & 1 & -1 \\ 2 & 0 & 0 \end{bmatrix} \cdot \begin{bmatrix} 1 & 0 & 0 \\ 0 & 1 & 0 \\ 0 & 0 & 1 \end{bmatrix}$

13. $\begin{bmatrix} 2 & -2 & -1 \\ 1 & 1 & -2 \\ 1 & 0 & -1 \end{bmatrix} \cdot \begin{bmatrix} -1 & -2 & 5 \\ -1 & -1 & 3 \\ -1 & -2 & 4 \end{bmatrix}$

14. $\begin{bmatrix} -1 & -2 & 5 \\ -1 & -1 & 3 \\ -1 & -2 & 4 \end{bmatrix} \cdot \begin{bmatrix} 2 & -2 & -1 \\ 1 & 1 & -2 \\ 1 & 0 & -1 \end{bmatrix}$

Let $A = \begin{bmatrix} 1 & -2 \\ 1 & 0 \end{bmatrix}$ and $B = \begin{bmatrix} -1 & 2 \\ -1 & 1 \end{bmatrix}$. Compute each of the following products.

15. AB
16. BA
17. $(AB)A$
18. $(BA)B$
19. A^tB
20. AB^t

Find a matrix X satisfying each matrix equation.

21. $3X + \begin{bmatrix} 1 & 0 \\ 2 & 1 \end{bmatrix} = \begin{bmatrix} -2 & 3 \\ -1 & -2 \end{bmatrix}$

22. $2X + 3\begin{bmatrix} 1 & 1 \\ 0 & 1 \end{bmatrix} = \begin{bmatrix} 7 & -1 \\ 3 & -5 \end{bmatrix}$

23. $X + 2I = \begin{bmatrix} 3 & -1 \\ 1 & 2 \end{bmatrix}$

24. $3X - 2I = \begin{bmatrix} 7 & 3 \\ 6 & 4 \end{bmatrix}$

25. Show that if $A = \begin{bmatrix} -1 & 2 \\ 0 & 1 \end{bmatrix}$ and $B = \begin{bmatrix} 1 & 0 \\ -1 & 2 \end{bmatrix}$, then

    a. $(A + B)(A + B) \neq A^2 + 2AB + B^2$,
    b. $(A + B)(A - B) \neq A^2 - B^2$.

26. a. Show that for each matrix $A_{2 \times 2}$,
    $$A_{2 \times 2} \cdot I_{2 \times 2} = I_{2 \times 2} \cdot A_{2 \times 2} = A_{2 \times 2}.$$
    b. Show that if, for a given matrix $B_{2 \times 2}$ and for *all* $A_{2 \times 2}$,
    $$A_{2 \times 2} \cdot B_{2 \times 2} = B_{2 \times 2} \cdot A_{2 \times 2} = A_{2 \times 2},$$
    then $B_{2 \times 2} = I_{2 \times 2}$.

27. Show that

$$\left(\begin{bmatrix} a_{11} & a_{12} \\ a_{21} & a_{22} \end{bmatrix} \cdot \begin{bmatrix} b_{11} & b_{12} \\ b_{21} & b_{22} \end{bmatrix}\right) \cdot \begin{bmatrix} c_{11} & c_{12} \\ c_{21} & c_{22} \end{bmatrix} = \begin{bmatrix} a_{11} & a_{12} \\ a_{21} & a_{22} \end{bmatrix} \cdot \left(\begin{bmatrix} b_{11} & b_{12} \\ b_{21} & b_{22} \end{bmatrix} \cdot \begin{bmatrix} c_{11} & c_{12} \\ c_{21} & c_{22} \end{bmatrix}\right).$$

28. Show that

$$\begin{bmatrix} a_{11} & a_{12} \\ a_{21} & a_{22} \end{bmatrix} \cdot \left(\begin{bmatrix} b_{11} & b_{12} \\ b_{21} & b_{22} \end{bmatrix} + \begin{bmatrix} c_{11} & c_{12} \\ c_{21} & c_{22} \end{bmatrix}\right)$$

$$= \begin{bmatrix} a_{11} & a_{12} \\ a_{21} & a_{22} \end{bmatrix} \cdot \begin{bmatrix} b_{11} & b_{12} \\ b_{21} & b_{22} \end{bmatrix} + \begin{bmatrix} a_{11} & a_{12} \\ a_{21} & a_{22} \end{bmatrix} \cdot \begin{bmatrix} c_{11} & c_{12} \\ c_{21} & c_{22} \end{bmatrix}.$$

29. Show that

$$\left(\begin{bmatrix} b_{11} & b_{12} \\ b_{21} & b_{22} \end{bmatrix} + \begin{bmatrix} c_{11} & c_{12} \\ c_{21} & c_{22} \end{bmatrix}\right) \cdot \begin{bmatrix} a_{11} & a_{12} \\ a_{21} & a_{22} \end{bmatrix}$$

$$= \begin{bmatrix} b_{11} & b_{12} \\ b_{21} & b_{22} \end{bmatrix} \cdot \begin{bmatrix} a_{11} & a_{12} \\ a_{21} & a_{22} \end{bmatrix} + \begin{bmatrix} c_{11} & c_{12} \\ c_{21} & c_{22} \end{bmatrix} \cdot \begin{bmatrix} a_{11} & a_{12} \\ a_{21} & a_{22} \end{bmatrix}.$$

30. Show that $\begin{bmatrix} 0 & a \\ a & 0 \end{bmatrix}^2 = a^2 I.$

31. Show that $(A_{2 \times 2} \cdot B_{2 \times 2})^t = B^t_{2 \times 2} \cdot A^t_{2 \times 2}.$

32. Show that $A^2_{2 \times 2} = (-A_{2 \times 2})^2.$

In Exercises 33–39, prove the specified part of Theorem 12.2 for 2×2 matrices.

33. Part I   34. Part II   35. Part III   36. Part V
37. Part VI   38. Part VII   39. Part VIII

40. Prove that if A and B are 2×2 matrices and a is a real number, then we have $a(AB) = (aA)B = A(aB).$

12.3   Solution of Linear Systems by Using Row-Equivalent Matrices

Elementary transformations

An **elementary transformation** of a matrix $A_{n \times m}$ is one of the following three operations upon the rows of the matrix:

1. Multiply the entries of any row of $A_{n \times m}$ by k, where $k \in R$, $k \neq 0$.
2. Interchange any two rows of $A_{n \times m}$.
3. Multiply the entries of any row of $A_{n \times m}$ by k, where $k \in R$, and add to the corresponding entries of any other row.

12.3 Solution of Linear Systems by Using Row-Equivalent Matrices

Notice that the inverse of an elementary transformation is an elementary transformation. That is, you can undo an elmentary transformation by means of an elementary transformation. For example, if A is transformed into B by interchanging two rows, then you can regain A from B by again interchanging the same two rows. It follows that if B results from performing a *succession* of elementary transformation on A, then A can similarly be obtained from B by performing the inverse operations in reverse order.

Row-equivalent matrices

Definition 12.12 *If B is a matrix resulting from a succession of a finite number of elementary transformations on a matrix A, then A and B are **row-equivalent** matrices. This is expressed by writing $A \sim B$ or $B \sim A$.*

Example Show that
$$\begin{bmatrix} 1 & -2 & -1 \\ 1 & 0 & 2 \\ -4 & 3 & 1 \end{bmatrix} \sim \begin{bmatrix} 3 & -6 & -3 \\ 1 & 0 & 2 \\ -4 & 3 & 1 \end{bmatrix}.$$

Solution Multiplying each entry of the first row of the left-hand matrix by 3, we obtain the right-hand matrix.

Example Show that
$$\begin{bmatrix} 1 & -2 & -1 \\ 1 & 0 & 2 \\ -4 & 3 & 1 \end{bmatrix} \sim \begin{bmatrix} -4 & 3 & 1 \\ 1 & 0 & 2 \\ 1 & -2 & -1 \end{bmatrix}.$$

Solution Interchanging the first and third row of the left-hand matrix, we obtain the right-hand matrix.

Sometimes it is convenient to make several elementary transformations on the same matrix.

Example Show that
$$\begin{bmatrix} 1 & -2 & -1 \\ 1 & 0 & 2 \\ -4 & 3 & 1 \end{bmatrix} \sim \begin{bmatrix} 1 & -2 & -1 \\ 0 & 2 & 3 \\ 0 & -5 & -3 \end{bmatrix}.$$

Solution Multiplying the entries of the first row of the left-hand matrix by -1 and adding these products to the corresponding entries of the second row, and then multiplying the entries of the first row by 4 and adding these products to the corresponding entries of the third row, we obtain the right-hand matrix.

The $n \times n$ matrices that are row equivalent to $I_{n \times n}$ are called **nonsingular** matrices.

Example Show that $\begin{bmatrix} 1 & 2 & 1 \\ -1 & 1 & 0 \\ 1 & 0 & 1 \end{bmatrix} \sim \begin{bmatrix} 1 & 0 & 0 \\ 0 & 1 & 0 \\ 0 & 0 & 1 \end{bmatrix}$.

Solution We first make appropriate transformations to obtain "0" elements in each entry (except for the principal diagonal) in columns 1, 2, and 3. Each reference to a row indicates the row of the preceding matrix.

$$\begin{bmatrix} 1 & 2 & 1 \\ -1 & 1 & 0 \\ 1 & 0 & 1 \end{bmatrix} \sim \begin{bmatrix} 1 & 2 & 1 \\ 0 & 3 & 1 \\ 0 & -2 & 0 \end{bmatrix} \begin{array}{l} \\ \text{Row 2 + Row 1} \\ \text{Row 3 + [(-1) \times Row 1]} \end{array}$$

$$\sim \begin{bmatrix} 1 & 0 & 1 \\ 0 & 3 & 1 \\ 0 & 0 & \frac{2}{3} \end{bmatrix} \begin{array}{l} \text{Row 1 + Row 3} \\ \\ \text{Row 3} + \left[\left(\frac{2}{3}\right) \times \text{Row 2}\right] \end{array}$$

$$\sim \begin{bmatrix} 1 & 0 & 0 \\ 0 & 3 & 0 \\ 0 & 0 & \frac{2}{3} \end{bmatrix} \begin{array}{l} \text{Row 1} + \left[\left(-\frac{3}{2}\right) \times \text{Row 3}\right] \\ \text{Row 2} + \left[\left(-\frac{3}{2}\right) \times \text{Row 3}\right] \\ \\ \end{array}$$

$$\sim \begin{bmatrix} 1 & 0 & 0 \\ 0 & 1 & 0 \\ 0 & 0 & 1 \end{bmatrix} \begin{array}{l} \\ \left(\frac{1}{3}\right) \times \text{Row 2} \\ \left(\frac{3}{2}\right) \times \text{Row 3} \end{array}$$

Matrix solution of linear systems

In a linear system of the form

$$a_{11}x + a_{12}y + a_{13}z = c_1$$
$$a_{21}x + a_{22}y + a_{23}z = c_2$$
$$a_{31}x + a_{32}y + a_{33}z = c_3$$

the matrices

$$\begin{bmatrix} a_{11} & a_{12} & a_{13} \\ a_{21} & a_{22} & a_{23} \\ a_{31} & a_{32} & a_{33} \end{bmatrix} \quad \text{and} \quad \begin{bmatrix} a_{11} & a_{12} & a_{13} & c_1 \\ a_{21} & a_{22} & a_{23} & c_2 \\ a_{31} & a_{32} & a_{33} & c_3 \end{bmatrix}$$

are called the **coefficient matrix** and the **augmented matrix**, respectively. Similar definitions hold for a system of n linear equations.

12.3 Solution of Linear Systems by Using Row-Equivalent Matrices

Starting with the augmented matrix of a linear system, and generating a sequence of row-equivalent matrices, we can obtain a matrix from which the solution set of the system is evident simply by inspection. The validity of the method, which is illustrated by example below, follows from the fact that performing elementary transformations on the augmented matrix of a system corresponds to performing the same sorts of operations on the equations of the system itself. Neither multiplying an equation by a nonzero constant nor interchanging two equations has any effect on the solution set of the system. Multiplying one equation by a constant and adding it to another equation in effect replaces the other by a linear combination of equations; a generalization of Theorem 11.1 and 11.2 ensures that the solution set remains the same.

Example Solve
$$x + 2y - 3z = -4$$
$$2x - y + z = 3$$
$$3x + 2y + z = 10$$

Solution The augmented matrix of the system is

$$\begin{bmatrix} 1 & 2 & -3 & | & -4 \\ 2 & -1 & 1 & | & 3 \\ 3 & 2 & 1 & | & 10 \end{bmatrix}.$$

We perform elementary transformations on this matrix as follows.

$$\begin{bmatrix} 1 & 2 & -3 & | & -4 \\ 0 & -5 & 7 & | & 11 \\ 3 & 2 & 1 & | & 10 \end{bmatrix} \quad -2 \times \text{Row 1} + \text{Row 2}$$

$$\begin{bmatrix} 1 & 2 & 3 & | & -4 \\ 0 & -5 & 7 & | & 11 \\ 0 & -4 & 10 & | & 22 \end{bmatrix} \quad -3 \times \text{Row 1} + \text{Row 3}$$

$$\begin{bmatrix} 1 & 2 & -3 & | & -4 \\ 0 & -5 & 7 & | & 11 \\ 0 & 0 & -22 & | & -66 \end{bmatrix} \quad -5 \times \text{Row 3, then } 4 \times \text{Row 2} + \text{Row 3}$$

The resulting system is
$$x + 2y - 3z = -4$$
$$-5y + 7z = 11$$
$$-22z = -66$$

which may be solved by reverse substitution. This example was solved in Section 11.2 and the augmented matrices here correspond exactly to the equivalent systems obtained there. Matrix notation, however, so facilitates the operations on the equations (rows) that it is easy to obtain even simpler equivalent systems. We can rework the above example; the system of equations corresponding to each successive matrix is shown on the right of the matrix in the solution on page 346.

$$\begin{array}{c} \text{Row } 2 + [-2 \times \text{Row } 1] \\ \text{Row } 3 + [-3 \times \text{Row } 1] \end{array} \begin{bmatrix} 1 & 2 & -3 & \vdots & -4 \\ 0 & -5 & 7 & \vdots & 11 \\ 0 & -4 & 10 & \vdots & 22 \end{bmatrix} \begin{array}{l} x + 2y - 3z = -4 \\ 0x - 5y + 7z = 11 \\ 0x - 4y + 10z = 22 \end{array}$$

$$\text{Row } 1 + \left[\frac{2}{5} \times \text{Row } 2\right] \begin{bmatrix} 1 & 0 & -\frac{1}{5} & \vdots & \frac{2}{5} \\ 0 & -5 & 7 & \vdots & 11 \\ 0 & 0 & \frac{22}{5} & \vdots & \frac{66}{5} \end{bmatrix} \begin{array}{l} x + 0y - \frac{1}{5}z = \frac{2}{5} \\ 0x - 5y + 7z = 11 \\ 0x + 0y + \frac{22}{5}z = \frac{66}{5} \end{array}$$

$$\begin{array}{c} 5 \times \text{Row } 1 \\ \\ \frac{5}{22} \times \text{Row } 3 \end{array} \begin{bmatrix} 5 & 0 & -1 & \vdots & 2 \\ 0 & -5 & 7 & \vdots & 11 \\ 0 & 0 & 1 & \vdots & 3 \end{bmatrix} \begin{array}{l} 5x + 0y - z = 2 \\ 0x - 5y + 7z = 11 \\ 0x + 0y + z = 3 \end{array}$$

$$\begin{array}{c} \text{Row } 1 + \text{Row } 3 \\ \text{Row } 2 + [-7 \times \text{Row } 3] \end{array} \begin{bmatrix} 5 & 0 & 0 & \vdots & 5 \\ 0 & -5 & 0 & \vdots & -10 \\ 0 & 0 & 1 & \vdots & 3 \end{bmatrix} \begin{array}{l} 5x + 0y + 0z = 5 \\ 0x - 5y + 0z = -10 \\ 0x + 0y + z = 3 \end{array}$$

$$\begin{array}{c} \frac{1}{5} \times \text{Row } 1 \\ -\frac{1}{5} \times \text{Row } 2 \\ \\ \end{array} \begin{bmatrix} 1 & 0 & 0 & \vdots & 1 \\ 0 & 1 & 0 & \vdots & 2 \\ 0 & 0 & 1 & \vdots & 3 \end{bmatrix} \begin{array}{l} x + 0y + 0z = 1 \\ 0x + y + 0z = 2 \\ 0x + 0y + z = 3 \end{array}$$

The last system is equivalent to

$$x = 1$$
$$y = 2$$
$$z = 3.$$

From this, the solution set, $\{(1, 2, 3)\}$, for the given system is evident by inspection.

For any given $n \times n$ linear system with nonsingular coefficient matrix, there are many sequences of row operations which will transform the augmented matrix of a system equivalently to one of the form

$$\begin{bmatrix} 1 & 0 & 0 & \cdots & 0 & \vdots & x_1 \\ 0 & 1 & 0 & & 0 & \vdots & \cdot \\ 0 & 0 & 1 & & 0 & \vdots & \cdot \\ \vdots & & & & \vdots & \vdots & \cdot \\ 0 & 0 & 0 & \cdots & 1 & \vdots & x_n \end{bmatrix},$$

12.3 Solution of Linear Systems by Using Row-Equivalent Matrices   347

from which the solution set, $\{(x_1, \ldots, x_n)\}$, of the original system is evident by inspection. Finding the most efficient sequence depends on experience and insight, but several good systematic procedures exist. The procedure used above was, briefly, the following one: obtain a 1 on the diagonal using type 1 or 2 transformations if necessary; then "clear out" the remainder of the column using type 3 transformations to obtain 0's; then proceed to the next column.

If the coefficient matrix for the $n \times n$ linear system is singular, that is, if it is not row-equivalent to $I_{n \times n}$, then the system has either no solution or an infinite number of solutions. The system will have no solution when the row reduction process yields a row of the form

$$0 \quad 0 \quad \cdots \quad 0 \quad a$$

where $a \neq 0$. The system will have an infinite number of solutions when the process does not yield such a row but does yield a row of zeros.

Exercise 12.3

Use row transformations of the augmented matrix to solve each system of equations. (Each coefficient matrix is nonsingular and is row-equivalent to $I_{n \times n}$.)

Example

$$\begin{aligned} x + 2y + z &= 4 \\ 2x + y - z &= -1 \\ -x + y + z &= 2 \end{aligned}$$

Solution

$\begin{bmatrix} 1 & 2 & 1 & | & 4 \\ 2 & 1 & -1 & | & -1 \\ -1 & 1 & 1 & | & 2 \end{bmatrix}$ (augmented matrix)

$\begin{bmatrix} 1 & 2 & 1 & | & 4 \\ 0 & -3 & -3 & | & -9 \\ 0 & 3 & 2 & | & 6 \end{bmatrix}$ $-2 \times$ Row 1 + Row 2
 Row 1 + Row 3

$\begin{bmatrix} 1 & 2 & 1 & | & 4 \\ 0 & 1 & 1 & | & 3 \\ 0 & 3 & 2 & | & 6 \end{bmatrix}$ $-\dfrac{1}{3} \times$ Row 2

$\begin{bmatrix} 1 & 0 & -1 & | & -2 \\ 0 & 1 & 1 & | & 3 \\ 0 & 0 & -1 & | & -3 \end{bmatrix}$ $-2 \times$ Row 2 + Row 1
 $-3 \times$ Row 2 + Row 3

$\begin{bmatrix} 1 & 0 & -1 & | & -2 \\ 0 & 1 & 1 & | & 3 \\ 0 & 0 & 1 & | & 3 \end{bmatrix}$ $-1 \times$ Row 3

$\begin{bmatrix} 1 & 0 & 0 & | & 1 \\ 0 & 1 & 0 & | & 0 \\ 0 & 0 & 1 & | & 3 \end{bmatrix}$ Row 3 + Row 1
 $-1 \times$ Row 3 + Row 2

The solution is (1, 0, 3). This same example was also solved in Exercise 11.2.

1. $x - 2y = 4$
   $x + 3y = -1$
2. $x + y = -1$
   $x - 4y = -14$
3. $3x - 2y = 13$
   $4x - y = 19$
4. $4x - 3y = 16$
   $2x + y = 8$
5. $x - 2y = 6$
   $3x + y = 25$
6. $x - y = -8$
   $x + 2y = 9$
7. $x + y - z = 0$
   $2x - y + z = -6$
   $x + 2y - 3z = 2$
8. $2x - y + 3z = 1$
   $x + 2y - z = -1$
   $3x + y + z = 2$
9. $2x - y = 0$
   $3y + z = 7$
   $2x + 3z = 1$
10. $3x - z = 7$
    $2x + y = 6$
    $3y - z = 7$
11. $2x - 5y + 3z = -1$
    $-3x - y + 2z = 11$
    $-2x + 7y + 5z = 9$
12. $2x + y + z = 4$
    $3x - z = 3$
    $2x + 3z = 13$

Use row transformations of the augmented matrix to solve each system of equations. (The coefficient matrix may be singular.)

Example

Solve.

$x - y + 2z = 3$

$2x + 3y - 6z = 1$

$4x + y - 2z = 7$

Solution

$\begin{bmatrix} 1 & -1 & 2 & | & 3 \\ 2 & 3 & -6 & | & 1 \\ 4 & 1 & -2 & | & 7 \end{bmatrix}$ (augmented matrix)

$\begin{bmatrix} 1 & -1 & 2 & | & 3 \\ 0 & 5 & -10 & | & -5 \\ 0 & 5 & -10 & | & -5 \end{bmatrix}$ $\begin{array}{l} -2 \times \text{Row 1} + \text{Row 2} \\ -4 \times \text{Row 1} + \text{Row 3} \end{array}$

$\begin{bmatrix} 1 & -1 & 2 & | & 3 \\ 0 & 1 & -2 & | & -1 \\ 0 & 5 & -10 & | & -1 \end{bmatrix}$ $\dfrac{1}{5} \times \text{Row 2}$

$\begin{bmatrix} 1 & 0 & 0 & | & 2 \\ 0 & 1 & -2 & | & -1 \\ 0 & 0 & 0 & | & 0 \end{bmatrix}$ $\begin{array}{l} \text{Row 2} + \text{Row 1} \\ \\ -5 \times \text{Row 2} + \text{Row 3} \end{array}$

The zero row indicates there are an infinite number of solutions. The final equivalent system is

$x = 2$ $x = 2$
           or
$y - 2z = -1$ $y = 2z - 1$

The solution set may be described as the set of all triples of the form $(2, 2z - 1, z)$.

13. $x + 2y + 2z = 3$
    $2x + 5y + 5z = 7$
    $y + z = 1$

14. $x - y + z = -1$
    $3x - 2y + 2z = 1$
    $2x - 4y + 4z = -10$

15. $x + 3y + 7z = 5$
    $-x + y + z = -1$
    $x + 11y + 22z = 14$

16. $x + 2y = -1$
    $3x + y - 5z = 5$
    $2x + 9y + 5z = -10$

17. $x + 2y + 2z = 5$
    $x + y + z = 4$
    $2x + 3y + 3z = 8$

18. $x - y + z = 4$
    $2x + 3y - z = 5$
    $x - 6y + 4z = 6$

19. $x + 2y - z = 1$
    $3x + y + z = 3$
    $2x - y + 2z = -1$

20. $x + 3y + z = 1$
    $2x + 4y - z = -1$
    $2y + 3z = 3$

Show that each product is a matrix that is row-equivalent to $\begin{bmatrix} a & b \\ c & d \end{bmatrix}$ if $k \neq 0$.

21. $\begin{bmatrix} k & 0 \\ 0 & 1 \end{bmatrix} \begin{bmatrix} a & b \\ c & d \end{bmatrix}$

22. $\begin{bmatrix} 1 & 0 \\ 0 & k \end{bmatrix} \begin{bmatrix} a & b \\ c & d \end{bmatrix}$

23. $\begin{bmatrix} 0 & 1 \\ 1 & 0 \end{bmatrix} \begin{bmatrix} a & b \\ c & d \end{bmatrix}$

24. $\begin{bmatrix} 1 & k \\ 0 & 1 \end{bmatrix} \begin{bmatrix} a & b \\ c & d \end{bmatrix}$

25. $\begin{bmatrix} 1 & 0 \\ k & 1 \end{bmatrix} \begin{bmatrix} a & b \\ c & d \end{bmatrix}$

12.4 The Determinant Function

Associated with each square matrix A having real-number entries is a real number called the **determinant** of A and denoted by δA or $\delta(A)$ (read "the determinant of A"). Thus we have a function, δ (delta), with domain the set of all square matrices having real-number entries, and with range the set of all real numbers; $\delta(A_{n \times n})$ is called a determinant of **order** n.

Let us begin by examining δ over the set $S_{2 \times 2}$ of 2×2 matrices.

Definition 12.13   The **determinant** of the matrix

$$\begin{bmatrix} a_{11} & a_{12} \\ a_{21} & a_{22} \end{bmatrix}$$

is the number $a_{11}a_{22} - a_{12}a_{21}$.

The determinant of a square matrix is customarily displayed in the same form as the matrix, but with vertical bars in lieu of brackets. Thus,

$$\delta \begin{bmatrix} a_{11} & a_{12} \\ a_{21} & a_{22} \end{bmatrix} = \begin{vmatrix} a_{11} & a_{12} \\ a_{21} & a_{22} \end{vmatrix} = a_{11}a_{22} - a_{12}a_{21}.$$

Example

$$\delta \begin{bmatrix} 3 & 1 \\ -2 & 3 \end{bmatrix} = \begin{vmatrix} 3 & 1 \\ -2 & 3 \end{vmatrix} = (3)(3) - (1)(-2) = 9 + 2 = 11.$$

Turning next to 3×3 matrices, we have the following definition.

Definition 12.14 The **determinant** of the matrix

$$\begin{bmatrix} a_{11} & a_{12} & a_{13} \\ a_{21} & a_{22} & a_{23} \\ a_{31} & a_{32} & a_{33} \end{bmatrix}, \text{ denoted by } \begin{vmatrix} a_{11} & a_{12} & a_{13} \\ a_{21} & a_{22} & a_{23} \\ a_{31} & a_{32} & a_{33} \end{vmatrix},$$

is the number

$$a_{11}a_{22}a_{33} - a_{11}a_{23}a_{32} + a_{12}a_{23}a_{31} - a_{12}a_{21}a_{33} + a_{13}a_{21}a_{32} - a_{13}a_{22}a_{31}.$$

Sign of a term in the expression of a determinant

An inspection of the subscripts of the factors of the products involved in this determinant (as well as those of the determinant of a 2×2 matrix) will show that each product is formed by taking one entry from each row and one entry from each column, with the restriction that no two factors be entries in the same row or column. The determinant consists of the sum of $\pm$ all such products as are possible. Whether a product or its negative is used in a determinant depends on the number of **inversions** in the second subscripts in the product when the first subscripts are in natural order, 1 2 3 .... An inversion occurs in a sequence of natural numbers each time a natural number is preceded by a greater natural number. For example, in the sequence 1 4 3 2, there are three inversions, because 4 precedes 3, 4 precedes 2, and 3 precedes 2. Now, if there is an odd number of inversions in the sequence formed by the *second subscripts of the factors in a product* when the first subscripts are in natural order, the negative of the product is used; otherwise, the product itself is used.

With this means of distinguishing products, we can generalize our definition of the determinant of a matrix.

Definition 12.15 The **determinant** of the square matrix

$$\begin{bmatrix} a_{11} & a_{12} & \cdots & a_{1n} \\ \vdots & \vdots & & \vdots \\ a_{n1} & a_{n2} & \cdots & a_{nn} \end{bmatrix}, \text{ denoted by } \begin{vmatrix} a_{11} & a_{12} & \cdots & a_{1n} \\ \vdots & \vdots & & \vdots \\ a_{n1} & a_{n2} & \cdots & a_{nn} \end{vmatrix},$$

is equal to the sum of all products, $\pm a_{1j_1} a_{2j_2} a_{3j_3} \cdots a_{nj_n}$, where each j takes on all values from 1 to n, and no j's in the same product have the same subscript. For each term in the sum, the negative sign is used if the number of inversions in the sequence formed by the j's is odd; otherwise, the positive sign is used.

Value of a determinant

It is evident from this definition that the determinant of a matrix with real entries is a real number. We shall refer to this real number as the **value** of the determinant; writing the determinant as a sum is called **expanding** the determinant. Note

12.4 The Determinant Function

that the arrays shown in the foregoing definition appear in print to be rectangular rather than square. The subscript on the lower right-hand entry, however, shows the dimension of the matrix or determinant. In any similar symbolism, this subscript should always be checked to determine the correct dimension. For example, in expanding the determinant of a 4×4 square matrix, one of the products is $a_{11}a_{23}a_{32}a_{44}$. The first subscripts are in natural order, 1 2 3 4, and the second subscripts are in the order 1 3 2 4. Since the only inversion is that 3 appears before 2, the negative of $a_{11}a_{23}a_{32}a_{44}$ is used in the expansion of the determinant.

Definition 12.16   The **minor** M_{ij} of the element a_{ij} in a given determinant is the determinant that remains after the ith row and jth column in the given determinant have been deleted.

For example, in the determinant

$$\begin{vmatrix} a_{11} & a_{12} & a_{13} \\ a_{21} & a_{22} & a_{23} \\ a_{31} & a_{32} & a_{33} \end{vmatrix}, \qquad (1)$$

the minor of a_{11} is

$$M_{11} = \begin{vmatrix} a_{22} & a_{23} \\ a_{32} & a_{33} \end{vmatrix},$$

the minor of a_{23} is

$$M_{23} = \begin{vmatrix} a_{11} & a_{12} \\ a_{31} & a_{32} \end{vmatrix},$$

the minor of a_{31} is

$$M_{31} = \begin{vmatrix} a_{12} & a_{13} \\ a_{22} & a_{23} \end{vmatrix}.$$

Definition 12.17   The **cofactor** A_{ij} of the element a_{ij} is the minor of a_{ij} if $i + j$ is an even integer, and the negative of the minor of a_{ij} if $i + j$ is an odd integer.

For example, in the determinant (1),

the cofactor of a_{11} is $A_{11} = \begin{vmatrix} a_{22} & a_{23} \\ a_{32} & a_{33} \end{vmatrix}$, because $1 + 1$ is 2, an even integer;

the cofactor of a_{23} is $A_{23} = -\begin{vmatrix} a_{11} & a_{12} \\ a_{31} & a_{32} \end{vmatrix}$, because $2 + 3$ is 5, an odd integer;

the cofactor of a_{31} is $A_{31} = \begin{vmatrix} a_{12} & a_{13} \\ a_{22} & a_{23} \end{vmatrix}$, because $3 + 1$ is 4, an even integer; etc.

The following sign array is a convenient means of determining whether the cofactor of a given element equals the minor or whether it equals the negative of the minor.

$$\begin{matrix} + & - & + & \cdots & (-)^{n+1} \\ - & + & - & \cdots & \cdot \\ + & - & + & \cdots & \cdot \\ \cdot & \cdot & \cdot & & \cdot \\ \cdot & \cdot & \cdot & & \cdot \\ \cdot & \cdot & \cdot & & \cdot \\ (-)^{n+1} & \cdot & & \cdots & + \end{matrix}$$

Expansion by cofactors

With the definition of a cofactor in mind, let us look again at Definition 12.14 for the determinant of the matrix

$$A = \begin{bmatrix} a_{11} & a_{12} & a_{13} \\ a_{21} & a_{22} & a_{23} \\ a_{31} & a_{32} & a_{33} \end{bmatrix}.$$

The value of this determinant is

$$\delta(A) = a_{11}a_{22}a_{33} - a_{11}a_{23}a_{32} + a_{12}a_{23}a_{31} - a_{12}a_{21}a_{33} + a_{13}a_{21}a_{32} - a_{13}a_{22}a_{31}.$$

By suitably factoring pairs of terms in the right-hand member, we obtain

$$\delta(A) = a_{11}(a_{22}a_{33} - a_{23}a_{32}) + a_{12}(a_{23}a_{31} - a_{21}a_{33}) + a_{13}(a_{21}a_{32} - a_{22}a_{31}).$$

If the binomial factor in the middle term is rewritten $-(a_{21}a_{33} - a_{23}a_{31})$ we have

$$\delta(A) = a_{11}(a_{22}a_{33} - a_{23}a_{32}) + a_{12}[-(a_{21}a_{33} - a_{23}a_{31})] + a_{13}(a_{21}a_{32} - a_{22}a_{31}),$$

which is equal to

$$a_{11}\begin{vmatrix} a_{22} & a_{23} \\ a_{32} & a_{33} \end{vmatrix} + a_{12}\left(-\begin{vmatrix} a_{21} & a_{23} \\ a_{31} & a_{33} \end{vmatrix}\right) + a_{13}\begin{vmatrix} a_{21} & a_{22} \\ a_{31} & a_{32} \end{vmatrix}.$$

Accordingly, we have

$$\delta(A) = a_{11}A_{11} + a_{12}A_{12} + a_{13}A_{13}.$$

Thus, the determinant

$$\delta(A) = \begin{vmatrix} a_{11} & a_{12} & a_{13} \\ a_{21} & a_{22} & a_{23} \\ a_{31} & a_{32} & a_{33} \end{vmatrix}$$

is equal to the sum formed by multiplying each entry in the first row by its cofactor and then adding these products.

Similar methods can be used to show that this determinant is also equal to the sum formed by multiplying each entry in *any* row (or column) by its cofactor and then adding the products.

Although we shall not show it here, Definition 12.15 is logically equivalent to the following.

12.4 The Determinant Function

Definition 12.18   The **determinant** of the square matrix

$$\begin{bmatrix} a_{11} & a_{12} & \cdots & a_{1n} \\ a_{21} & a_{22} & \cdots & a_{2n} \\ \vdots & & & \vdots \\ a_{n1} & a_{n2} & \cdots & a_{nn} \end{bmatrix}$$

is the sum of the n products formed by multiplying each entry in any single row (or any single column) by its cofactor.

When this latter definition is used to rewrite a determinant, the determinant is said to be **expanded** about whatever row (or column) is chosen.

Example   If $A = \begin{bmatrix} 3 & 2 & 1 \\ 0 & 1 & -2 \\ 1 & 3 & 4 \end{bmatrix}$, find $\delta(A)$ by expansion about the first column.

Solution   Noting that $a_{11} = 3$, $a_{21} = 0$, and $a_{31} = 1$, we have

$$\delta(A) = 3 \begin{vmatrix} 1 & -2 \\ 3 & 4 \end{vmatrix} + 0\left(-\begin{vmatrix} 2 & 1 \\ 3 & 4 \end{vmatrix}\right) + 1 \begin{vmatrix} 2 & 1 \\ 1 & -2 \end{vmatrix}$$

$$= 3(10) + 0(-5) + 1(-5) = 25.$$

Exercise 12.4

Let $A = \begin{bmatrix} 2 & 1 & -2 & 0 \\ 1 & 0 & 3 & -1 \\ -2 & 1 & 2 & 2 \\ 1 & -1 & 3 & 1 \end{bmatrix}$.

Each of the following products or its negative is a term in an expansion of $\delta(A)$. Determine the product, and write the product or its negative in accordance with the number of inversions in the second subscripts.

Example $a_{13}a_{21}a_{34}a_{42}$

Solution   The product is $(-2) \cdot 1 \cdot 2 \cdot (-1) = 4$.

The second subscripts are in order 3142; the number of inversions is 3 since 3 precedes 1 and 2, and 4 precedes 2. Therefore the negative of the product, that is, -4, is used.

1. $a_{11}a_{22}a_{33}a_{44}$
2. $a_{11}a_{23}a_{34}a_{42}$
3. $a_{11}a_{24}a_{32}a_{43}$
4. $a_{12}a_{21}a_{33}a_{44}$
5. $a_{13}a_{24}a_{32}a_{41}$
6. $a_{14}a_{23}a_{32}a_{41}$

Let A be the matrix given above. Determine the minor M_{ij} and cofactor A_{ij} (in determinant form) of each of the following entries.

Example a_{21}

Solution The minor is $\begin{vmatrix} 1 & -2 & 0 \\ 1 & 2 & 2 \\ -1 & 3 & 1 \end{vmatrix}$.

Since $2 + 1$ is odd, the cofactor is

$$-\begin{vmatrix} 1 & -2 & 0 \\ 1 & 2 & 2 \\ -1 & 3 & 1 \end{vmatrix}.$$

7. a_{11} 8. a_{13} 9. a_{23} 10. a_{41}
11. a_{31} 12. a_{33} 13. a_{44} 14. a_{14}

Evaluate each determinant.

Examples a. $\begin{vmatrix} 2 & -3 \\ 1 & 4 \end{vmatrix}$ b. $\begin{vmatrix} 1 & 2 & 0 \\ 3 & -1 & 4 \\ -2 & 1 & 3 \end{vmatrix}$

Solutions a. $\begin{vmatrix} 2 & -3 \\ 1 & 4 \end{vmatrix} = (2)(4) - (-3)(1) = 11.$

b. Expand about any row or column; the first row is used here:

$$\begin{vmatrix} 1 & 2 & 0 \\ 3 & -1 & 4 \\ -2 & 1 & 3 \end{vmatrix} = 1\begin{vmatrix} -1 & 4 \\ 1 & 3 \end{vmatrix} - 2\begin{vmatrix} 3 & 4 \\ -2 & 3 \end{vmatrix} + 0\begin{vmatrix} 3 & -1 \\ -2 & 1 \end{vmatrix}$$

$$= 1[(-1)(3) - (4)(1)] - 2[(3)(3) - (4)(-2)] + 0$$

$$= -41.$$

15. $\begin{vmatrix} 1 & 0 \\ 2 & 1 \end{vmatrix}$ 16. $\begin{vmatrix} 3 & -2 \\ 4 & 1 \end{vmatrix}$ 17. $\begin{vmatrix} -3 & -1 \\ 3 & 1 \end{vmatrix}$ 18. $\begin{vmatrix} -1 & 6 \\ 0 & -2 \end{vmatrix}$

19. $\begin{vmatrix} -4 & 2 \\ 1 & 7 \end{vmatrix}$ 20. $\begin{vmatrix} 8 & -1 \\ 3 & 2 \end{vmatrix}$ 21. $\begin{vmatrix} x & -1 \\ 1 & 3 \end{vmatrix}$ 22. $\begin{vmatrix} -3 & x \\ -1 & 3 \end{vmatrix}$

23. $\begin{vmatrix} 2 & 0 & 1 \\ 1 & 1 & 2 \\ -1 & 0 & 1 \end{vmatrix}$ 24. $\begin{vmatrix} 1 & 3 & 1 \\ -1 & 2 & 1 \\ 0 & 2 & 0 \end{vmatrix}$ 25. $\begin{vmatrix} 1 & 2 & 3 \\ 3 & -1 & 2 \\ 2 & 0 & 2 \end{vmatrix}$

12.5 Properties of Determinants

26. $\begin{vmatrix} 1 & 0 & 0 \\ 0 & 1 & 2 \\ 0 & 3 & 4 \end{vmatrix}$

27. $\begin{vmatrix} -1 & 0 & 2 \\ -2 & 1 & 0 \\ 0 & 1 & -3 \end{vmatrix}$

28. $\begin{vmatrix} 2 & 1 & 4 \\ 3 & 2 & 6 \\ 5 & -3 & 10 \end{vmatrix}$

29. $\begin{vmatrix} a & b & 1 \\ a & b & 1 \\ 1 & 1 & 1 \end{vmatrix}$

30. $\begin{vmatrix} a & a & a \\ 1 & 2 & 3 \\ 4 & 5 & 6 \end{vmatrix}$

31. $\begin{vmatrix} x & 0 & 0 \\ 0 & x & 0 \\ 0 & 0 & x \end{vmatrix}$

32. $\begin{vmatrix} 0 & 0 & x \\ 0 & x & 0 \\ x & 0 & 0 \end{vmatrix}$

Solve for x.

33. $\begin{vmatrix} x & -1 \\ 2 & 3 \end{vmatrix} = 17$

34. $\begin{vmatrix} 1 & -5 \\ x & 3 \end{vmatrix} = -7$

35. $\begin{vmatrix} x & 0 & 0 \\ 2 & 1 & 3 \\ 0 & 1 & 4 \end{vmatrix} = 3$

36. $\begin{vmatrix} x^2 & x & 1 \\ 0 & 2 & 1 \\ 3 & 1 & 4 \end{vmatrix} = 28$

Expand by cofactors and verify.

37. $\begin{vmatrix} 0 & 1 & 0 & 0 \\ 1 & 0 & 3 & 2 \\ 5 & -1 & 2 & 1 \\ 1 & 0 & 1 & 1 \end{vmatrix} = 5$

38. $\begin{vmatrix} 1 & 2 & 0 & -1 \\ 1 & 0 & -1 & 2 \\ 0 & 1 & 1 & 1 \\ 2 & -1 & 0 & 1 \end{vmatrix} = 17$

39. In accordance with Definition 12.18, the determinant of an $n \times n$ matrix is the sum of a certain number of products. What is this number for $n = 2$? For $n = 3$? For $n = 4$?

40. Generalize the results of Exercise 39, making a conjecture about the number of such products in the determinant of an $n \times n$ matrix.

41. Show that for any 2×2 matrix A, $\delta(aA) = a^2 \delta(A)$.

42. Show that for any 2×2 matrix A, $\delta(A^t) = \delta(A)$.

43. Show that for any 2×2 matrices A and B, $\delta(AB) = \delta(A) \cdot \delta(B)$.

12.5 Properties of Determinants

Determinants have some properties that are useful by virtue of the fact that they permit us to generate equal determinants with different and simpler configurations of entries. This, in turn, helps us find values for determinants.

Theorem 12.7   *If each entry in any row, or each entry in any column, of a determinant is 0, then the determinant is equal to 0.*

Proof   By Definition 12.15, a determinant is a sum of products, each product having an entry from each row and each column of the determinant. This means that if any row or any column has only zero entries, then each product will contain 0 as a factor. Thus each product, and therefore the sum of all such products, is 0.

For example,

$$\begin{vmatrix} 0 & 0 \\ 1 & 2 \end{vmatrix} = 0, \quad \begin{vmatrix} 1 & 1 & 0 \\ 3 & 5 & 0 \\ 2 & 7 & 0 \end{vmatrix} = 0, \quad \text{and} \quad \begin{vmatrix} 0 & 1 & 0 & 0 \\ 1 & 0 & 0 & 0 \\ 0 & 0 & 0 & 1 \\ 0 & 0 & 0 & 1 \end{vmatrix} = 0.$$

Theorem 12.8   *If any two rows (or any two columns) of a determinant are interchanged, then the resulting determinant is the negative of the original determinant.*

Proof   Let us look first at the case in which two *adjacent* rows of a determinant D of order n are interchanged. If the ith and $(i+1)$st rows of D are interchanged to give us D' with entries a'_{ij}, then a_{ij} in D is equal to $a'_{i+1,j}$ in D' and the minors of a_{ij} in D are identical to those of $a'_{i+1,j}$ in D'. An expansion of D about the ith row leads to

$$D = a_{i1}A_{i1} + a_{i2}A_{i2} + \cdots + a_{in}A_{in},$$

and an expansion of D' about its $(i+1)$st row leads to

$$D' = a'_{i+1,1}A'_{i+1,1} + a'_{i+1,2}A'_{i+1,2} + \cdots + a'_{i+1,n}A'_{i+1,n}.$$

Since $a_{ij} = a'_{i+1,j}$, and since, from the definition of cofactor, $A_{ij} = -A'_{i+1,j}$, it follows immediately that $D = -D'$. The same argument can be applied to an interchange of *adjacent* columns.

Now, let us turn to the case in which *any* two rows are interchanged. Let the interchange of the ith and $(i+k)$th rows of D lead to D'. This change can be viewed as the result of making k adjacent-row interchanges to bring the ith row into the $(i+k)$th position, and then $k-1$ similar interchanges to bring the $(i+k)$th row into the ith position. Since each such interchange results in a sign reversal, and since $k+k-1$ such reversals occur, $D' = (-1)^{2k-1}D$. But $2k-1$ is an odd number for any natural number k, and hence $D' = -D$. A similar argument can be made for the interchange of the jth and $(j+k)$th columns.

For example,

$$\begin{vmatrix} 1 & 2 \\ 3 & 4 \end{vmatrix} = -\begin{vmatrix} 3 & 4 \\ 1 & 2 \end{vmatrix}, \quad \text{and} \quad \begin{vmatrix} 1 & 2 & 3 \\ 4 & 5 & 6 \\ 7 & 8 & 9 \end{vmatrix} = -\begin{vmatrix} 3 & 2 & 1 \\ 6 & 5 & 4 \\ 9 & 8 & 7 \end{vmatrix}.$$

In the first example, rows 1 and 2 were interchanged. In the second example, columns 1 and 3 were interchanged.

12.5 Properties of Determinants

Theorem 12.9 *If two rows (or two columns) in a determinant have corresponding entries that are equal, the determinant is equal to 0.*

Proof Let the determinant D have two rows (or columns) with corresponding entries equal. Then by Theorem 12.8, an interchange of these rows (or columns) will produce D', which is the negative of D. That is, $D' = -D$. Since, however, the interchanged rows (or columns) have identical elements, $D' = D$. Thus, $D = -D$, $2D = 0$, and D must be 0.

For example,

$$\begin{vmatrix} 1 & 1 \\ 3 & 3 \end{vmatrix} = 0, \quad \begin{vmatrix} 1 & 2 & 1 \\ 3 & 1 & 0 \\ 1 & 2 & 1 \end{vmatrix} = 0, \quad \text{and} \quad \begin{vmatrix} 1 & 2 & 3 & 4 \\ 5 & 6 & 7 & 8 \\ 0 & 0 & 1 & 0 \\ 1 & 2 & 3 & 4 \end{vmatrix} = 0.$$

Theorem 12.10 *If each of the entries of one row (or column) of a determinant is multiplied by k, the determinant is multiplied by k.*

Proof Let the ith row of the determinant D be multipled by k to yield D'. Then an expansion of D' about the ith row leads to

$$\begin{aligned} D' &= ka_{i1}A_{i1} + ka_{i2}A_{i2} + \cdots + ka_{in}A_{in} \\ &= k(a_{i1}A_{i1} + a_{i2}A_{i2} + \cdots + a_{in}A_{in}) \\ &= kD. \end{aligned}$$

A similar argument holds for columns. Thus,

$$\begin{vmatrix} 1 & 0 & 0 \\ 2 & 1 & 3 \\ 1 \times 2 & 3 \times 2 & 4 \times 2 \end{vmatrix} = 2 \begin{vmatrix} 1 & 0 & 0 \\ 2 & 1 & 3 \\ 1 & 3 & 4 \end{vmatrix}, \quad \text{and} \quad \begin{vmatrix} 4 & 5 & 8 \\ 1 & 1 & 2 \\ 3 & 1 & 6 \end{vmatrix} = 2 \begin{vmatrix} 4 & 5 & 4 \\ 1 & 1 & 1 \\ 3 & 1 & 3 \end{vmatrix}.$$

Note that this process is different from that of the multiplication of a matrix by a real number. In the latter, each entry in the matrix is multiplied by the real number, rather than, as here, only the entries in a single row or column being so multiplied.

Theorem 12.11 *If each entry in a row (or column) of a determinant is written as the sum of two terms, the determinant can be written as the sum of two determinants as follows: If*

$$D = \begin{vmatrix} a_{11} & a_{12} & \cdots & a_{1n} \\ \vdots & \vdots & & \vdots \\ b_{i1} + c_{i1} & b_{i2} + c_{i2} & \cdots & b_{in} + c_{in} \\ \vdots & \vdots & & \vdots \\ a_{n1} & a_{n2} & \cdots & a_{nn} \end{vmatrix},$$

Theorem Continued on Overleaf

then
$$D = \begin{vmatrix} a_{11} & a_{12} & \cdots & a_{1n} \\ \vdots & \vdots & & \vdots \\ b_{i1} & b_{i2} & \cdots & b_{in} \\ \vdots & \vdots & & \vdots \\ a_{n1} & a_{n2} & \cdots & a_{nn} \end{vmatrix} + \begin{vmatrix} a_{11} & a_{12} & \cdots & a_{1n} \\ \vdots & \vdots & & \vdots \\ c_{i1} & c_{i2} & \cdots & c_{in} \\ \vdots & \vdots & & \vdots \\ a_{n1} & a_{n2} & \cdots & a_{nn} \end{vmatrix},$$

and if
$$D = \begin{vmatrix} a_{11} & \cdots & b_{1j} + c_{1j} & \cdots & a_{1n} \\ a_{21} & \cdots & b_{2j} + c_{2j} & \cdots & a_{2n} \\ \vdots & & \vdots & & \vdots \\ a_{n1} & \cdots & b_{nj} + c_{nj} & \cdots & a_{nn} \end{vmatrix},$$

then
$$D = \begin{vmatrix} a_{11} & \cdots & b_{1j} & \cdots & a_{1n} \\ a_{21} & \cdots & b_{2j} & \cdots & a_{2n} \\ \vdots & & \vdots & & \vdots \\ a_{n1} & \cdots & b_{nj} & \cdots & a_{nn} \end{vmatrix} + \begin{vmatrix} a_{11} & \cdots & c_{1j} & \cdots & a_{1n} \\ a_{21} & \cdots & c_{2j} & \cdots & a_{2n} \\ \vdots & & \vdots & & \vdots \\ a_{n1} & \cdots & c_{nj} & \cdots & a_{nn} \end{vmatrix}.$$

Proof Let
$$D = \begin{vmatrix} a_{11} & \cdots & a_{1n} \\ \vdots & & \vdots \\ b_{i1} + c_{i1} & \cdots & b_{in} + c_{in} \\ \vdots & & \vdots \\ a_{n1} & \cdots & a_{nn} \end{vmatrix},$$

and expand about the ith row. This leads to
$$\begin{aligned} D &= (b_{i1} + c_{i1})A_{i1} + (b_{i2} + c_{i2})A_{i2} + \cdots + (b_{in} + c_{in})A_{in} \\ &= b_{i1}A_{i1} + c_{i1}A_{i1} + b_{i2}A_{i2} + c_{i2}A_{i2} + \cdots + b_{in}A_{in} + c_{in}A_{in} \\ &= b_{i1}A_{i1} + b_{i2}A_{i2} + \cdots + b_{in}A_{in} + c_{i1}A_{i1} + c_{i2}A_{i2} + \cdots + c_{in}A_{in} \\ &= \begin{vmatrix} a_{11} & \cdots & a_{1n} \\ \vdots & & \vdots \\ b_{i1} & \cdots & b_{in} \\ \vdots & & \vdots \\ a_{n1} & \cdots & a_{nn} \end{vmatrix} + \begin{vmatrix} a_{11} & \cdots & a_{1n} \\ \vdots & & \vdots \\ c_{i1} & \cdots & c_{in} \\ \vdots & & \vdots \\ a_{n1} & \cdots & a_{nn} \end{vmatrix}. \end{aligned}$$

A similar argument proves the result for columns.

For example,
$$\begin{vmatrix} 1 & 3 \\ 2 & 5 \end{vmatrix} = \begin{vmatrix} 1 & 1 \\ 2 & 4 \end{vmatrix} + \begin{vmatrix} 1 & 2 \\ 2 & 1 \end{vmatrix}$$

and
$$\begin{vmatrix} 4 & 0 & 0 \\ 0 & 4 & 0 \\ 0 & 0 & 4 \end{vmatrix} = \begin{vmatrix} 4 & 0 & 0 \\ 0 & 2 & 0 \\ 0 & 0 & 4 \end{vmatrix} + \begin{vmatrix} 4 & 0 & 0 \\ 0 & 2 & 0 \\ 0 & 0 & 4 \end{vmatrix}.$$

12.5 Properties of Determinants

Theorem 12.12 *If each entry of one row (or column) of a determinant is multiplied by a real number k and the resulting product is added to the corresponding entry in another row (or column, respectively) in the determinant, the resulting determinant is equal to the original determinant.*

Proof Let

$$D = \begin{vmatrix} a_{11} & \cdots & a_{1n} \\ \vdots & & \vdots \\ a_{n1} & \cdots & a_{nn} \end{vmatrix}.$$

Then, if the ith row of D is multiplied by k and added to another row, say the pth row, we have

$$\begin{vmatrix} a_{11} & \cdots & a_{1n} \\ \vdots & & \vdots \\ a_{i1} & \cdots & a_{in} \\ \vdots & & \vdots \\ a_{p1} + ka_{i1} & \cdots & a_{pn} + ka_{in} \\ \vdots & & \vdots \\ a_{n1} & \cdots & a_{nn} \end{vmatrix},$$

which, by Theorem 12.11, is equal to

$$\begin{vmatrix} a_{11} & \cdots & a_{1n} \\ \vdots & & \vdots \\ a_{i1} & \cdots & a_{in} \\ \vdots & & \vdots \\ a_{p1} & \cdots & a_{pn} \\ \vdots & & \vdots \\ a_{n1} & \cdots & a_{nn} \end{vmatrix} + \begin{vmatrix} a_{11} & \cdots & a_{1n} \\ \vdots & & \vdots \\ a_{i1} & \cdots & a_{in} \\ \vdots & & \vdots \\ ka_{i1} & \cdots & ka_{in} \\ \vdots & & \vdots \\ a_{n1} & \cdots & a_{nn} \end{vmatrix}.$$

The last determinant in this sum can, by Theorem 12.10, be written

$$k \begin{vmatrix} a_{11} & \cdots & a_{1n} \\ \vdots & & \vdots \\ a_{i1} & \cdots & a_{in} \\ \vdots & & \vdots \\ a_{i1} & \cdots & a_{in} \\ \vdots & & \vdots \\ a_{n1} & \cdots & a_{nn} \end{vmatrix},$$

and since this determinant contains two rows that have their corresponding entries equal, it follows from Theorem 12.9 that it is 0. Hence, the sum is D, and the theorem is proved for rows. A similar argument shows that it is true also for columns.

For example,

$$\begin{vmatrix} 1 & 1 \\ 2 & 1 \end{vmatrix} = \begin{vmatrix} 1 & 1 \\ 2 + 3(1) & 1 + 3(1) \end{vmatrix} = \begin{vmatrix} 1 & 1 \\ 5 & 4 \end{vmatrix}$$

and

$$\begin{vmatrix} 1 & 2 & 3 \\ 4 & 5 & 6 \\ 7 & 8 & 9 \end{vmatrix} = \begin{vmatrix} 1+2(3) & 2 & 3 \\ 4+2(6) & 5 & 6 \\ 7+2(9) & 8 & 9 \end{vmatrix} = \begin{vmatrix} 7 & 2 & 3 \\ 16 & 5 & 6 \\ 25 & 8 & 9 \end{vmatrix}.$$

Evaluation of determinants

The preceding theorems can be used to write sequences of equal determinants, leading from one form of a determinant to another and more useful form.

Example

Evaluate

$$D = \begin{vmatrix} 2 & -1 & 1 & -3 \\ 1 & 3 & -4 & 2 \\ 1 & 0 & -2 & 1 \\ 3 & -1 & 5 & 2 \end{vmatrix}.$$

Solution

As a step toward evaluating the determinant, we shall use Theorem 12.12 to produce an equal determinant with a row or a column containing zero entries in all but one place. Let us arbitrarily select the second column for this role, because one entry is already zero. Multiplying a_{1j} by 3 and adding the result to a_{2j}, we obtain

$$D = \begin{vmatrix} 2 & -1 & 1 & -3 \\ 1+3(2) & 3+3(-1) & -4+3(1) & 2+3(-3) \\ 1 & 0 & -2 & 1 \\ 3 & -1 & 5 & 2 \end{vmatrix} = \begin{vmatrix} 2 & -1 & 1 & -3 \\ 7 & 0 & -1 & -7 \\ 1 & 0 & -2 & 1 \\ 3 & -1 & 5 & 2 \end{vmatrix}.$$

Next, multiplying a_{1j} by -1 and adding the result to a_{4j}, we find that

$$D = \begin{vmatrix} 2 & -1 & 1 & -3 \\ 7 & 0 & -1 & -7 \\ 1 & 0 & -2 & 1 \\ 3-1(2) & -1-1(-1) & 5-1(1) & 2-1(-3) \end{vmatrix} = \begin{vmatrix} 2 & -1 & 1 & -3 \\ 7 & 0 & -1 & -7 \\ 1 & 0 & -2 & 1 \\ 1 & 0 & 4 & 5 \end{vmatrix}.$$

If we now expand the determinant about the second column, we have

$$D = \begin{vmatrix} 2 & -1 & 1 & -3 \\ 7 & 0 & -1 & -7 \\ 1 & 0 & -2 & 1 \\ 1 & 0 & 4 & 5 \end{vmatrix} = -(-1)\begin{vmatrix} 7 & -1 & -7 \\ 1 & -2 & 1 \\ 1 & 4 & 5 \end{vmatrix} + 0A_{22} + 0A_{32} + 0A_{42}.$$

From this point, we can reduce the third-order determinant to a second-order determinant by a similar procedure or, alternatively, expand directly about the elements in any row or column. Expanding about the elements of the first row, we obtain

$$D = \begin{vmatrix} 7 & -1 & -7 \\ 1 & -2 & 1 \\ 1 & 4 & 5 \end{vmatrix} = 7\begin{vmatrix} -2 & 1 \\ 4 & 5 \end{vmatrix} - (-1)\begin{vmatrix} 1 & 1 \\ 1 & 5 \end{vmatrix} + (-7)\begin{vmatrix} 1 & -2 \\ 1 & 4 \end{vmatrix},$$

from which

$$D = 7(-14) + (4) - 7(6) = -98 + 4 - 42 = -136.$$

12.5 Properties of Determinants

Exercise 12.5

Without evaluating, state why each statement is true. Verify selected examples by expansion.

Example

$$\begin{vmatrix} 1 & 0 & 3 \\ 2 & 1 & 7 \\ 1 & 0 & 2 \end{vmatrix} = \begin{vmatrix} 1 & 0 & 3 \\ 2 & 1 & 7 \\ 3 & 1 & 9 \end{vmatrix}$$

Solution The right-hand determinant is obtained from the left by adding row 2 to row 3.

1. $\begin{vmatrix} 2 & 3 & 1 \\ 0 & 0 & 0 \\ -1 & 2 & 0 \end{vmatrix} = 0$

2. $\begin{vmatrix} 3 & 1 & 3 \\ 0 & 1 & 0 \\ 1 & 2 & 1 \end{vmatrix} = 0$

3. $\begin{vmatrix} 2 & 3 & 1 & 1 \\ 2 & 0 & 1 & 2 \\ 2 & 3 & 1 & 1 \\ 0 & 1 & 2 & 0 \end{vmatrix} = 0$

4. $\begin{vmatrix} 7 & 3 & 2 & 0 \\ 2 & 1 & 2 & 0 \\ 4 & 1 & 1 & 0 \\ 0 & 2 & 1 & 0 \end{vmatrix} = 0$

5. $\begin{vmatrix} 4 & 2 & 1 \\ 0 & -1 & -2 \\ 1 & 0 & 2 \end{vmatrix} = - \begin{vmatrix} 4 & 2 & 1 \\ 0 & 1 & 2 \\ 1 & 0 & 2 \end{vmatrix}$

6. $\begin{vmatrix} -2 & 3 & 1 \\ -1 & 0 & 1 \\ -2 & 1 & 0 \end{vmatrix} = - \begin{vmatrix} 2 & 3 & 1 \\ 1 & 0 & 1 \\ 2 & 1 & 0 \end{vmatrix}$

7. $2\begin{vmatrix} 1 & 0 & 2 \\ -1 & 2 & 0 \\ 1 & 1 & 1 \end{vmatrix} = \begin{vmatrix} 1 & 0 & 2 \\ -1 & 2 & 0 \\ 2 & 2 & 2 \end{vmatrix}$

8. $\begin{vmatrix} 3 & -4 & 2 \\ 1 & -2 & 0 \\ 0 & 8 & 1 \end{vmatrix} = -2 \begin{vmatrix} 3 & 2 & 2 \\ 1 & 1 & 0 \\ 0 & -4 & 1 \end{vmatrix}$

9. $\begin{vmatrix} 1 & 2 \\ 3 & 4 \end{vmatrix} = \begin{vmatrix} 1+2 & 2 \\ 3+4 & 4 \end{vmatrix}$

10. $\begin{vmatrix} 1 & 2 \\ 3 & 4 \end{vmatrix} = \begin{vmatrix} 1+4 & 2 \\ 3+8 & 4 \end{vmatrix}$

11. $\begin{vmatrix} 1 & 2 & 1 \\ 0 & 2 & 3 \\ 2 & -1 & 2 \end{vmatrix} = \begin{vmatrix} 1 & 2 & 1 \\ 0 & 2 & 3 \\ 0 & -5 & 0 \end{vmatrix}$

12. $\begin{vmatrix} -1 & 1 & 0 \\ 2 & 3 & -1 \\ 2 & 1 & 2 \end{vmatrix} = \begin{vmatrix} 0 & 1 & 0 \\ 5 & 3 & -1 \\ 3 & 1 & 2 \end{vmatrix}$

Theorem 12.12 was used on the left-hand member of each of the following equalities to produce the elements in the right-hand member. Complete the entries.

Example

$$\begin{vmatrix} 1 & 5 \\ 4 & 3 \end{vmatrix} = \begin{vmatrix} 1 & 5 \\ 0 & \end{vmatrix}$$

Solution To obtain the 0 in the 2, 1 position using Theorem 12.12, it must be that -4 times row 1 was added to row 2. The missing entry is therefore -17.

13. $\begin{vmatrix} 1 & 3 \\ 2 & 2 \end{vmatrix} = \begin{vmatrix} 1 & 3 \\ 0 & 0 \end{vmatrix}$

14. $\begin{vmatrix} 2 & -1 \\ 3 & 1 \end{vmatrix} = \begin{vmatrix} 0 \\ 3 & 1 \end{vmatrix}$

15. $\begin{vmatrix} 1 & -2 & 1 \\ 3 & 1 & 4 \\ 0 & 2 & 1 \end{vmatrix} = \begin{vmatrix} 1 & -2 & 1 \\ 0 & 7 \\ 0 & 2 & 1 \end{vmatrix}$

16. $\begin{vmatrix} 3 & -1 & 0 \\ 1 & 2 & 1 \\ 2 & 3 & 1 \end{vmatrix} = \begin{vmatrix} 3 & -1 & 0 \\ 1 & 2 & 1 \\ 1 & & 0 \end{vmatrix}$

17. $\begin{vmatrix} 2 & 3 & 1 & 4 \\ 0 & 2 & 1 & 2 \\ 1 & 1 & 2 & 3 \\ 0 & 1 & 1 & 1 \end{vmatrix} = \begin{vmatrix} 0 & 1 & & -2 \\ 0 & 2 & 1 & 2 \\ 1 & 1 & 2 & 3 \\ 0 & 1 & 1 & 1 \end{vmatrix}$

18. $\begin{vmatrix} 2 & 1 & 1 & 0 \\ 1 & 2 & 0 & 2 \\ 3 & 1 & 0 & 3 \\ 2 & 1 & 4 & 2 \end{vmatrix} = \begin{vmatrix} 2 & 1 & 1 & 0 \\ 1 & 2 & 0 & 2 \\ 3 & 1 & 0 & 3 \\ & -3 & 0 & 2 \end{vmatrix}$

First reduce each determinant to an equal 2×2 determinant and then evaluate.

Example

$\begin{vmatrix} 0 & 3 & 2 \\ 1 & 7 & 8 \\ 0 & 5 & 4 \end{vmatrix}$

Solution

Expanding by column 1, we see that the determinant is equal to

$$-1 \cdot \begin{vmatrix} 3 & 2 \\ 5 & 4 \end{vmatrix} = -1(12 - 10) = -2.$$

19. $\begin{vmatrix} 2 & 1 & 0 \\ 3 & 2 & 1 \\ -1 & 2 & 0 \end{vmatrix}$

20. $\begin{vmatrix} 1 & 2 & 1 \\ 2 & -1 & 2 \\ 0 & 1 & 0 \end{vmatrix}$

21. $\begin{vmatrix} 1 & 0 & 3 \\ 2 & -1 & 1 \\ 1 & 2 & 1 \end{vmatrix}$

22. $\begin{vmatrix} 1 & 2 & -1 \\ 2 & 1 & 3 \\ 0 & 1 & 2 \end{vmatrix}$

23. $\begin{vmatrix} 1 & 2 & 1 \\ -1 & 2 & 3 \\ 2 & -1 & 1 \end{vmatrix}$

24. $\begin{vmatrix} 3 & -1 & 2 \\ 1 & 2 & 1 \\ -2 & 1 & 3 \end{vmatrix}$

25. $\begin{vmatrix} 11 & -10 & 32 \\ 12 & -11 & 35 \\ 12 & -6 & 33 \end{vmatrix}$

26. $\begin{vmatrix} 9 & 31 & 16 \\ 11 & 38 & 19 \\ 10 & 34 & 21 \end{vmatrix}$

27. $\begin{vmatrix} 28 & 27 & 25 \\ 31 & 30 & 26 \\ 36 & 35 & 30 \end{vmatrix}$

28. $\begin{vmatrix} 26 & 29 & 29 \\ 25 & 27 & 30 \\ 25 & 26 & 28 \end{vmatrix}$

29. $\begin{vmatrix} 13 & 16 & 19 \\ 28 & 34 & 40 \\ 27 & 33 & 39 \end{vmatrix}$

30. $\begin{vmatrix} 19 & 54 & 17 \\ 20 & 57 & 18 \\ 19 & 55 & 19 \end{vmatrix}$

31. $\begin{vmatrix} 0 & 0 & 1 & 2 \\ 6 & 0 & 0 & 1 \\ 6 & 1 & 0 & -1 \\ 6 & 1 & 0 & 2 \end{vmatrix}$

32. $\begin{vmatrix} 4 & 2 & 0 & 2 \\ -1 & 0 & 2 & 1 \\ 3 & 0 & -1 & 1 \\ 0 & 0 & 2 & 1 \end{vmatrix}$

33. $\begin{vmatrix} 0 & 1 & 0 & 2 \\ 0 & 2 & 0 & 3 \\ 2 & -1 & 1 & 0 \\ 0 & 0 & 8 & 8 \end{vmatrix}$

12.6 The Inverse of a Square Matrix

34. $\begin{vmatrix} 0 & 2 & -1 & 3 \\ 0 & 0 & 2 & 1 \\ 3 & 0 & 1 & 0 \\ -6 & 6 & 0 & 0 \end{vmatrix}$

35. $\begin{vmatrix} 1 & 2 & 3 & -1 \\ 0 & 4 & 8 & 4 \\ -2 & 0 & 1 & 1 \\ 2 & 1 & 0 & 1 \end{vmatrix}$

36. $\begin{vmatrix} 1 & 2 & 1 & 1 \\ 2 & -1 & 0 & 1 \\ 0 & 6 & 3 & 9 \\ 2 & 0 & -1 & 1 \end{vmatrix}$

37. Show that $\begin{vmatrix} x & y & 1 \\ x_1 & y_1 & 1 \\ x_2 & y_2 & 1 \end{vmatrix} = 0$ represents an equation of the line through the points (x_1, y_1) and (x_2, y_2).

38. Use the results in Exercise 37 to find an equation of the line through the points $(3, -1)$ and $(-2, 5)$.

39. Show that $\begin{vmatrix} 1 & a & a^2 \\ 1 & b & b^2 \\ 1 & c & c^2 \end{vmatrix} = (b-c)(c-a)(a-b)$.

40. Show that $\begin{vmatrix} a_{11} & a_{12} & a_{13} & a_{14} \\ a_{21} & a_{22} & a_{23} & a_{24} \\ 0 & 0 & a_{33} & a_{34} \\ 0 & 0 & a_{43} & a_{44} \end{vmatrix} = \begin{vmatrix} a_{11} & a_{12} \\ a_{21} & a_{22} \end{vmatrix} \cdot \begin{vmatrix} a_{33} & a_{34} \\ a_{43} & a_{44} \end{vmatrix}$.

12.6 The Inverse of a Square Matrix

In the field of real numbers, every element a except 0 has a multiplicative inverse $1/a$ with the property that $a \cdot 1/a = 1$. The question should (and does) arise, "Does every square matrix A have a multiplicative inverse A^{-1}?"

Definition 12.19 For a given square matrix A of order n, if there is a square matrix A^{-1} of order n such that

$$AA^{-1} = I \quad \text{and} \quad A^{-1}A = I,$$

where I is the multiplicative identity matrix of order n, then A^{-1} is the **multiplicative inverse** of A.

To answer the question about the existence of a multiplicative inverse for a matrix, we shall begin by considering the simple case of 2×2 matrices. If we let

$$A = \begin{bmatrix} a_{11} & a_{12} \\ a_{21} & a_{22} \end{bmatrix},$$

then we must see whether or not there exists a 2×2 matrix A^{-1} such that $AA^{-1} = I$. If so, let $A^{-1} = \begin{bmatrix} b & c \\ d & e \end{bmatrix}$. We wish to have

$$\begin{bmatrix} a_{11} & a_{12} \\ a_{21} & a_{22} \end{bmatrix} \begin{bmatrix} b & c \\ d & e \end{bmatrix} = \begin{bmatrix} 1 & 0 \\ 0 & 1 \end{bmatrix}.$$

This leads to

$$\begin{bmatrix} a_{11}b + a_{12}d & a_{11}c + a_{12}e \\ a_{21}b + a_{22}d & a_{21}c + a_{22}e \end{bmatrix} = \begin{bmatrix} 1 & 0 \\ 0 & 1 \end{bmatrix},$$

which is true if and only if

$$\begin{aligned} a_{11}b + a_{12}d = 1, & \quad a_{11}c + a_{12}e = 0, \\ a_{21}b + a_{22}d = 0, & \quad a_{21}c + a_{22}e = 1. \end{aligned} \tag{1}$$

Solving these equations for b, c, d, and e, we have

$$(a_{11}a_{22} - a_{12}a_{21})b = a_{22}, \quad (a_{11}a_{22} - a_{12}a_{21})c = -a_{12},$$
$$(a_{11}a_{22} - a_{12}a_{21})d = -a_{21}, \quad (a_{11}a_{22} - a_{12}a_{21})e = a_{11},$$

from which

$$b = \frac{a_{22}}{a_{11}a_{22} - a_{12}a_{21}}, \quad c = \frac{-a_{12}}{a_{11}a_{22} - a_{12}a_{21}},$$

$$d = \frac{-a_{21}}{a_{11}a_{22} - a_{12}a_{21}}, \quad e = \frac{a_{11}}{a_{11}a_{22} - a_{12}a_{21}},$$

provided $a_{11}a_{22} - a_{12}a_{21} \neq 0$. Now the denominator of each of these fractions is just $\delta(A)$, so that

$$A^{-1} = \begin{bmatrix} b & c \\ d & e \end{bmatrix} = \begin{bmatrix} \dfrac{a_{22}}{\delta(A)} & \dfrac{-a_{12}}{\delta(A)} \\ \dfrac{-a_{21}}{\delta(A)} & \dfrac{a_{11}}{\delta(A)} \end{bmatrix} = \frac{1}{\delta(A)} \begin{bmatrix} a_{22} & -a_{12} \\ -a_{21} & a_{11} \end{bmatrix}.$$

By direct multiplication, it can be verified not only that

$$AA^{-1} = I,$$

but also (surprisingly, since matrix multiplication is not always commutative) that

$$A^{-1}A = I.$$

Inverse of a 2 × 2 matrix Thus, to write the inverse of a 2 × 2 square matrix A for which $\delta(A) \neq 0$, we interchange the entries on the principal diagonal, replace each of the other two entries with its negative, and multiply the result by $1/\delta(A)$.

Example If $A = \begin{bmatrix} 1 & 3 \\ 2 & -1 \end{bmatrix}$, find A^{-1}.

12.6 The Inverse of a Square Matrix

Solution We first observe that $\delta(A) = (1)(-1) - (3)(2) = -7$. Hence,

$$A^{-1} = -\frac{1}{7}\begin{bmatrix} -1 & -3 \\ -2 & 1 \end{bmatrix} = \begin{bmatrix} \frac{1}{7} & \frac{3}{7} \\ \frac{2}{7} & -\frac{1}{7} \end{bmatrix}.$$

It is a good idea always to check the result when finding A^{-1}, because there is much room for blundering in the process of determining the inverse. In the present example, we have

$$A^{-1}A = -\frac{1}{7}\begin{bmatrix} -1 & -3 \\ -2 & 1 \end{bmatrix}\begin{bmatrix} 1 & 3 \\ 2 & -1 \end{bmatrix} = -\frac{1}{7}\begin{bmatrix} -7 & 0 \\ 0 & -7 \end{bmatrix} = \begin{bmatrix} 1 & 0 \\ 0 & 1 \end{bmatrix}.$$

Matrices with no inverse

We have now arrived at a position where we can answer the question, "Does every 2×2 square matrix A have an inverse?" The answer is "No," for if $\delta(A)$ is 0, then the foregoing equations (1) for b, c, d, e would have no solution.

Example The matrix $\begin{bmatrix} 3 & 5 \\ 6 & 10 \end{bmatrix}$ has no inverse because

$$\delta(A) = 3(10) - 6(5) = 0.$$

More generally, and without proving it, we have the following result.

Theorem 12.13 If

$$A = \begin{bmatrix} a_{11} & a_{12} & \cdots & a_{1n} \\ a_{21} & a_{22} & \cdots & a_{2n} \\ \vdots & \vdots & & \vdots \\ a_{n1} & a_{n2} & \cdots & a_{nn} \end{bmatrix},$$

and if $\delta(A) \neq 0$, then

$$A^{-1} = \frac{1}{\delta(A)}\begin{bmatrix} A_{11} & A_{21} & \cdots & A_{n1} \\ A_{12} & A_{22} & \cdots & A_{n2} \\ \vdots & \vdots & & \vdots \\ A_{1n} & A_{2n} & \cdots & A_{nn} \end{bmatrix},$$

where A_{ij} is the cofactor of a_{ij} in A. If $\delta(A) = 0$, then A has no inverse.

Square matrices A for which $\delta(A) \neq 0$ are *nonsingular*, for it can be shown that A is row-equivalent to the identity if and only if $\delta(A) \neq 0$. Thus by Theorem 12.13, A has an inverse if and only if A is nonsingular.

	Inverse of an $n \times n$ matrix

Observe that A^{-1} is the matrix having as its entries the cofactors of the entries in A multiplied by $1/\delta(A)$, but that the cofactors of the *row* entries in A are the *column* entries in A^{-1}. One way to obtain A^{-1} is to replace each entry in A with its cofactor and multiply the *transpose* of the resulting matrix by $1/\delta(A)$.

Example If $A = \begin{bmatrix} 1 & 0 & 1 \\ 2 & 1 & 0 \\ 1 & -1 & 1 \end{bmatrix}$, find A^{-1}.

Solution We first observe that $\delta(A) = -2$, and since $\delta(A)$ is not zero, A has an inverse. Next, replacing each entry in A with its cofactor, we obtain the matrix

$$\begin{bmatrix} 1 & -2 & -3 \\ -1 & 0 & 1 \\ -1 & 2 & 1 \end{bmatrix}, \text{ whose transpose is } \begin{bmatrix} 1 & -1 & -1 \\ -2 & 0 & 2 \\ -3 & 1 & 1 \end{bmatrix},$$

so that

$$A^{-1} = -\frac{1}{2} \begin{bmatrix} 1 & -1 & -1 \\ -2 & 0 & 2 \\ -3 & 1 & 1 \end{bmatrix}$$

As a check, we have

$$A^{-1}A = -\frac{1}{2} \begin{bmatrix} 1 & -1 & -1 \\ -2 & 0 & 2 \\ -3 & 1 & 1 \end{bmatrix} \begin{bmatrix} 1 & 0 & 1 \\ 2 & 1 & 0 \\ 1 & -1 & 1 \end{bmatrix}$$

$$= -\frac{1}{2} \begin{bmatrix} -2 & 0 & 0 \\ 0 & -2 & 0 \\ 0 & 0 & -2 \end{bmatrix} = \begin{bmatrix} 1 & 0 & 0 \\ 0 & 1 & 0 \\ 0 & 0 & 1 \end{bmatrix}$$

Theorem 12.13 is applicable to $n \times n$ square matrices, although, clearly, the process of actually determining A^{-1} by the formula given in that theorem becomes very laborious for matrices much larger than 3×3.

Properties of matrices and their inverses

There are a number of useful properties associated with matrices and their inverses. The following theorem gives one example.

Theorem 12.14 If A and B are $n \times n$ nonsingular square matrices, then AB has an inverse, namely

$$(AB)^{-1} = B^{-1}A^{-1}.$$

Proof If we right-multiply AB by $B^{-1}A^{-1}$, and apply the associative law for the multiplication of matrices, we have

$$AB \cdot B^{-1}A^{-1} = A \cdot I \cdot A^{-1} = A \cdot A^{-1} = I.$$

12.6 The Inverse of a Square Matrix

Moreover, if we left-multiply AB by $B^{-1}A^{-1}$, we have

$$B^{-1}A^{-1} \cdot AB = B^{-1} \cdot I \cdot B = B^{-1} \cdot B = I.$$

Thus, since $(AB)(B^{-1}A^{-1}) = (B^{-1}A^{-1})(AB) = I$, by the definition of the inverse of a matrix, we have

$$(AB)^{-1} = B^{-1}A^{-1}.$$

This theorem can be used to find the inverse of products of any number of nonsingular matrices. For example, if there are three factors A, B, and C in a product,

$$(ABC)^{-1} = [(AB)C]^{-1} = C^{-1}(AB)^{-1} = C^{-1}B^{-1}A^{-1}.$$

Exercise 12.6

Find the inverse of each matrix if one exists.

Example

$$B = \begin{bmatrix} 1 & 0 & -1 \\ 1 & 3 & 1 \\ 0 & 1 & 2 \end{bmatrix}$$

Solution

The determinant $\delta(B)$ is given by

$$\delta\begin{bmatrix} 1 & 0 & -1 \\ 1 & 3 & 1 \\ 0 & 1 & 2 \end{bmatrix} = 1(5) - 0 - 1(1) = 4.$$

Replacing each entry of B with its cofactor gives

$$\begin{bmatrix} 5 & -2 & 1 \\ -1 & 2 & -1 \\ 3 & -2 & 3 \end{bmatrix}; \quad \begin{bmatrix} 5 & -2 & 1 \\ -1 & 2 & -1 \\ 3 & -2 & 3 \end{bmatrix}^t = \begin{bmatrix} 5 & -1 & 3 \\ -2 & 2 & -2 \\ 1 & -1 & 3 \end{bmatrix};$$

$$B^{-1} = \frac{1}{\delta(B)} \begin{bmatrix} \text{each } b_{ij} \text{ of} \\ B \text{ replaced} \\ \text{by } B_{ij} \end{bmatrix}^t = \frac{1}{4} \begin{bmatrix} 5 & -1 & 3 \\ -2 & 2 & -2 \\ 1 & -1 & 3 \end{bmatrix}$$

$$= \begin{bmatrix} \frac{5}{4} & -\frac{1}{4} & \frac{3}{4} \\ -\frac{2}{4} & \frac{2}{4} & -\frac{2}{4} \\ \frac{1}{4} & -\frac{1}{4} & \frac{3}{4} \end{bmatrix}$$

1. $\begin{bmatrix} 1 & 2 \\ 1 & 3 \end{bmatrix}$
2. $\begin{bmatrix} 3 & 1 \\ 2 & -1 \end{bmatrix}$
3. $\begin{bmatrix} 2 & -3 \\ 1 & 1 \end{bmatrix}$

4. $\begin{bmatrix} 3 & -2 \\ 2 & 1 \end{bmatrix}$
5. $\begin{bmatrix} -2 & -1 \\ 4 & 2 \end{bmatrix}$
6. $\begin{bmatrix} 3 & 1 \\ 9 & 3 \end{bmatrix}$

7. $\begin{bmatrix} 5 & 7 \\ 3 & 4 \end{bmatrix}$
8. $\begin{bmatrix} 5 & -4 \\ 4 & -3 \end{bmatrix}$
9. $\begin{bmatrix} 7 & 4 \\ -4 & -2 \end{bmatrix}$

10. $\begin{bmatrix} -9 & 5 \\ -4 & 2 \end{bmatrix}$
11. $\begin{bmatrix} -2 & -6 \\ -3 & -9 \end{bmatrix}$
12. $\begin{bmatrix} 21 & 7 \\ 9 & 3 \end{bmatrix}$

13. $\begin{bmatrix} 1 & -1 & 2 \\ 2 & 1 & 3 \\ 0 & 0 & 2 \end{bmatrix}$
14. $\begin{bmatrix} 0 & 4 & 2 \\ 1 & 0 & 2 \\ 0 & -1 & 1 \end{bmatrix}$
15. $\begin{bmatrix} 2 & -1 & 1 \\ 3 & 0 & 1 \\ 2 & 2 & 1 \end{bmatrix}$

16. $\begin{bmatrix} 1 & 2 & 1 \\ 0 & 2 & 1 \\ -2 & 2 & 3 \end{bmatrix}$
17. $\begin{bmatrix} 2 & 1 & 1 \\ 1 & 0 & 2 \\ 4 & 2 & 2 \end{bmatrix}$
18. $\begin{bmatrix} -3 & 1 & -6 \\ 2 & 1 & 4 \\ 2 & 0 & 4 \end{bmatrix}$

19. $\begin{bmatrix} 1 & 2 & -3 \\ 3 & -1 & 0 \\ 5 & 3 & -6 \end{bmatrix}$
20. $\begin{bmatrix} 2 & 4 & -1 \\ 1 & 6 & 2 \\ 5 & 14 & 0 \end{bmatrix}$
21. $\begin{bmatrix} 2 & -1 & -5 \\ 1 & 3 & 4 \\ 0 & 1 & 2 \end{bmatrix}$

22. $\begin{bmatrix} 2 & 1 & -8 \\ 1 & 1 & -2 \\ 1 & 2 & 3 \end{bmatrix}$
23. $\begin{bmatrix} 0 & 0 & 1 \\ 0 & 1 & 0 \\ 1 & 0 & 0 \end{bmatrix}$
24. $\begin{bmatrix} 1 & 0 & 1 \\ 0 & 1 & 0 \\ 1 & 0 & 0 \end{bmatrix}$

25. Verify that

$$\left(\begin{bmatrix} 2 & 3 \\ 1 & -1 \end{bmatrix} \cdot \begin{bmatrix} 0 & 1 \\ 3 & 1 \end{bmatrix} \right)^{-1} = \begin{bmatrix} 0 & 1 \\ 3 & 1 \end{bmatrix}^{-1} \cdot \begin{bmatrix} 2 & 3 \\ 1 & -1 \end{bmatrix}^{-1}.$$

26. Verify that

$$\left(\begin{bmatrix} 1 & 2 \\ -1 & 0 \end{bmatrix} \cdot \begin{bmatrix} 1 & 1 \\ 2 & 0 \end{bmatrix} \cdot \begin{bmatrix} 2 & -1 \\ 0 & 1 \end{bmatrix} \right)^{-1} = \begin{bmatrix} 2 & -1 \\ 0 & 1 \end{bmatrix}^{-1} \cdot \begin{bmatrix} 1 & 1 \\ 2 & 0 \end{bmatrix}^{-1} \cdot \begin{bmatrix} 1 & 2 \\ -1 & 0 \end{bmatrix}^{-1}.$$

27. Verify that

$$\left(\begin{bmatrix} 3 & 0 & 1 \\ 2 & 1 & 0 \\ 0 & 1 & 2 \end{bmatrix} \cdot \begin{bmatrix} 2 & 1 & 0 \\ 1 & 1 & 2 \\ 0 & 1 & 0 \end{bmatrix} \right)^{-1} = \begin{bmatrix} 2 & 1 & 0 \\ 1 & 1 & 2 \\ 0 & 1 & 0 \end{bmatrix}^{-1} \cdot \begin{bmatrix} 3 & 0 & 1 \\ 2 & 1 & 0 \\ 0 & 1 & 2 \end{bmatrix}^{-1}.$$

28. Show that $[A^t]^{-1} = [A^{-1}]^t$ for each nonsingular 2×2 matrix.

29. Show that $\delta(A^{-1}) = 1/\delta(A)$ for each nonsingular 2×2 matrix.

30. Prove that if a and b are real numbers, then $\delta(aA^2 + bA) = \delta(aA + bI) \times \delta(A)$ for all 2×2 matrices A.

31. Prove that $\delta(B^{-1}AB) = \delta(A)$ for all nonsingular 2×2 matrices A and B.

32. Prove that if A is a 2×2 matrix and a, b, and c are real numbers, with $c \neq 0$, and if $aA^2 + bA + cI = \mathbf{0}$, then A has an inverse.

12.7   Solution of Linear Systems Using Inverses of Matrices

In Section 12.3 we solved linear systems using row-equivalent matrices. The solution for a linear system can also be found by using the inverse of a matrix.

We first verify the matrix product equation

$$\begin{bmatrix} a_{11} & a_{12} & \cdots & a_{1n} \\ \vdots & \vdots & & \vdots \\ a_{n1} & a_{n2} & \cdots & a_{nn} \end{bmatrix} \begin{bmatrix} x_1 \\ \vdots \\ x_n \end{bmatrix} = \begin{bmatrix} a_{11}x_1 + a_{12}x_2 + \cdots + a_{1n}x_n \\ \vdots \\ a_{n1}x_1 + a_{n2}x_2 + \cdots + a_{nn}x_n \end{bmatrix},$$

and hence note that the linear system

$$\begin{aligned} a_{11}x_1 + a_{12}x_2 + \cdots + a_{1n}x_n &= c_1 \\ a_{21}x_1 + a_{22}x_2 + \cdots + a_{2n}x_n &= c_2 \\ \vdots \quad\quad \vdots \quad\quad\quad\quad \vdots \quad\quad &\vdots \\ a_{n1}x_1 + a_{n2}x_2 + \cdots + a_{nn}x_n &= c_n \end{aligned} \quad (1)$$

can be written as the matrix equation

$$\begin{bmatrix} a_{11} & a_{12} & \cdots & a_{1n} \\ \vdots & \vdots & & \vdots \\ a_{n1} & a_{n2} & \cdots & a_{nn} \end{bmatrix} \begin{bmatrix} x_1 \\ \vdots \\ x_n \end{bmatrix} = \begin{bmatrix} c_1 \\ \vdots \\ c_n \end{bmatrix},$$

where the first factor in the left-hand member is the coefficient matrix for the system. In more concise notation, this latter equation can be written

$$AX = B,$$

where A is an $n \times n$ square matrix, and X and B are $n \times 1$ column matrices.

Solution of systems

If A in the foregoing equation is nonsingular, we can left-multiply both members of this equation by A^{-1} to obtain the equivalent matrices

$$A^{-1}AX = A^{-1}B,$$
$$IX = A^{-1}B,$$
$$X = A^{-1}B,$$

Example Use matrices to find the solution set of

$$2x + y + z = 1$$
$$x - 2y - 3z = 1$$
$$3x + 2y + 4z = 5.$$

Solution We first write this as a matrix equation of the form $AX = B$, thus:

$$\begin{bmatrix} 2 & 1 & 1 \\ 1 & -2 & -3 \\ 3 & 2 & 4 \end{bmatrix} \begin{bmatrix} x \\ y \\ z \end{bmatrix} = \begin{bmatrix} 1 \\ 1 \\ 5 \end{bmatrix}.$$

We next determine $\delta(A)$, obtaining

$$\delta \begin{bmatrix} 2 & 1 & 1 \\ 1 & -2 & -3 \\ 3 & 2 & 4 \end{bmatrix} = 2(-2) - 1(13) + 1(8) = -9,$$

observe that A is nonsingular, and then find A^{-1} by the method of Section 12.6, as shown below.

$$A^{-1} = \begin{bmatrix} 2 & 1 & 1 \\ 1 & -2 & -3 \\ 3 & 2 & 4 \end{bmatrix}^{-1} = -\frac{1}{9} \begin{bmatrix} -2 & -2 & -1 \\ -13 & 5 & 7 \\ 8 & -1 & -5 \end{bmatrix}.$$

As a matter of routine, we check the latter by verifying that $A^{-1}A = I$.

$$A^{-1}A = -\frac{1}{9} \begin{bmatrix} -2 & -2 & -1 \\ -13 & 5 & 7 \\ 8 & -1 & -5 \end{bmatrix} \begin{bmatrix} 2 & 1 & 1 \\ 1 & -2 & -3 \\ 3 & 2 & 4 \end{bmatrix}$$

$$= -\frac{1}{9} \begin{bmatrix} -9 & 0 & 0 \\ 0 & -9 & 0 \\ 0 & 0 & -9 \end{bmatrix}$$

$$= \begin{bmatrix} 1 & 0 & 0 \\ 0 & 1 & 0 \\ 0 & 0 & 1 \end{bmatrix}.$$

Now, since $X = A^{-1}B$, we have

$$\begin{bmatrix} x \\ y \\ z \end{bmatrix} = -\frac{1}{9} \begin{bmatrix} -2 & -2 & -1 \\ -13 & 5 & 7 \\ 8 & -1 & -5 \end{bmatrix} \begin{bmatrix} 1 \\ 1 \\ 5 \end{bmatrix} = -\frac{1}{9} \begin{bmatrix} -9 \\ 27 \\ -18 \end{bmatrix} = \begin{bmatrix} 1 \\ -3 \\ 2 \end{bmatrix}.$$

Hence, $x = 1$, $y = -3$, and $z = 2$, and the solution set of the system is $\{(1, -3, 2)\}$.

12.7 Solutions of Linear Systems Using Inverses of Matrices

The computation of A^{-1} is laborious when A is a square matrix containing many rows and columns. The foregoing method is not always the easiest to use in solving systems. It is, however, very useful when solving several systems of the form $AX = B$ having the same coefficient matrix A.

Example Solve the systems

$$\begin{bmatrix} 2 & 1 & 1 \\ 1 & -2 & -3 \\ 3 & 2 & 4 \end{bmatrix} \begin{bmatrix} x \\ y \\ z \end{bmatrix} = \begin{bmatrix} 8 \\ 5 \\ 10 \end{bmatrix} \quad \text{and} \quad \begin{bmatrix} 2 & 1 & 1 \\ 1 & -2 & -3 \\ 3 & 2 & 4 \end{bmatrix} \begin{bmatrix} x \\ y \\ z \end{bmatrix} = \begin{bmatrix} 0 \\ 10 \\ -11 \end{bmatrix}.$$

Solution Here $A = \begin{bmatrix} 2 & 1 & 1 \\ 1 & -2 & -3 \\ 3 & 2 & 4 \end{bmatrix}$,

and from the previous example,

$$A^{-1} = -\frac{1}{9} \begin{bmatrix} -2 & -2 & -1 \\ -13 & 5 & 7 \\ 8 & -1 & -5 \end{bmatrix}.$$

Therefore the solution to the first system is

$$\begin{bmatrix} x \\ y \\ z \end{bmatrix} = A^{-1} \begin{bmatrix} 8 \\ 5 \\ 10 \end{bmatrix} = \begin{bmatrix} 4 \\ 1 \\ -1 \end{bmatrix},$$

and the solution to the second is

$$\begin{bmatrix} x \\ y \\ z \end{bmatrix} = A^{-1} \begin{bmatrix} 0 \\ 10 \\ -11 \end{bmatrix} = \begin{bmatrix} 1 \\ 3 \\ -5 \end{bmatrix}.$$

Exercise 12.7

Find the solution set of the given system by means of matrices. If the system has no unique solution, so state.

1. $2x - 3y = -1$
   $x + 4y = 5$

2. $3x - 4y = -2$
   $x - 2y = 0$

3. $3x - 4y = -2$
   $6x + 12y = 36$

4. $2x - 4y = 7$
   $x - 2y = 1$

5. $2x - 3y = 0$
   $2x + y = 16$

6. $2x + 3y = 3$
   $3x - 4y = 0$

7. $x + y = 2$
   $2x - z = 1$
   $2y - 3z = -1$

8. $2x - 6y + 3z = -12$
   $3x - 2y + 5z = -4$
   $4x + 5y - 2z = 10$

9. $x - 2y + z = -1$
   $3x + y - 2z = 4$
   $y - z = 1$

10. $2x + 5z = 9$
    $4x + 3y = -1$
    $3y - 4z = -13$

11. $2x + 2y + z = 1$
    $x - y + 6z = 21$
    $3x + 2y - z = -4$

12. $4x + 8y + z = -6$
    $2x - 3y + 2z = 0$
    $x + 7y - 3z = -8$

13. $\quad x + y + z = 0$
$\quad 2x - y - 4z = 15$
$\quad x - 2y - z = 7$

14. $\quad x + y - 2z = 3$
$\quad 3x - y + z = 5$
$\quad 3x + 3y - 6z = 9$

12.8 Cramer's Rule

In Section 12.7, we obtained the solution set of the linear system (1) on page 369 with nonsingular coefficient matrix by first expressing the system in the matrix form $AX = B$ and then left-multiplying both members of the equation by A^{-1} to obtain

$$A^{-1}AX = X = A^{-1}B.$$

If, now, this technique is viewed in terms of determinants, we arrive at a general solution for such systems.

If the coefficient matrix A is nonsingular, then its inverse, A^{-1}, is

$$A^{-1} = \frac{1}{\delta(A)} \begin{bmatrix} A_{11} & A_{21} & \cdots & A_{n1} \\ \vdots & \vdots & & \vdots \\ A_{1n} & A_{2n} & \cdots & A_{nn} \end{bmatrix}.$$

Now, since $B = \begin{bmatrix} c_1 \\ c_2 \\ \vdots \\ c_n \end{bmatrix}$, we have

$$X = A^{-1}B = \frac{1}{\delta(A)} \begin{bmatrix} c_1 A_{11} + c_2 A_{21} + \cdots + c_n A_{n1} \\ c_1 A_{12} + c_2 A_{22} + \cdots + c_n A_{n2} \\ \vdots & \vdots \\ c_1 A_{1n} + c_2 A_{2n} + \cdots + c_n A_{nn} \end{bmatrix}.$$

Each entry in $X = A^{-1}B$ can be seen to be of the form

$$\frac{c_1 A_{1j} + c_2 A_{2j} + \cdots + c_n A_{nj}}{\delta(A)}.$$

But $c_1 A_{1j} + c_2 A_{2j} + \cdots + c_n A_{nj}$ is just the expansion of the determinant

$$\begin{vmatrix} a_{11} & a_{12} & \cdots & \overset{\underset{\text{jth}}{\downarrow}}{c_1} & \cdots & a_{1n} \\ a_{21} & a_{22} & \cdots & c_2 & \cdots & a_{2n} \\ \vdots & \vdots & & \vdots & & \vdots \\ a_{n1} & a_{n2} & \cdots & c_n & \cdots & a_{nn} \end{vmatrix}$$

about the jth column, which has entries $c_1, c_2, c_3, \ldots, c_n$ in place of $a_{1j}, a_{2j}, \ldots, a_{nj}$. Thus, each entry x_j in the matrix $X = \begin{bmatrix} x_1 \\ x_2 \\ \vdots \\ x_n \end{bmatrix} = A^{-1}B$ is

12.8 Cramer's Rule

$$x_j = \frac{\delta(A_j)}{\delta(A)} = \frac{\begin{vmatrix} a_{11} & a_{12} & \cdots & c_1 & \cdots & a_{1n} \\ a_{21} & a_{22} & \cdots & c_2 & \cdots & a_{2n} \\ \vdots & \vdots & & \vdots & & \vdots \\ a_{n1} & a_{n2} & \cdots & c_n & \cdots & a_{nn} \end{vmatrix}}{\begin{vmatrix} a_{11} & a_{12} & & & & a_{1n} \\ a_{21} & a_{22} & & \cdots & & a_{2n} \\ \vdots & \vdots & & & & \vdots \\ a_{n1} & a_{n2} & & \cdots & & a_{nn} \end{vmatrix}}$$

with the jth column indicated.

Application of Cramer's rule

This relationship expresses **Cramer's rule**. Cramer's rule is the assertion that if the determinant of the coefficient matrix of an $n \times n$ linear system *is not* 0, then the equations are consistent (the system has a solution) and have a unique solution which can be found as follows.

To find x_j in solving the matrix equation $AX = B$:

1. Write the determinant of the coefficient matrix for the system.
2. Replace each entry in the jth column of the coefficient matrix A with the corresponding entry from the column matrix B, and find the determinant of the resulting matrix.
3. Divide the result in Step 2 by the result in Step 1.

Hence, in a nonsingular 3×3 system:

$$x = \frac{\delta(A_x)}{\delta(A)}, \quad y = \frac{\delta(A_y)}{\delta(A)}, \quad \text{and} \quad z = \frac{\delta(A_z)}{\delta(A)}.$$

Example   Use Cramer's rule to solve the system

$$-4x + 2y - 9z = 2$$
$$3x + 4y + z = 5$$
$$x - 3y + 2z = 8.$$

Solution   By inspection,

$$\delta(A) = \begin{vmatrix} -4 & 2 & -9 \\ 3 & 4 & 1 \\ 1 & -3 & 2 \end{vmatrix},$$

$$= -4(11) - 2(5) - 9(-13)$$
$$= -44 - 10 + 117 = 63.$$

Solution Continued on Overleaf

Replacing the entries in the first column of A with corresponding constants 2, 5, and 8, we have

$$\delta(A_x) = \begin{vmatrix} 2 & 2 & -9 \\ 5 & 4 & 1 \\ 8 & -3 & 2 \end{vmatrix}$$

$$= 2(11) - 2(2) - 9(-47)$$

$$= 22 - 4 + 423 = 441.$$

Hence, by Cramer's rule,

$$x = \frac{\delta(A_x)}{\delta(A)} = \frac{441}{63} = 7.$$

Similarly, by replacing, in turn, the entries of the second and third columns of A with the corresponding constants, 2, 5, and 8, we have

$$\delta(A_y) = \begin{vmatrix} -4 & 2 & -9 \\ 3 & 5 & 1 \\ 1 & 8 & 2 \end{vmatrix} \quad \text{and} \quad \delta(A_z) = \begin{vmatrix} -4 & 2 & 2 \\ 3 & 4 & 5 \\ 1 & -3 & 8 \end{vmatrix}.$$

Now,

$$\delta(A_y) = -4(2) - 2(5) - 9(19)$$
$$= -8 - 10 - 171 = -189$$

and

$$\delta(A_z) = -4(47) - 2(19) + 2(-13)$$
$$= -188 - 38 - 26 = -252,$$

so that

$$y = \frac{\delta(A_y)}{\delta(A)} = \frac{-189}{63} = -3$$

and

$$z = \frac{\delta(A_z)}{\delta(A)} = \frac{-252}{63} = -4,$$

and the solution set of the system is $\{(7, -3, -4)\}$.

Test for consistency

If $\delta(A) = 0$ for a linear system, then the system either has infinitely many members in its solution set (the equations are consistent and one of them can be obtained from the others by linear combinations) or has an empty solution set (the equations are inconsistent). The distinction can be determined as follows: Consider the matrix of coefficients

$$\begin{bmatrix} a_{11} & \cdots & a_{1n} \\ \vdots & & \vdots \\ a_{n1} & \cdots & a_{nn} \end{bmatrix}$$

12.8 Cramer's Rule

and the augmented matrix

$$\begin{bmatrix} a_{11} & \cdots & a_{1n} & c_1 \\ \vdots & & \vdots & \vdots \\ a_{n1} & \cdots & a_{nn} & c_n \end{bmatrix},$$

and in each find a determinant (obtained by striking out certain rows and columns) of order as great as possible with value not 0. The order of such a nonvanishing determinant is called the **rank** of the matrix. The rank of the augmented matrix is either the same as, or 1 greater than, that of the matrix of coefficients. The equations are consistent if and only if the two ranks are the same.

For example, the coefficient matrix C for the system

$$x + 2y + 3z = 2$$
$$2x + 4y + 2z = -1$$
$$x + 2y - 2z = 5$$

is given by

$$C = \begin{bmatrix} 1 & 2 & 3 \\ 2 & 4 & 2 \\ 1 & 2 & -2 \end{bmatrix},$$

while the augmented matrix C_A for the system is given by

$$C_A = \begin{bmatrix} 1 & 2 & 3 & 2 \\ 2 & 4 & 2 & -1 \\ 1 & 2 & -2 & 5 \end{bmatrix}.$$

Since $\delta(C) = 0$, the rank of C is not 3, and we therefore know that the system does not have a unique solution. If the first column and third row are deleted, the remaining determinant

$$\begin{vmatrix} 2 & 3 \\ 4 & 2 \end{vmatrix}$$

is not zero (check this), so C has rank 2.

Now, if the first column of C_A is deleted, the remaining entries form the determinant

$$\delta \begin{bmatrix} 2 & 3 & 2 \\ 4 & 2 & -1 \\ 2 & -2 & 5 \end{bmatrix},$$

which is also not zero (check this). Hence C_A has rank 3. Therefore, the equations in the system are inconsistent.

Exercise 12.8

Find the solution set of each of the following systems by Cramer's rule. If $\delta(A) = 0$ in any of the systems, use the ranks of the coefficient matrix and the augmented matrix to determine whether or not the equations in the system are consistent.

1. $x - y = 2$
   $x + 4y = 5$

2. $x + y = 4$
   $x - 2y = 0$

3. $3x - 4y = -2$
   $x + y = 6$

4. $2x - 4y = 7$
   $x - 2y = 1$

5. $\frac{1}{3}x - \frac{1}{2}y = 0$
   $\frac{1}{2}x + \frac{1}{4}y = 4$

6. $\frac{2}{3}x + y = 1$
   $x - \frac{4}{3}y = 0$

7. $x - 2y = 6$
   $\frac{2}{3}x - \frac{4}{3}y = 6$

8. $\frac{1}{2}x + y = 3$
   $-\frac{1}{4}x - y = -3$

9. $x - 3y = 1$
   $y = 1$

10. $2x - 3y = 12$
    $x = 4$

11. $ax + by = 1$
    $bx + ay = 1$

12. $x + y = a$
    $x - y = b$

13. $x - 2y + z = -1$
    $3x + y - 2z = 4$
    $y - z = 1$

14. $2x + 5z = 9$
    $4x + 3y = -1$
    $3y - 4z = -13$

15. $2x + 2y + z = 1$
    $x - y + 6z = 21$
    $3x + 2y - z = -4$

16. $4x + 8y + z = -6$
    $2x - 3y + 2z = 0$
    $x + 7y - 3z = -8$

17. $x + y + z = 0$
    $2x - y - 4z = 15$
    $x - 2y - z = 7$

18. $x + y - 2z = 2$
    $3x - y + z = 5$
    $3x + 3y - 6z = 6$

19. $x - 2y - 2z = 3$
    $2x - 4y + 4z = 1$
    $3x - 3y - 3z = 4$

20. $3x - 2y + 5z = 6$
    $4x - 4y + 3z = 0$
    $5x - 4y + z = -5$

21. $x - 4z = -1$
    $3x + 3y = 2$
    $3x + 4z = 5$

22. $2x - \frac{2}{3}y + z = 2$
    $\frac{1}{2}x - \frac{1}{3}y - \frac{1}{4}z = 0$
    $4x + 5y - 3z = -1$

23. $x + y + z = 0$
    $w + 2y - z = 4$
    $2w - y + 2z = 3$
    $-2w + 2y - z = -2$

24. $x + y + z = 0$
    $x + z + w = 0$
    $x + y + w = 0$
    $y + z + w = 0$

25. For the system
$$a_1 x + b_1 y + c_1 = 0$$
$$a_2 x + b_2 y + c_2 = 0,$$
show that if both $\delta(A_y) = 0$ and $\delta(A_x) = 0$, and if c_1 and c_2 are not both 0, then $\delta(A) = 0$, and the equations are consistent. *Hint:* Show that the first two determinant equations imply that $a_1 c_2 = a_2 c_1$ and $b_1 c_2 = b_2 c_1$ and that the rest follows from the formation of a proportion with these equations.

26. Show that if $\delta(A) = 0$ and $\delta(A_x) = 0$, and if a_1 and a_2 are not both 0, then $\delta(A_y) = 0$, where $\delta(A)$ is the determinant of the coefficient matrix of the system in Exercise 25.

Chapter Review

[12.1] Write each sum or difference as a single matrix.

1. $\begin{bmatrix} 4 & -7 \\ 2 & 1 \end{bmatrix} + \begin{bmatrix} -3 & 6 \\ -1 & 0 \end{bmatrix}$

2. $\begin{bmatrix} 3 & -1 & 7 \\ 6 & 2 & 5 \end{bmatrix} + \begin{bmatrix} -1 & 6 & -9 \\ 8 & -3 & 7 \end{bmatrix}$

3. $\begin{bmatrix} 2 & -4 & 3 \\ 6 & 1 & 7 \\ 2 & 8 & 0 \end{bmatrix} - \begin{bmatrix} 4 & -1 & 2 \\ 3 & 8 & 1 \\ 7 & 6 & -5 \end{bmatrix}$

4. $\begin{bmatrix} -11 & 2 & -6 \\ 7 & 1 & 2 \\ -3 & 4 & 8 \end{bmatrix} - \begin{bmatrix} -3 & 5 & 0 \\ 1 & 4 & 2 \\ 6 & -1 & 3 \end{bmatrix}$

[12.2] Write each product as a single matrix.

5. $-7 \begin{bmatrix} 3 & -1 \\ 2 & 0 \\ 1 & 1 \end{bmatrix}$

6. $\begin{bmatrix} 3 & -1 & 2 \end{bmatrix} \begin{bmatrix} 4 \\ -1 \\ 0 \end{bmatrix}$

7. $\begin{bmatrix} 3 & -1 \\ 6 & 5 \end{bmatrix} \cdot \begin{bmatrix} -4 & 2 \\ 1 & 3 \end{bmatrix}$

8. $\begin{bmatrix} -1 & 7 & 6 \\ 3 & 1 & 2 \\ 1 & 0 & 1 \end{bmatrix} \cdot \begin{bmatrix} 1 & -1 & 2 \\ 1 & 0 & 3 \\ 2 & 1 & 1 \end{bmatrix}$

[12.3] Use row transformations on the augmented matrix to solve each system of equations.

9. $2x - y = 5$
   $x + 3y = -1$

10. $2x - y + z = 4$
    $x + 3y - z = 4$
    $x + 2y + z = 5$

[12.4] Evaluate each determinant.

11. $\begin{vmatrix} -3 & 0 \\ 2 & 1 \end{vmatrix}$

12. $\begin{vmatrix} 1 & 5 \\ -1 & 2 \end{vmatrix}$

13. $\begin{vmatrix} 3 & 1 & 0 \\ 2 & 0 & 1 \\ 1 & 2 & -1 \end{vmatrix}$

14. $\begin{vmatrix} 3 & 1 & -2 \\ -1 & 2 & 1 \\ 1 & -2 & 1 \end{vmatrix}$

[12.5] Reduce each determinant to an equal 2 × 2 determinant and evaluate.

15. $\begin{vmatrix} 3 & -1 & 2 \\ 1 & -2 & 0 \\ 2 & 1 & -1 \end{vmatrix}$

16. $\begin{vmatrix} 1 & 7 & -11 \\ 12 & 10 & 15 \\ 5 & 11 & 14 \end{vmatrix}$

[12.6] Find the inverse of each nonsingular matrix.

17. $\begin{bmatrix} -4 & 2 \\ 11 & 3 \end{bmatrix}$

18. $\begin{bmatrix} 1 & -1 & 2 \\ 3 & 1 & 0 \\ 2 & 1 & 1 \end{bmatrix}$

[12.7] Use matrices to solve each system.

19. $x - y = -3$
    $2x + 3y = -1$

20. $2x + z = 7$
    $y + 2z = 1$
    $3x + y + z = 9$

[12.8] Use Cramer's rule to solve each system.

21. $3x - y = -5$
    $x + 2y = -6$

22. $x - y + 2z = 3$
    $2x + y - z = 3$
    $x - 2y + 2z = 4$

13 Sequences and Series

A function with domain $\{1, 2, 3, \ldots\}$ will relate some object with 1, some object with 2, and so forth. Thus such a function puts objects in sequence, something goes first, second, and so on. The notion of a sequence is formalized in this chapter. The chapter concludes with a discussion of mathematical induction, a most useful technique for proving results about sequences.

13.1 Sequences

Let us consider a class of functions in which each function has as its domain either the set N of positive integers or a subset of successive members of N.

Definition 13.1 *A **sequence function** is a function having as its domain the set N of positive integers $1, 2, 3, \ldots$. A **finite-sequence function** has as its domain the set of positive integers $1, 2, 3, \ldots, n$, for some fixed n.*

For example, the function defined by

$$s(n) = n + 3, \quad n \in \{1, 2, 3, \ldots\}, \tag{1}$$

is a sequence function. The elements in the range of such a function, considered in the order

$$s(1), s(2), s(3), s(4), \ldots,$$

are said to form a **sequence**. Similarly, the elements of a finite-sequence function, considered in order, constitute a **finite sequence**.

For example, the sequence associated with (1) is found by successively substituting the numbers $1, 2, 3, \ldots,$ for n:

$$s(1) = 1 + 3 = 4,$$
$$s(2) = 2 + 3 = 5,$$
$$s(3) = 3 + 3 = 6,$$
$$s(4) = 4 + 3 = 7,$$

and so on. Thus the first four terms of (1) are 4, 5, 6, and 7. The nth term, or general term, is $n + 3$. As another example, the first five terms of the sequence defined by the equation

$$s(n) = \frac{3}{2n - 1}, \quad n \in \{1, 2, 3, \ldots\},$$

are 3/1, 3/3, 3/5, 3/7, and 3/9; and the twenty-fifth term is

$$s(25) = \frac{3}{2(25) - 1} = \frac{3}{49}.$$

Given several terms in a sequence, we are often able to construct an expression for a general term of a sequence to which they belong. Such a sequence is not unique. Thus, if the first three terms in a sequence are 2, 4, 6, ..., we may *surmise* that the general term is $s(n) = 2n$. Note, however, that the sequences for both

$$s(n) = 2n$$

and

$$s(n) = 2n + (n - 1)(n - 2)(n - 3)$$

start with 2, 4, 6, but that the two sequences differ for terms following the third.

Sequence notation

The notation ordinarily used for the terms in a sequence is not function notation as such. It is customary to denote a term in a sequence by means of a subscript. Thus, the sequence $s(1), s(2), s(3), s(4), \ldots$ would appear as $s_1, s_2, s_3, s_4, \ldots$

Arithmetic progressions

Let us next consider two special kinds of sequences that have many applications. The first kind can be defined as follows:

Definition 13.2   An **arithmetic progression** *is a sequence defined by equations of the form*

$$s_1 = a, \quad s_{n+1} = s_n + d,$$

where $a, d \in R$, and $n \in N$.

Since each term in such a sequence is obtained from the preceding term by adding d, d is called the **common difference**. Thus, 3, 7, 11, 15, ... is an arithmetic progression, in which $s_1 = 3$ and $d = 4$.

In Definition 13.2, each term of the sequence, after the first term, is defined by its relation to previous terms. Such a formulation is called a **recursive** definition. The general term is obtained as follows.

13.1 Sequences

Theorem 13.1   The nth term in the sequence defined by

$$s_1 = a, \quad s_{n+1} = s_n + d,$$

where $a, d \in R$, and $n \in N$, is

$$s_n = a + (n-1)d. \tag{2}$$

Proof   Since the sequence progresses from term to term by adding the common difference, any term is obtained by adding an appropriate number of differences to a. The second term requires one difference, the third requires two differences and in general, s_n requires $n - 1$ differences added to a. A complete proof requires the technique of mathematical induction which is provided in Section 13.5.

Geometric progressions

The second kind of sequence we shall consider can be defined as follows:

Definition 13.3   *A **geometric progression** is a sequence defined by equations of the form*

$$s_1 = a, \quad s_{n+1} = rs_n,$$

where $a, r \in R$, $a, r \neq 0$, and $n \in N$.

Thus, $3, 9, 27, 81, \ldots$ is a geometric progression in which each term except the first is obtained by multiplying the preceding term by 3. Since the effect of multiplying the terms in this way is to produce a fixed ratio between any two successive terms, the multiplier, r, is called the **common ratio**.

The general term for a geometric progression is that established by the following theorem. The proof by mathematical induction is left as an exercise in Section 13.5.

Theorem 13.2   The nth term in the sequence defined by

$$s_1 = a, \quad s_{n+1} = rs_n,$$

where $a, r \in R$, $a \neq 0$, $r \neq 0$, and $n \in N$, is

$$s_n = ar^{n-1}.$$

Exercise 13.1

Find the first four terms in the sequence with the general term as given.

Examples

a. $s_n = \dfrac{n(n+1)}{2}$ b. $s_n = (-1)^n 2^n$

Solutions on Overleaf

Solutions

a. $s_1 = \dfrac{1(1+1)}{2} = 1$

$s_2 = \dfrac{2(2+1)}{2} = 3$

$s_3 = \dfrac{3(3+1)}{2} = 6$

$s_4 = \dfrac{4(4+1)}{2} = 10;$

1, 3, 6, 10

b. $s_1 = (-1)^1 2^1 = -2$

$s_2 = (-1)^2 2^2 = 4$

$s_3 = (-1)^3 2^3 = -8$

$s_4 = (-1)^4 2^4 = 16;$

$-2, 4, -8, 16$

1. $s_n = n - 5$
2. $s_n = 2n - 3$
3. $s_n = \dfrac{n^2 - 2}{2}$
4. $s_n = \dfrac{3}{n^2 + 1}$
5. $s_n = 1 + \dfrac{1}{n}$
6. $s_n = \dfrac{n}{2n - 1}$
7. $s_n = \dfrac{n(n-1)}{2}$
8. $s_n = \dfrac{5}{n(n+1)}$
9. $s_n = (-1)^n$
10. $s_n = (-1)^{n+1}$
11. $s_n = \dfrac{(-1)^n(n-2)}{n}$
12. $s_n = (-1)^{n-1} 3^{n+1}$

Write the next three terms in each of the following arithmetic progressions.

Examples

a. 5, 9, ...

b. $x, x - a, \ldots$

Solutions

Find the common difference and then continue the sequence.

a. $d = 9 - 5 = 4;$

13, 17, 21

b. $d = (x - a) - x = -a;$

$x - 2a, x - 3a, x - 4a$

13. 3, 7, ...
14. $-6, -1, \ldots$
15. $x, x + 1, \ldots$
16. $a, a + 5, \ldots$
17. $2x + 1, 2x + 4, \ldots$
18. $3a, 5a, \ldots$

Write the next four terms in each of the following geometric progressions.

Examples

a. 3, 6, ...

b. $x, 2, \ldots$

Solutions

Find the common ratio, and then continue the sequence.

a. $r = \dfrac{6}{3} = 2;$

12, 24, 48, 96

b. $r = \dfrac{2}{x} \quad (x \neq 0);$

$\dfrac{4}{x}, \dfrac{8}{x^2}, \dfrac{16}{x^3}, \dfrac{32}{x^4}$

13.1 Sequences

19. 2, 8, ...   **20.** 4, 8, ...   **21.** $\dfrac{2}{3}, \dfrac{4}{3}, \ldots$

22. $\dfrac{1}{2}, -\dfrac{3}{2}, \ldots$ **23.** $\dfrac{a}{x}, -1, \ldots$ **24.** $\dfrac{a}{b}, \dfrac{a}{bc}, \ldots$

Example Find the general term and the fourteenth term of the arithmetic progression $-6, -1, \ldots$.

Solution Find the common difference.
$$d = -1 - (-6) = 5$$
Use $s_n = a + (n-1)d$.
$$s_n = -6 + (n-1)5 = 5n - 11$$
$$s_{14} = 5(14) - 11 = 59$$

25. Find the general term and the seventh term in the arithmetic progression 7, 11, ...

26. Find the general term and the twelfth term in the arithmetic progression $2, \dfrac{5}{2}, \ldots$.

27. Find the general term and the twentieth term in the arithmetic progression $3, -2, \ldots$.

28. Find the general term and the ninth term in the arithmetic progression $\dfrac{3}{4}, 2, \ldots$.

Example Find the general term and also the ninth term of the geometric progression $-24, 12, \ldots$.

Solution Find the common ratio.
$$r = \dfrac{12}{-24} = -\dfrac{1}{2}$$
Use $s_n = ar^{n-1}$.
$$s_n = -24\left(-\dfrac{1}{2}\right)^{n-1} \qquad s_9 = -24\left(-\dfrac{1}{2}\right)^8 = -\dfrac{3}{32}$$

29. Find the general term and the sixth term in the geometric progression 48, 96, ...

30. Find the general term and the eighth term in the geometric progression $-3, \dfrac{3}{2}, \ldots$.

31. Find the general term and the seventh term in the geometric progression $-\frac{1}{3}, 1, \ldots$.

32. Find the general term and the ninth term in the geometric progression $-81, 27, \ldots$.

Example Find the first term in an arithmetic progression in which the third term is 7 and the eleventh term is 55.

Solution Find the common difference by considering an arithmetic progression with first term 7 and with ninth term 55. Use $s_n = a + (n-1)d$.

$$s_9 = 7 + (9-1)d$$
$$55 = 7 + 8d$$
$$d = 6$$

Use the difference to find the first term in an arithmetic progression in which the third term is 7. Again use $s_n = a + (n-1)d$.

$$s_3 = a + (3-1)6$$
$$7 = a + 12$$
$$a = -5$$

33. If the third term in an arithmetic progression is 7 and the eighth term is 17, find the common difference. What are the first and the twentieth terms?
34. If the fifth term of an arithmetic progression is -16 and the twentieth term is -46, what is the twelfth term?
35. Which term in the arithmetic progression $4, 1, \ldots$ is -77?
36. What is the twelfth term in an arithmetic progression in which the second term is x and the third term is y?
37. Find the first term of a geometric progression with fifth term 48 and ratio 2.
38. Find two different values for x so that $-\frac{3}{2}, x, -\frac{8}{27}$ will be in geometric progression.

13.2 Series

Associated with any sequence is a *series*.

Definition 13.4   *A series is the indicated sum of the terms in a sequence.*

We shall ordinarily denote a series of n terms by S_n. For example, with the finite sequence

$$4, 7, 10, \ldots, 3n+1,$$

for a given counting number n, there is associated the finite series

$$S_n = 4 + 7 + 10 + \cdots + (3n+1);$$

13.2 Series

similarly, with the finite sequence

$$x, x^2, x^3, x^4, \ldots, x^n,$$

there is associated the finite series

$$S_n = x + x^2 + x^3 + x^4 + \cdots + x^n.$$

Since the terms in the series are the same as those in the sequence, we can refer to the first term or the second term or the general term of a series in the same manner as we do for a sequence.

Sum of the first n terms of an arithmetic progression

Consider the series S_n of the first n terms of the general arithmetic progression,

$$S_n = a + (a + d) + (a + 2d) + \cdots + [a + (n-1)d], \tag{1}$$

and then consider the same series written as

$$S_n = s_n + (s_n - d) + (s_n - 2d) + \cdots + [s_n - (n-1)d], \tag{2}$$

where the terms are displayed in reverse order. Adding (1) and (2) term by term, we have

$$S_n + S_n = (a + s_n) + (a + s_n) + (a + s_n) + \cdots + (a + s_n),$$

where the term $(a + s_n)$ occurs n times. Then

$$2S_n = n(a + s_n),$$

$$S_n = \frac{n}{2}(a + s_n). \tag{3}$$

If (3) is rewritten as

$$S_n = n\left(\frac{a + s_n}{2}\right),$$

we observe that the sum is given by the product of the number of terms in the series and the average of the first and last terms. The validity of (3) can be established by mathematical induction and is deferred until Section 13.5.

An alternate form for (3) is obtained by substituting $a + (n-1)d$ for s_n in (3) to obtain

$$S_n = \frac{n}{2}(a + [a + (n-1)d]),$$

$$S_n = \frac{n}{2}[2a + (n-1)d], \tag{3'}$$

where the sum is now expressed in terms of a, n, and d.

Sum of the first n terms of a geometric progression

To find an explicit representation for the sum of a given number of terms in a geometric progression in terms of a, r, and n, we employ a device somewhat similar to the one used in finding the sum in an arithmetic progression. Consider the geometric series (4) containing n terms, and the series (5) obtained by multiplying both members of (4) by r:

$$S_n = a + ar + ar^2 + ar^3 + \cdots + ar^{n-2} + ar^{n-1}, \tag{4}$$

$$rS_n = ar + ar^2 + ar^3 + ar^4 + \cdots + ar^{n-1} + ar^n. \tag{5}$$

When we subtract (5) from (4), all terms in the right-hand members except the first term in (4) and the last term in (5) vanish, yielding

$$S_n - rS_n = a - ar^n.$$

Factoring S_n from the left-hand member gives

$$(1 - r)S_n = a - ar^n,$$

$$S_n = \frac{a - ar^n}{1 - r}, \tag{6}$$

if $r \neq 1$, and we have a formula for the sum of the first n terms of a geometric progression. Establishing the validity of Equation (6), which can be accomplished by mathematical induction, is left to the exercises in Section 13.5.

An alternative expression for (6) can be obtained by first writing

$$S_n = \frac{a - r(ar^{n-1})}{1 - r},$$

and then, since $s_n = ar^{n-1}$, expressing this as

$$S_n = \frac{a - rs_n}{1 - r}, \tag{7}$$

where the sum is now given in terms of a, s_n, and r.

Sigma notation

A series for which the general term is known can be represented in a very convenient, compact way by means of what is called **sigma**, or **summation, notation**. The Greek letter $\sum$ (sigma) is used to denote a sum. For example,

$$S_n = 4 + 7 + 10 + \cdots + (3n + 1)$$

can be written

$$S_n = \sum_{j=1}^{n} (3j + 1),$$

where we understand that S_n is the series having terms obtained by replacing j in the expression $3j + 1$ with the numbers $1, 2, 3, \ldots, n$, successively. Similarly,

$$S = \sum_{j=3}^{6} j^2$$

13.2 Series

appears in expanded form as

$$S = 3^2 + 4^2 + 5^2 + 6^2,$$

where the first value for j is 3 and the last is 6.

The variable used in conjunction with summation notation is called the **index of summation**, and the set of integers over which we sum (in this case, $\{3, 4, 5, 6\}$) is called the **range of summation**.

Notation for an infinite sum

To indicate that a series has an infinite number of terms, we cannot use the notation S_n for the sum, because there is no value to substitute for n. We therefore adopt notation such as

$$S_\infty = \sum_{j=1}^{\infty} \frac{1}{2^j} \tag{8}$$

to indicate that there is no last term in a series. In expanded form, the infinite series (8) is given by

$$S_\infty = \frac{1}{2} + \frac{1}{4} + \frac{1}{8} + \cdots.$$

The meaning, if any, of such an infinite sum will be discussed in Section 13.3.

Exercise 13.2

Write each series in expanded form.

Examples

a. $\sum_{j=2}^{5} (j^2 + 1)$ b. $\sum_{k=1}^{\infty} (-1)^k 2^{k+1}$

Solutions

a. $j = 2$, $2^2 + 1 = 5$;
   $j = 3$, $3^2 + 1 = 10$;
   $j = 4$, $4^2 + 1 = 17$;
   $j = 5$, $5^2 + 1 = 26$.

   $\sum_{j=2}^{5} (j^2 + 1) = 5 + 10 + 17 + 26$

b. $k = 1$, $(-1)^1 2^{1+1} = (-1)(4) = -4$;
   $k = 2$, $(-1)^2 2^{2+1} = (1)(8) = 8$;
   $k = 3$, $(-1)^3 2^{3+1} = (-1)(16) = -16$.

   $\sum_{k=1}^{\infty} (-1)^k 2^{k+1} = -4 + 8 - 16 + \cdots$

1. $\sum_{j=1}^{4} j^2$ 2. $\sum_{j=1}^{4} (3j - 2)$ 3. $\sum_{j=1}^{3} \frac{(-1)^j}{2^j}$

4. $\sum_{k=3}^{5} \frac{(-1)^{k+1}}{k - 2}$ 5. $\sum_{k=0}^{\infty} \frac{1}{2^k}$ 6. $\sum_{k=0}^{\infty} \frac{k}{1 + k}$

Write each series in sigma notation.

Examples a. $5 + 8 + 11 + 14$ b. $x^2 + x^4 + x^6$ c. $\dfrac{3}{5} + \dfrac{5}{7} + \dfrac{7}{9} + \cdots$

Solutions Find an expression for the general term and write in sigma notation.

a. $3j + 2$ b. x^{2j} c. $\dfrac{2j+1}{2j+3}$

$\sum_{j=1}^{4}(3j+2)$ $\sum_{j=1}^{3}x^{2j}$ $\sum_{j=1}^{\infty}\dfrac{2j+1}{2j+3}$

7. $x + x^3 + x^5 + x^7$

8. $x^3 + x^5 + x^7 + x^9 + x^{11}$

9. $1 + 4 + 9 + 16 + 25$

10. $\dfrac{1}{3} + \dfrac{1}{9} + \dfrac{1}{27} + \dfrac{1}{81}$

11. $1 \cdot 2 + 2 \cdot 3 + 3 \cdot 4 + 4 \cdot 5 + \cdots$

12. $\dfrac{1}{2} + \dfrac{2}{3} + \dfrac{3}{4} + \dfrac{4}{5} + \cdots$

13. $\dfrac{2}{1} + \dfrac{3}{2} + \dfrac{4}{3} + \dfrac{5}{4} + \cdots$

14. $\dfrac{1}{1} + \dfrac{2}{3} + \dfrac{3}{5} + \dfrac{4}{7} + \cdots$

Find each of the following sums.

Example $\sum_{j=1}^{12}(4j + 1)$

Solution Write the first two or three terms in expanded form:
$$5 + 9 + 13 + \cdots.$$

This is an arithmetic series. The first term is 5 and the common difference is 4. Therefore we can use
$$S_n = \dfrac{n}{2}[2a + (n - 1)d]$$
to obtain
$$S_{12} = \dfrac{12}{2}[2(5) + (12 - 1)4] = 324.$$

15. $\sum_{j=1}^{7}(2j+1)$ 16. $\sum_{j=1}^{21}(3j-2)$ 17. $\sum_{j=3}^{15}(7j-1)$

18. $\sum_{j=10}^{20}(2j-3)$ 19. $\sum_{k=1}^{8}\left(\dfrac{1}{2}k-3\right)$ 20. $\sum_{k=1}^{100}k$

Example $\quad \sum_{j=2}^{5} \left(\frac{1}{3}\right)^j$

Solution Write the first two terms in expanded form:

$$\left(\frac{1}{3}\right)^2 + \left(\frac{1}{3}\right)^3 + \cdots.$$

This is a geometric series in which the first term is $\frac{1}{9}$, the ratio is $\frac{1}{3}$, and $n = 4$. Therefore we can use $S_n = \dfrac{a - ar^n}{1 - r}$ to obtain

$$S_4 = \frac{\frac{1}{9} - \frac{1}{9}\left(\frac{1}{3}\right)^4}{1 - \frac{1}{3}} = \frac{40}{243}.$$

21. $\sum_{j=1}^{6} 3^j$
22. $\sum_{j=1}^{4} (-2)^j$
23. $\sum_{k=3}^{7} \left(\frac{1}{2}\right)^{k-2}$

24. $\sum_{j=3}^{12} 2^{j-5}$
25. $\sum_{j=1}^{6} \left(\frac{1}{3}\right)^j$
26. $\sum_{j=1}^{4} (3 + 2^j)$

27. Find the sum of all even integers n, for $13 < n < 29$.

28. Find the sum of all integral multiples of 7 between 8 and 110.

29. How many bricks will there be in a wall one brick in thickness if there are 27 bricks in the bottom row, 25 in the second row, and so forth, to the top row, which has one brick?

30. If there are a total of 256 bricks in a wall arranged in the manner of those in Exercise 29, how many bricks are there in the tenth row from the bottom?

31. Find $\sum_{j=1}^{n} \left(\frac{1}{2}\right)^j$ for $n = 2, 3, 4,$ and 5. What value do you think $\sum_{j=1}^{n} \left(\frac{1}{2}\right)^j$ approximates as n becomes greater and greater?

32. Find p if $\sum_{j=1}^{5} pj = 14$.

33. Find p and q if $\sum_{j=1}^{4} (pj + q) = 28$ and $\sum_{j=2}^{5} (pj + q) = 44$.

34. Consider $S_n = \sum_{j=1}^{n} f(j)$. Explain why this equation defines a sequence function. What is the variable denoting an element in the domain? The range?

35. Show that the sequence formed by adding the corresponding terms in two arithmetic progressions is an arithmetic progression.

36. Show that the sum of the terms in two series with terms in arithmetic progression can be written as a series with terms in arithmetic progression.

13.3 Limits of Sequences and Series

A sequence that is strictly increasing but bounded

Consider the sequence function defined by

$$s_n = \frac{n}{n+1}, \quad n \in N. \tag{1}$$

If we write the range of (1) in the form

$$\frac{1}{2}, \frac{2}{3}, \frac{3}{4}, \frac{4}{5}, \ldots, \frac{n}{n+1}, \ldots,$$

then it is clear that each of the terms is greater than the preceding term; indeed, the difference of consecutive terms is

$$\frac{n+1}{n+2} - \frac{n}{n+1} = \frac{(n^2 + 2n + 1) - (n^2 + 2n)}{(n+1)(n+2)} = \frac{1}{(n+1)(n+2)} > 0.$$

Such a sequence is said to be **strictly increasing**. On the other hand, it is also clear that, no matter how large a value is assigned to n, we have

$$\frac{n}{n+1} < 1,$$

because the denominator is one larger than the numerator; in fact, we have

$$1 - \frac{n}{n+1} = \frac{(n+1) - n}{n+1} = \frac{1}{n+1} > 0.$$

Thus we have a sequence in which each term is greater than the preceding term and yet no term is equal to or greater than 1.

Limit of a sequence

We note, however—and this is a very basic consideration—that the value of $n/(n+1)$ is as close to 1 as we please if n is large enough. For example, the difference satisfies

$$1 - \frac{n}{n+1} = \frac{1}{n+1}, \quad \text{and we have} \quad \frac{1}{n+1} < \frac{1}{1000}$$

provided $n + 1 > 1000$—that is, $n > 999$. If it is true that the nth term in a sequence differs from the number L by as little as we please for all sufficiently large n, we

13.3 Limits of Sequences and Series

say that **the sequence approaches the number L as a limit**. The symbolism

$$\lim_{n \to \infty} s_n = L$$

(read "the limit, as n increases without bound, of s_n is L") is used to denote this situation. A thorough discussion of the notion of a limit is included in courses in calculus, and will not be attempted here. A few elementary ideas, however, are in order.

A sequence in which the nth term approaches a number L as $n \to \infty$ is said to be a **convergent sequence**, and the sequence is said to **converge** to L. It is not necessary for convergence that a sequence be strictly increasing. For example,

$$1, \frac{1}{2}, \frac{1}{3}, \frac{1}{4}, \ldots, \frac{1}{n}, \ldots$$

converges to 0, but each term in the sequence is less than, instead of greater than, the term that precedes it. Again, the sequence

$$-1, \frac{1}{2}, -\frac{1}{3}, \frac{1}{4}, \ldots, \frac{(-1)^n}{n}, \ldots$$

converges to 0 but is neither increasing nor decreasing. We can rephrase the definition of convergence of a sequence as follows:

Definition 13.5 A sequence $s_1, s_2, \ldots, s_n, \ldots$ **converges** to the number L,

$$\lim_{n \to \infty} s_n = L,$$

if and only if the absolute value of the difference between the nth term in the sequence and the number L is as small as we please for all sufficiently large n. Thus the sequence converges to the number L if and only if

$$\lim_{n \to \infty} |L - s_n| = 0.$$

For example, the **alternating** (because the signs alternate) **sequence**

$$-\frac{1}{3}, \frac{1}{9}, -\frac{1}{27}, \ldots, \frac{(-1)^n}{3^n}, \ldots$$

converges to 0 since the absolute value of the difference between $\frac{(-1)^n}{3^n}$ and 0 is as small as we please for n large enough. We express this by writing

$$\lim_{n \to \infty} \left(\frac{-1}{3}\right)^n = 0.$$

On the other hand, the alternating sequence

$$\frac{1}{2}, -\frac{2}{3}, \frac{3}{4}, -\frac{4}{5}, \ldots, (-1)^{n+1} \frac{n}{n+1}, \ldots$$

does not converge. As n increases, the nth term oscillates back and forth from the neighborhood of $+1$ to the neighborhood of -1, and we cannot find a number L such that $\lim_{n \to \infty} |L - s_n| = 0$. Such a sequence is said to **diverge**. A sequence such as

$$1, 2, 3, \ldots, n, \ldots$$

also is said to diverge. An answer to the logical question of what we mean by "enough" when we say "n large enough" requires a more precise definition of limit than we have given here. As remarked earlier, a course in the calculus will treat this in detail.

Sequence of partial sums of a series

For an infinite series,

$$S_\infty = \sum_{j=1}^{\infty} s_j,$$

we can consider the infinite sequence of **partial sums**:

$$S_1 = s_1$$
$$S_2 = s_1 + s_2$$
$$\cdot \quad \cdot \quad \cdot$$
$$S_n = s_1 + s_2 + \cdots + s_n$$
$$\cdot \quad \cdot \quad \cdot$$

Definition 13.6 *An infinite series*

$$S_\infty = \sum_{j=1}^{\infty} s_j$$

converges if and only if $S_1, S_2, \ldots, S_n, \ldots$, the corresponding sequence of partial sums, converges.

If the sequence of partial sums converges to the number L,

$$\lim_{n \to \infty} S_n = L,$$

then L is said to be the **sum** of the infinite series, and we write

$$S_\infty = \sum_{j=1}^{\infty} s_j = L.$$

If the sequence of partial sums diverges, then the series is said to **diverge**.

Sum of an infinite geometric progression

We recall from Section 13.2 that the sum of n terms (the nth partial sum) of a geometric progression is given, for $r \neq 1$, by

$$S_n = \frac{a - ar^n}{1 - r}. \tag{2}$$

If $|r| < 1$, that is, if $-1 < r < 1$, then $|r|^n$ becomes smaller and smaller for increasingly large n. For example, if $r = 1/2$, then

$$r^2 = \frac{1}{4}, \quad r^3 = \frac{1}{8}, \quad r^4 = \frac{1}{16},$$

and so forth; and $(1/2)^n$ is as small as we please if n is sufficiently large. Writing (2) as

$$S_n = \frac{a}{1 - r}(1 - r^n), \tag{3}$$

we see that the value of the factor $(1 - r^n)$ is as close as we please to 1 provided $|r| < 1$ and n is taken large enough. Since this argument shows that the sequence of partial sums (3) converges to

$$\frac{a}{1 - r},$$

we have the following result.

Theorem 13.3   *The sum of an infinite geometric progression,*

$$a + ar + ar^2 + \cdots + ar^n + \cdots,$$

with $|r| < 1$, is

$$S_\infty = \lim_{n \to \infty} S_n = \frac{a}{1 - r}.$$

An interesting application of this sum arises in connection with repeating decimals—that is, decimal numerals that, after a finite number of decimal places, have endlessly repeating groups of digits. For example,

$$0.21212\overline{1},$$

$$0.138512512\overline{512}$$

are repeating decimals. The bar denotes that the numerals appearing under it are repeated endlessly. Consider the problem of expressing such a decimal fraction as an arithmetic fraction. We illustrate the process involved with the first example above. The decimal $0.21212\overline{1}$ can be written as

$$0.21 + 0.0021 + 0.000021 + \cdots, \tag{4}$$

which is a geometric progression with ratio $r = 0.01$. Since the ratio is less than 1 in absolute value, we can use Theorem 13.3 to find the sum of the infinite series (4).

Thus

$$S_\infty = \frac{a}{1-r} = \frac{0.21}{1-0.01} = \frac{21}{99} = \frac{7}{33},$$

and the given decimal fraction is equivalent to 7/33.

Exercise 13.3

Discuss the limiting behavior of each expression as $n \to \infty$.

Example $\dfrac{n^2 + 3}{n^2}$

Solution By writing $\dfrac{n^2+3}{n^2}$ as $\dfrac{n^2}{n^2} + \dfrac{3}{n^2}$ and then as $1 + \dfrac{3}{n^2}$, we observe that

$$\lim_{n\to\infty} \frac{n^2+3}{n^2} = \lim_{n\to\infty}\left(1 + \frac{3}{n^2}\right) = 1 + 0 = 1.$$

1. $\dfrac{1}{n}$ 2. $1 + \dfrac{1}{n^2}$ 3. $\dfrac{n+1}{n}$ 4. $\dfrac{n+3}{n^2}$

5. $2n$ 6. $(-1)^n$ 7. $\dfrac{1}{2^n}$ 8. $(-1)^n \dfrac{1}{n}$

State which of the following sequences are convergent.

Example $1, \dfrac{3}{2}, \dfrac{7}{4}, \dfrac{15}{8}, \ldots, \dfrac{2^n - 1}{2^{n-1}}, \ldots$

Solution Writing the general term as $\dfrac{2^n}{2^{n-1}} - \dfrac{1}{2^{n-1}}$, or $2 - \dfrac{1}{2^{n-1}}$, we observe that

$$\lim_{n\to\infty} \frac{2^n - 1}{2^{n-1}} = \lim_{n\to\infty}\left(2 - \frac{1}{2^{n-1}}\right) = 2 - 0 = 2.$$

The sequence is convergent.

9. $\dfrac{1}{2}, \dfrac{1}{4}, \dfrac{1}{8}, \dfrac{1}{16}, \ldots, \dfrac{1}{2^n}, \ldots$ 10. $2, \dfrac{3}{2}, \dfrac{4}{3}, \dfrac{5}{4}, \ldots, \dfrac{n+1}{n}, \ldots$

11. $1, 2, 3, 4, 5, \ldots, n, \ldots$ 12. $2, 4, 6, 8, \ldots, 2n, \ldots$

13.3 Limits of Sequences and Series

13. $1, -\dfrac{1}{2}, \dfrac{1}{4}, -\dfrac{1}{8}, \ldots, (-1)^{n-1}\dfrac{1}{2^{n-1}}, \ldots$

14. $1, -1, 1, -1, \ldots, (-1)^{n+1}, \ldots$

Find the sum of each of the following infinite geometric series. If the series has no sum, so state.

Examples

a. $3 + 2 + \cdots$

b. $\dfrac{1}{81} - \dfrac{1}{54} + \cdots$

Solutions

a. $r = \dfrac{2}{3}$; series has a sum since $|r| < 1$. Using Theorem 13.3, we have

$$S_\infty = \dfrac{a}{1-r} = \dfrac{3}{1-\dfrac{2}{3}} = 9.$$

b. $r = -\dfrac{1}{54} \div \dfrac{1}{81} = -\dfrac{3}{2}$; series does not have a sum since $|r| > 1$.

15. $12 + 6 + \cdots$

16. $2 + 1 + \cdots$

17. $\dfrac{1}{36} + \dfrac{1}{30} + \cdots$

18. $\dfrac{1}{16} - \dfrac{1}{8} + \cdots$

19. $\sum_{j=1}^{\infty} \left(\dfrac{2}{3}\right)^j$

20. $\sum_{j=1}^{\infty} \left(-\dfrac{1}{4}\right)^j$

Find an arithmetic fraction equal to each of the given decimal numerals.

Example

$0.81\overline{81}$

Solution

Rewrite as a series: $0.81 + 0.0081 + 0.000081 + \cdots$.

Find the common ratio: $r = 0.01$. Use $S_\infty = \dfrac{a}{1-r}$.

$$S_\infty = \dfrac{0.81}{1 - 0.01} = \dfrac{81}{99} = \dfrac{9}{11}.$$

21. $0.31\overline{31}$

22. $0.45\overline{45}$

23. $2.41\overline{10}$

24. $3.02\overline{027}$

25. $0.12\overline{8888}$

26. $0.8\overline{3333}$

27. A force is applied to a particle moving in a straight line in such a fashion that each second it moves only one half of the distance it moved the preceding second. If the particle moves ten centimeters the first second, approximately how far will it move before coming to rest?

28. The arc length through which the bob on a pendulum moves is nine-tenths of its preceding arc length. Approximately how far will the bob move before coming to rest if the first arc length is 12 inches?

13.4 The Binomial Theorem

Factorial notation

There are situations in which it is necessary to write the product of several consecutive positive integers. To facilitate writing products of this type, we use a special symbol $n!$ (read "n factorial" or "factorial n"), which is defined by

$$n! = n(n-1)(n-2)\ldots(3)(2)(1).$$

Thus
$$5! = 5 \cdot 4 \cdot 3 \cdot 2 \cdot 1 \quad \text{(read "five factorial")},$$
and
$$8! = 8 \cdot 7 \cdot 6 \cdot 5 \cdot 4 \cdot 3 \cdot 2 \cdot 1 \quad \text{(read "eight factorial")}.$$

Factorial notation can also be used to represent products of consecutive positive integers, beginning with integers different from 1. For example,

$$8 \cdot 7 \cdot 6 \cdot 5 = \frac{8!}{4!},$$

because

$$\frac{8!}{4!} = \frac{8 \cdot 7 \cdot 6 \cdot 5 \cdot 4 \cdot 3 \cdot 2 \cdot 1}{4 \cdot 3 \cdot 2 \cdot 1} = 8 \cdot 7 \cdot 6 \cdot 5.$$

Since
$$n! = n(n-1)(n-2)(n-3) \cdots 5 \cdot 4 \cdot 3 \cdot 2 \cdot 1$$
and
$$(n-1)! = (n-1)(n-2)(n-3) \cdots 5 \cdot 4 \cdot 3 \cdot 2 \cdot 1,$$

for $n > 1$ we can write the recursive relationship

$$n! = n(n-1)!. \tag{1}$$

For example,
$$7! = 7 \cdot 6!,$$
$$27! = 27 \cdot 26!,$$
$$(n+2)! = (n+2)(n+1)!.$$

13.4 The Binomial Theorem

If (1) is to hold also for $n = 1$, then we must have

$$1! = 1 \cdot (1 - 1)!$$

or

$$1! = 1 \cdot 0!.$$

Therefore, for consistency, we define $0!$ by

$0! = 1.$

A special case of the use of factorial notation occurs in the **binomial coefficient**

$$\binom{n}{r} = \frac{n!}{r!(n-r)!}. \tag{2}$$

The symbol $\binom{n}{r}$ is read "the binomial coefficient n, r," or simply, "n, r." Some examples are

$$\binom{5}{3} = \frac{5!}{3!(5-3)!} = \frac{5!}{3!2!} = \frac{5 \cdot 4 \cdot 3!}{3!2 \cdot 1} = 10,$$

$$\binom{5}{1} = \frac{5!}{1!(5-1)!} = \frac{5!}{4!} = \frac{5 \cdot 4!}{4!} = 5,$$

and

$$\binom{5}{0} = \frac{5!}{0!(5-0)!} = \frac{5!}{5!} = 1.$$

Binomial expansions

The series obtained by expanding a binomial of the form

$$(a + b)^n$$

is particularly useful in certain branches of mathematics. Starting with familiar examples, where n takes the values 1, 2, 3, 4, and 5 in turn, we can show by direct multiplication that

$$(a + b)^1 = a + b$$
$$(a + b)^2 = a^2 + 2ab + b^2$$
$$(a + b)^3 = a^3 + 3a^2b + 3ab^2 + b^3$$
$$(a + b)^4 = a^4 + 4a^3b + 6a^2b^2 + 4ab^3 + b^4$$
$$(a + b)^5 = a^5 + 5a^4b + 10a^3b^2 + 10a^2b^3 + 5ab^4 + b^5.$$

We observe that in each case:

1. The first term is a^n.

2. The variable factors of the second term are $a^{n-1}b^1$, and the coefficient is n, which can be written in the form

$$\frac{n}{1!}.$$

3. The variable factors of the third term are $a^{n-2}b^2$, and the coefficient can be written in the form

$$\frac{n(n-1)}{2!}.$$

4. The variable factors of the fourth term are $a^{n-3}b^3$, and the coefficient can be written in the form

$$\frac{n(n-1)(n-2)}{3!}.$$

The foregoing expansions suggest the following result, known as the **binomial theorem**. Since its proof, which requires the use of mathematical induction, is quite lengthy, it is omitted.

Theorem 13.4   For each natural number n,

$$(a+b)^n = a^n + \frac{n}{1!}a^{n-1}b + \frac{n(n-1)}{2!}a^{n-2}b^2 + \frac{n(n-1)(n-2)}{3!}a^{n-3}b^3$$

$$+ \cdots + \frac{n(n-1)(n-2)\cdots(n-r+2)}{(r-1)!}a^{n-r+1}b^{r-1} + \cdots + b^n, \qquad (3)$$

where r is the number of the term.

For example,

$$(x-2)^4 = x^4 + \frac{4}{1!}x^3(-2)^1 + \frac{4\cdot 3}{2!}x^2(-2)^2 + \frac{4\cdot 3\cdot 2}{3!}x(-2)^3 + \frac{4\cdot 3\cdot 2\cdot 1}{4!}(-2)^4$$

$$= x^4 - 8x^3 + 24x^2 - 32x + 16.$$

In this case, $a = x$ and $b = -2$ in the binomial expansion.

Observe that the coefficients of the terms in the binomial expansion (3) can be represented as follows.

1st term: $1 = \dfrac{n!}{0!\,n!} = \dbinom{n}{0}$,

2nd term: $\dfrac{n}{1!} = \dfrac{n\cdot(n-1)!}{1!\,(n-1)!} = \dfrac{n!}{1!\,(n-1)!} = \dbinom{n}{1}$,

3rd term: $\dfrac{n(n-1)}{2!} = \dfrac{n(n-1)(n-2)!}{2!\,(n-2)!} = \dfrac{n!}{2!\,(n-2)!} = \dbinom{n}{2}$,

13.4 The Binomial Theorem

rth term: $$\frac{n(n-1)(n-2)\cdots(n-r+2)}{(r-1)!}$$

$$= \frac{n(n-1)(n-2)\cdots(n-r+2)(n-r+1)!}{(r-1)!(n-r+1)!}$$

$$= \frac{n!}{(r-1)!(n-r+1)!} = \binom{n}{r-1}.$$

Hence, the binomial expansion (3) can be represented by the expression

$$(a+b)^n = \binom{n}{0}a^n + \binom{n}{1}a^{n-1}b + \binom{n}{2}a^{n-2}b^2 + \binom{n}{3}a^{n-3}b^3 + \cdots$$

$$+ \binom{n}{r-1}a^{n-r+1}b^{r-1} + \cdots + \binom{n}{n}b^n. \quad (4)$$

For example,

$$(x-2)^4 = \binom{4}{0}x^4 + \binom{4}{1}x^3(-2)^1 + \binom{4}{2}x^2(-2)^2 + \binom{4}{3}x(-2)^3 + \binom{4}{4}(-2)^4.$$

Using the formula for $\binom{n}{r}$ to find the coefficients, we obtain the same result as above,

$$(x-2)^4 = x^4 - 8x^3 + 24x^2 - 32x + 16.$$

rth term in a binomial expansion

Note that the rth term in a binomial expansion is given by

$$\binom{n}{r-1}a^{n-r+1}b^{r-1} = \frac{n!}{(r-1)!(n-r+1)!}a^{n-r+1}b^{r-1}$$

$$= \frac{n(n-1)(n-2)\cdots(n-r+2)}{(r-1)!}a^{n-r+1}b^{r-1}. \quad (5)$$

For example, by the left-hand member of (5), the seventh term of $(x-2)^{10}$ is

$$\binom{10}{6}x^4(-2)^6 = \frac{10!}{6!4!}x^4(-2)^6 = \frac{10\cdot 9\cdot 8\cdot 7\cdot 6!}{6!\cdot 4\cdot 3\cdot 2\cdot 1}x^4(64)$$

$$= 13440\,x^4,$$

while, by the right-hand member of (5), we have

$$\frac{10\cdot 9\cdot 8\cdot 7\cdot 6\cdot 5}{6\cdot 5\cdot 4\cdot 3\cdot 2\cdot 1}x^4(64) = 13440\,x^4.$$

Exercise 13.4

1. Write $(2n)!$ in expanded form for $n = 4$.
2. Write $2n!$ in expanded form for $n = 4$.
3. Write $n(n-1)!$ in expanded form for $n = 6$.
4. Write $2n(2n-1)!$ in expanded form for $n = 2$.

Write in expanded form and simplify.

Examples a. $\dfrac{7!}{4!}$ b. $\dfrac{4!\,6!}{8!}$

Solutions a. $\dfrac{7\cdot 6\cdot 5\cdot 4!}{4!}$ b. $\dfrac{4\cdot 3\cdot 2\cdot 1\cdot 6!}{8\cdot 7\cdot 6!}$

$\qquad\qquad\quad 210 \qquad\qquad\qquad\qquad\qquad \dfrac{3}{7}$

5. $5!$ **6.** $7!$ **7.** $\dfrac{9!}{7!}$ **8.** $\dfrac{12!}{11!}$

9. $\dfrac{5!\,7!}{8!}$ **10.** $\dfrac{12!\,8!}{16!}$ **11.** $\dfrac{8!}{2!(8-2)!}$ **12.** $\dfrac{10!}{4!(10-4)!}$

Write each product in factorial notation.

Examples a. $1\cdot 2\cdot 3\cdot 4\cdot 5\cdot 6$ b. $11\cdot 12\cdot 13\cdot 14$ c. 150

Solutions a. $6!$ b. $\dfrac{14!}{10!}$ c. $\dfrac{150!}{149!}$

13. $1\cdot 2\cdot 3$ **14.** $1\cdot 2\cdot 3\cdot 4\cdot 5$ **15.** $3\cdot 4\cdot 5\cdot 6$
16. 7 **17.** $8\cdot 7\cdot 6$ **18.** $28\cdot 27\cdot 26\cdot 25\cdot 24$

Write each expression in factorial notation and simplify.

Examples a. $\binom{6}{2}$ b. $\binom{4}{4}$

Solutions a. $\binom{6}{2} = \dfrac{6!}{2!(6-2)!} = \dfrac{6!}{2!\,4!}$ b. $\binom{4}{4} = \dfrac{4!}{4!(4-4)!}$

$\qquad\qquad\qquad = \dfrac{6\cdot 5\cdot 4!}{2\cdot 1\cdot 4!} = 15 \qquad\qquad = \dfrac{4!}{4!\,0!} = 1$

19. $\binom{6}{5}$ **20.** $\binom{4}{2}$ **21.** $\binom{3}{3}$ **22.** $\binom{5}{5}$

23. $\binom{7}{0}$ **24.** $\binom{2}{0}$ **25.** $\binom{5}{2}$ **26.** $\binom{5}{3}$

13.4 The Binomial Theorem

Write each expression in factored form and show the first three factors and the last three factors.

Example $(2n + 1)!$

Solution $(2n + 1)(2n)(2n - 1) \cdot \cdots \cdot 3 \cdot 2 \cdot 1$

27. $n!$
28. $(n + 4)!$
29. $(3n)!$
30. $3n!$
31. $(n - 2)!$
32. $(3n - 2)!$

Simplify each expression.

Examples a. $\dfrac{(n - 1)!}{(n - 3)!}$ b. $\dfrac{(n - 1)!(2n)!}{2n!(2n - 2)!}$

Solutions a. $\dfrac{(n - 1)(n - 2)(n - 3)!}{(n - 3)!}$ b. $\dfrac{(n - 1)!(2n)(2n - 1)(2n - 2)!}{2n(n - 1)!(2n - 2)!}$

$(n - 1)(n - 2)$ $2n - 1$

33. $\dfrac{(n + 2)!}{n!}$
34. $\dfrac{(n + 2)!}{(n - 1)!}$
35. $\dfrac{(n + 1)(n + 2)!}{(n + 3)!}$

36. $\dfrac{(2n + 4)!}{(2n + 2)!}$
37. $\dfrac{(2n)!(n - 2)!}{4(2n - 2)!\ n!}$
38. $\dfrac{(2n + 1)!(2n - 1)!}{[(2n)!]^2}$

Expand.

Example $(a - 3b)^4$

Solution From the binomial expansion (3) on page 398,

$$(a - 3b)^4 = a^4 + \frac{4}{1!}a^3(-3b) + \frac{4 \cdot 3}{2!}a^2(-3b)^2 + \frac{4 \cdot 3 \cdot 2}{3!}a(-3b)^3$$

$$+ \frac{4 \cdot 3 \cdot 2 \cdot 1}{4!}(-3b)^4$$

$$= a^4 - 12a^3b + 54a^2b^2 - 108ab^3 + 81b^4.$$

Alternatively, from the abbreviation of the binomial expansion (4) on page 399, we have

$$(a - 3b)^4 = \binom{4}{0}a^4 + \binom{4}{1}a^3(-3b) + \binom{4}{2}a^2(-3b)^2 + \binom{4}{3}a(-3b)^3 + \binom{4}{4}(-3b)^4,$$

which also simplifies to the expression obtained above.

39. $(x + 3)^5$ 40. $(2x + y)^4$ 41. $(x - 3)^4$ 42. $(2x - 1)^5$

43. $\left(2x - \dfrac{y}{2}\right)^3$ 44. $\left(\dfrac{x}{3} + 3\right)^5$ 45. $\left(\dfrac{x}{2} + 2\right)^6$ 46. $\left(\dfrac{2}{3} - a^2\right)^4$

Write the first four terms in each expansion. Do not simplify the terms.

Example $(x + 2y)^{15}$

Solution From the binomial expansion (3), we have

$$x^{15} + \frac{15}{1!}x^{14}(2y) + \frac{15 \cdot 14}{2!}x^{13}(2y)^2 + \frac{15 \cdot 14 \cdot 13}{3!}x^{12}(2y)^3$$

or alternatively from the form (4),

$$\binom{15}{0}x^{15} + \binom{15}{1}x^{14}(2y) + \binom{15}{2}x^{13}(2y)^2 + \binom{15}{3}x^{12}(2y)^3.$$

47. $(x + y)^{20}$ 48. $(x - y)^{15}$ 49. $(a - 2b)^{12}$

50. $(2a - b)^{12}$ 51. $(x - \sqrt{2})^{10}$ 52. $\left(\dfrac{x}{2} + 2\right)^8$

Find each power to the nearest hundredth.

Example $(0.97)^7$

Solution Either form of the binomial expansion (3) or (4) can be used. We shall use (3).

$(0.97)^7 = (1 - 0.03)^7$

$$= 1^7 + \frac{7}{1!}(1)^6(-0.03)^1 + \frac{7 \cdot 6}{2!}(1)^5(-0.03)^2 + \frac{7 \cdot 6 \cdot 5}{3!}(1)^4(-0.03)^3 + \cdots$$

$$= 1 - 0.21 + 0.0189 - 0.000945 + \cdots$$

$$= 0.807955^+$$

Hence, to the nearest hundredth, $(0.97)^7 = 0.81$.

53. $(1.02)^{10}$ *Hint:* $1.02 = (1 + 0.02)$
54. $(1.01)^{15}$
55. If an amount of money (P) is invested at 4% compounded annually, the amount (A) present at the end of (n) years is given by $A = P(1 + 0.04)^n$. Find the amount A (to the nearest dollar) if $1000 was invested for 10 years.
56. In Exercise 55, find the amount present at the end of 20 years.

Find each specified term.

Example $(x - 2y)^{12}$, the seventh term.

Solution In Equation (5), page 399, use $n = 12$ and $r = 7$.

$$\frac{12!}{6!\,(12-6)!} x^6(-2y)^6 = \frac{12 \cdot 11 \cdot 10 \cdot 9 \cdot 8 \cdot 7}{6!} x^6(-2y)^6 = 59{,}136 x^6 y^6$$

57. $(a - b)^{15}$, the sixth term
58. $(x + 2)^{12}$, the fifth term
59. $(x - 2y)^{10}$, the fifth term
60. $(a^3 - b)^9$, the seventh term

61. Given that the binomial formula holds for $(1 + x)^n$ where n is a negative integer:
   a. Write the first four terms of $(1 + x)^{-1}$.
   b. Find the first four terms of the quotient $1/(1 + x)$, by dividing 1 by $(1 + x)$. Compare the results of (a) and (b).

62. Given that the binomial formula holds as an infinite "sum" for $(1 + x)^n$, where n is a rational number and $|x| < 1$, find to two decimal places.
   a. $\sqrt{1.02}$
   b. $\sqrt{0.99}$

13.5 Mathematical Induction

The material in the present section depends on a special property of the set of natural numbers, or positive integers, $N = \{1, 2, 3, \ldots\}$:

a. $1 \in N$

b. If $k \in N$, then $k + 1 \in N$.

In addition, *N contains no elements not implied by properties a and b*. These properties underlie the following theorem, called the **principle of mathematical induction**, which we state without proof.

Theorem 13.5 *If a given sentence involving natural numbers n is true for $n = 1$, and if its truth for $n = k$ implies its truth for $n = k + 1$, then it is true for every natural number n.*

Requirements of a proof by mathematical induction

We can exploit Theorem 13.5 to prove a number of assertions. Although the technique we shall use is called **proof by mathematical induction**, the argument we shall employ is deductive, as have been all of the other arguments in this book. Proofs by mathematical induction require two things:

a. A demonstration that the assertion to be proved is true for the natural number 1.
b. A demonstration that the truth of the assertion for a natural number k implies its truth for $k + 1$.

When these two demonstrations have been made, the principle of mathematical induction assures us that the assertion is true for every natural number.

Example Prove that the sum of the first n natural numbers is $\dfrac{n(n+1)}{2}$.

Solution We wish to show that
$$1 + 2 + 3 + \cdots + n = \frac{n(n+1)}{2}.$$

We must do two things:

1. We must first show that the assertion is true for $n = 1$, that is, that
$$1 = \frac{1(1+1)}{2},$$
which is true.

2. We must next show that the truth of the assertion for $n = k$ implies its truth for $n = k + 1$. That is, we must show that the truth of
$$1 + 2 + 3 + \cdots + k = \frac{k(k+1)}{2} \tag{1}$$
implies the truth of
$$1 + 2 + 3 + \cdots + k + (k+1) = \frac{(k+1)[(k+1)+1]}{2}. \tag{2}$$

Assuming the truth of (1) we add $k + 1$ to each member of this equation, obtaining
$$1 + 2 + 3 + \cdots + k + (k+1) = \frac{k(k+1)}{2} + (k+1)$$
$$= (k+1)\left(\frac{k}{2} + 1\right)$$
$$= (k+1)\left(\frac{k+2}{2}\right)$$
$$= \frac{(k+1)[(k+1)+1]}{2},$$
which is (2).

13.5 Mathematical Induction

Thus the second fact necessary for our proof is established. By the principle of mathematical induction, the assertion is true for every natural number n.

Example

Prove that for any arithmetic progression,

$$S_n = \frac{n}{2}(2a + (n-1)d).$$

Solution

For $n = 1$, the assertion is that

$$S_1 = \frac{1}{2}(2a + 0 \cdot d) = a,$$

which is true. We next must show that

$$S_n = \frac{n}{2}(2a + (n-1)d)$$

implies that

$$S_{n+1} = \frac{n+1}{2}(2a + nd).$$

Now since S_{n+1} is the sum of the first $n+1$ terms, S_{n+1} is the sum of the first n terms and the $n+1$st term. That is,

$$S_{n+1} = S_n + s_{n+1}.$$

Therefore

$$S_{n+1} = \frac{n}{2}(2a + (n-1)d) + a + nd$$

$$= na + a + \frac{n(n-1)}{2}d + nd$$

$$= (n+1)a + \frac{n(n+1)}{2} \cdot d$$

$$= \frac{n+1}{2}(2a + nd).$$

This method of proof is often compared to lining up a row of dominoes, with the assumption that whenever one domino is toppled, the one following will topple. One then needs only to topple the first domino ($n = 1$) and the whole row behind it will topple, as indicated in Figure 13.1 on page 406.

Figure 13.1

Exercise 13.5

By mathematical induction, prove the validity of the formulas in Exercises 1–10 for all positive integral values of n.

1. $\dfrac{1}{2} + \dfrac{2}{2} + \dfrac{3}{2} + \cdots + \dfrac{n}{2} = \dfrac{n(n+1)}{4}$

2. $1 + 3 + 5 + \cdots + (2n-1) = n^2$

3. $2 + 4 + 6 + \cdots + 2n = n(n+1)$

4. $2 + 6 + 10 + \cdots + (4n-2) = 2n^2$

5. $1^2 + 2^2 + 3^2 + \cdots + n^2 = \dfrac{n(n+1)(2n+1)}{6}$

6. $2 + 2^2 + 2^3 + \cdots + 2^n = 2^{n+1} - 2$

7. $1^3 + 3^3 + 5^3 + \cdots + (2n-1)^3 = n^2(2n^2 - 1)$

8. $\dfrac{1}{1 \cdot 2} + \dfrac{1}{2 \cdot 3} + \dfrac{1}{3 \cdot 4} + \cdots + \dfrac{1}{n(n+1)} = \dfrac{n}{n+1}$

9. $1 \cdot 2 + 2 \cdot 3 + 3 \cdot 4 + \cdots + n(n+1) = \dfrac{n(n+1)(n+2)}{3}$

10. $1 \cdot 4 + 2 \cdot 9 + 3 \cdot 16 + \cdots + n(n+1)^2 = \dfrac{1}{12} n(n+1)(n+2)(3n+5)$

11. Show that $n^3 + 2n$ is divisible by 3 for every $n \in N$.

12. Show that $3^{2n} - 1$ is divisible by 8 for every $n \in N$.

13. Show that if $2 + 4 + 6 + \cdots + 2n = n(n+1) + 2$ is true for $n = k$, then it is true for $n = k+1$. Is it true for every $n \in N$?

14. Show that $n^3 + 11n = 6(n^2 + 1)$ is true for $n = 1, 2,$ and 3. Is it true for every $n \in N$?

15. Use mathematical induction to prove Theorem 13.1.
16. Use mathematical induction to prove Theorem 13.2.
17. Use mathematical induction to prove that for a geometric progression,
$$S_n = \frac{a - ar^n}{1 - r} \text{ for all } n \in N, \text{ provided } r \neq 1.$$

13.6 Basic Counting Principles; Permutations

Associated with each finite set A is a nonnegative integer n, namely the number of elements in A. Hence, we have a function from the set of all finite sets to the set of nonnegative integers. The symbolism $n(A)$ is used to denote elements in the range of this set function n. For example, if

$$A = \{5, 7, 9\}, \quad B = \{1/2, 0, 3, -5, 7\}, \quad C = \emptyset,$$

then

$$n(A) = 3, \quad n(B) = 5, \quad \text{and} \quad n(C) = 0.$$

Counting properties

All the sets with which we shall hereafter be concerned are assumed to be finite sets. We then have the following properties, called **counting properties**, for the function n.

I $\quad n(A \cup B) = n(A) + n(B), \quad \text{if} \quad A \cap B = \emptyset$.

Thus, if A and B are disjoint sets, then the number of elements in their union is the sum of the number of elements in A and the number of elements in B.

For example, suppose there are five roads from town R to town S, and two railroads from town R to town S. If A is the set of roads and B the set of railroads from R to S, then $n(A) = 5$, $n(B) = 2$, and $n(A \cup B) = 5 + 2 = 7$; thus there are seven ways one can go from town R to town S by driving or riding on a train.

II $\quad n(A \cup B) = n(A) + n(B) - n(A \cap B), \quad \text{if} \quad A \cap B \neq \emptyset$.

That is, if A and B have a nonnull intersection, then to count the number of elements in $A \cup B$, we might add the number of elements in A to the number of elements in B. But since any elements in the intersection of A and B are counted twice in this process (once in A and once in B), we must subtract the number of such elements from the sum $n(A) + n(B)$ to obtain the number of elements in $A \cup B$, as suggested in Figure 13.2.

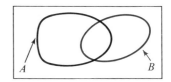

Figure 13.2

For example, suppose there are fifteen unrelated girls and seventeen unrelated boys in a mathematics class, and suppose that there are precisely two brother-sister pairs in the class. If A denotes the set of different families represented by the girls and B denotes the set of different families represented by the boys, then the number of different families represented by all of the members of the class is

$$n(A) + n(B) - n(A \cap B) = 15 + 17 - 2 = 30.$$

Actually, Property II is a consequence of Property I.

III $n(A \times B) = n(A) \cdot n(B)$.

This asserts that the number of elements in the Cartesian product of sets A and B is the product of the number of elements in A and the number of elements in B.

For example, suppose that there are five roads from town R to town S (set A), and further suppose that there are three roads from town S to town T (set B). Then for each element of A there are three elements of B, and the total possible ways one can drive from R to T via S is

$$n(A \times B) = n(A) \cdot n(B) = 5 \cdot 3 = 15.$$

Property III can be stated in an equivalent way as follows. Suppose the first of two operations (for example choosing a road from R to S) can be done in a ways. Suppose the second operation (choosing a road from S to T) can be done in b ways and each possible second operation can succeed every first operation. Then the two operations in sequence can be done in $a \cdot b$ ways. This formulation has the following very useful generalization.

IV Suppose the first of several operations can be done in a ways, the second in b ways, no matter what came first, the third in c ways, no matter what came prior, and so on. Then the number of ways the operations can be done in sequence is $a \cdot b \cdot c \cdot \cdots$.

Permutations Given the set of digits $A = \{1, 2, 3\}$, how many different three-digit numerals can be constructed from the members of A if no member is used more than once? The answer to this question can be obtained by simply listing the different three-digit numerals, 1 2 3, 1 3 2, 2 1 3, 2 3 1, 3 1 2, 3 2 1, and counting them. Such a procedure would be quite impracticable, however, if the number of members of the given set of digits were very large. Another way to arrive at the same conclusion is by applying the fourth counting property. The first digit can be any of the three numerals 1, 2, or 3. No matter what comes first, 2 choices remain for the second digit and then 1 choice remains for the third digit. By the fourth counting property, the number of ways of choosing the three digits in order is $3 \cdot 2 \cdot 1 = 6$.

Definition 13.7 *A **permutation** of a set A is an ordering (first, second, etc.) of the members of A.*

With Definition 13.7, we can state the following result.

13.6 Basic Counting Principles; Permutations

Theorem 13.6 Let $P_{n,n}$ denote the number of distinct permutations of the members of a set A containing n members. Then

$$P_{n,n} = n! \qquad (1)$$

The symbol $P_{n,n}$ (or sometimes $_nP_n$ or P_n^n) is read "the number of permutations of n things taken n at a time."

Proof We have n choices for the first member in the permutation, $n - 1$ for the second, and so on until only 1 remains for the last. By Property IV, the total number of possible permutations is

$$n(n - 1) \cdot (n - 2) \cdot \cdots \cdot 1 = n!$$

as was to be proved.

Example In how many ways can nine men be assigned positions to form distinct baseball teams?

Solution Let A denote the set of men, so that $n(A) = 9$. The total number of ways in which 9 men can be assigned 9 positions on a team, or, in other words, the number of possible permutations of the members of a 9-element set, is, by Equation (1),

$$P_{9,9} = 9! = 9 \cdot 8 \cdot 7 \cdots 1 = 362{,}880.$$

In many applications there are more members from which to choose than there are places to fill, as the following theorem and example illustrate.

Theorem 13.7 Let $P_{n,r}$ denote the number of permutations of the members, taken r at a time, of a set A containing n members; that is, let $P_{n,r}$ be the number of distinct orderings of r elements when there is a set A of n elements from which to choose. Then

$$P_{n,r} = n(n-1)(n-2) \cdots (n-r+1). \qquad (2)$$

The proof follows the proof of Theorem 13.6 except that only r choices are made so the last choice involves $n - (r - 1) = n - r + 1$ possibilities.

Example In how many ways can a basketball team be formed by choosing players for the five positions from a set of ten players?

Solution Let A denote the set of players, so that $n(A) = 10$. Then from (2) and the fact that a basketball team consists of 5 players, we have

$$P_{10,5} = 10 \cdot 9 \cdot 8 \cdots (10 - 5 + 1) = 10 \cdot 9 \cdot 8 \cdot 7 \cdot 6 = 30{,}240.$$

An alternative expression for $P_{n,r}$ can be obtained by observing that

$$P_{n,r} = n(n-1)(n-2) \cdots (n-r+1)$$

$$= \frac{n(n-1)(n-2)\cdots(n-r+1)(n-r)!}{(n-r)!},$$

so that

$$P_{n,r} = \frac{n!}{(n-r)!}. \tag{3}$$

Distinguishable permutations

The problem of finding the number of distinguishable permutations of n objects taken n at a time, if some of the objects are identical, requires a little more careful analysis. As an example, consider the number of permutations of the letters of the word *DIVISIBLE*. We can make a distinction between the three *I*'s by assigning subscripts to each so that we have nine distinct letters,

$$D, I_1, V, I_2, S, I_3, B, L, E.$$

The number of permutations of these nine letters is of course 9!. If the letters other than I_1, I_2, and I_3 are retained in the positions they occupy in a permutation of the above nine letters, I_1, I_2, and I_3 can be permuted among themselves 3! ways. Thus, if P is the number of *distinguishable* permutations of the letters

$$D, I, V, I, S, I, B, L, E,$$

then, since for each of these there are 3! ways in which the *I*'s can be permuted without otherwise changing the order of the other letters, it follows that

$$3! \cdot P = 9!,$$

from which

$$P = \frac{9!}{3!}.$$

As another example, consider the letters of the word *MISSISSIPPI*. There would exist 11! distinguishable permutations of the letters in this word if each letter were distinct. Note, however, that the letters *S* and *I* each appear four times and the letter *P* appears twice. Reasoning as we did in the previous example, we see that the number P of distinguishable permutations of the letters in *MISSISSIPPI* is given by

$$4! \, 4! \, 2! \, P = 11!,$$

from which

$$P = \frac{11!}{4! \, 4! \, 2!}.$$

In general, if there are n_1 identical objects of a first kind, n_2 of a second, ..., and n_k of a kth, with $n_1 + n_2 + \cdots + n_k = n$, then the number of distinguishable permutations of the n objects is given by

$$P = \frac{n!}{n_1! \, n_2! \cdots n_k!}.$$

13.6 Basic Counting Principles; Permutations

Exercise 13.6

Given the following sets, find $n(A \cap B)$, $n(A \cup B)$, and $n(A \times B)$.

Example $A = \{a, b, c\}$, $B = \{c, d\}$

Solution

$A \cap B = \{c\}$. Therefore $n(A \cap B) = 1$.
$n(A \cup B) = n(A) + n(B) - n(A \cap B) = 3 + 2 - 1 = 4$.
$n(A \times B) = n(A) \cdot n(B) = 3 \cdot 2 = 6$.

1. $A = \{d, e\}$, $B = \{e, f, g, h\}$
2. $A = \{e\}$, $B = \{a, b, c, d\}$
3. $A = \{1, 2, 3, 4\}$, $B = \{3, 4, 5, 6\}$
4. $A = \{1, 2\}$, $B = \{3, 4, 5\}$
5. $A = \{1, 2\}$, $B = \{1, 2\}$
6. $A = \emptyset$, $B = \{2, 3, 4\}$

Example   In how many ways can three members of a class be assigned a grade of A, B, C, or D?

Solution   Sometimes a simple diagram, such as ——, ——, ——, designating a sequence, is a helpful preliminary device. Since each of the students may receive any one of four different grades, the sequence would appear as

$$\underline{4}, \underline{4}, \underline{4}.$$

From counting Property IV, there are $4 \cdot 4 \cdot 4$, or 64, possible ways the grades may be assigned.

Example   In how many different ways can three members of a class be assigned a grade of A, B, C, or D so that no two members receive the same grade?

Solution   Since the first student may receive any one of four different grades, the second student may then receive any one of three different grades, and the third student may then receive any one of two different grades, the sequence would appear as

$$\underline{4}, \underline{3}, \underline{2}.$$

From counting Property IV, there are $4 \cdot 3 \cdot 2$, or 24, possible ways the grades may be assigned. In this case, we could have obtained the same result directly from Theorem 13.7, since $P_{4,3} = 4 \cdot 3 \cdot 2 = 24$.

In each problem, a digit or letter may be used more than once unless stated otherwise.

7. How many different two-digit numerals can be formed from the digits 5 and 6?
8. How many different two-digit numerals can be formed from the digits 7, 8, and 9?

9. In how many different ways can four students be seated in a row?
10. In how many different ways can five students be seated in a row?
11. In how many different ways can four questions on a true-false test be answered?
12. In how many different ways can five questions on a true-false test be answered?
13. In how many ways can you write different three-digit numerals from {2, 3, 4, 5}?
14. How many different seven-digit telephone numbers can be formed from the set of digits {0, 1, 2, 3, 4, 5, 6, 7, 8, 9}?
15. In how many ways can you write different three-digit numerals, using {2, 3, 4, 5}, if no digit is to be used more than once in each numeral?
16. How many different seven-digit telephone numbers can be formed from the set of digits {0, 1, 2, 3, 4, 5, 6, 7, 8, 9} if no digit is to be used more than once in any number?
17. How many three-letter arrangements can be formed from {A, N, S, W, E, R}?
18. If no letter is to be used more than once in any arrangement, how many different three-letter arrangements can be formed from {A, N, S, W, E, R}?
19. How many four-digit numerals for positive odd integers can be formed from {1, 2, 3, 4, 5}?
20. How many four-digit numerals for positive even integers can be formed from {1, 2, 3, 4, 5}?
21. How many numerals for positive integers less than 500 can be formed from {3, 4, 5}?
22. How many numerals for positive odd integers less than 500 can be formed from {3, 4, 5}?
23. How many numerals for positive even integers less than 500 can be formed from {3, 4, 5}?
24. How many numerals for positive even integers between 400 and 500, inclusive, can be formed from {3, 4, 5}?
25. How many permutations of the elements of {P, R, I, M, E} end in a vowel?
26. How many permutations of the elements of {P, R, O, D, U, C, T} end in a vowel?
27. Find the number of distinguishable permutations of the letters in the word

    LIMIT.

28. Find the number of distinguishable permutations of the letters in the word

    COMBINATION.

29. Find the number of distinguishable permutations of the letters in the word

    COLORADO.

30. Find the number of distinguishable permutations of the letters in the word

    TALLAHASSEE.

31. Show that $P_{5,3} = 5(P_{4,2})$.
32. Show that $P_{5,r} = 5(P_{4,r-1})$.
33. Show that $P_{n,3} = n(P_{n-1,2})$.
34. Show that $P_{n,3} - P_{n,2} = (n-3)(P_{n,2})$.
35. Solve for n: $P_{n,5} = 5(P_{n,4})$.
36. Solve for n: $P_{n,5} = 9(P_{n-1,4})$.

Example

In how many ways can four students be seated around a circular table?

Solution

In any such arrangements (which is called a **circular permutation**), there is no first position. Each person can take four different initial positions without affecting the arrangement. Thus, there are

$$\frac{4!}{4} = 6 \quad \text{arrangements.}$$

(In general, there are $n!/n$, or $(n-1)!$, circular permutations of n things taken n at a time).

37. In how many ways can five students be seated around a circular table?
38. In how many ways can six students be seated around a circular table?
39. In how many ways can six students be seated around a circular table if a certain two must be seated together?
40. In how many ways can three different keys be arranged on a key ring? *Hint:* Arrangements should be considered identical if one can be obtained from the other by turning the ring over. In general, there are only $(1/2)(n-1)!$ distinct arrangements of n keys on a ring $(n \geq 3)$.

13.7 Combinations

An additional counting concept is used frequently in computing probabilities—namely, finding the number of distinct r-element subsets of an n-element set with no reference to relative order of the elements in the subset. For example, five different cards can be arranged in $5!$ permutations, but to a poker player they represent the same hand. The set of five cards (with no reference to the arrangement of the cards) is called a *combination*.

Definition 13.8 *A subset of an n-element set A is called a **combination**.*

The counting of combinations is related to the counting of permutations. From Theorem 13.7, we know that the number of distinct permutations of n elements of a set A taken r at a time is given by

$$P_{n,r} = \frac{n!}{(n-r)!}.$$

With this in mind, consider the following result concerning the number $C_{n,r}$ of combinations of n things taken r at a time. The symbol $C_{n,r}$ is read "the number of combinations of n things taken r at a time."

Theorem 13.8 *The number $C_{n,r}$ of distinct combinations of the members, taken r at a time, of a set containing n members is given by*

$$C_{n,r} = \frac{P_{n,r}}{r!} \qquad (1)$$

Proof There are, by definition, $C_{n,r}$ r-element subsets of the set A, where $n(A) = n$. Also, from Theorem 13.6, each of these subsets has $r!$ permutations of its members. There are therefore $C_{n,r} r!$ permutations of n elements of A taken r at a time. Thus

$$P_{n,r} = C_{n,r} r!,$$

from which we obtain

$$C_{n,r} = \frac{P_{n,r}}{r!},$$

as was to be shown.

Thus, to find the number of r-element subsets of an n-element set A, we count the number of permutations of the elements of A taken r at a time, and then divide by the number of possible permutations of an r-element set. This seems very much like counting a set of people by counting the number of arms and legs and dividing the result by 4, but this approach gives us a very useful expression for the number we seek, $C_{n,r}$. Since

$$P_{n,r} = n(n-1)(n-2) \cdots (n-r+1),$$

it follows that

$$C_{n,r} = \frac{P_{n,r}}{r!} = \frac{n(n-1)(n-2) \cdots (n-r+1)}{r!} \qquad (2)$$

Example In how many ways can a committee of five be selected from a set of twelve persons?

Solution What we wish here is the number of 5-element subsets of a 12-element set. From (2), we have

$$C_{12,5} = \frac{12 \cdot 11 \cdot 10 \cdot 9 \cdot 8}{5 \cdot 4 \cdot 3 \cdot 2 \cdot 1} = 792.$$

13.7 Combinations

By Equation (3) on page 410, we have the alternative expression

$$C_{n,r} = \frac{P_{n,r}}{r!} = \frac{n!}{r!(n-r)!}. \tag{3}$$

Notice that the value of $C_{n,r}$ given in the right-hand member of (3) is the same as the value of the binomial coefficient $\binom{n}{r}$ given in the right-hand member of (2) on page 397. Thus $C_{n,r}$ and $\binom{n}{r}$ are equal and therefore are interchangeable in mathematical formulas:

$$C_{n,r} = \binom{n}{r}.$$

Since the numbers $\binom{n}{r}$, or $C_{n,r}$, are the coefficients in the binomial expansion, and since these coefficients are symmetric, we have the following plausible assertion.

Theorem 13.9 $C_{n,r} = C_{n,n-r}.$

Proof From (3), we have both

$$C_{n,r} = \frac{n!}{r!(n-r)!}$$

and

$$C_{n,n-r} = \frac{n!}{(n-r)![n-(n-r)]!} = \frac{n!}{(n-r)!r!},$$

and the theorem is proved.

Theorem 13.9 is plausible also since each time a distinct set of r objects is chosen, a distinct set of $n - r$ objects remains unchosen.

Exercise 13.7

Example How many different amounts of money can be formed from a penny, a nickel, a dime and a quarter?

Solution We want to find the total number of combinations that can be formed by taking the coins 1, 2, 3, and 4 at a time. By Equation (3) we have,

$$C_{4,1} = \frac{4!}{1!\,3!} = 4, \quad C_{4,2} = \frac{4!}{2!\,2!} = 6, \quad C_{4,3} = \frac{4!}{3!\,1!} = 4, \quad C_{4,4} = \frac{4!}{4!\,0!} = 1,$$

Solution Continued on Overleaf

and the total number of combinations is 15. Clearly each combination gives a different amount.

1. How many different amounts of money can be formed from a penny, a nickel, and a dime?
2. How many different amounts of money can be formed from a penny, a nickel, a dime, a quarter, and a half-dollar?
3. How many different committees of four persons each can be chosen from a group of six persons?
4. How many different committees of four persons each can be chosen from a group of ten persons?
5. In how many different ways can a set of five cards be selected from a standard bridge deck containing 52 cards?
6. In how many different ways can a set of 13 cards be selected from a standard bridge deck of 52 cards?
7. In how many different ways can a hand consisting of five spades, five hearts, and three diamonds be selected from a standard bridge deck of 52 cards?
8. In how many different ways can a hand consisting of ten spades, one heart, one diamond, and one club be selected from a standard bridge deck of 52 cards?
9. In how many different ways can a hand consisting of either five spades, five hearts, five diamonds, or five clubs be selected from a standard bridge deck?
10. In how many different ways can a hand consisting of three aces and two cards that are not aces be selected from a standard bridge deck?
11. A combination of three balls is picked at random from a box containing five red, four white, and three blue balls. In how many ways can the set chosen contain at least one white ball?
12. In Exercise 11, in how many ways can the set chosen contain at least one white and one blue ball?
13. A set of five distinct points lies on a circle. How many inscribed triangles can be drawn having all their vertices in this set?
14. A set of ten distinct points lies on a circle. How many inscribed quadrilaterals can be drawn having all their vertices in this set?
15. A set of ten distinct points lies on a circle. How many inscribed hexagons can be drawn having all their vertices in this set?
16. Given $C_{n,3} = C_{50,47}$, find n.
17. Given $C_{n,7} = C_{n,5}$, find n.

Chapter Review

[13.1] *Write the first three terms of the sequence with the general term as given.*

1. $s_n = n^2 + 1$
2. $s_n = \dfrac{1}{n+1}$

Write the next three terms of each of the following arithmetic progressions.

3. 7, 10, ...
4. $a, a - 2, \ldots$

Write the next three terms in each of the following geometric progressions.

5. $-2, 6, \ldots$
6. $\dfrac{2}{3}, 1, \ldots$

7. Find the general term and the seventh term of the arithmetic progression $-3, 2, \ldots$.

8. Find the general term and the fifth term of the geometric progression $-2, \dfrac{2}{3}, \ldots$.

9. If the fourth term of an arithmetic progression is 13 and the ninth term is 33, find the seventh term.

10. Which term in a geometric progression $-\dfrac{2}{9}, \dfrac{2}{3}, \ldots$ is 54?

[13.2] 11. Write $\sum\limits_{k=2}^{5} k(k-1)$ in expanded form.

12. Write $x^2 + x^3 + x^4 + \cdots$ in sigma notation.

13. Find the value for $\sum\limits_{j=3}^{9} (3j - 1)$.

14. Find the value for $\sum\limits_{j=1}^{5} \left(\dfrac{1}{3}\right)^j$.

[13.3] 15. Specify the limit of $\dfrac{3n^2 - 1}{n^2}$ as $n \to \infty$.

16. Find the value of the geometric series $4 - 2 + 1 - \dfrac{1}{2} + \cdots$

17. Find the value of $\sum\limits_{i=1}^{\infty} \left(\dfrac{1}{3}\right)^i$.

18. Find a fraction equivalent to $0.\overline{444}$.

[13.4] 19. Write $n(n-3)!$ in expanded form for $n = 5$.

Write each of the following expressions in expanded form and simplify.

20. $\dfrac{8!\,3!}{7!}$ 21. $\dbinom{7}{2}$ 22. $\dfrac{(n-1)!}{n!(n+1)!}$

23. Write the first four terms of the binomial expansion of $(x - 2y)^{10}$.
24. Find the eighth term in the expansion of $(x - 2y)^{10}$.

[13.5] *By mathematical induction, prove each formula for all positive integral values of n.*

25. $3 + 6 + 9 + \cdots + 3n = \dfrac{3n(n+1)}{2}$

26. $\dfrac{1}{2} + \dfrac{1}{4} + \dfrac{1}{8} + \cdots + \dfrac{1}{2^n} = 1 - \dfrac{1}{2^n}$

[13.6] 27. How many different two-digit numerals can be formed from $\{6, 7, 8, 9\}$?
28. How many different ways can six questions on a true-false test be answered?
29. How many four-digit numerals for positive odd integers can be formed from $\{3, 4, 5, 6\}$?
30. How many distinguishable permutations can be formed using the letters in the word *TENNIS*?

[13.7] 31. How many different committees of 5 persons can be formed from a group of 12 persons?
32. In how many different ways can a hand consisting of 3 spades, 5 hearts, 4 diamonds, and 1 club be selected from a standard bridge deck of 52 cards?
33. A box contains 4 red, 6 white, and 2 blue marbles. In how many ways can you select 3 marbles from the box if at least one of the marbles chosen is white?
34. How many hexagons can be drawn whose vertices are members of a set of 9 fixed points on the circle?

14 Theory of Equations

14.1 Synthetic Division and the Factor Theorem Over C

In Section 2.5, a synthetic process was developed for the division of $P(x)$ by $(x - r)$, where $P(x)$ was a real polynomial over R, and r was any real number. Then, in Section 5.5, the remainder theorem was developed, and these two concepts were combined to help us find ordered pairs $(x, P(x))$ in real polynomial functions.

Now let us reexamine these ideas for polynomials over the field C of complex numbers. To begin with, let us examine the problem of finding $P(r)$ for any given real or complex number r by the method that we employed in Section 5.5.

Remainder theorem

Because no properties other than the field properties are necessary for its validity, the remainder theorem also holds in the set of complex numbers.

Theorem 14.1   *If $P(x)$ is a polynomial over the field C of complex numbers, and $c \in C$, then there exists a unique polynomial $Q(x)$ and a complex number r, such that*

$$P(x) = (x - c)Q(x) + r,$$

and $r = P(c)$.

Example   If $P(x) = 2x^4 - x^3 + 2x^2 - x + 1,$ find $P(1 - i)$.

Solution   Using synthetic division in the field of complex numbers, we obtain:

$$\begin{array}{r|rrrrr} 1-i & 2 & -1 & 2 & -1 & 1 \\ & & 2-2i & -1-3i & -2-4i & -7-i \\ \hline & 2 & 1-2i & 1-3i & -3-4i & -6-i \end{array}$$

Thus $P(1 - i) = -6 - i$.

Factor theorem

Another theorem, called the **factor theorem**, follows immediately from the remainder theorem.

Theorem 14.2 *If $P(x)$ is a polynomial over the field C of complex numbers, and $P(r) = 0$, then $(x - r)$ is a factor of $P(x)$.*

Proof If $P(r) = 0$, then, by the remainder theorem,
$$P(x) = (x - r)Q(x) + P(r) = (x - r)Q(x) + 0,$$
and by the definition of a factor, $(x - r)$ is a factor of $P(x)$.

Converse of factor theorem

It might be noted, as a converse of the factor theorem, that if $(x - r)$ is a factor of $P(x)$, so that $P(x) = (x - r)Q(x)$, then $P(r) = 0$, since
$$P(r) = (r - r)Q(r) = 0 \cdot Q(r) = 0.$$

Example Show that $(x - 2i)$ is a factor of $P(x) = x^3 - x^2 + 4x - 4$.

Solution If $P(2i) = 0$, then, by the factor theorem, $x - 2i$ must be a factor of $P(x)$. Dividing $P(x)$ by $x - 2i$ synthetically, we have

```
2i | 1   -1        4        -4
   |     2i    -4 - 2i       4
     ─────────────────────────
     1  -1 + 2i   -2i        0
```

and, since the remainder is 0, we have shown that $(x - 2i)$ is a factor of $P(x)$.

As a last observation, note that, in the synthetic division of $P(x)$ by $x - r$, the terms occurring in the bottom row of the division process are the coefficients in the polynomial $Q(x)$, except that the final one is $P(r)$. Thus, in the preceding example, we showed that $x - 2i$ is a factor of $P(x) = x^3 - x^2 + 4x - 4$. By looking at the last row in the synthetic-division process, we see that, since $P(2i) = 0$, the polynomial $P(x)$ can be expressed as
$$P(x) = (x - 2i)[x^2 + (-1 + 2i)x - 2i].$$

Exercise 14.1

Find the value of the polynomial for the given value of x.

Example $P(x) = 3x^3 + 2x^2 - x + 1;\ 2i$

Solution Using synthetic division, we obtain

```
2i | 3   2        -1          1
   |      6i   -12 + 4i    -8 - 26i
     ──────────────────────────────
     3  2 + 6i  -13 + 4i   -7 - 26i .
```

14.2 Complex Zeros of Polynomial Functions

By the remainder theorem, $P(2i) = -7 - 26i$

1. $P(x) = 4x^3 - x^2 + 5x + 2$; 2
2. $Q(x) = 3x^3 + 2x^2 - x - 1$; -3
3. $Q(x) = x^3 + 4x^2 - 2x - 5$; $-3i$
4. $G(x) = 3x^3 - 2x^2 + 7x - 1$; i
5. $P(x) = x^4 - 3x^3 + x - 2$; -2.
6. $P(x) = 2x^4 - 3x^2 - x$; 4
7. $Q(x) = x^5 - x^3 + 3x^2 + 1$; $2i$
8. $Q(x) = 3x^5 - x^4 + 2x - 1$; $-i$
9. $P(x) = x^3 + 3x^2 - x + 1$; $1 - i$
10. $P(x) = 2x^3 - x^2 + 3x - 5$; $2 + i$

For each expression, find the quotient and the remainder.

Example $(3x^3 - x^2 + 1) \div (x + 2)$

Solution Using synthetic division, we obtain

$$\begin{array}{r|rrrr} -2 & 3 & -1 & 0 & 1 \\ & & -6 & 14 & -28 \\ \hline & 3 & -7 & 14 & -27 \end{array}.$$

The quotient is $3x^2 - 7x + 14$ and the remainder is -27.

11. $(2x^3 - x^2 + 3x + 4) \div (x - 3)$
12. $(x^3 - 4x^2 - 2x - 2) \div (x + 1)$
13. $(x^3 + x^2 - 3x + 3) \div (x - 2i)$
14. $(2x^3 - 2x^2 + 4x + 1) \div (x + i)$
15. $(x^4 - 3x^2 + x - 1) \div (x + i)$
16. $(x^4 - x^3 + 2x^2 + 2) \div (x - 2i)$
17. $(x^3 - 2x^2 + x + 1) \div [x - (1 + i)]$
18. $(2x^3 + x^2 - 3x) \div [x + (1 - i)]$
19. Show that $x - 2$ is a factor of $x^3 + 2x^2 - 5x - 6$.
20. Show that $x + 1$ is a factor of $x^4 - 5x^3 - 13x^2 + 53x + 60$.
21. Show that $-2i$ is a root of $x^3 - x^2 + 4x - 4 = 0$.
22. Show that $2 + 4i$ is a root of $x^4 - 4x^3 + 18x^2 + 8x - 40 = 0$.
23. Find $Q(x)$ if $x^3 - 3x^2 + x - 3 = (x - i)Q(x)$.
24. Find $Q(x)$ if $2x^3 - 3x^2 - 4x + 4 = (x - 2)Q(x)$.

Find a value for k so that the second polynomial is a factor of the first.

25. $x^3 - 5x^2 - 16x + k$; $x - 5$
26. $x^3 - 9x^2 + 14x + k$; $x + 1$
27. $x^4 + 2x^3 - 21x^2 + kx + 40$; $x - 4$
28. $3x^4 - 40x^3 + 130x^2 + kx + 27$; $x - 9$

14.2 Complex Zeros of Polynomial Functions

Conjugate complex roots

Recalling from Section 1.5 that the conjugate of the complex number $z = a + bi$ is $\bar{z} = a - bi$, where $a, b \in R$, we state without proof an important property of a polynomial over R.

Theorem 14.3 *If $P(z)$ is a polynomial over the field R of real numbers, and $P(z) = 0$ for some $z \in C$, then $P(\bar{z}) = 0$.*

One implication this theorem has for the zeros of a function defined by a polynomial with *real coefficients* is that *complex zeros always occur in conjugate pairs*.

Example Given that $2 - i$ is a zero of

$$P(x) = x^3 - 6x^2 + 13x - 10,$$

find all zeros of P.

Solution The zeros of P are the solutions of $P(x) = 0$. By synthetic division, we have

$$\begin{array}{r|rrrr} 2-i & 1 & -6 & 13 & -10 \\ & & 2-i & -9+2i & 10 \\ \hline & 1 & -4-i & 4+2i & 0 \end{array},$$

and the quotient is $x^2 + (-4 - i)x + 4 + 2i$. By Theorem 14.3, we know that $2 + i$ must also be a root of $P(x) = 0$. We can now write

$$P(x) = [x - (2 - i)][x^2 + (-4 - i)x + 4 + 2i] = 0,$$

and since, by inspection, $2 + i$ is not a root of $x - (2 - i) = 0$, it must be a root of

$$x^2 + (-4 - i)x + 4 + 2i = 0.$$

Using synthetic division again, we have

$$\begin{array}{r|rrr} 2+i & 1 & -4-i & 4+2i \\ & & 2+i & -4-2i \\ \hline & 1 & -2 & 0 \end{array},$$

where the quotient is $x - 2$. Then

$$P(x) = [x - (2 - i)][x - (2 + i)](x - 2) = 0$$

and, by inspection, the solutions of this equation—and hence the zeros of P—are $2 - i$, $2 + i$, and 2.

Fundamental theorem of algebra

When Theorem 14.3 is coupled with the following theorem, which is called the **fundamental theorem of algebra**, a great deal of information relative to the zeros of polynomial functions becomes readily available.

Theorem 14.4 *Every polynomial function of degree $n \geq 1$ over the field C of complex numbers has at least one zero in the field C.*

The proof of this theorem involves concepts beyond those available to us, and is omitted.

14.2 Complex Zeros of Polynomial Functions

As an extension of Theorem 14.4, we have the following consequent result.

Theorem 14.5 *Every polynomial of degree $n \geq 1$ over the field C can be expressed as a product of a constant and n linear factors of the form $x - x_j$, where $x_j \in C$.*

Proof Let $P(x) = a_0 x^n + a_1 x^{n-1} + \cdots + a_n$, $a_i \in C$, $a_0 \neq 0$, $x \in C$, and $n \in N$. Then, since this equation defines a polynomial function, by the fundamental theorem of algebra there is at least one real or complex number x, say x_1, such that

$$P(x_1) = a_0 x_1^n + a_1 x_1^{n-1} + \cdots + a_n = 0.$$

By the factor theorem,

$$P(x) = (x - x_1)Q_{n-1}(x),$$

where $Q_{n-1}(x)$ is of degree $n - 1$. Again, by the fundamental theorem, if $n - 1 \geq 1$, there must exist an $x_2 \in C$ such that $Q_{n-1}(x_2) = 0$. Hence, we can write

$$P(x) = (x - x_1)(x - x_2)Q_{n-2}(x),$$

where $Q_{n-2}(x)$ is a polynomial of degree $n - 2$. If this factoring process is performed n times, the result is

$$P(x) = (x - x_1)(x - x_2) \cdots (x - x_n)Q_0(x),$$

where $Q_0(x)$ consists solely of a_0, and the theorem is proved.

An nth-degree polynomial function has n zeros

If a factor $(x - x_i)$ occurs k times in such a linear factorization of $P(x)$, then x_i is said to be a zero of *multiplicity k*. With this agreement, Theorem 14.5 shows that every polynomial function defined by a polynomial $P(x)$ of degree n with complex coefficients has exactly n zeros.

Note that any theorem stated in terms of zeros of polynomial functions applies to solutions of polynomial equations and vice versa; a *zero* of

$$P(x) = a_0 x^n + a_1 x^{n-1} + \cdots + a_n$$

is a *solution* of $P(x) = 0$.

Exercise 14.2

In Exercises 1–8, one or more zeros are given for each of the polynomial functions; find the other zeros. Verify by synthetic division.

1. $P(x) = x^2 + 4$; $2i$ is one zero.
2. $P(x) = 3x^2 + 27$; $-3i$ is one zero.
3. $P(x) = x^3 - 3x^2 + x - 3$; 3 and i are zeros.
4. $Q(x) = x^3 - 5x^2 + 7x + 13$; -1 and $3 - 2i$ are zeros.
5. $Q(x) = x^4 + 5x^2 + 4$; $-i$ and $2i$ are zeros.
6. $P(x) = x^4 + 11x^2 + 18$; $3i$ and $\sqrt{2}i$ are zeros.

7. $Q(x) = x^4 + 3x^3 + 4x^2 + 27x - 45$; $-3i$ is a zero. *(hint)*

8. $Q(x) = x^5 - 2x^4 + 8x^3 - 16x^2 + 16x - 32$; $2i$ (multiplicity 2) and 2 are zeros.

9. A cubic equation with real coefficients has roots -2 and $1 + i$. What is the third root? Write the equation in the form $P(x) = 0$, given that the leading coefficient (the coefficient of the highest power of x) is 1.

10. A cubic equation with real coefficients has roots 4 and $2 - i$. What is the third root? Write the equation in the form $P(x) = 0$, given that the leading coefficient is 1.

11. One zero of $P(x) = 2x^3 - 11x^2 + 28x - 24$ is $2 - 2i$. Factor $P(x)$ over the complex numbers.

12. One zero of $Q(x) = 3x^3 - 10x^2 + 7x + 10$ is $2 + i$. Factor $Q(x)$ over the complex numbers.

13. One root of $x^4 - 10x^3 + 35x^2 - 50x + 34 = 0$ is $4 - i$. Find the remaining roots.

14. One root of $5x^4 + 34x^3 + 40x^2 - 78x + 51 = 0$ is $-4 - i$. Find the remaining roots.

15. Argue that every polynomial equation with real coefficients and of odd degree has at least one real root.

16. One zero of a polynomial P is i. Is $-i$ necessarily a zero of P? Why or why not? The number i is a zero of $P(x) = x^3 + i$. Determine by synthetic division whether or not $-i$ is a zero.

17. Determine the three cube roots of -1. *Hint:* Consider the roots of the equation $x^3 + 1 = 0$.

18. Show that there are four fourth roots of -1. *Hint:* Consider the roots of the equation $x^4 + 1 = 0$, that is, of $(x^2 + i)(x^2 - i) = 0$.

14.3 Real Zeros of Polynomial Functions

Since we have adopted the convention that $R \subset C$, if $\{z_1, z_2, z_3, \ldots z_n\}$ is the set of zeros for a complex polynomial function of degree n, any one or more of these zeros, or, indeed, all of them, may be real numbers. Now we wish to find some ways of identifying real zeros of these polynomials and we shall, for the time being, restrict replacements for x in $P(x)$ to real numbers.

Intermediate values

For real-number replacements for x, we know that for the $a_j \in R$, the graph of

$$y = P(x) = a_0 x^n + a_1 x^{n-1} + \cdots + a_n$$

lies entirely in the real plane (see Section 5.1). Moreover, although we shall not prove it here, we have the following theorem.

14.3 Real Zeros of Polynomial Functions

Theorem 14.6 *If $P(x) = a_0 x^n + a_1 x^{n-1} + \cdots + a_n$, $a_j, x \in R$, and if $k \in R$ is between $P(x_1)$ and $P(x_2)$, then there exists at least one $c \in R$ between x_1 and x_2 such that $P(c) = k$.*

More intuitively, we can show that there are no breaks or jumps in the values of $P(x)$, so we know that $P(x)$ must assume all values between any two of its values. Thus, the graph of P must be a continuous, unbroken curve.

Descartes' Rule of Signs

Another useful theorem, which again we shall not prove here, is that which establishes **Descartes' Rule of Signs.**

Theorem 14.7 *If $P(x)$ is a polynomial over the field R of real numbers, then the number of positive real solutions of $P(x) = 0$ either is equal to the number of variations in sign occurring in the coefficients of $P(x)$, or else is less than this number by an even natural number. Moreover, the number of negative real solutions of $P(x) = 0$ either is equal to the number of variations in sign occurring in $P(-x)$, or else is less than this number by an even natural number.*

Variations in sign

A **variation in sign** occurs in a polynomial with real coefficients if, as the polynomial is viewed from left to right, successive coefficients are opposite in sign. For example, in the polynomial

$$P(x) = 3x^5 - 2x^4 - 2x^2 + x - 1,$$

there are three variations in sign, and, in

$$P(-x) = -3x^5 - 2x^4 - 2x^2 - x - 1,$$

there are no variations in sign.

Example

Find an upper bound on the number of positive real solutions and an upper bound on the number of negative real solutions for the equation

$$3x^4 + 3x^3 - 2x^2 + x + 1 = 0. \tag{1}$$

Solution

Since $P(x) = 3x^4 + 3x^3 - 2x^2 + x + 1$ has but two variations in sign, the equation can have no more than two positive real solutions. Since

$$P(-x) = 3x^4 - 3x^3 - 2x^2 - x + 1$$

has two variations in sign, the number of negative real solutions of (1) cannot exceed two.

Isolation of real zeros

The following theorem on page 426 is sometimes helpful in isolating real zeros of a polynomial function with real coefficients.

Theorem 14.8 Let $P(x)$ be a polynomial over the field R of real numbers.

I   If $r_1 \geq 0$, and the coefficients of the terms in $Q(x)$ and the term $P(r_1)$ are all of the same sign in the right-hand member of

$$P(x) = (x - r_1)Q(x) + P(r_1),$$

then $P(x) = 0$ can have no solution greater than r_1.

II If $r_2 \leq 0$, and the coefficients of the terms in $Q(x)$ and the term $P(r_2)$ alternate in sign (zero suitably denoted by $+0$ or -0) in the right-hand member of

$$P(x) = (x - r_2)Q(x) + P(r_2),$$

then $P(x) = 0$ can have no solution less than r_2.

Proof We shall prove only the first part of the theorem here. The second part follows from the first by considering $P(-x) = (-x - r_2)Q(-x) + P(r_2)$.

For all $x > r_1$, $(x - r_1) > 0$. Moreover, if all the coefficients in $Q(x)$ are of the same sign, say positive, then, since $x > r_1 \geq 0$, we have $Q(x) > 0$ and $(x - r_1)Q(x) > 0$. Since $P(r_1) \geq 0$ by hypothesis, we have

$$P(x) = (x - r_1)Q(x) + P(r_1) > 0,$$

and the first part is proved.

This theorem permits us to place upper and lower bounds on the set of real zeros of the polynomial function

$$P(x) = a_0 x^n + \cdots + a_n,$$

and consequently on the real members of the solution set of $P(x) = 0$.

Example Show that 2 and -2 are upper and lower bounds, respectively, for the location of the zeros of

$$P(x) = 18x^3 - 12x^2 - 11x + 10.$$

Solution Using synthetic division to divide $P(x)$ by $x - 2$, we have

$$\underline{2|} \quad \begin{array}{cccc} 18 & -12 & -11 & 10 \\ & 36 & 48 & 74 \\ \hline 18 & 24 & 37 & 84 \end{array}.$$

Thus $Q(x) = 18x^2 + 24x + 37$ and $P(2) = 84 > 0$. Hence, by Theorem 14.8-I, 2 is an upper bound for the zeros of P. Next dividing $P(x)$ by $x + 2$, we have

$$\underline{-2|} \quad \begin{array}{cccc} 18 & -12 & -11 & 10 \\ & -36 & 96 & -170 \\ \hline 18 & -48 & 85 & -160 \end{array}.$$

Here $Q(x) = 18x^2 - 48x + 85$ and $P(-2) = -160$. Thus, by Theorem 14.8-II, -2 is a lower bound for the zeros of P.

14.3 Real Zeros of Polynomial Functions

Example Find the least nonnegative integer and the greatest nonpositive integer that are, by Theorem 14.8, upper and lower bounds, respectively, for the real zeros of

$$P(x) = x^4 - x^3 - 10x^2 - 2x + 12.$$

Solution We shall first seek an upper bound by dividing $P(x)$ successively by $x - 1$, $x - 2$, and so on. Each row after the first in the following array is the bottom row in the respective synthetic division involved.

	1	-1	-10	-2	12
1	1	0	-10	-12	0
2	1	1	-8	-18	-24
3	1	2	-4	-14	-30
4	1	3	2	6	36

Since the numbers in the last row are all positive, 4 is an upper bound. Next, we divide by $x + 1$, $x + 2$, and so on, in search of a lower bound.

	1	-1	-10	-2	12
-1	1	-2	-8	6	6
-2	1	-3	-4	6	0
-3	1	-4	2	-8	36

Since the numbers in the row following -3 alternate from positive to negative, etc., -3 is a lower bound. (Had the numbers in the row been 1, 0, 2, -8, 36, then the sign "$-$" could arbitrarily have been assigned to 0 to give the desired pattern of alternating signs.)

The location theorem To narrow the search for real zeros of a polynomial function still further, we have the following **location theorem**.

Theorem 14.9 Let $P(x)$ be a polynomial over the field R of real numbers. If $x_1, x_2 \in R$, with $x_1 < x_2$, and if $P(x_1)$ and $P(x_2)$ are opposite in sign, then there exists at least one $c \in R$, $x_1 < c < x_2$, such that $P(c) = 0$.

Proof First, let $P(x_1) < 0 < P(x_2)$, and $x_1 < x_2$. Then, by Theorem 14.6 with $k = 0$, there must exist $c \in R$, $x_1 < c < x_2$, such that $P(c) = 0$. Next, let $P(x_2) < 0 < P(x_1)$ and $x_1 < x_2$. Then, also by Theorem 14.6, there must exist $c \in R$, $x_1 < c < x_2$, such that $P(c) = 0$. Since x_1 and x_2 are arbitrary, the theorem is proved.

Example

Discuss the possibilities for real zeros of
$$P(x) = 32x^4 - 8x^3 - 148x^2 + 162x - 45.$$

Solution

We observe first that, by Descartes' Rule of Signs, Theorem 14.7, we can have at most three positive and one negative real zeros. Next, we apply synthetic division to obtain the following array:

	32	-8	-148	162	-45
0	32	-8	-148	162	-45
1	32	24	-124	38	-7
2	32	56	-36	90	135
3	32	88	116	510	1485
-1	32	-40	-108	270	-315
-2	32	-72	-4	170	-385
-3	32	-104	164	-330	945

Examining this array, we find that, by Theorem 14.8, 3 is an upper and -3 is a lower bound for real zeros. By the remainder theorem, $P(1) = -7$ and $P(2) = 135$, and since these are of opposite sign, Theorem 14.9 assures us that there is at least one real zero between 1 and 2. Similarly, we find there is at least one zero between -2 and -3, because $P(-2) = -385$ while $P(-3) = 945$. Furthermore, by Theorem 14.5, P must have precisely 4 real or complex zeros, not necessarily all distinct, because $P(x)$ is of degree four. By Theorem 14.3, any nonreal (imaginary) complex roots must occur in conjugate pairs, so we can assert, as a conclusion, the following possibilities:

1. P has two real zeros, one between 1 and 2, and one between -2 and -3;

2. P has four real zeros:
   a. three between 1 and 2 and one between -2 and -3, or vice versa,
   b. two between 2 and 3, one between 1 and 2, and one between -2 and -3,
   c. two between -2 and 1, one between 1 and 2, and one between -2 and -3.

As a matter of fact, the zeros of P are $1/2$, $3/4$, $3/2$, and $-5/2$, so that case 2(c) is the true situation; but in order to have detected this, we would have had either to find these zeros or to conduct a rather extensive search for sign changes in $P(x)$ over the interval $0 \leq x \leq 1$.

14.4 Rational Zeros of Polynomial Functions

Exercise 14.3

Use Theorem 14.7 to discuss the nature of the roots of each polynomial equation.

1. $x^4 - 2x^3 + 2x + 1 = 0$
2. $3x^4 + 3x^3 + 2x^2 - x + 1 = 0$
3. $2x^5 + 3x^3 + 2x + 1 = 0$
4. $4x^5 - 2x^3 - 3x - 2 = 0$
5. $3x^4 + 1 = 0$
6. $2x^5 - 1 = 0$

Find an upper bound and a lower bound for the real roots of each polynomial equation.

7. $x^3 + 2x^2 - 7x - 8 = 0$
8. $x^3 - 8x + 5 = 0$
9. $x^4 - 2x^3 - 7x^2 + 10x + 10 = 0$
10. $x^3 - 4x^2 - 4x + 12 = 0$
11. $x^5 - 3x^3 + 24 = 0$
12. $x^5 - 3x^4 - 1 = 0$
13. $2x^5 + x^4 - 2x - 1 = 0$
14. $2x^5 - 2x^2 + x - 2 = 0$

Use Theorem 14.9 to verify each statement in Exercises 15–20.

15. $f(x) = x^3 - 3x + 1$ has a zero between 0 and 1.
16. $f(x) = 2x^3 + 7x^2 + 2x - 6$ has a zero between -2 and -1.
17. $g(x) = x^4 - 2x^2 + 12x - 17$ has a zero between -3 and -2.
18. $g(x) = 2x^4 + 3x^3 - 14x^2 - 15x + 9$ has a zero between -2 and -1.
19. $P(x) = 2x^2 + 4x - 4$ has one zero between -3 and -2, and one between 0 and 1.
20. $P(x) = x^3 - x^2 - 2x + 1$ has one zero between -2 and -1, one between 0 and 1, and one between 1 and 2.

14.4 Rational Zeros of Polynomial Functions

Prime and composite numbers

The results of the present section depend on the notion of a **prime number**. If a is an element of the set N of natural numbers, and $a \neq 1$, then a is a prime number if and only if a has no factor in N other than itself and 1; otherwise, a is a **composite number**. For example, 2 and 3 are prime numbers, and $6 = 2 \cdot 3$ is composite; but 1 is considered to be neither prime nor composite.

The **unique-factorization theorem**, or **fundamental theorem of arithmetic** (which we shall not prove), follows.

Theorem 14.10 *If a is a composite number, then a has only one set of prime factors; that is, there are unique primes $p_1, \ldots, p_n$ and natural numbers $\alpha_1, \ldots, \alpha_n$ so that*

$$a = p_1^{\alpha_1} \cdot p_2^{\alpha_2} \cdot \ldots \cdot p_n^{\alpha_n}.$$

Two integers a and b are said to be **relatively prime** if and only if they have no prime factors in common. For example, -6 and 35 are relatively prime, since $-6 = -2 \cdot 3$ and $35 = 5 \cdot 7$; but 6 and 8 are not, since they have the prime factor 2 in common. The fraction a/b is said to express a rational number in **lowest terms** if and only if a and b are relatively prime.

Identifying possible rational zeros

If all the coefficients of the defining equation

$$P(x) = a_0 x^n + a_1 x^{n-1} + \cdots + a_n$$

of a polynomial function P are integers, then we can identify all possible rational zeros of P by means of the following theorem.

Theorem 14.11 *If the rational number p/q, in lowest terms, is a solution of*

$$P(x) = a_0 x^n + a_1 x^{n-1} + \cdots + a_n = 0,$$

where $a_j \in J$, then p is an integral factor of a_n and q is an integral factor of a_0.

Proof Since p/q is a solution of $P(x) = 0$, we have

$$a_0 \left(\frac{p}{q}\right)^n + a_1 \left(\frac{p}{q}\right)^{n-1} + \cdots + a_n = 0,$$

and we can multiply each member here by q^n to obtain

$$a_0 p^n + a_1 p^{n-1} q + \cdots + a_n q^n = 0.$$

Adding $-a_n q^n$ to each member and factoring p from each term in the left-hand member of the resulting equation, we have

$$p(a_0 p^{n-1} + a_1 p^{n-2} q + \cdots + a_{n-1} q^{n-1}) = -a_n q^n.$$

Since J is closed with respect to addition and multiplication, the expression in parentheses in the left-hand member here represents an integer, say r, so that we have

$$pr = -a_n q^n,$$

where pr is an integer having p as a factor. Hence, p is a factor of $-a_n q^n$. But, by Theorem 14.10, p and q^n have no factor in common, because, by hypothesis, p/q is in lowest terms; hence p must be a factor of a_n. In a similar manner, by writing the equation

$$a_0 p^n + a_1 p^{n-1} q + \cdots + a_n q^n = 0$$

in the form

$$-a_0 p^n = a_1 p^{n-1} q + \cdots + a_n q^n,$$

we can factor q from each term in the right-hand member and show that q must be a factor of a_0.

14.4 Rational Zeros of Polynomial Functions

Example List all possible rational zeros of
$$P(x) = 2x^3 - 4x^2 + 3x + 9.$$

Solution Rational zeros, p/q, must, by Theorem 14.11, be such that p is an integral factor of 9 and q is an integral factor of 2. Hence
$$p \in \{-9, -3, -1, 1, 3, 9\}, \quad q \in \{-2, -1, 1, 2\},$$
and the set of possible rational zeros of P is
$$\left\{-9, -\frac{9}{2}, -3, -\frac{3}{2}, -1, -\frac{1}{2}, \frac{1}{2}, 1, \frac{3}{2}, 3, \frac{9}{2}, 9\right\}.$$

Test for rational zeros It is important to observe that Theorem 14.11 does not assure us that a polynomial function with integral coefficients indeed has a rational zero; it simply enables us to identify possibilities for rational zeros. These can then be checked by synthetic division. The identification of the zeros of P in the previous example is left as an exercise.

As a special case of Theorem 14.11, it is evident that if a function P is defined by the equation
$$P(x) = x^n + a_1 x^{n-1} + \cdots + a_n,$$
in which $a_i \in J$ and $a_0 = 1$, then any rational zero of P must be an integer, and, moreover, must be an integral factor of a_n.

Example Find all rational zeros of $P(x) = x^3 - 4x^2 + x + 6.$

Solution The only possible rational zeros of P are $-6, -3, -2, -1, 1, 2, 3,$ and 6. Using synthetic division, we set up the following array:

	1	−4	1	6
1	1	−3	−2	4
−1	1	−5	6	0

We can cease our trials with -1, since by the remainder theorem, -1 is a zero of P, and the remaining zeros can be obtained from the equation $x^2 - 5x + 6 = 0$, whose coefficients are taken from the last line in the array. We can write the equation as $(x-2)(x-3) = 0$ and observe that 2 and 3 are also zeros. Note that it is often useful to start with possible roots of *least* absolute value, since the synthetic-division testing might reveal an upper or lower bound and thus eliminate the necessity of testing some of the numbers.

Exercise 14.4

Find all integral zeros of each function.

1. $f(x) = 3x^3 - 13x^2 + 6x - 8$

2. $f(x) = x^4 - x^2 - 4x + 4$
3. $f(x) = x^4 + x^3 + 2x - 4$
4. $f(x) = 5x^3 + 11x^2 - 2x - 8$
5. $P(x) = 2x^4 - 3x^3 - 8x^2 - 5x - 3$
6. $P(x) = 3x^4 - 40x^3 + 130x^2 - 120x + 27$

Find all rational zeros of each function.

7. $f(x) = 2x^3 + 3x^2 - 14x - 21$
8. $f(x) = 3x^4 - 11x^3 + 9x^2 + 13x - 10$
9. $P(x) = 4x^4 - 13x^3 - 7x^2 + 41x - 14$
10. $P(x) = 2x^3 - 4x^2 + 3x + 9$
11. $Q(x) = 2x^3 - 7x^2 + 10x - 6$
12. $Q(x) = x^3 + 3x^2 - 4x - 12$

Find all complex zeros of each function. Hint: First find all rational zeros.

13. $P(x) = 3x^3 - 5x^2 - 14x - 4$
14. $P(x) = x^3 - 4x^2 - 5x + 14$
15. $P(x) = 2x^4 + 3x^3 + 2x^2 - 1$
16. $P(x) = 8x^4 - 22x^3 + 29x^2 - 66x + 15$
17. $P(x) = 12x^4 + 7x^3 + 7x - 12$
18. $P(x) = 6x^4 - 13x^3 + 2x^2 - 4x + 15$
19. Factor the polynomial $2x^3 + 3x^2 - 2x - 3$ over C.
20. Factor the polynomial $x^4 - 6x^3 - 3x^2 - 24x - 28$ over C.
21. Show that $\sqrt{3}$ is irrational. *Hint:* Consider the equation $x^2 - 3 = 0$.
22. Show that $\sqrt{2}$ is irrational.

Chapter Review

[14.1] Find the value of the polynomial for the given value of x.

1. $P(x) = 2x^3 - 3x^2 + 15x - 6;\quad -3$
2. $Q(x) = x^4 - 3x^3 + 2x^2 + 15x - 12;\quad i$

3. Find the quotient and remainder when $x^3 - 2x^2 + 3x - 5$ is divided by $x - i$.
4. Show that $x - i$ is a factor of $2x^3 - 3x^2 + 2x - 3$.

Chapter Review

[14.2] *Find the other zeros of the given polynomial function if one zero is as given.*

5. $P(x) = x^3 - 2x^2 + 4x - 8$; $2i$ is one zero.
6. $Q(x) = 2x^3 - 11x^2 + 28x - 24$; $2 + 2i$ is one zero.

[14.3]
7. Discuss the nature of the roots of $x^4 + 3x^3 - 2x^2 - x + 2 = 0$ in accordance with Descartes' Rule of Signs.
8. Find an upper bound and a lower bound for the real roots of
$$x^4 + 2x^3 - 7x^2 - 10x + 10 = 0.$$
9. Use synthetic division and Theorem 14.9 to show that
$$P(x) = 2x^3 + x^2 - 4x - 2$$
has a zero between 1 and 2.

[14.4]
10. Find all integral zeros of $Q(x) = x^4 - 3x^3 - x^2 - 11x - 4$.
11. Find all rational zeros of $P(x) = 2x^3 - 11x^2 + 12x + 9$.
12. Find all complex zeros of $P(x) = x^3 + 2x^2 + 2x + 4$.

Appendices

A De Moivre's Theorem; Polar Coordinates

A.1   Trigonometric Form of Complex Numbers

Graphs of complex numbers

In Chapter 5, we used Cartesian coordinates to establish a one-to-one correspondence between the set of ordered pairs (a, b) in R^2 and the set of points P in the geometric plane. By pairing each complex number $a + bi$ with the ordered pair (a, b) (we call a the **real part** and b the **imaginary part** of the complex number), we can establish a one-to-one correspondence between the set of all complex numbers and R^2, and hence each point in the plane can be viewed as the graph of a complex number (see Figure A.1). Since the real part a of $a + bi$ is taken as the abscissa, or x-coordinate, of P, in this context the x-axis is called the **real axis.** Similarly, since the imaginary part b is taken as the ordinate or y-coordinate of P, the y-axis is called the **imaginary axis.** Just as we sometimes speak, for instance, of the point $(2, 3)$, meaning of course the point having coordinates 2 and 3, we

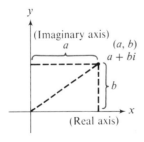

Figure A.1

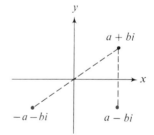

Figure A.2

likewise speak of the point $2 + 3i$, meaning the point $(2, 3)$ representing the complex number $2 + 3i$.

A plane on which complex numbers are thus represented is often called a **complex plane.** It is also sometimes called an **Argand plane,** after the French mathematician

A.1 Trigonometric Form of Complex Numbers

Jean Robert Argand (1768–1822), who systematically used it, or a **Gauss plane,** after the great German mathematician Carl Friedrich Gauss (1777–1855).

A complex number $z = a + bi$, its conjugate $\bar{z} = a - bi$, and also its negative $-z = -a - bi$ are represented in Figure A.2. It is evident that $\bar{z}$ is the reflection of z in the real axis, and $-z$ is the reflection of z in the origin as well as the reflection of $\bar{z}$ in the imaginary axis.

Modulus and argument

Since each nonzero complex number $z = a + bi$ lies on a ray with the origin as endpoint (Figure A.3), we can associate with each z two useful concepts.

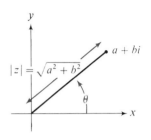

Definition A.1 The **absolute value**, or **modulus**, of the complex number $z = a + bi$ is denoted by r, $|z|$, or $|a + bi|$, and is given by

$$r = |z| = |a + bi| = \sqrt{a^2 + b^2}.$$

Figure A.3

Thus the modulus $|a + bi|$ is just the distance from the origin to the point $a + bi$.

Definition A.2 An **argument**, or an **amplitude**, of the complex number $z = a + bi$ is an angle θ with initial side the positive x-axis and terminal side the ray from the origin containing $a + bi$.

Note that if θ is an argument of $a + bi$, then so is $\theta + 2k\pi^R$, or $\theta + k360°$, for each $k \in J$. For $a + bi = 0$, that is, for $a^2 + b^2 = 0$, any angle θ might be used as an argument of $a + bi$. Also if θ is an argument of $a + bi$, then $b/a = \tan \theta$ when $a \neq 0$.

Trigonometric form

As seen in Figure A.4, any ordered pair (a, b) can be written in the form $(r \cos \theta, r \sin \theta)$, where $r = \sqrt{a^2 + b^2}$ is the modulus of $a + bi$ and θ is an argument of $a + bi$, it follows that any complex number $a + bi$ can be written in the form

$$r \cos \theta + ir \sin \theta, \quad \text{or} \quad r(\cos \theta + i \sin \theta),$$

which is called the **trigonometric form**, or **polar form,** for a complex number. More generally, using degree measure for angles, we have

$$a + bi = r[\cos(\theta + k360°) + i \sin(\theta + k360°)], \quad k \in J.$$

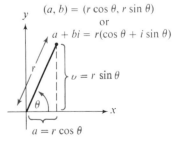

Figure A.4

A convenient abbreviation that is used for the expression $\cos \theta + i \sin \theta$ is **cis θ** (read "cosine θ plus i sine θ"), so that we can write

$$a + bi = r \operatorname{cis}(\theta + k360°), \quad k \in J.$$

We ordinarily use for θ the angle of least nonnegative measure that is a solution of $a + bi = r \text{ cis } \theta$.

Example Represent $1 + \sqrt{3}i$ in trigonometric form.

Solution $r = |1 + \sqrt{3}\, i| = \sqrt{1 + 3} = 2$

Noting that the graph of the complex number is in Quadrant I and that

$$\tan \theta = \frac{\sqrt{3}}{1} = \sqrt{3},$$

we find that $\theta = 60°$. Hence,

$$1 + \sqrt{3}\, i = 2\,(\cos 60° + i \sin 60°) = 2 \text{ cis } 60°$$

Example Represent $4 \text{ cis } 225°$ graphically, and write the number in rectangular form.

Solution
$$a = r \cos \theta = 4\left(-\frac{\sqrt{2}}{2}\right) = -2\sqrt{2},$$
$$b = r \sin \theta = 4\left(-\frac{\sqrt{2}}{2}\right) = -2\sqrt{2},$$

and
$$a + bi = -2\sqrt{2} - 2\sqrt{2}i.$$

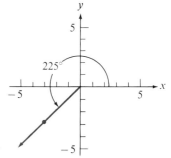

Products and quotients of complex numbers can be found quite easily when the complex numbers are in trigonometric form, as shown by the following results.

Theorem A.1 If $z_1, z_2 \in C$, with $z_1 = r_1 \text{ cis } \theta_1$ and $z_2 = r_2 \text{ cis } \theta_2$, then

I $z_1 \cdot z_2 = r_1 r_2 \text{ cis } (\theta_1 + \theta_2)$,

II $\dfrac{z_1}{z_2} = \dfrac{r_1}{r_2} \text{ cis } (\theta_1 - \theta_2) \quad (z_2 \neq 0 + 0i)$.

We shall prove only Part I here and leave the proof of Part II as an exercise.

Proof of Theorem A.1-I We have
$$z_1 = r_1\,(\cos \theta_1 + i \sin \theta_1) \quad \text{and} \quad z_2 = r_2\,(\cos \theta_2 + i \sin \theta_2),$$
from which
$$z_1 \cdot z_2 = r_1\,(\cos \theta_1 + i \sin \theta_1) \cdot r_2(\cos \theta_2 + i \sin \theta_2)$$
$$= r_1 \cdot r_2 \cdot [\cos \theta_1 \cos \theta_2 + i \cos \theta_1 \sin \theta_2 + i \sin \theta_1 \cos \theta_2 + i^2 \sin \theta_1 \sin \theta_2]$$
$$= r_1 \cdot r_2 \cdot [(\cos \theta_1 \cos \theta_2 - \sin \theta_1 \sin \theta_2) + i(\cos \theta_1 \sin \theta_2 + \sin \theta_1 \cos \theta_2)].$$

A.1 Trigonometric Form of Complex Numbers

By using the sum formula derived in Sections 10.2 and 10.3, the right-hand member can be written

$$r_1 r_2 [\cos(\theta_1 + \theta_2) + i \sin(\theta_1 + \theta_2)],$$

so that

$$z_1 \cdot z_2 = r_1 r_2 \operatorname{cis}(\theta_1 + \theta_2),$$

as was to be shown.

Example Write the product $3 \operatorname{cis} 80° \cdot 5 \operatorname{cis} 40°$ in the form $a + bi$.

Solution By Theorem A.1-I,

$$3 \operatorname{cis} 80° \cdot 5 \operatorname{cis} 40° = 15 \operatorname{cis} 120° = 15 (\cos 120° + i \sin 120°).$$

Since $\cos 120° = -\dfrac{1}{2}$ and $\sin 120° = \dfrac{\sqrt{3}}{2}$, we have

$$15(\cos 120° + i \sin 120°) = 15\left(-\frac{1}{2} + \frac{\sqrt{3}}{2} i\right) = -\frac{15}{2} + \frac{15\sqrt{3}}{2} i.$$

Example Write the quotient $\dfrac{8 \operatorname{cis} 540°}{2 \operatorname{cis} 225°}$ as a complex number in the form $a + bi$.

Solution By Theorem A.1-II,

$$\frac{8 \operatorname{cis} 540°}{2 \operatorname{cis} 225°} = \frac{8}{2} \operatorname{cis}(540° - 225°) = 4 \operatorname{cis} 315° = 4(\cos 315° + i \sin 315°).$$

Since $\cos 315° = \dfrac{1}{\sqrt{2}} = \dfrac{\sqrt{2}}{2}$ and $\sin 315° = \dfrac{-1}{\sqrt{2}} = \dfrac{-\sqrt{2}}{2}$, we have

$$4(\cos 315° + i \sin 315°) = 4\left(\frac{\sqrt{2}}{2} - \frac{\sqrt{2}}{2} i\right) = 2\sqrt{2} - 2\sqrt{2}\, i.$$

Exercise A.1

Graph the given complex number, its conjugate, its negative, and the negative of its conjugate. Draw line segments joining each pair of these four points.

1. $2 + 3i$
2. $-3 + 4i$
3. $4 - i$
4. $-2 - i$
5. $4i$
6. $-3i$

Write without absolute-value notation.

Examples a. $|-3|$ b. $|2 + 5i|$ c. $|(-3, 2)|$

Solutions on Overleaf

Solutions   By Definition A.1,

a. $\sqrt{(-3)^2} = 3$ b. $\sqrt{2^2 + 5^2} = \sqrt{29}$ c. $\sqrt{(-3)^2 + (2)^2} = \sqrt{13}$

7. $|4|$ 8. $|-2|$ 9. $|3 + 2i|$
10. $|4 - i|$ 11. $|2|$ 12. $|3 - 5i|$
13. $|-2 - i|$ 14. $|-7 - i|$

Write the complex number in the form r cis θ.

Example $3\sqrt{3} - 3i$

Solution $r = |3\sqrt{3} - 3i| = \sqrt{27 + 9} = 6$.

Noting that the graph of the complex number is in Quadrant IV and also that $\tan\theta = \dfrac{-3}{3\sqrt{3}} = \dfrac{-1}{\sqrt{3}}$, we find that $\theta = 330°$. Hence

$$3\sqrt{3} - 3i = 6(\cos 330° + i \sin 330°)$$
$$= 6 \text{ cis } 330°.$$

15. $3 + 3i$ 16. $2 - 2i$ 17. 5
18. $-7i$ 19. $2\sqrt{3} - 2i$ 20. $-3\sqrt{3} - 3i$

Write the complex number in the form a + bi.

Example   2 cis 120°

Solution   Since $r = 2$ and $\theta = 120°$, we have

$$a = r\cos\theta = 2\cos 120° = 2\left(-\frac{1}{2}\right) = -1,$$

and

$$b = r\sin\theta = 2\sin 120° = 2\left(\frac{\sqrt{3}}{2}\right) = \sqrt{3}.$$

Hence, $a + bi = -1 + \sqrt{3}\, i$.

21. 4 cis 240° 22. 3 cis 300° 23. 6 cis (−30°)
24. 5 cis 180° 25. 12 cis 420° 26. 10 cis (−480°)

A.2 De Moivre's Theorem—Powers and Roots

For the given pair of complex numbers z_1 and z_2, find (a) $z_1 \cdot z_2$, and (b) z_1/z_2. Express each result in the form $a + bi$. Use Table VI as needed.

27. $z_1 = 3 \text{ cis } 90°$ and $z_2 = \sqrt{2} \text{ cis } 45°$
28. $z_1 = 4 \text{ cis } 30°$ and $z_2 = 2 \text{ cis } 60°$
29. $z_1 = 6 \text{ cis } 150°$ and $z_2 = 18 \text{ cis } 570°$
30. $z_1 = 14 \text{ cis } 210°$ and $z_2 = 2 \text{ cis } 120°$
31. $z_1 = -3 + i$ and $z_2 = -2 - 4i$
32. $z_1 = 2 + 3i$ and $z_2 = 2 + 3i$
33. Prove that the sum and product of two conjugate complex numbers are both real.
34. Show that if $a + bi = r \text{ cis } \theta$, then $(a + bi)^2 = r^2 \text{ cis } 2\theta$.
35. Use the result of Exercise 34 to show that if $a + bi = r \text{ cis } \theta$, then
$$(a + bi)^3 = r^3 \text{ cis } 3\theta.$$

A.2 De Moivre's Theorem—Powers and Roots

Powers of complex numbers

Since $a + bi = r \text{ cis } \theta$, an application of Theorem A.1-I to $(a + bi)^2$ results in

$$(a + bi)^2 = (r \text{ cis } \theta)(r \text{ cis } \theta) = r^2 \text{ cis } 2\theta. \tag{1}$$

Because

$$(a + bi)^3 = (a + bi)^2(a + bi),$$

from (1) we have

$$(a + bi)^3 = (r^2 \text{ cis } 2\theta)(r \text{ cis } \theta) = r^3 \text{ cis } 3\theta.$$

In a similar way, we can show that

$$(a + bi)^4 = (r^3 \text{ cis } 3\theta)(r \text{ cis } \theta) = r^4 \text{ cis } 4\theta,$$

and it seems plausible to make the following assertion, known as **De Moivre's theorem**. The proof is omitted.

Theorem A.2 If $z \in C$, $z = r \text{ cis } \theta$ and $n \in N$, then

$$z^n = r^n \text{ cis } n\theta.$$

Example Write $(\sqrt{3} + i)^7$ in the form $a + bi$.

Solution For the modulus, we have $r = \sqrt{(\sqrt{3})^2 + 1^2} = 2$. For an argument, $\tan \theta = 1/\sqrt{3}$ and from the fact that the graph of $\sqrt{3} + i$ is in Quadrant I, we obtain $\theta = 30°$.

Solution Continued on Overleaf

Thus, $(\sqrt{3} + i)^7 = (2 \text{ cis } 30°)^7$. Then, by De Moivre's theorem,
$$(2 \text{ cis } 30°)^7 = 2^7 \text{ cis } (7 \cdot 30°) = 128 \text{ cis } 210°.$$

Converting to the form $a + bi$, we find that
$$128(\cos 210° + i \sin 210°) = 128\left(-\frac{\sqrt{3}}{2} - \frac{1}{2}i\right)$$
$$= -64\sqrt{3} - 64i,$$

so
$$(\sqrt{3} + i)^7 = -64\sqrt{3} - 64i.$$

By appropriately defining z^0 and z^{-n}, we can use Theorem A.1 to extend De Moivre's theorem to include as exponents all $n \in J$. The simple proof is omitted.

Definition A.3 If $z \neq 0 + 0i$, then

I $z^0 = 1 + 0i$,   II $z^{-n} = \dfrac{1}{z^n}$, for $n \in J$.

Theorem A.3 If $z \in C$, $z \neq 0 + 0i$, $z = r \text{ cis } \theta$ and $n \in J$, then
$$z^n = r^n \text{ cis } n\theta.$$

Example Write $(1 + i)^{-6}$ in the form $a + bi$.

Solution $r = \sqrt{1^2 + 1^2} = \sqrt{2}.$

Noting that the graph of $1 + i$ is in Quadrant I and $\tan \theta = 1/1$, we have $\theta = 45°$. Hence,
$$(1 + i)^{-6} = (\sqrt{2} \text{ cis } 45°)^{-6}.$$

By Theorem A.3,
$$(\sqrt{2} \text{ cis } 45°)^{-6} = (\sqrt{2})^{-6} \text{ cis } (-6 \cdot 45°)$$
$$= \frac{1}{8} \text{ cis } (-270°)$$
$$= \frac{1}{8} [\cos(-270°) + i \sin(-270°)].$$

Since $\cos(-270°) = 0$ and $\sin(-270°) = 1$, we obtain
$$(1 + i)^{-6} = \frac{1}{8}(0 + i) = \frac{1}{8}i.$$

A.2 De Moivre's Theorem—Powers and Roots

Roots of complex numbers

Yet another extension of De Moivre's theorem is possible if we make the following definition.

Definition A.4   For $z \in C$, $n \in N$, w is an nth root of z provided
$$w^n = z.$$

Theorem A.4   If $z \in C$, $z = r \operatorname{cis} \theta$ and $n \in N$, then
$$w = r^{1/n} \operatorname{cis}\left(\frac{\theta}{n}\right)$$
is an nth root of z.

This theorem follows directly from De Moivre's theorem. The fact that
$$\operatorname{cis} \theta = \operatorname{cis}(\theta + k360°),$$
for $k \in J$, enables us to find n distinct complex nth roots for each $z \in C$, $z \neq 0 + 0i$ as illustrated in the following example.

Example   Write each of the four fourth roots of $z = 2 + 2\sqrt{3}\,i$ in the form $r \operatorname{cis} \theta$.

Solution $r = \sqrt{2^2 + (2\sqrt{3})^2} = 4.$

The graph of $2 + 2\sqrt{3}\,i$ is in Quadrant I and $\tan \theta = \sqrt{3}$. Hence, $\theta = 60°$ and
$$z = 2 + 2\sqrt{3}\,i = 4 \operatorname{cis} 60°.$$

From Theorem A.4 and the periodic property of cosine and sine, each number
$$4^{1/4} \operatorname{cis}\left(\frac{60° + k360°}{4}\right),$$
for $k \in J$, is a fourth root of z. Taking $k = 0, 1, 2,$ and 3, in turn, gives the roots
$$w_0 = \sqrt{2} \operatorname{cis} 15°, \quad w_1 = \sqrt{2} \operatorname{cis} 105°, \quad w_2 = \sqrt{2} \operatorname{cis} 195°, \quad w_3 = \sqrt{2} \operatorname{cis} 285°.$$
The substitution of any other integer for k will produce one of these four complex numbers.

Exercise A.2

Use Theorem A.2 or A.3 as appropriate to write the given expression as a complex number of the form $a + bi$. Use Table VI as necessary.

1. $[2 \operatorname{cis}(-30°)]^7$
2. $(4 \operatorname{cis} 36°)^5$
3. $-\left(\frac{1}{2} + \frac{1}{2}\sqrt{3}\,i\right)^3$
4. $(1 + i)^{12}$
5. $(\sqrt{3} \operatorname{cis} 5°)^{12}$
6. $(\sqrt{2} \operatorname{cis} 30°)^{-7}$

7. $(\sqrt{3} - i)^{-5}$ 8. $(1 - i)^{-6}$ 9. $\dfrac{(1 + i)^3}{(1 + \sqrt{3}\, i)^5(1 - i)^2}$

10. $\dfrac{4(\sqrt{3} + i)^3}{(1 - i)^3}$ 11. $\dfrac{(1 - i)^5}{(1 + i)^6}$ 12. $\dfrac{(1 + \sqrt{3}\, i)^{-4}}{(\sqrt{3} + i)^{-6}}$

Find the nth roots of z by applying Theorem A.4. Leave the results in trigonometric form and list all n of the nth roots.

13. $z = 32 \text{ cis } 45°, \quad n = 5$ 14. $z = 27, \quad n = 3$
15. $z = -16\sqrt{3} + 16i, \quad n = 5$ 16. $z = 1 - i, \quad n = 4$
17. $z = -i, \quad n = 6$ 18. $z = 2 + 2\sqrt{3}\, i, \quad n = 3$

Solve the given equation over C.

19. $x^5 = 16 - 16\sqrt{3}\, i$ 20. $x^3 + 4i = 4\sqrt{3}$
21. $x^7 + 1 = 0$ 22. $x^7 - 1 = 0$

23. Factor $x^4 + 16$ into linear factors.
24. Factor $x^5 - 1$ into linear factors.
25. Show that the sum of the four fourth roots of 1 is $0 + 0i$.
26. Explain why the sum of the *n*th roots of any complex number is zero for all $n > 1$.

A.3 Polar Coordinates

The Cartesian coordinates that we have been using specify the location of a point in the plane by giving the directed distances of the point from a pair of fixed perpendicular lines, the axes. There is an alternative coordinate system that is frequently used in the plane, in which the location of a point is specified in a different manner.

In the plane, consider a fixed ray $\overrightarrow{PB}$ and any point *A*. We can describe the location of *A* by giving the distance *r* from *P* to *A* and specifying the angle *BPA* (Figure A.5), which is customarily designated by θ. By stating the ordered pair $(r, \theta°)$ or (r, θ^R), we clearly identify the location of *A*. We ordinarily write (r, θ) for either of these ordered pairs, where the meaning should be clear from the context. The components of such an ordered pair are called **polar coordinates** of *A*. The fixed ray $\overrightarrow{PB}$ is called the **polar axis,** and the initial point *P* of the polar axis is called the **pole** of the system.

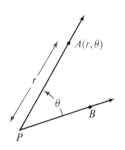

Figure A.5

A.3 Polar Coordinates

Notice that while there is a one-to-one correspondence between the set of ordered pairs in a Cartesian coordinate system and the points in the geometric plane, each point in the plane has infinitely many pairs of polar coordinates. In the first place, if (r, θ) are polar coordinates of A, then so are

$$(r, \theta + k360°), \quad k \in J$$

(see Figure A.6-a). In the second place, if we let $\overrightarrow{PA'}$ denote the ray in a direction opposite that of $\overrightarrow{PA}$ (Figure A.6-b), then we see that $-r \leq 0$ denotes the directed

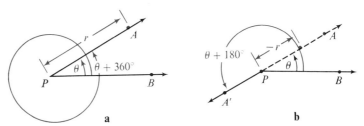

Figure A.6

distance from P to A along the negative extension of $\overrightarrow{PA'}$. Thus we see that also $(-r, \theta + 180°)$, and more generally

$$(-r, \theta + 180° + k360°), \quad k \in J,$$

are polar coordinates of A. The pole P itself is represented by $(0, \theta)$ for any θ.

Example Write four additional sets of polar coordinates for the point having polar coordinates $(3, 30°)$.

Solution With positive values for r, two more pairs of polar coordinates for $(3, 30°)$ are $(3, 390°)$ and $(3, -330°)$. Using negative values for r, we have $(-3, 210°)$ and $(-3, -150°)$. The figures below show these cases. Of course there are infinitely many other possible polar coordinates for the same point.

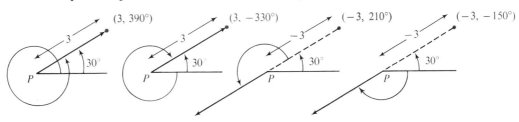

Relationships between polar and rectangular coordinates

Cartesian and polar coordinates of a point can be related by means of the trigonometric functions. If the pole P in a polar coordinate system is also the origin O in a Cartesian coordinate system, and if the polar axis coincides with the positive x-axis of the Cartesian system, as in Figure A.7 on page 446, then the coordinates (x, y) can be expressed in terms of the polar coordinates (r, θ) by the following equations:

$$x = r \cos \theta \quad \text{and} \quad y = r \sin \theta. \qquad (1)$$

Conversely, we have

$$r = \pm\sqrt{x^2 + y^2}, \quad \cos \theta = \frac{x}{\pm\sqrt{x^2 + y^2}}, \qquad (2)$$

$$\sin \theta = \frac{y}{\pm\sqrt{x^2 + y^2}} \quad [(x, y) \neq (0, 0)].$$

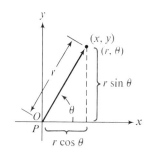

Figure A.7

The sets of equations (1) and (2) enable us to find rectangular coordinates for a point with a given pair of polar coordinates, and vice versa. Often, in determining θ, it is simplest first to determine the quadrant from the signs of x and y, and then to use the relation

$$\tan \theta = \frac{y}{x}.$$

Example Find the rectangular coordinates of the point with polar coordinates (4, 30°). Show the graph of the point on a combined polar and rectangular coordinate system.

Solution Using (1), we obtain

$$x = 4 \cos 30° = 4 \cdot \frac{\sqrt{3}}{2} = 2\sqrt{3},$$

$$y = 4 \sin 30° = 4 \cdot \frac{1}{2} = 2.$$

The rectangular coordinates are $(2\sqrt{3}, 2)$.

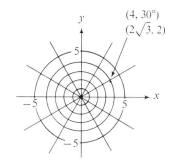

Example Find a pair of polar coordinates for the point with Cartesian coordinates $(7, -2)$. Show the graph of the point on a combined polar and rectangular coordinate system.

Solution By (2), $r = \pm\sqrt{x^2 + y^2} = \pm\sqrt{49 + 4} = \pm\sqrt{53}$. Choosing the positive sign, and noting that the point $(7, -2)$ is in the fourth quadrant, we have

$$\tan \theta = \frac{-2}{7},$$

from which

$$\theta \approx -15.9°$$

A pair of polar coordinates is therefore

$$(\sqrt{53}, -15.9°),$$

where the given angle measure is an approximation.

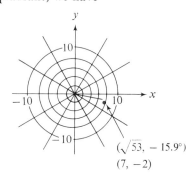

A.3 Polar Coordinates

Equations in polar form and in rectangular form

Equations (1) on page 446 can be used to transform Cartesian equations to polar form, and Equations (2) can be used to transform polar equations to Cartesian form. Several examples are shown in the exercise set.

Exercise A.3

Find four additional sets of polar coordinates $(-360° < \theta \le 360°)$ for the point with polar coordinates as given.

1. $(6, 485°)$
2. $(-3, 518°)$
3. $(-2, -450°)$
4. $(5, 720°)$
5. $(6, -600°)$
6. $(3, -395°)$

Find the Cartesian coordinates of a point with polar coordinates as given.

7. $(5, 45°)$
8. $(-3, 30°)$
9. $\left(\frac{1}{2}, 330°\right)$
10. $\left(\frac{3}{4}, 225°\right)$
11. $(10, -135°)$
12. $(-6, -240°)$

Find two sets of polar coordinates, one involving an angle of positive measure and one an angle of negative measure, for the point with Cartesian coordinates as given.

13. $(3\sqrt{2}, 3\sqrt{2})$
14. $\left(-\frac{\sqrt{3}}{2}, \frac{1}{2}\right)$
15. $(-1, -\sqrt{3})$
16. $(0, -4)$
17. $(0, 0)$
18. $(-6, 0)$

Transform the given equation to an equation in polar form.

Example

$x^2 + y^2 - 2x + 3 = 0$

Solution

From Equations (1) on page 446, we find that $x^2 = r^2 \cos^2 \theta$ and $y^2 = r^2 \sin^2 \theta$. Thus we have

$$r^2 \cos^2 \theta + r^2 \sin^2 \theta - 2r \cos \theta + 3 = 0,$$

from which

$$r^2 (\cos^2 \theta + \sin^2 \theta) - 2r \cos \theta + 3 = 0,$$

$$r^2 - 2r \cos \theta + 3 = 0.$$

19. $x^2 + y^2 = 25$
20. $x = 3$
21. $y = -4$
22. $x^2 + y^2 - 4y = 0$
23. $x^2 + 9y^2 = 9$
24. $x^2 - 4y^2 = 4$

Transform the given equation to an equation in Cartesian form free from radicals.

Example $r(1 - 2 \cos \theta) = 3$

Solution Using Equations (2) on page 446, we have

$$\pm\sqrt{x^2 + y^2}\left[1 - 2\left(\frac{x}{\pm\sqrt{x^2 + y^2}}\right)\right] = 3,$$

from which

$$\pm\sqrt{x^2 + y^2} - 2x = 3,$$
$$\pm\sqrt{x^2 + y^2} = 2x + 3.$$

Upon squaring each member, we have the equation

$$x^2 + y^2 = 4x^2 + 12x + 9,$$

from which

$$3x^2 - y^2 + 12x + 9 = 0.$$

25. $r = 5$ **26.** $r = 4 \sin \theta$ **27.** $r = 9 \cos \theta$
28. $r \cos \theta = 3$ **29.** $r(1 - \cos \theta) = 2$ **30.** $r(1 + \sin \theta) = 2$

31. Show by transformation of coordinates that the graph of $r = \sec^2 (\theta/2)$ is a parabola.
32. Show that the graph of $r = \csc^2 (\theta/2)$ is a parabola.

Graph the given equation.

Example $r = 1 + \sin \theta$

Solution Using selected values for θ ($0° \leq \theta \leq 360°$) we first obtain the following ordered pairs (r, θ), where values for $r = 1 + \sin \theta$ are approximated. The graph is then sketched as shown on page 449. Since the sine function is periodic, with period 2π, any other values of θ less than $0°$ or greater than $360°$ would simply yield ordered pairs whose graphs would also be on the curve.

A.3 Polar Coordinates

θ	$\sin\theta$	$r = 1 + \sin\theta$
0°	0	1
30°	0.5	1.5
60°	0.9	1.9
90°	1	2
120°	0.9	1.9
150°	0.5	1.5
180°	0	1
210°	−0.5	0.5
240°	−0.9	0.1
270°	−1	0
300°	−0.9	0.1
330°	−0.5	0.5
360°	0	1

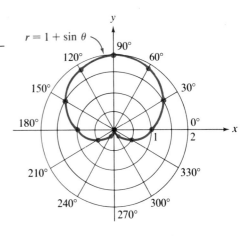

33. $r = 4\sin\theta$
34. $r = 9\cos\theta$
35. $r = 1 + \cos\theta$
36. $r = 1 - \sin\theta$
37. $r\cos\theta = 3$
38. $r\sin\theta = -3$
39. $r = 4\sin 2\theta$
40. $r = 4\cos 2\theta$
41. $r = 3$ *Hint:* All solutions are of the form $(3, \theta)$.
42. $\theta = 30°$ *Hint:* All solutions are of the form $(r, 30°)$.

B Mathematical Structure

Here we present the basic substance of the text in compact form. It is a handy guide for following the development of the course and for reviewing material in preparation for tests. Also, you may find it a convenient reference for mathematics courses that you take in the future.

Postulates for Real Numbers

[1.2]
Equality postulates

E-1 $a = a$.

E-2   If $a = b$, then $b = a$.

E-3   If $a = b$ and $b = c$, then $a = c$.

E-4   If $a = b$, then a may be replaced by b and b by a in any mathematical statement without altering the truth or falsity of the statement.

[1.2]
Field postulates

F-1 $a + b$ is a unique element of R.

F-2 $(a + b) + c = a + (b + c)$.

F-3   There exists an element $0 \in R$ with the property

$$a + 0 = a \quad \text{and} \quad 0 + a = a$$

for all $a \in R$.

F-4   For each $a \in R$, there exists an element $-a \in R$ with the property

$$a + (-a) = 0 \quad \text{and} \quad (-a) + a = 0.$$

F-5 $a + b = b + a$.

F-6 $a \cdot b$ is a unique element of R.

F-7 $(a \cdot b) \cdot c = a \cdot (b \cdot c)$.

Mathematical Structure

F-8   For all $a \in R$, there exists an element $1 \in R$, $1 \neq 0$, with the property

$$a \cdot 1 = a \quad \text{and} \quad 1 \cdot a = a$$

F-9   For each element $a \in R$ $(a \neq 0)$, there exists an element $a^{-1} \in R$ with the property

$$a \cdot (a^{-1}) = 1 \quad \text{and} \quad (a^{-1}) \cdot a = 1.$$

F-10 $a \cdot b = b \cdot a$.

F-11 $a \cdot (b + c) = a \cdot b + a \cdot c \quad \text{and} \quad (b + c) \cdot a = b \cdot a + c \cdot a$.

Properties of Numbers

[1.3]   *Assume all variables represent arbitrary real numbers unless otherwise noted.*

If $a = b$, then
$a + c = b + c$ and $a \cdot c = b \cdot c$.

If $a + b = 0$, then $a = -b$.

If $a \cdot b = 1$ $(a,b \neq 0)$, then $a = \dfrac{1}{b}$.

If $a + c = b + c$, then $a = b$.

If $a \cdot c = b \cdot c$ $(c \neq 0)$, then $a = b$.

For every a, $a \cdot 0 = 0$.

If $a \cdot b = 0$, then
either $a = 0$ or $b = 0$, or both.

$-(-a) = a$.

$(-a) + (-b) = -(a + b)$.

$(-a)(b) = -(ab)$.

$(-a)(-b) = ab$.

$\dfrac{-a}{b} = \dfrac{a}{-b} = -\dfrac{a}{b}$ $(b \neq 0)$.

$\dfrac{-a}{-b} = \dfrac{a}{b}$ $(b \neq 0)$.

$\dfrac{a}{b} = \dfrac{c}{d}$ $(b,d \neq 0)$ if and only if $ad = bc$.

$\dfrac{ac}{bc} = \dfrac{a}{b}$ $(b,c \neq 0)$.

$\dfrac{1}{a} \cdot \dfrac{1}{b} = \dfrac{1}{ab}$ $(a,b \neq 0)$.

$\dfrac{a}{b} \cdot \dfrac{c}{d} = \dfrac{ac}{bd}$ $(b,d \neq 0)$.

$\dfrac{a}{c} + \dfrac{b}{c} = \dfrac{a + b}{c}$ $(c \neq 0)$.

$\dfrac{a}{b} + \dfrac{c}{d} = \dfrac{ad + bc}{bd}$ $(b,d \neq 0)$.

$\dfrac{a}{b} - \dfrac{c}{d} = \dfrac{ad - bc}{bd}$ $(b,d \neq 0)$.

$1 \div \dfrac{a}{b} = \dfrac{b}{a}$ $(a,b \neq 0)$.

$\dfrac{a}{b} \div \dfrac{c}{d} = \dfrac{ad}{bc}$ $(b,c,d \neq 0)$.

If $a < b$ and $b < c$, then $a < c$.

If $a < b$, then $a + c < b + c$.

If $a < b$ and $c > 0$, then $ac < bc$.

If $a < b$ and $c < 0$, then $ac > bc$.

[1.4]

Order postulates

O-1   If a is a real number, then exactly one of the following statements is true:
a is positive, a is zero, or $-a$ is positive.

O-2   If a and b are positive real numbers, then $a + b$ and $a \cdot b$ are positive.

O-3 There is a one-to-one correspondence between the set of real numbers and the set of points on a geometric line.

[1.5] The complex numbers satisfy the field postulates F-1 through F-11 and the properties listed above.

[2.2] $a^m \cdot a^n = a^{m+n}.$ $(a^m)^n = a^{mn}.$ $(ab)^n = a^n b^n.$

[2.4] $\dfrac{a^m}{a^n} = a^{m-n} \quad (a \neq 0).$ $\left(\dfrac{a}{b}\right)^m = \dfrac{a^m}{b^m} \quad (b \neq 0).$

[2.5] If $P(x)$ is a real polynomial and c is any real number, then there exists a unique real polynomial $Q(x)$ and a real number r such that

$$P(x) = (x - c)Q(x) + r.$$

[2.9] If $r_1 \neq r_2$, then there exist constants $c_1, c_2 \in R$ such that

$$\frac{ax + b}{(x - r_1)(x - r_2)} = \frac{c_1}{x - r_1} + \frac{c_2}{x - r_2}.$$

If $P(x)$ is of degree less than $Q(x)$, and if $Q(x) = (x - r_1)(x - r_2) \cdots (x - r_n)$, where no two factors are identical, then there exist constants $c_1, c_2, \ldots c_n \in R$ such that

$$\frac{P(x)}{Q(x)} = \frac{c_1}{x - r_1} + \frac{c_2}{x - r_2} + \cdots + \frac{c_n}{x - r_n}.$$

If $P(x)$ is of degree less than $Q(x)$, and if $Q(x) = (x - r_1)^n$, then there exist constants $c_1, c_2, \ldots c_n \in R$ such that

$$\frac{P(x)}{Q(x)} = \frac{c_1}{x - r_1} + \frac{c_2}{(x - r_1)^2} + \cdots + \frac{c_n}{(x - r_1)^n}.$$

[3.1] $a^0 = 1 \quad (a \neq 0).$ $a^{-n} = \dfrac{1}{a^n} \quad (a \neq 0).$

[3.2] $(a^{1/n})^n = a$ (n an odd natural number).
$(a^{1/n})^n = |a|$ (n an even natural number).
$(a^{1/n})^m = (a^m)^{1/n}$ ($n \in N, m \in J$).
$(a^{1/np})^{mp} = (a^{1/n})^m$ ($m \in J, n, p \in N$).
$a^{m/n} = (a^{1/n})^m$ ($m \in J, n \in N$).

[3.3]
$\sqrt[n]{a} = a^{1/n} \quad (a \geq 0)$.

$\sqrt[n]{a^n} = a \quad$ (n an odd natural number).

$\sqrt[n]{a^n} = |a| \quad$ (n an even natural number).

$\sqrt[n]{a^m} = (\sqrt[n]{a})^m \quad (n \in N, m \in J)$.

$\sqrt[n]{a} \sqrt[n]{b} = \sqrt[n]{ab} \quad (n \in N)$.

$\dfrac{\sqrt[n]{a}}{\sqrt[n]{b}} = \sqrt[n]{\dfrac{a}{b}} \quad (b \neq 0, n \in N)$.

$\sqrt[cn]{a^{cm}} = \sqrt[n]{a^m} \quad (m \in J, n, c \in N)$.

[4.1] If $P(x)$, $Q(x)$, and $R(x)$ are expressions, then for all values of x for which $P(x)$, $Q(x)$, and $R(x)$ are real numbers, the sentence
$$P(x) = Q(x)$$
is equivalent to each of the following:

    I $P(x) + R(x) = Q(x) + R(x)$,

    II $P(x) \cdot R(x) = Q(x) \cdot R(x)$ for $x \in \{x \mid R(x) \neq 0\}$.

[4.2] $ab = 0$ if and only if $a = 0$ or $b = 0$ or both.

[4.3] If $ax^2 + bx + c = 0 \quad (a \neq 0)$, then $x = \dfrac{-b \pm \sqrt{b^2 - 4ac}}{2a}$.

[4.4] The solution set of $U(x) = V(x)$ is a subset of the solution set of $[U(x)]^n = [V(x)]^n$, for each natural number n.

[4.6] If $P(x)$, $Q(x)$, and $R(x)$ are expressions, then for all values of x for which $P(x)$, $Q(x)$, and $R(x)$ are real numbers, the sentence
$$P(x) < Q(x)$$
is equivalent to each of the following.

    I $P(x) + R(x) < Q(x) + R(x)$,

    II $P(x) \cdot R(x) < Q(x) \cdot R(x)$ for $x \in \{x \mid R(x) > 0\}$,

    III $P(x) \cdot R(x) > Q(x) \cdot R(x)$ for $x \in \{x \mid R(x) < 0\}$.

Similarly, the sentence
$$P(x) \leq Q(x)$$
is equivalent to sentences of the form I-III, with $<$ (or $>$) replaced by $\leq$ (or $\geq$) under the same conditions, $R(x) > 0$ and $R(x) < 0$, as above.

[4.8] $|x| = a$ is equivalent to $x = a$ or $x = -a$.
$|x| < a$ is equivalent to $-a < x$ and $x < a$, i.e., $-a < x < a$.
$|x| > a$ is equivalent to $x < -a$ or $x > a$.
$|x| \leq a$ is equivalent to $-a \leq x$ and $x \leq a$, i.e., $-a \leq x \leq a$.
$|x| \geq a$ is equivalent to $x \leq -a$ or $x \geq a$.

[5.2]   The length of the line segment between two points (x_1, y_1) and (x_2, y_2) is given by
$$d = \sqrt{(x_2 - x_1)^2 + (y_2 - y_1)^2},$$
and the slope is given by
$$m = \frac{y_2 - y_1}{x_2 - x_1} \quad (x_2 \neq x_1).$$

[5.3]   Forms for linear equations:
$$y - y_1 = m(x - x_1) \quad \text{point-slope form,}$$
$$y = mx + b \quad \text{slope-intercept form,}$$
$$\frac{x}{a} + \frac{y}{b} = 1 \quad \text{intercept form.}$$

[5.5]   If $P(x) = a_0 x^n + a_1 x^{n-1} + \cdots + a_n$ is a real polynomial equation of degree n, then the graph of
$$P(x) = a_0 x^n + a_1 x^{n-1} + \cdots + a_n, \quad a_0 \neq 0$$
is a smooth curve that has at most $n - 1$ turning points.

If $P(x)$ is a real polynomial, then for every real number c there exists a unique real polynomial $Q(x)$ such that
$$P(x) = (x - c)Q(x) + P(c).$$

[5.6]   The graph of the rational function over R defined by $y = P(x)/Q(x)$ has a vertical asymptote at $x = a$ for each value a at which $Q(x)$ vanishes and $P(x)$ does not vanish.

The graph of the rational function over R defined by
$$y = \frac{a_0 x^n + a_1 x^{n-1} + \cdots + a_n}{b_0 x^m + b_1 x^{m-1} + \cdots + b_m},$$
where $a_0, b_0 \neq 0$ and n, m are nonnegative integers, has

I   a horizontal asymptote at $y = 0$ if $n < m$,

II a horizontal asymptote at $y = a_0/b_0$ if $n = m$,

III no horizontal asymptotes if $n > m$.

Mathematical Structure

[6.3]
$$y = kx \quad \text{direct variation,}$$
$$y = \frac{k}{x} \quad \text{inverse variation,}$$
$$y = kuvw \quad \text{joint variation.}$$

[7.1] Let $x, y \in Q$, and let $x > y > 0$. Then
$$b^x > b^y \text{ if } b > 1, \quad b^x = b^y \text{ if } b = 1, \quad \text{and} \quad b^x < b^y \text{ if } 0 < b < 1.$$

[7.2]
$$\log_b (x_1 x_2) = \log_b x_1 + \log_b x_2.$$
$$\log_b \frac{x_2}{x_1} = \log_b x_2 - \log_b x_1.$$
$$\log_b (x_1)^m = m \log_b x_1.$$

[8.1] When measuring arcs on the unit circle, the counterclockwise direction is considered positive and the clockwise direction is considered negative.

$$\sin^2 s + \cos^2 s = 1$$

[8.3] For any arc A on the unit circle, with initial point $(1, 0)$ and with measure x, the reference arc is the arc on the unit circle with least nonnegative measure between the terminal point of A and the horizontal axis. The length of the reference arc is denoted by $\bar{x}$.

For any arc of measure x,
$$|\cos x| = \cos \bar{x} \quad \text{and} \quad |\sin x| = \sin \bar{x}.$$

[8.4] If f is a function such that for some $p \in R$, $p \neq 0$,
$$f(x + p) = f(x - p) = f(x)$$
for all x in the domain of f, then f is periodic with period p.

If p is the least positive period of a periodic function f, then p is called the fundamental period of f.

The fundamental period of the cosine function and of the sine function is 2π.

[8.5] The fundamental period of the tangent function is π.

[9.1]
$$1^R = \left(\frac{180}{\pi}\right)^\circ \quad \text{and} \quad 1^\circ = \left(\frac{\pi}{180}\right)^R$$

If α is the radian measure of a central angle of a circle of radius r, then the length s of the intercepted arc of the circle is
$$s = r\alpha.$$

[9.2] For any angle α in standard position, the reference angle $\bar{\alpha}$ is the least nonnegative angle which the terminal ray of α forms with the x-axis.

If T is any trigonometric function and $\bar{\alpha}$ is the reference angle for the angle α, then
$$|T(\alpha)| = T(\bar{\alpha}).$$

[9.3] If α is an angle of a right triangle, then
$$\sin \alpha = \frac{\text{length of side opposite } \alpha}{\text{length of hypotenuse}},$$
$$\cos \alpha = \frac{\text{length of side adjacent to } \alpha}{\text{length of hypotenuse}},$$
$$\tan \alpha = \frac{\text{length of side opposite } \alpha}{\text{length of side adjacent to } \alpha}.$$

[9.4] If α, β, and γ are the angles of a triangle, and a, b, and c are the lengths of the sides opposite α, β, and γ, respectively, then
$$\frac{\sin \alpha}{a} = \frac{\sin \beta}{b} = \frac{\sin \gamma}{c}. \qquad \textbf{Law of Sines}$$

[9.5] If α, β, and γ are the angles of a triangle, and a, b, and c are the lengths of the sides opposite α, β, and γ, respectively, then
$$c^2 = a^2 + b^2 - 2ab \cos \gamma,$$
$$b^2 = a^2 + c^2 - 2ac \cos \beta, \qquad \textbf{Law of Cosines}$$
$$a^2 = b^2 + c^2 - 2bc \cos \alpha.$$

[10.1] The following basic identities are useful:
$$\tan^2 x + 1 = \sec^2 x \qquad 1 + \cot^2 x = \csc^2 x$$
$$\sin(-x) = -\sin x \qquad \cos(-x) = \cos x \qquad \tan(-x) = -\tan x$$

[10.2]
$$\cos(x_1 + x_2) = \cos x_1 \cos x_2 - \sin x_1 \sin x_2$$
$$\cos(x_1 - x_2) = \cos x_1 \cos x_2 + \sin x_1 \sin x_2$$
$$\cos 2x = \cos^2 x - \sin^2 x$$
$$= 1 - 2\sin^2 x$$
$$= 2\cos^2 x - 1$$

$$\cos \frac{x}{2} = \begin{cases} \sqrt{\dfrac{1 + \cos x}{2}} & \text{when } \dfrac{x}{2} \text{ terminates in Quadrant I or IV;} \\ -\sqrt{\dfrac{1 + \cos x}{2}} & \text{when } \dfrac{x}{2} \text{ terminates in Quadrant II or III.} \end{cases}$$

[10.3]
$$\sin(x_1 + x_2) = \sin x_1 \cos x_2 + \cos x_1 \sin x_2$$
$$\sin(x_1 - x_2) = \sin x_1 \cos x_2 - \cos x_1 \sin x_2$$
$$\sin 2x = 2 \sin x \cos x$$

$$\sin \frac{x}{2} = \begin{cases} \sqrt{\dfrac{1 - \cos x}{2}} & \text{when } \dfrac{x}{2} \text{ terminates in Quadrant I or II;} \\ -\sqrt{\dfrac{1 - \cos x}{2}} & \text{when } \dfrac{x}{2} \text{ terminates in Quadrant III or IV.} \end{cases}$$

[10.4]
$$\tan(x_1 + x_2) = \frac{\tan x_1 + \tan x_2}{1 - \tan x_1 \tan x_2} \qquad \tan(x_1 - x_2) = \frac{\tan x_1 - \tan x_2}{1 + \tan x_1 \tan x_2}$$

$$\tan 2x = \frac{2 \tan x}{1 - \tan^2 x} \qquad \tan \frac{x}{2} = \frac{1 - \cos x}{\sin x}$$

[11.1] Any ordered pair that satisfies both the equations

$$f(x, y) = 0 \quad \text{and} \quad g(x, y) = 0$$

will also satisfy the equation

$$a \cdot f(x, y) + b \cdot g(x, y) = 0,$$

for all real numbers a and b.

[11.2] If any equation in the system

$$f(x, y, z) = 0, \quad g(x, y, z) = 0, \quad \text{and} \quad h(x, y, z) = 0$$

is replaced by a linear combination, with nonzero coefficients, of itself and any one of the other equations in the system, then the result is an equivalent system.

[11.5] Any half-plane is a convex set.

The intersection of two convex sets is a convex set.

[12.1] If A, B, and C are $m \times n$ matrices with real-number entries, then:

    I $(A + B)_{m \times n}$ is a matrix with real-number entries.
    II $(A + B) + C = A + (B + C)$.
    III The matrix $\mathbf{0}_{m \times n}$ has the property that for every matrix $A_{m \times n}$,

$$A + \mathbf{0} = A \quad \text{and} \quad \mathbf{0} + A = A.$$

    IV For every matrix $A_{m \times n}$, the matrix $-A_{m \times n}$ has the property that

$$A + (-A) = \mathbf{0} \quad \text{and} \quad (-A) + A = \mathbf{0}.$$

    V $A + B = B + A$

[12.2] If A and B are $m \times n$ matrices and $c, d \in R$ then:

    I cA is an $m \times n$ matrix,
    II $c(dA) = (cd)A$,
    III $(c + d)A = cA + dA$,
    IV $c(A + B) = cA + cB$,
    V $1A = A$,
    VI $(-1)A = -A$,
    VII $0A = \mathbf{0}$,
    VIII $c\mathbf{0} = \mathbf{0}$.

If A, B, and C are $n \times n$ square matrices, then
$$(AB)C = A(BC).$$

If A, B, and C are $n \times n$ square matrices, then
$$A(B + C) = AB + AC \quad \text{and} \quad (B + C)A = BA + CA.$$

For each matrix $A_{n \times n}$, we have
$$A_{n \times n} I_{n \times n} = I_{n \times n} A_{n \times n} = A_{n \times n}.$$

Furthermore, $I_{n \times n}$ is the unique matrix having this property for all matrices $A_{n \times n}$.

If A and B are $n \times n$ square matrices, and a is a real number, then
$$a(AB) = (aA)B = A(aB).$$

[12.5] If each entry in any row, or each entry in any column, of a determinant is 0, then the determinant is equal to 0.

If any two rows (or any two columns) of a determinant are interchanged, the resulting determinant is the negative of the original determinant.

If two rows (or two columns) in a determinant have corresponding entries that are equal, the determinant is equal to 0.

If each of the entries of one row (or column) of a determinant is multiplied by k, the determinant is multiplied by k.

If each entry in a row (or column) of a determinant is written as the sum of two terms, the determinant can be written as the sum of two determinants as follows: If

$$D = \begin{vmatrix} a_{11} & a_{12} & \cdots & a_{1n} \\ \vdots & \vdots & & \vdots \\ b_{i1} + c_{i1} & b_{i2} + c_{i2} & \cdots & b_{in} + c_{in} \\ \vdots & \vdots & & \vdots \\ a_{n1} & a_{n2} & \cdots & a_{nn} \end{vmatrix},$$

then

$$D = \begin{vmatrix} a_{11} & a_{12} & \cdots & a_{1n} \\ \vdots & \vdots & & \vdots \\ b_{i1} & b_{i2} & \cdots & b_{in} \\ \vdots & \vdots & & \vdots \\ a_{n1} & a_{n2} & \cdots & a_{nn} \end{vmatrix} + \begin{vmatrix} a_{11} & a_{12} & \cdots & a_{1n} \\ \vdots & \vdots & & \vdots \\ c_{i1} & c_{i2} & \cdots & c_{in} \\ \vdots & \vdots & & \vdots \\ a_{n1} & a_{n2} & \cdots & a_{nn} \end{vmatrix},$$

and if
$$D = \begin{vmatrix} a_{11} & \cdots & b_{1j}+c_{1j} & \cdots & a_{1n} \\ a_{21} & \cdots & b_{2j}+c_{2j} & \cdots & a_{2n} \\ \vdots & & \vdots & & \vdots \\ a_{n1} & \cdots & b_{nj}+c_{nj} & \cdots & a_{nn} \end{vmatrix},$$

then

$$D = \begin{vmatrix} a_{11} & \cdots & b_{1j} & \cdots & a_{1n} \\ a_{21} & \cdots & b_{2j} & \cdots & a_{2n} \\ \vdots & & \vdots & & \vdots \\ a_{n1} & \cdots & b_{nj} & \cdots & a_{nn} \end{vmatrix} + \begin{vmatrix} a_{11} & \cdots & c_{1j} & \cdots & a_{1n} \\ a_{21} & \cdots & c_{2j} & \cdots & a_{2n} \\ \vdots & & \vdots & & \vdots \\ a_{n1} & \cdots & c_{nj} & \cdots & a_{nn} \end{vmatrix}.$$

If each entry of one row (or column) of a determinant is multiplied by a real number k and the resulting product is added to the corresponding entry in another row (or column, respectively) in the determinant, the resulting determinant is equal to the original determinant.

[12.6] If

$$A = \begin{bmatrix} a_{11} & a_{12} & \cdots & a_{1n} \\ a_{21} & a_{22} & \cdots & a_{2n} \\ \vdots & \vdots & & \vdots \\ a_{n1} & a_{n2} & \cdots & a_{nn} \end{bmatrix},$$

and if $\delta(A) \neq 0$, then

$$A^{-1} = \frac{1}{\delta(A)} \begin{bmatrix} A_{11} & A_{21} & \cdots & A_{n1} \\ A_{12} & A_{22} & \cdots & A_{n2} \\ \vdots & \vdots & & \vdots \\ A_{1n} & A_{2n} & \cdots & A_{nn} \end{bmatrix},$$

where A_{ij} is the cofactor of a_{ij} in A. If $\delta(A) = 0$, then A has no inverse.

If A and B are $n \times n$ nonsingular square matrices, then AB has an inverse, namely,

$$(AB)^{-1} = B^{-1}A^{-1}.$$

[12.8] Cramer's Rule:

$$x = \frac{\delta(A_x)}{\delta(A)}, \quad y = \frac{\delta(A_y)}{\delta(A)}, \quad \text{and} \quad z = \frac{\delta(A_z)}{\delta(A)}, \quad (\delta(A) \neq 0).$$

[13.1] The nth term in the sequence defined by

$$s_1 = a, \quad s_{n+1} = s_n + d,$$

where $a, d \in R$, and $n \in N$, is

$$s_n = a + (n-1)d.$$

The nth term in the sequence defined by

$$s_1 = a, \qquad s_{n+1} = rs_n,$$

where $a, r \in R$, $a \neq 0$, $r \neq 0$, and $n \in N$, is

$$s_n = ar^{n-1}.$$

[13.2] The sum of the first n terms of an arithmetic progression is

$$S_n = \frac{n}{2}(a + s_n), \quad \text{or} \quad S_n = \frac{n}{2}[2a + (n-1)d].$$

The sum of the first n terms of a geometric progression is

$$S_n = \frac{a - ar^n}{1 - r}, \quad \text{or} \quad S_n = \frac{a - rs_n}{1 - r} \quad (r \neq 1).$$

[13.3] The sum of an infinite geometric progression, $a + ar + ar^2 + \cdots + ar^n + \cdots$, with $|r| < 1$, is

$$S_\infty = \lim_{n \to \infty} S_n = \frac{a}{1 - r}.$$

[13.4] For each natural number n,

$$(a+b)^n = a^n + \frac{n}{1!}a^{n-1}b + \frac{n(n-1)}{2!}a^{n-2}b^2 + \frac{n(n-1)(n-2)}{3!}a^{n-3}b^3 + \cdots$$

$$+ \frac{n(n-1)(n-2)\cdots(n-r+2)}{(r-1)!}a^{n-r+1}b^{r-1} + \cdots + b^n,$$

where r is the number of the term. Alternatively,

$$(a+b)^n = \binom{n}{0}a^n + \binom{n}{1}a^{n-1}b + \binom{n}{2}a^{n-2}b^2 + \binom{n}{3}a^{n-3}b^3 + \cdots$$

$$+ \binom{n}{r-1}a^{n-r+1}b^{r-1} + \cdots + \binom{n}{n}b^n.$$

The rth term in a binomial expansion is given by

$$\binom{n}{r-1}a^{n-r+1}b^{r-1} = \frac{n!}{(r-1)!(n-r+1)!}a^{n-r+1}b^{r-1}$$

$$= \frac{n(n-1)(n-2)(n-r+2)}{(r-1)!}a^{n-r+1}b^{r-1}.$$

[13.5] If a given open sentence involving n is true for $n = 1$, and if its truth for $n = k$ implies its truth for $n = k + 1$, then it is true for every natural number n.

[13.6] Counting properties:

I $n(A \cup B) = n(A) + n(B)$, if $A \cap B = \emptyset$,

II $n(A \cup B) = n(A) + n(B) - n(A \cap B)$,

III $n(A \times B) = n(A) \cdot n(B)$.

IV Suppose the first of several operations can be done in a ways, the second in b ways, no matter what came first, the third in c ways, no matter what came prior, and so on. Then the number of ways the operations can be done in sequence is $a \cdot b \cdot c \cdots$.

Let $P_{n,n}$ denote the number of distinct permutations of a set A, where $n(A) = n$. Then

$$P_{n,n} = n!.$$

[13.7] Let $P_{n,r}$ denote the number of permutations of the members, taken r at a time, of a set containing n members; that is, let $P_{n,r}$ be the number of distinct orderings of r elements when there is a set of n elements from which to choose. Then

$$P_{n,r} = n(n-1)(n-2)\cdots[n-(r-1)]$$
$$= n(n-1)(n-2)\cdots(n-r+1)$$
$$= \frac{n!}{(n-r)!}.$$

Let $\binom{n}{r}$ denote the number of distinct combinations of the members, taken r at a time, of a set containing n members. Then

$$\binom{n}{r} = \frac{P_{n,r}}{r!} = \frac{n!}{r!(n-r)!};$$

$$\binom{n}{r} = \binom{n}{n-r}.$$

[14.1] If $P(x)$ is a polynomial over the field C of complex numbers and $c \in C$, then there exists a unique polynomial $Q(x)$ and a complex number r, such that

$$P(x) = (x-c)Q(x) + r \quad \text{and} \quad r = P(c).$$

If $P(x)$ is a polynomial over the field C of complex numbers, and $P(r) = 0$, then $(x - r)$ is a factor of $P(x)$.

[14.2] If $P(z)$ is a polynomial over the field R of real numbers, and $P(z) = 0$ for some $z \in C$, then $P(\bar{z}) = 0$.

Every polynomial function of degree $n \geq 1$ over the field C of complex numbers has at least one complex zero.

If $P(z)$ is a polynomial of degree $n \geq 1$ over the field C of complex numbers, then $P(z)$ can be expressed as a product of a constant and n linear factors of the form $(z - z_k)$.

[14.3]   If $P(x) = a_0 x^n + a_1 x^{n-1} + \cdots + a_n$, $a_j, x \in R$, and if $k \in R$ is between $P(x_1)$ and $P(x_2)$, then there exists at least one $c \in R$ between x_1 and x_2 such that $P(c) = k$.

If $P(x)$ is a polynomial over the field R of real numbers, then the number of positive real zeros of $P(x)$ either is equal to the number of variations in sign occurring in the coefficients of $P(x)$ or else is less than this number by an even natural number. Moreover, the number of negative real solutions of $P(x) = 0$ either is equal to the number of variations in sign occurring in $P(-x)$ or else is less than this number by an even natural number.

Let $P(x)$ be a polynomial over the field R of real numbers.

I   If $r_1 \geq 0$ and the coefficients of the terms in $Q(x)$ and the term $P(r_1)$ are all of the same sign in the right-hand member of

$$P(x) = (x - r_1)Q(x) + P(r_1),$$

then $P(x) = 0$ can have no solution greater than r_1.

II   If $r_2 \leq 0$ and the coefficients of the terms in $Q(x)$ and the term $P(r_2)$ alternate in sign (zero suitably denoted by $+0$ and -0) in the right-hand member of

$$P(x) = (x - r_2)Q(x) + P(r_2),$$

then $P(x) = 0$ can have no solution less than r_2.

Let $P(x)$ be a polynomial over the field R of real numbers. If $x_1, x_2 \in R$, with $x_1 < x_2$, and $P(x_1)$ and $P(x_2)$ are opposite in sign, then there exists at least one $c \in R$, $x_1 < c < x_2$, such that $P(c) = 0$.

[14.4]   If a is a composite number, then a is the product of only one set of prime factors; that is, the prime factorization of a is unique except for the ordering of the factors.

If the rational number p/q, in lowest terms, is a solution of

$$P(x) = a_0 x^n + a_1 x^{n-1} + \cdots + a_n = 0,$$

where $a_j \in J$, then p is an integral factor of a_n and q is an integral factor of a_0.

[A.1]   If r is the modulus and θ is an argument of the complex number $a + bi$, then

$$a + bi = r[\cos(\theta + k360°) + i \sin(\theta + k360°)], \quad k \in J.$$

If $z_1 = r_1 \text{ cis } \theta_1$ and $z_2 = r_2 \text{ cis } \theta_2$, then

I. $z_1 \cdot z_2 = r_1 r_2 \text{ cis } (\theta_1 + \theta_2)$

II. $\dfrac{z_1}{z_2} = \dfrac{r_1}{r_2} \text{ cis } (\theta_1 - \theta_2) \quad (z_2 \neq 0 + 0i).$

[A.2]   If $z \in C$, $z = r \text{ cis } \theta$ and $n \in J$, then

$$z^n = r^n \text{ cis } n\theta. \quad \text{(De Moivre's Theorem)}$$

Mathematical Structure

Mathematical Systems

Three operations appear over and over in the algebraic systems of this text, operations called "addition," "multiplication," and "multiplication by a scalar." Operations having the same properties as these occur also in many other systems in pure and applied mathematics.

As a consequence, theorems and the relevant skills developed for one of the systems carry over to any other system whose operations have properties that include the basic properties of the analogous operations in the first system.

Postulates F-1 to F-11 listed on pages 450, 451 are the basic properties for addition and multiplication. Some important systems for which various ones of these postulates hold appear in the following table.

System	Binary Operations	Postulates
Group	one	1, 2, 3, 4
Abelian group	one	1, 2, 3, 4, 5
Ring	two	1, 2, 3, 4, 5, 6, 7, 11
Ring with identity	two	1, 2, 3, 4, 5, 6, 7, 11, 8
Commutative ring with identity	two	1, 2, 3, 4, 5, 6, 7, 11, 8, 10
Integral domain	two	1, 2, 3, 4, 5, 6, 7, 11, 8, 10, 9'
Field	two	1, 2, 3, 4, 5, 6, 7, 11, 8, 10, 9

Postulate 9', which applies to an integral domain, is a corollary of Postulate 9. It is stated as follows:

If $a \cdot c = b \cdot c$ (or $c \cdot a = c \cdot b$) and $c \neq 0$, then $a = b$.

The above listing of systems and the postulates which hold in the respective system shows that every field is an integral domain, every integral domain is a ring, every ring is an Abelian group, and every Albelian group is a group.

Some of the above systems also satisfy one or more of the order postulates listed on page 451. A system satisfying Postulates O-1 and O-2 is said to be *ordered*. If it also satisfies O-3 it is said to be *completely ordered*. The **set of complex numbers** constitutes a field but does not satisfy the order postulates; it is not an ordered field. The **set of rational numbers** is a field satisfying Postulates O-1 and O-2 but not O-3; it is an ordered field. The **set of real numbers** is also an ordered field, and it satisfies Postulate O-3; it is a complete ordered field, and is the only complete ordered field in the sense that any other complete ordered field would have to be "just like" (isomorphic with) the complete field of real numbers.

There are still other algebraic systems, with different sets of postulates, which we have not considered in this text. For example, in Boolean algebra, which is important in logic and in the theory of switching circuits, not only does multiplication distribute over addition,

$$A \times (B + C) = (A \times B) + (A \times C),$$

but also addition distributes over multiplication,

$$A + (B \times C) = (A + B) \times (A + C).$$

C *Tables*

Table I Common Logarithms

x	0	1	2	3	4	5	6	7	8	9
1.0	.0000	.0043	.0086	.0128	.0170	.0212	.0253	.0294	.0334	.0374
1.1	.0414	.0453	.0492	.0531	.0569	.0607	.0645	.0682	.0719	.0755
1.2	.0792	.0828	.0864	.0899	.0934	.0969	.1004	.1038	.1072	.1106
1.3	.1139	.1173	.1206	.1239	.1271	.1303	.1335	.1367	.1399	.1430
1.4	.1461	.1492	.1523	.1553	.1584	.1614	.1644	.1673	.1703	.1732
1.5	.1761	.1790	.1818	.1847	.1875	.1903	.1931	.1959	.1987	.2014
1.6	.2041	.2068	.2095	.2122	.2148	.2175	.2201	.2227	.2253	.2279
1.7	.2304	.2330	.2355	.2380	.2405	.2430	.2455	.2480	.2504	.2529
1.8	.2553	.2577	.2601	.2625	.2648	.2672	.2695	.2718	.2742	.2765
1.9	.2788	.2810	.2833	.2856	.2878	.2900	.2923	.2945	.2967	.2989
2.0	.3010	.3032	.3054	.3075	.3096	.3118	.3139	.3160	.3181	.3201
2.1	.3222	.3243	.3263	.3284	.3304	.3324	.3345	.3365	.3385	.3404
2.2	.3424	.3444	.3464	.3483	.3502	.3522	.3541	.3560	.3579	.3598
2.3	.3617	.3636	.3655	.3674	.3692	.3711	.3729	.3747	.3766	.3784
2.4	.3802	.3820	.3838	.3856	.3874	.3892	.3909	.3927	.3945	.3962
2.5	.3979	.3997	.4014	.4031	.4048	.4065	.4082	.4099	.4116	.4133
2.6	.4150	.4166	.4183	.4200	.4216	.4232	.4249	.4265	.4281	.4298
2.7	.4314	.4330	.4346	.4362	.4378	.4393	.4409	.4425	.4440	.4456
2.8	.4472	.4487	.4502	.4518	.4533	.4548	.4564	.4579	.4594	.4609
2.9	.4624	.4639	.4654	.4669	.4683	.4698	.4713	.4728	.4742	.4757
3.0	.4771	.4786	.4800	.4814	.4829	.4843	.4857	.4871	.4886	.4900
3.1	.4914	.4928	.4942	.4955	.4969	.4983	.4997	.5011	.5024	.5038
3.2	.5051	.5065	.5079	.5092	.5105	.5119	.5132	.5145	.5159	.5172
3.3	.5185	.5198	.5211	.5224	.5237	.5250	.5263	.5276	.5289	.5302
3.4	.5315	.5328	.5340	.5353	.5366	.5378	.5391	.5403	.5416	.5428
3.5	.5441	.5453	.5465	.5478	.5490	.5502	.5514	.5527	.5539	.5551
3.6	.5563	.5575	.5587	.5599	.5611	.5623	.5635	.5647	.5658	.5670
3.7	.5682	.5694	.5705	.5717	.5729	.5740	.5752	.5763	.5775	.5786
3.8	.5798	.5809	.5821	.5832	.5843	.5855	.5866	.5877	.5888	.5899
3.9	.5911	.5922	.5933	.5944	.5955	.5966	.5977	.5988	.5999	.6010
4.0	.6021	.6031	.6042	.6053	.6064	.6075	.6085	.6096	.6107	.6117
4.1	.6128	.6138	.6149	.6160	.6170	.6180	.6191	.6201	.6212	.6222
4.2	.6232	.6243	.6253	.6263	.6274	.6284	.6294	.6304	.6314	.6325
4.3	.6335	.6345	.6355	.6365	.6375	.6385	.6395	.6405	.6415	.6425
4.4	.6435	.6444	.6454	.6464	.6474	.6484	.6493	.6503	.6513	.6522
4.5	.6532	.6542	.6551	.6561	.6571	.6580	.6590	.6599	.6609	.6618
4.6	.6628	.6637	.6646	.6656	.6665	.6675	.6684	.6693	.6702	.6712
4.7	.6721	.6730	.6739	.6749	.6758	.6767	.6776	.6785	.6794	.6803
4.8	.6812	.6821	.6830	.6839	.6848	.6857	.6866	.6875	.6884	.6893
4.9	.6902	.6911	.6920	.6928	.6937	.6946	.6955	.6964	.6972	.6981
5.0	.6990	.6998	.7007	.7016	.7024	.7033	.7042	.7050	.7059	.7067
5.1	.7076	.7084	.7093	.7101	.7110	.7118	.7126	.7135	.7143	.7152
5.2	.7160	.7168	.7177	.7185	.7193	.7202	.7210	.7218	.7226	.7235
5.3	.7243	.7251	.7259	.7267	.7275	.7284	.7292	.7300	.7308	.7316
5.4	.7324	.7332	.7340	.7348	.7356	.7364	.7372	.7380	.7388	.7396
x	0	1	2	3	4	5	6	7	8	9

Table I *(Continued)*

x	0	1	2	3	4	5	6	7	8	9
5.5	.7404	.7412	.7419	.7427	.7435	.7443	.7451	.7459	.7466	.7474
5.6	.7482	.7490	.7497	.7505	.7513	.7520	.7528	.7536	.7543	.7551
5.7	.7559	.7566	.7574	.7582	.7589	.7597	.7604	.7612	.7619	.7627
5.8	.7634	.7642	.7649	.7657	.7664	.7672	.7679	.7686	.7694	.7701
5.9	.7709	.7716	.7723	.7731	.7738	.7745	.7752	.7760	.7767	.7774
6.0	.7782	.7789	.7796	.7803	.7810	.7818	.7825	.7832	.7839	.7846
6.1	.7853	.7860	.7868	.7875	.7882	.7889	.7896	.7903	.7910	.7917
6.2	.7924	.7931	.7938	.7945	.7952	.7959	.7966	.7973	.7980	.7987
6.3	.7993	.8000	.8007	.8014	.8021	.8028	.8035	.8041	.8048	.8055
6.4	.8062	.8069	.8075	.8082	.8089	.8096	.8102	.8109	.8116	.8122
6.5	.8129	.8136	.8142	.8149	.8156	.8162	.8169	.8176	.8182	.8189
6.6	.8195	.8202	.8209	.8215	.8222	.8228	.8235	.8241	.8248	.8254
6.7	.8261	.8267	.8274	.8280	.8287	.8293	.8299	.8306	.8312	.8319
6.8	.8325	.8331	.8338	.8344	.8351	.8357	.8363	.8370	.8376	.8382
6.9	.8388	.8395	.8401	.8407	.8414	.8420	.8426	.8432	.8439	.8445
7.0	.8451	.8457	.8463	.8470	.8476	.8482	.8488	.8494	.8500	.8506
7.1	.8513	.8519	.8525	.8531	.8537	.8543	.8549	.8555	.8561	.8567
7.2	.8573	.8579	.8585	.8591	.8597	.8603	.8609	.8615	.8621	.8627
7.3	.8633	.8639	.8645	.8651	.8657	.8663	.8669	.8675	.8681	.8686
7.4	.8692	.8698	.8704	.8710	.8716	.8722	.8727	.8733	.8739	.8745
7.5	.8751	.8756	.8762	.8768	.8774	.8779	.8785	.8791	.8797	.8802
7.6	.8808	.8814	.8820	.8825	.8831	.8837	.8842	.8848	.8854	.8859
7.7	.8865	.8871	.8876	.8882	.8887	.8893	.8899	.8904	.8910	.8915
7.8	.8921	.8927	.8932	.8938	.8943	.8949	.8954	.8960	.8965	.8971
7.9	.8976	.8982	.8987	.8993	.8998	.9004	.9009	.9015	.9020	.9025
8.0	.9031	.9036	.9042	.9047	.9053	.9058	.9063	.9069	.9074	.9079
8.1	.9085	.9090	.9096	.9101	.9106	.9112	.9117	.9122	.9128	.9133
8.2	.9138	.9143	.9149	.9154	.9159	.9165	.9170	.9175	.9180	.9186
8.3	.9191	.9196	.9201	.9206	.9212	.9217	.9222	.9227	.9232	.9238
8.4	.9243	.9248	.9253	.9258	.9263	.9269	.9274	.9279	.9284	.9289
8.5	.9294	.9299	.9304	.9309	.9315	.9320	.9325	.9330	.9335	.9340
8.6	.9345	.9350	.9355	.9360	.9365	.9370	.9375	.9380	.9385	.9390
8.7	.9395	.9400	.9405	.9410	.9415	.9420	.9425	.9430	.9435	.9440
8.8	.9445	.9450	.9455	.9460	.9465	.9469	.9474	.9479	.9484	.9489
8.9	.9494	.9499	.9504	.9509	.9513	.9518	.9523	.9528	.9533	.9538
9.0	.9542	.9547	.9552	.9557	.9562	.9566	.9571	.9576	.9581	.9586
9.1	.9590	.9595	.9600	.9605	.9609	.9614	.9619	.9624	.9628	.9633
9.2	.9638	.9643	.9647	.9652	.9657	.9661	.9666	.9671	.9675	.9680
9.3	.9685	.9689	.9694	.9699	.9703	.9708	.9713	.9717	.9722	.9727
9.4	.9731	.9736	.9741	.9745	.9750	.9754	.9759	.9763	.9768	.9773
9.5	.9777	.9782	.9786	.9791	.9795	.9800	.9805	.9809	.9814	.9818
9.6	.9823	.9827	.9832	.9836	.9841	.9845	.9850	.9854	.9859	.9863
9.7	.9868	.9872	.9877	.9881	.9886	.9890	.9894	.9899	.9903	.9908
9.8	.9912	.9917	.9921	.9926	.9930	.9934	.9939	.9943	.9948	.9952
9.9	.9956	.9961	.9965	.9969	.9974	.9978	.9983	.9987	.9991	.9996
x	0	1	2	3	4	5	6	7	8	9

Table II Exponential Functions

x	e^x	e^{-x}	x	e^x	e^{-x}
0.00	1.0000	1.0000	1.5	4.4817	0.2231
0.01	1.0101	0.9901	1.6	4.9530	0.2019
0.02	1.0202	0.9802	1.7	5.4739	0.1827
0.03	1.0305	0.9705	1.8	6.0496	0.1653
0.04	1.0408	0.9608	1.9	6.6859	0.1496
0.05	1.0513	0.9512	2.0	7.3891	0.1353
0.06	1.0618	0.9418	2.1	8.1662	0.1225
0.07	1.0725	0.9324	2.2	9.0250	0.1108
0.08	1.0833	0.9331	2.3	9.9742	0.1003
0.09	1.0942	0.9139	2.4	11.023	0.0907
0.10	1.1052	0.9048	2.5	12.182	0.0821
0.11	1.1163	0.8958	2.6	13.464	0.0743
0.12	1.1275	0.8869	2.7	14.880	0.0672
0.13	1.1388	0.8781	2.8	16.445	0.0608
0.14	1.1503	0.8694	2.9	18.174	0.0550
0.15	1.1618	0.8607	3.0	20.086	0.0498
0.16	1.1735	0.8521	3.1	22.198	0.0450
0.17	1.1853	0.8437	3.2	24.533	0.0408
0.18	1.1972	0.8353	3.3	27.113	0.0369
0.19	1.2092	0.8270	3.4	29.964	0.0334
0.20	1.2214	0.8187	3.5	33.115	0.0302
0.21	1.2337	0.8106	3.6	36.598	0.0273
0.22	1.2461	0.8025	3.7	40.447	0.0247
0.23	1.2586	0.7945	3.8	44.701	0.0224
0.24	1.2712	0.7866	3.9	49.402	0.0202
0.25	1.2840	0.7788	4.0	54.598	0.0183
0.30	1.3499	0.7408	4.1	60.340	0.0166
0.35	1.4191	0.7047	4.2	66.686	0.0150
0.40	1.4918	0.6703	4.3	73.700	0.0136
0.45	1.5683	0.6376	4.4	81.451	0.0123
0.50	1.6487	0.6065	4.5	90.017	0.0111
0.55	1.7333	0.5769	4.6	99.484	0.0101
0.60	1.8221	0.5488	4.7	109.95	0.0091
0.65	1.9155	0.5220	4.8	121.51	0.0082
0.70	2.0138	0.4966	4.9	134.29	0.0074
0.75	2.1170	0.4724	5.0	148.41	0.0067
0.80	2.2255	0.4493	5.5	244.69	0.0041
0.85	2.3396	0.4274	6.0	403.43	0.0025
0.90	2.4596	0.4066	6.5	665.14	0.0015
0.95	2.5857	0.3867	7.0	1096.6	0.0009
1.0	2.7183	0.3679	7.5	1808.0	0.0006
1.1	3.0042	0.3329	8.0	2981.0	0.0003
1.2	3.3201	0.3012	8.5	4914.8	0.0002
1.3	3.6693	0.2725	9.0	8103.1	0.0001
1.4	4.0552	0.2466	10.0	22026	0.00005

Table III   Natural Logarithms of Numbers

n	$\log_e n$	n	$\log_e n$	n	$\log_e n$
	*	4.5	1.5041	9.0	2.1972
0.1	7.6974	4.6	1.5261	9.1	2.2083
0.2	8.3906	4.7	1.5476	9.2	2.2192
0.3	8.7960	4.8	1.5686	9.3	2.2300
0.4	9.0837	4.9	1.5892	9.4	2.2407
0.5	9.3069	5.0	1.6094	9.5	2.2513
0.6	9.4892	5.1	1.6292	9.6	2.2618
0.7	9.6433	5.2	1.6487	9.7	2.2721
0.8	9.7769	5.3	1.6677	9.8	2.2824
0.9	9.8946	5.4	1.6864	9.9	2.2925
1.0	0.0000	5.5	1.7047	10	2.3026
1.1	0.0953	5.6	1.7228	11	2.3979
1.2	0.1823	5.7	1.7405	12	2.4849
1.3	0.2624	5.8	1.7579	13	2.5649
1.4	0.3365	5.9	1.7750	14	2.6391
1.5	0.4055	6.0	1.7918	15	2.7081
1.6	0.4700	6.1	1.8083	16	2.7726
1.7	0.5306	6.2	1.8245	17	2.8332
1.8	0.5878	6.3	1.8405	18	2.8904
1.9	0.6419	6.4	1.8563	19	2.9444
2.0	0.6931	6.5	1.8718	20	2.9957
2.1	0.7419	6.6	1.8871	25	3.2189
2.2	0.7885	6.7	1.9021	30	3.4012
2.3	0.8329	6.8	1.9169	35	3.5553
2.4	0.8755	6.9	1.9315	40	3.6889
2.5	0.9163	7.0	1.9459	45	3.8067
2.6	0.9555	7.1	1.9601	50	3.9120
2.7	0.9933	7.2	1.9741	55	4.0073
2.8	1.0296	7.3	1.9879	60	4.0943
2.9	1.0647	7.4	2.0015	65	4.1744
3.0	1.0986	7.5	2.0149	70	4.2485
3.1	1.1314	7.6	2.0281	75	4.3175
3.2	1.1632	7.7	2.0412	80	4.3820
3.3	1.1939	7.8	2.0541	85	4.4427
3.4	1.2238	7.9	2.0669	90	4.4998
3.5	1.2528	8.0	2.0794	100	4.6052
3.6	1.2809	8.1	2.0919	110	4.7005
3.7	1.3083	8.2	2.1041	120	4.7875
3.8	1.3350	8.3	2.1163	130	4.8676
3.9	1.3610	8.4	2.1282	140	4.9416
4.0	1.3863	8.5	2.1401	150	5.0106
4.1	1.4110	8.6	2.1518	160	5.0752
4.2	1.4351	8.7	2.1633	170	5.1358
4.3	1.4586	8.8	2.1748	180	5.1930
4.4	1.4816	8.9	2.1861	190	5.2470

* Subtract 10 for $n < 1$. Thus $\log_e 0.1 = 7.6974 - 10 = -2.3026$.

Table IV Squares, Square Roots, and Prime Factors

No.	Sq.	Sq. Root	Prime Factors	No.	Sq.	Sq. Root	Prime Factors
1	1	1.000		51	2,601	7.141	$3 \cdot 17$
2	4	1.414	2	52	2,704	7.211	$2^2 \cdot 13$
3	9	1.732	3	53	2,809	7.280	53
4	16	2.000	2^2	54	2,916	7.348	$2 \cdot 3^3$
5	25	2.236	5	55	3,025	7.416	$5 \cdot 11$
6	36	2.449	$2 \cdot 3$	56	3,136	7.483	$2^3 \cdot 7$
7	49	2.646	7	57	3,249	7.550	$3 \cdot 19$
8	64	2.828	2^3	58	3,364	7.616	$2 \cdot 29$
9	81	3.000	3^2	59	3,481	7.681	59
10	100	3.162	$2 \cdot 5$	60	3,600	7.746	$2^2 \cdot 3 \cdot 5$
11	121	3.317	11	61	3,721	7.810	61
12	144	3.464	$2^2 \cdot 3$	62	3,844	7.874	$2 \cdot 31$
13	169	3.606	13	63	3,969	7.937	$3^2 \cdot 7$
14	196	3.742	$2 \cdot 7$	64	4,096	8.000	2^6
15	225	3.873	$3 \cdot 5$	65	4,225	8.062	$5 \cdot 13$
16	256	4.000	2^4	66	4,356	8.124	$2 \cdot 3 \cdot 11$
17	289	4.123	17	67	4,489	8.185	67
18	324	4.243	$2 \cdot 3^2$	68	4,624	8.246	$2^2 \cdot 17$
19	361	4.359	19	69	4,761	8.307	$3 \cdot 23$
20	400	4.472	$2^2 \cdot 5$	70	4,900	8.367	$2 \cdot 5 \cdot 7$
21	441	4.583	$3 \cdot 7$	71	5,041	8.426	71
22	484	4.690	$2 \cdot 11$	72	5,184	8.485	$2^3 \cdot 3^2$
23	529	4.796	23	73	5,329	8.544	73
24	576	4.899	$2^3 \cdot 3$	74	5,476	8.602	$2 \cdot 37$
25	625	5.000	5^2	75	5,625	8.660	$3 \cdot 5^2$
26	676	5.099	$2 \cdot 13$	76	5,776	8.718	$2^2 \cdot 19$
27	729	5.196	3^3	77	5,929	8.775	$7 \cdot 11$
28	784	5.292	$2^2 \cdot 7$	78	6,084	8.832	$2 \cdot 3 \cdot 13$
29	841	5.385	29	79	6,241	8.888	79
30	900	5.477	$2 \cdot 3 \cdot 5$	80	6,400	8.944	$2^4 \cdot 5$
31	961	5.568	31	81	6,561	9.000	3^4
32	1,024	5.657	2^5	82	6,724	9.055	$2 \cdot 41$
33	1,089	5.745	$3 \cdot 11$	83	6,889	9.110	83
34	1,156	5.831	$2 \cdot 17$	84	7,056	9.165	$2^2 \cdot 3 \cdot 7$
35	1,225	5.916	$5 \cdot 7$	85	7,225	9.220	$5 \cdot 17$
36	1,296	6.000	$2^2 \cdot 3^2$	86	7,396	9.274	$2 \cdot 43$
37	1,369	6.083	37	87	7,569	9.327	$3 \cdot 29$
38	1,444	6.164	$2 \cdot 19$	88	7,744	9.381	$2^3 \cdot 11$
39	1,521	6.245	$3 \cdot 13$	89	7,921	9.434	89
40	1,600	6.325	$2^3 \cdot 5$	90	8,100	9.487	$2 \cdot 3^2 \cdot 5$
41	1,681	6.403	41	91	8,281	9.539	$7 \cdot 13$
42	1,764	6.481	$2 \cdot 3 \cdot 7$	92	8,464	9.592	$2^2 \cdot 23$
43	1,849	6.557	43	93	8,649	9.644	$3 \cdot 31$
44	1,936	6.633	$2^2 \cdot 11$	94	8,836	9.695	$2 \cdot 47$
45	2,025	6.708	$3^2 \cdot 5$	95	9,025	9.747	$5 \cdot 19$
46	2,116	6.782	$2 \cdot 23$	96	9,216	9.798	$2^5 \cdot 3$
47	2,209	6.856	47	97	9,409	9.849	97
48	2,304	6.928	$2^4 \cdot 3$	98	9,604	9.899	$2 \cdot 7^2$
49	2,401	7.000	7^2	99	9,801	9.950	$3^2 \cdot 11$
50	2,500	7.071	$2 \cdot 5^2$	100	10,000	10.000	$2^2 \cdot 5^2$

Table V Values of Circular Functions

Real Number x or θ radians	θ degrees	sin x or sin θ	csc x or csc θ	tan x or tan θ	cot x or cot θ	sec x or sec θ	cos x or cos θ
0.00	0° 00′	0.0000	No value	0.0000	No value	1.000	1.000
.01	0° 34′	.0100	100.0	.0100	100.0	1.000	1.000
.02	1° 09′	.0200	50.00	.0200	49.99	1.000	0.9998
.03	1° 43′	.0300	33.34	.0300	33.32	1.000	0.9996
.04	2° 18′	.0400	25.01	.0400	24.99	1.001	0.9992
0.05	2° 52′	0.0500	20.01	0.0500	19.98	1.001	0.9988
.06	3° 26′	.0600	16.68	.0601	16.65	1.002	.9982
.07	4° 01′	.0699	14.30	.0701	14.26	1.002	.9976
.08	4° 35′	.0799	12.51	.0802	12.47	1.003	.9968
.09	5° 09′	.0899	11.13	.0902	11.08	1.004	.9960
0.10	5° 44′	0.0998	10.02	0.1003	9.967	1.005	0.9950
.11	6° 18′	.1098	9.109	.1104	9.054	1.006	.9940
.12	6° 53′	.1197	8.353	.1206	8.293	1.007	.9928
.13	7° 27′	.1296	7.714	.1307	7.649	1.009	.9916
.14	8° 01′	.1395	7.166	.1409	7.096	1.010	.9902
0.15	8° 36′	0.1494	6.692	0.1511	6.617	1.011	0.9888
.16	9° 10′	.1593	6.277	.1614	6.197	1.013	.9872
.17	9° 44′	.1692	5.911	.1717	5.826	1.015	.9856
.18	10° 19′	.1790	5.586	.1820	5.495	1.016	.9838
.19	10° 53′	.1889	5.295	.1923	5.200	1.018	.9820
0.20	11° 28′	0.1987	5.033	0.2027	4.933	1.020	0.9801
.21	12° 02′	.2085	4.797	.2131	4.692	1.022	.9780
.22	12° 36′	.2182	4.582	.2236	4.472	1.025	.9759
.23	13° 11′	.2280	4.386	.2341	4.271	1.027	.9737
.24	13° 45′	.2377	4.207	.2447	4.086	1.030	.9713
0.25	14° 19′	0.2474	4.042	0.2553	3.916	1.032	0.9689
.26	14° 54′	.2571	3.890	.2660	3.759	1.035	.9664
.27	15° 28′	.2667	3.749	.2768	3.613	1.038	.9638
.28	16° 03′	.2764	3.619	.2876	3.478	1.041	.9611
.29	16° 37′	.2860	3.497	.2984	3.351	1.044	.9582
0.30	17° 11′	0.2955	3.384	0.3093	3.233	1.047	0.9553
.31	17° 46′	.3051	3.278	.3203	3.122	1.050	.9523
.32	18° 20′	.3146	3.179	.3314	3.018	1.053	.9492
.33	18° 54′	.3240	3.086	.3425	2.920	1.057	.9460
.34	19° 29′	.3335	2.999	.3537	2.827	1.061	.9428
0.35	20° 03′	0.3429	2.916	0.3650	2.740	1.065	0.9394
.36	20° 38′	.3523	2.839	.3764	2.657	1.068	.9359
.37	21° 12′	.3616	2.765	.3879	2.578	1.073	.9323
.38	21° 46′	.3709	2.696	.3994	2.504	1.077	.9287
.39	22° 21′	.3802	2.630	.4111	2.433	1.081	.9249
0.40	22° 55′	0.3894	2.568	0.4228	2.365	1.086	0.9211
.41	23° 29′	.3986	2.509	.4346	2.301	1.090	.9171
.42	24° 04′	.4078	2.452	.4466	2.239	1.095	.9131
.43	24° 38′	.4169	2.399	.4586	2.180	1.100	.9090
.44	25° 13′	.4259	2.348	.4708	2.124	1.105	.9048
0.45	25° 47′	0.4350	2.299	0.4831	2.070	1.111	0.9004

Table V (*continued*)

Real Number x or θ radians	θ degrees	$\sin x$ or $\sin \theta$	$\csc x$ or $\csc \theta$	$\tan x$ or $\tan \theta$	$\cot x$ or $\cot \theta$	$\sec x$ or $\sec \theta$	$\cos x$ or $\cos \theta$
0.45	25° 47′	0.4350	2.299	0.4831	2.070	1.111	0.9004
.46	26° 21′	.4439	2.253	.4954	2.018	1.116	.8961
.47	26° 56′	.4529	2.208	.5080	1.969	1.122	.8916
.48	27° 30′	.4618	2.166	.5206	1.921	1.127	.8870
.49	28° 04′	.4706	2.125	.5334	1.875	1.133	.8823
0.50	28° 39′	0.4794	2.086	0.5463	1.830	1.139	0.8776
.51	29° 13′	.4882	2.048	.5594	1.788	1.146	.8727
.52	29° 48′	.4969	2.013	.5726	1.747	1.152	.8678
.53	30° 22′	.5055	1.978	.5859	1.707	1.159	.8628
.54	30° 56′	.5141	1.945	.5994	1.668	1.166	.8577
0.55	31° 31′	0.5227	1.913	0.6131	1.631	1.173	0.8525
.56	32° 05′	.5312	1.883	.6269	1.595	1.180	.8473
.57	32° 40′	.5396	1.853	.6410	1.560	1.188	.8419
.58	33° 14′	.5480	1.825	.6552	1.526	1.196	.8365
.59	33° 48′	.5564	1.797	.6696	1.494	1.203	.8309
0.60	34° 23′	0.5646	1.771	0.6841	1.462	1.212	0.8253
.61	34° 57′	.5729	1.746	.6989	1.431	1.220	.8196
.62	35° 31′	.5810	1.721	.7139	1.401	1.229	.8139
.63	36° 06′	.5891	1.697	.7291	1.372	1.238	.8080
.64	36° 40′	.5972	1.674	.7445	1.343	1.247	.8021
0.65	37° 15′	0.6052	1.652	0.7602	1.315	1.256	0.7961
.66	37° 49′	.6131	1.631	.7761	1.288	1.266	.7900
.67	38° 23′	.6210	1.610	.7923	1.262	1.276	.7838
.68	38° 58′	.6288	1.590	.8087	1.237	1.286	.7776
.69	39° 32′	.6365	1.571	.8253	1.212	1.297	.7712
0.70	40° 06′	0.6442	1.552	0.8423	1.187	1.307	0.7648
.71	40° 41′	.6518	1.534	.8595	1.163	1.319	.7584
.72	41° 15′	.6594	1.517	.8771	1.140	1.330	.7518
.73	41° 50′	.6669	1.500	.8949	1.117	1.342	.7452
.74	42° 24′	.6743	1.483	.9131	1.095	1.354	.7385
0.75	42° 58′	0.6816	1.467	0.9316	1.073	1.367	0.7317
.76	43° 33′	.6889	1.452	.9505	1.052	1.380	.7248
.77	44° 07′	.6961	1.436	.9697	1.031	1.393	.7179
.78	44° 41′	.7033	1.422	.9893	1.011	1.407	.7109
.79	45° 16′	.7104	1.408	1.009	.9908	1.421	.7038
0.80	45° 50′	0.7174	1.394	1.030	0.9712	1.435	0.6967
.81	46° 25′	.7243	1.381	1.050	.9520	1.450	.6895
.82	46° 59′	.7311	1.368	1.072	.9331	1.466	.6822
.83	47° 33′	.7379	1.355	1.093	.9146	1.482	.6749
.84	48° 08′	.7446	1.343	1.116	.8964	1.498	.6675
0.85	48° 42′	0.7513	1.331	1.138	0.8785	1.515	0.6600
.86	49° 16′	.7578	1.320	1.162	.8609	1.533	.6524
.87	49° 51′	.7643	1.308	1.185	.8437	1.551	.6448
.88	50° 25′	.7707	1.297	1.210	.8267	1.569	.6372
.89	51° 00′	.7771	1.287	1.235	.8100	1.589	.6294
0.90	51° 34′	0.7833	1.277	1.260	0.7936	1.609	0.6216
.91	52° 08′	.7895	1.267	1.286	.7774	1.629	.6137
.92	52° 43′	.7956	1.257	1.313	.7615	1.651	.6058
.93	53° 17′	.8016	1.247	1.341	.7458	1.673	.5978
.94	53° 51′	.8076	1.238	1.369	.7303	1.696	.5898
0.95	54° 26′	0.8134	1.229	1.398	0.7151	1.719	0.5817

Table V (continued)

Real Number x or θ radians	θ degrees	sin x or sin θ	csc x or csc θ	tan x or tan θ	cot x or cot θ	sec x or sec θ	cos x or cos θ
0.95	54° 26′	0.8134	1.229	1.398	0.7151	1.719	0.5817
.96	55° 00′	.8192	1.221	1.428	.7001	1.744	.5735
.97	55° 35′	.8249	1.212	1.459	.6853	1.769	.5653
.98	56° 09′	.8305	1.204	1.491	.6707	1.795	.5570
.99	56° 43′	.8360	1.196	1.524	.6563	1.823	.5487
1.00	57° 18′	0.8415	1.188	1.557	0.6421	1.851	0.5403
1.01	57° 52′	.8468	1.181	1.592	.6281	1.880	.5319
1.02	58° 27′	.8521	1.174	1.628	.6142	1.911	.5234
1.03	59° 01′	.8573	1.166	1.665	.6005	1.942	.5148
1.04	59° 35′	.8624	1.160	1.704	.5870	1.975	.5062
1.05	60° 10′	0.8674	1.153	1.743	0.5736	2.010	0.4976
1.06	60° 44′	.8724	1.146	1.784	.5604	2.046	.4889
1.07	61° 18′	.8772	1.140	1.827	.5473	2.083	.4801
1.08	61° 53′	.8820	1.134	1.871	.5344	2.122	.4713
1.09	62° 27′	.8866	1.128	1.917	.5216	2.162	.4625
1.10	63° 02′	0.8912	1.122	1.965	0.5090	2.205	0.4536
1.11	63° 36′	.8957	1.116	2.014	.4964	2.249	.4447
1.12	64° 10′	.9001	1.111	2.066	.4840	2.295	.4357
1.13	64° 45′	.9044	1.106	2.120	.4718	2.344	.4267
1.14	65° 19′	.9086	1.101	2.176	.4596	2.395	.4176
1.15	65° 53′	0.9128	1.096	2.234	0.4475	2.448	0.4085
1.16	66° 28′	.9168	1.091	2.296	.4356	2.504	.3993
1.17	67° 02′	.9208	1.086	2.360	.4237	2.563	.3902
1.18	67° 37′	.9246	1.082	2.427	.4120	2.625	.3809
1.19	68° 11′	.9284	1.077	2.498	.4003	2.691	.3717
1.20	68° 45′	0.9320	1.073	2.572	0.3888	2.760	0.3624
1.21	69° 20′	.9356	1.069	2.650	.3773	2.833	.3530
1.22	69° 54′	.9391	1.065	2.733	.3659	2.910	.3436
1.23	70° 28′	.9425	1.061	2.820	.3546	2.992	.3342
1.24	71° 03′	.9458	1.057	2.912	.3434	3.079	.3248
1.25	71° 37′	0.9490	1.054	3.010	0.3323	3.171	0.3153
1.26	72° 12′	.9521	1.050	3.113	.3212	3.270	.3058
1.27	72° 46′	.9551	1.047	3.224	.3102	3.375	.2963
1.28	72° 20′	.9580	1.044	3.341	.2993	3.488	.2867
1.29	73° 55′	.9608	1.041	3.467	.2884	3.609	.2771
1.30	74° 29′	0.9636	1.038	3.602	0.2776	3.738	0.2675
1.31	75° 03′	.9662	1.035	3.747	.2669	3.878	.2579
1.32	75° 38′	.9687	1.032	3.903	.2562	4.029	.2482
1.33	76° 12′	.9711	1.030	4.072	.2456	4.193	.2385
1.34	76° 47′	.9735	1.027	4.256	.2350	4.372	.2288
1.35	77° 21′	0.9757	1.025	4.455	0.2245	4.566	0.2190
1.36	77° 55′	.9779	1.023	4.673	.2140	4.779	.2092
1.37	78° 30′	.9799	1.021	4.913	.2035	5.014	.1994
1.38	79° 04′	.9819	1.018	5.177	.1931	5.273	.1896
1.39	79° 38′	.9837	1.017	5.471	.1828	5.561	.1798
1.40	80° 13′	0.9854	1.015	5.798	0.1725	5.883	0.1700
1.41	80° 47′	.9871	1.013	6.165	.1622	6.246	.1601
1.42	81° 22′	.9887	1.011	6.581	.1519	6.657	.1502
1.43	81° 56′	.9901	1.010	7.055	.1417	7.126	.1403
1.44	82° 30′	.9915	1.009	7.602	.1315	7.667	.1304
1.45	83° 05′	0.9927	1.007	8.238	0.1214	8.299	0.1205

Table V *(continued)*

Real Number x or θ radians	θ degrees	$\sin x$ or $\sin \theta$	$\csc x$ or $\csc \theta$	$\tan x$ or $\tan \theta$	$\cot x$ or $\cot \theta$	$\sec x$ or $\sec \theta$	$\cos x$ or $\cos \theta$
1.45	83° 05′	0.9927	1.007	8.238	0.1214	8.299	0.1205
1.46	83° 39′	.9939	1.006	8.989	.1113	9.044	.1106
1.47	84° 13′	.9949	1.005	9.887	.1011	9.938	.1006
1.48	84° 48′	.9959	1.004	10.98	.0910	11.03	.0907
1.49	85° 22′	.9967	1.003	12.35	.0810	12.39	.0807
1.50	85° 57′	0.9975	1.003	14.10	0.0709	14.14	0.0707
1.51	86° 31′	.9982	1.002	16.43	.0609	16.46	.0608
1.52	87° 05′	.9987	1.001	19.67	.0508	19.69	.0508
1.53	87° 40′	.9992	1.001	24.50	.0408	24.52	.0408
1.54	88° 14′	.9995	1.000	32.46	.0308	32.48	.0308
1.55	88° 49′	0.9998	1.000	48.08	0.0208	48.09	0.0208
1.56	89° 23′	.9999	1.000	92.62	.0108	92.63	.0108
1.57	89° 57′	1.000	1.000	1256	.0008	1256	.0008

Appendix C

Table VI Values of Trigonometric Functions

θ deg	θ deg-min	sin θ	cos θ	tan θ	csc θ	sec θ	cot θ		
0.0°	0°00′	0.0000	1.0000	0.0000	no value	1.0000	no value	90°0′	90.0°
0.1	0 06	0.0017	1.0000	0.0017	572.96	1.0000	572.96	89 54	89.9
0.2	0 12	0.0035	1.0000	0.0035	286.48	1.0000	286.48	89 48	89.8
0.3	0 18	0.0052	1.0000	0.0052	190.99	1.0000	190.98	89 42	89.7
0.4	0 24	0.0070	1.0000	0.0070	143.24	1.0000	143.24	89 36	89.6
0.5	0 30	0.0087	1.0000	0.0087	114.59	1.0000	114.59	89 30	89.5
0.6	0 36	0.0105	0.9999	0.0105	95.495	1.0001	95.490	89 24	89.4
0.7	0 42	0.0122	0.9999	0.0122	81.853	1.0001	81.847	89 18	89.3
0.8	0 48	0.0140	0.9999	0.0140	71.622	1.0001	71.615	89 12	89.2
0.9	0 54	0.0157	0.9999	0.0157	63.665	1.0001	63.657	89 06	89.1
1.0°	1°00′	0.0175	0.9998	0.0175	57.299	1.0002	57.290	89°00′	89.0°
1.1	1 06	0.0192	0.9998	0.0192	52.090	1.0002	52.081	88 54	88.9
1.2	1 12	0.0209	0.9998	0.0209	47.750	1.0002	47.740	88 48	88.8
1.3	1 18	0.0227	0.9997	0.0227	44.077	1.0003	44.066	88 42	88.7
1.4	1 24	0.0244	0.9997	0.0244	40.930	1.0003	40.917	88 36	88.6
1.5	1 30	0.0262	0.9997	0.0262	38.202	1.0003	38.188	88 30	88.5
1.6	1 36	0.0279	0.9996	0.0279	35.815	1.0004	35.801	88 24	88.4
1.7	1 42	0.0297	0.9996	0.0297	33.708	1.0004	33.694	88 18	88.3
1.8	1 48	0.0314	0.9995	0.0314	31.836	1.0005	31.821	88 12	88.2
1.9	1 54	0.0332	0.9995	0.0332	30.161	1.0005	30.145	88 06	88.1
2.0°	2°00′	0.0349	0.9994	0.0349	28.654	1.0006	28.636	88°00′	88.0°
2.1	2 06	0.0366	0.9993	0.0367	27.290	1.0007	27.271	87 54	87.9
2.2	2 12	0.0384	0.9993	0.0384	26.050	1.0007	26.031	87 48	87.8
2.3	2 18	0.0401	0.9992	0.0402	24.918	1.0008	24.898	87 42	87.7
2.4	2 24	0.0419	0.9991	0.0419	23.880	1.0009	23.859	87 36	87.6
2.5	2 30	0.0436	0.9990	0.0437	22.926	1.0010	22.904	87 30	87.5
2.6	2 36	0.0454	0.9990	0.0454	22.044	1.0010	22.022	87 24	87.4
2.7	2 42	0.0471	0.9989	0.0472	21.229	1.0011	21.205	87 18	87.3
2.8	2 48	0.0488	0.9988	0.0489	20.471	1.0012	20.446	87 12	87.2
2.9	2 54	0.0506	0.9987	0.0507	19.766	1.0013	19.740	87 06	87.1
3.0°	3°00′	0.0523	0.9986	0.0524	19.107	1.0014	19.081	87°00′	87.0°
3.1	3 06	0.0541	0.9985	0.0542	18.492	1.0015	18.464	86 54	86.9
3.2	3 12	0.0558	0.9984	0.0559	17.914	1.0016	17.886	86 48	86.8
3.3	3 18	0.0576	0.9983	0.0577	17.372	1.0017	17.343	86 42	86.7
3.4	3 24	0.0593	0.9982	0.0594	16.862	1.0018	16.832	86 36	86.6
3.5	3 30	0.0610	0.9981	0.0612	16.380	1.0019	16.350	86 30	86.5
3.6	3 36	0.0628	0.9980	0.0629	15.926	1.0020	15.895	86 24	86.4
3.7	3 42	0.0645	0.9979	0.0647	15.496	1.0021	15.464	86 18	86.3
3.8	3 48	0.0663	0.9978	0.0664	15.089	1.0022	15.056	86 12	86.2
3.9	3 54	0.0680	0.9977	0.0682	14.703	1.0023	14.669	86 06	86.1
4.0°	4°00′	0.0698	0.9976	0.0699	14.336	1.0024	14.301	86°00′	86.0°
4.1	4 06	0.0715	0.9974	0.0717	13.987	1.0026	13.951	85 54	85.9
4.2	4 12	0.0732	0.9973	0.0734	13.654	1.0027	13.617	85 48	85.8
4.3	4 18	0.0750	0.9972	0.0752	13.337	1.0028	13.300	85 42	85.7
4.4	4 24	0.0767	0.9971	0.0769	13.035	1.0030	12.996	85 36	85.6
4.5	4 30	0.0785	0.9969	0.0787	12.746	1.0031	12.706	85 30	85.5
4.6	4 36	0.0802	0.9968	0.0805	12.469	1.0032	12.429	85 24	85.4
4.7	4 42	0.0819	0.9966	0.0822	12.204	1.0034	12.163	85 18	85.3
4.8	4 48	0.0837	0.9965	0.0840	11.951	1.0035	11.909	85 12	85.2
4.9	4 54	0.0854	0.9963	0.0857	11.707	1.0037	11.665	85°06′	85.1°
		cos θ	sin θ	cot θ	sec θ	csc θ	tan θ	θ deg-min	θ deg

Table VI (continued)

θ deg	θ deg-min	sin θ	cos θ	tan θ	csc θ	sec θ	cot θ		
5.0°	5°00′	0.0872	0.9962	0.0875	11.474	1.0038	11.430	85°00′	85.0°
5.1	5 06	0.0889	0.9960	0.0892	11.249	1.0040	11.205	84 54	84.9
5.2	5 12	0.0906	0.9959	0.0910	11.034	1.0041	10.988	84 48	84.8
5.3	5 18	0.0924	0.9957	0.0928	10.826	1.0043	10.780	84 42	84.7
5.4	5 24	0.0941	0.9956	0.0945	10.626	1.0045	10.579	84 36	84.6
5.5	5 30	0.0958	0.9954	0.0963	10.433	1.0046	10.385	84 30	84.5
5.6	5 36	0.0976	0.9952	0.0981	10.248	1.0048	10.199	84 24	84.4
5.7	5 42	0.0993	0.9951	0.0998	10.069	1.0050	10.019	84 18	84.3
5.8	5 48	0.1011	0.9949	0.1016	9.8955	1.0051	9.8448	84 12	84.2
5.9	5 54	0.1028	0.9947	0.1033	9.7283	1.0053	9.6768	84 06	84.1
6.0°	6°00′	0.1045	0.9945	0.1051	9.5668	1.0055	9.5144	84°00′	84.0°
6.1	6 06	0.1063	0.9943	0.1069	9.4105	1.0057	9.3573	83 54	83.9
6.2	6 12	0.1080	0.9942	0.1086	9.2593	1.0059	9.2052	83 48	83.8
6.3	6 18	0.1097	0.9940	0.1104	9.1129	1.0061	9.0579	83 42	83.7
6.4	6 24	0.1115	0.9938	0.1122	8.9711	1.0063	8.9152	83 36	83.6
6.5	6 30	0.1132	0.9936	0.1139	8.8337	1.0065	8.7769	83 30	83.5
6.6	6 36	0.1149	0.9934	0.1157	8.7004	1.0067	8.6428	83 24	83.4
6.7	6 42	0.1167	0.9932	0.1175	8.5711	1.0069	8.5126	83 18	83.3
6.8	6 48	0.1184	0.9930	0.1192	8.4457	1.0071	8.3863	83 12	83.2
6.9	6 54	0.1201	0.9928	0.1210	8.3238	1.0073	8.2636	83 06	83.1
7.0°	7°00′	0.1219	0.9925	0.1228	8.2055	1.0075	8.1444	83°00′	83.0°
7.1	7 06	0.1236	0.9923	0.1246	8.0905	1.0077	8.0285	82 54	82.9
7.2	7 12	0.1253	0.9921	0.1263	7.9787	1.0079	7.9158	82 48	82.8
7.3	7 18	0.1271	0.9919	0.1281	7.8700	1.0082	7.8062	82 42	82.7
7.4	7 24	0.1288	0.9917	0.1299	7.7642	1.0084	7.6996	82 36	82.6
7.5	7 30	0.1305	0.9914	0.1317	7.6613	1.0086	7.5958	82 30	82.5
7.6	7 36	0.1323	0.9912	0.1334	7.5611	1.0089	7.4947	82 24	82.4
7.7	7 42	0.1340	0.9910	0.1352	7.4635	1.0091	7.3962	82 18	82.3
7.8	7 48	0.1357	0.9907	0.1370	7.3684	1.0093	7.3002	82 12	82.2
7.9	7 54	0.1374	0.9905	0.1388	7.2757	1.0096	7.2066	82 06	82.1
8.0°	8°00′	0.1392	0.9903	0.1405	7.1853	1.0098	7.1154	82°00′	82.0°
8.1	8 06	0.1409	0.9900	0.1423	7.0972	1.0101	7.0264	81 54	81.9
8.2	8 12	0.1426	0.9898	0.1441	7.0112	1.0103	6.9395	81 48	81.8
8.3	8 18	0.1444	0.9895	0.1459	6.9273	1.0106	6.8548	81 42	81.7
8.4	8 24	0.1461	0.9893	0.1477	6.8454	1.0108	6.7720	81 36	81.6
8.5	8 30	0.1478	0.9890	0.1495	6.7655	1.0111	6.6912	81 30	81.5
8.6	8 36	0.1495	0.9888	0.1512	6.6874	1.0114	6.6122	81 24	81.4
8.7	8 42	0.1513	0.9885	0.1530	6.6111	1.0116	6.5350	81 18	81.3
8.8	8 48	0.1530	0.9882	0.1548	6.5366	1.0119	6.4596	81 12	81.2
8.9	8 54	0.1547	0.9880	0.1566	6.4637	1.0122	6.3859	81 06	81.1
9.0°	9°00′	0.1564	0.9877	0.1584	6.3925	1.0125	6.3138	81°00′	81.0°
9.1	9 06	0.1582	0.9874	0.1602	6.3228	1.0127	6.2432	80 54	80.9
9.2	9 12	0.1599	0.9871	0.1620	6.2547	1.0130	6.1742	80 48	80.8
9.3	9 18	0.1616	0.9869	0.1638	6.1880	1.0133	6.1066	80 42	80.7
9.4	9 24	0.1633	0.9866	0.1655	6.1227	1.0136	6.0405	80 36	80.6
9.5	9 30	0.1650	0.9863	0.1673	6.0589	1.0139	5.9758	80 30	80.5
9.6	9 36	0.1668	0.9860	0.1691	5.9963	1.0142	5.9124	80 24	80.4
9.7	9 42	0.1685	0.9857	0.1709	5.9351	1.0145	5.8502	80 18	80.3
9.8	9 48	0.1702	0.9854	0.1727	5.8751	1.0148	5.7894	80 12	80.2
9.9	9 54	0.1719	0.9851	0.1745	5.8164	1.0151	5.7297	80°06′	80.1°
		cos θ	sin θ	cot θ	sec θ	csc θ	tan θ	θ deg-min	θ deg

Appendix C

Table VI (continued)

θ deg	θ deg-min	sin θ	cos θ	tan θ	csc θ	sec θ	cot θ		
10.0°	10°00′	0.1736	0.9848	0.1763	5.7588	1.0154	5.6713	80°00′	80.0°
10.1	10 06	0.1754	0.9845	0.1781	5.7023	1.0157	5.6140	79 54	79.9
10.2	10 12	0.1771	0.9842	0.1799	5.6470	1.0161	5.5578	79 48	79.8
10.3	10 18	0.1788	0.9839	0.1817	5.5928	1.0164	5.5027	79 42	79.7
10.4	10 24	0.1805	0.9836	0.1835	5.5396	1.0167	5.4486	79 36	79.6
10.5	10 30	0.1822	0.9833	0.1853	5.4874	1.0170	5.3955	79 30	79.5
10.6	10 36	0.1840	0.9829	0.1871	5.4362	1.0174	5.3435	79 24	79.4
10.7	10 42	0.1857	0.9826	0.1890	5.3860	1.0177	5.2924	79 18	79.3
10.8	10 48	0.1874	0.9823	0.1908	5.3367	1.0180	5.2422	79 12	79.2
10.9	10 54	0.1891	0.9820	0.1926	5.2883	1.0184	5.1929	79 06	79.1
11.0°	11°00′	0.1908	0.9816	0.1944	5.2408	1.0187	5.1446	79°00′	79.0°
11.1	11 06	0.1925	0.9813	0.1962	5.1942	1.0191	5.0970	78 54	78.9
11.2	11 12	0.1942	9.9810	0.1980	5.1484	1.0194	5.0504	78 48	78.8
11.3	11 18	0.1959	0.9806	0.1998	5.1034	1.0198	5.0045	78 42	78.7
11.4	11 24	0.1977	0.9803	0.2016	5.0593	1.0201	4.9595	78 36	78.6
11.5	11 30	0.1994	0.9799	0.2035	5.0159	1.0205	4.9152	78 30	78.5
11.6	11 36	0.2011	0.9796	0.2053	4.9732	1.0209	4.8716	78 24	78.4
11.7	11 42	0.2028	0.9792	0.2071	4.9313	1.0212	4.8288	78 18	78.3
11.8	11 48	0.2045	0.9789	0.2089	4.8901	1.0216	4.7867	78 12	78.2
11.9	11 54	0.2062	0.9785	0.2107	4.8496	1.0220	4.7453	78 06	78.1
12.0°	12°00′	0.2079	0.9781	0.2126	4.8097	1.0223	4.7046	78°00′	78.0°
12.1	12 06	0.2096	0.9778	0.2144	4.7706	1.0227	4.6646	77 54	77.9
12.2	12 12	0.2113	0.9774	0.2162	4.7321	1.0231	4.6252	77 48	77.8
12.3	12 18	0.2130	0.9770	0.2180	4.6942	1.0235	4.5864	77 42	77.7
12.4	12 24	0.2147	0.9767	0.2199	4.6569	1.0239	4.5483	77 36	77.6
12.5	12 30	0.2164	0.9763	0.2217	4.6202	1.0243	4.5107	77 30	77.5
12.6	12 36	0.2181	0.9759	0.2235	4.5841	1.0247	4.4737	77 24	77.4
12.7	12 42	0.2198	0.9755	0.2254	4.5486	1.0251	4.4374	77 18	77.3
12.8	12 48	0.2215	0.9751	0.2272	4.5137	1.0255	4.4015	77 12	77.2
12.9	12 54	0.2232	0.9748	0.2290	4.4793	1.0259	4.3662	77 06	77.1
13.0°	13°00′	0.2250	0.9744	0.2309	4.4454	1.0263	4.3315	77°00′	77.0°
13.1	13 06	0.2267	0.9740	0.2327	4.4121	1.0267	4.2972	76 54	76.9
13.2	13 12	0.2284	0.9736	0.2345	4.3792	1.0271	4.2635	76 48	76.8
13.3	13 18	0.2300	0.9732	0.2364	4.3469	1.0276	4.2303	76 42	76.7
13.4	13 24	0.2317	0.9728	0.2382	4.3150	1.0280	4.1976	76 36	76.6
13.5	13 30	0.2334	0.9724	0.2401	4.2837	1.0284	4.1653	76 30	76.5
13.6	13 36	0.2351	0.9720	0.2419	4.2528	1.0288	4.1335	76 24	76.4
13.7	13 42	0.2368	0.9715	0.2438	4.2223	1.0293	4.1022	76 18	76.3
13.8	13 48	0.2385	0.9711	0.2456	4.1923	1.0297	4.0713	76 12	76.2
13.9	13 54	0.2402	0.9707	0.2475	4.1627	1.0302	4.0408	76 06	76.1
14.0°	14°00′	0.2419	0.9703	0.2493	4.1336	1.0306	4.0108	76°00′	76.0°
14.1	14 06	0.2436	0.9699	0.2512	4.1048	1.0311	3.9812	75 54	75.9
14.2	14 12	0.2453	0.9694	0.2530	4.0765	1.0315	3.9520	75 48	75.8
14.3	14 18	0.2470	0.9690	0.2549	4.0486	1.0320	3.9232	75 42	75.7
14.4	14 24	0.2487	0.9686	0.2568	4.0211	1.0324	3.8947	75 36	75.6
14.5	14 30	0.2504	0.9681	0.2586	3.9939	1.0329	3.8667	75 30	75.5
14.6	14 36	0.2521	0.9677	0.2605	3.9672	1.0334	3.8391	75 24	75.4
14.7	14 42	0.2538	0.9673	0.2623	3.9408	1.0338	3.8118	75 18	75.3
14.8	14 48	0.2554	0.9668	0.2642	3.9147	1.0343	3.7849	75 12	75.2
14.9	14 54	0.2571	0.9664	0.2661	3.8890	1.0348	3.7583	75°06′	75.1°
		cos θ	sin θ	cot θ	sec θ	csc θ	tan θ	θ deg-min	θ deg

Table VI (*continued*)

θ deg	θ deg-min	sin θ	cos θ	tan θ	csc θ	sec θ	cot θ		
15.0°	15°00′	0.2588	0.9659	0.2679	3.8637	1.0353	3.7321	75°00′	75.0°
15.1	15 06	0.2605	0.9655	0.2698	3.8387	1.0358	3.7062	74 54	74.9
15.2	15 12	0.2622	0.9650	0.2717	3.8140	1.0363	3.6806	74 48	74.8
15.3	15 18	0.2639	0.9646	0.2736	3.7897	1.0367	3.6554	74 42	74.7
15.4	15 24	0.2656	0.9641	0.2754	3.7657	1.0372	3.6305	74 36	74.6
15.5	15 30	0.2672	0.9636	0.2773	3.7420	1.0377	3.6059	74 30	74.5
15.6	15 36	0.2689	0.9632	0.2792	3.7186	1.0382	3.5816	74 24	74.4
15.7	15 42	0.2706	0.9627	0.2811	3.6955	1.0388	3.5576	74 18	74.3
15.8	15 48	0.2723	0.9622	0.2830	3.6727	1.0393	3.5339	74 12	74.2
15.9	15 54	0.2740	0.9617	0.2849	3.6502	1.0398	3.5105	74 06	74.1
16.0°	16°00′	0.2756	0.9613	0.2867	3.6280	1.0403	3.4874	74°00′	74.0°
16.1	16 06	0.2773	0.9608	0.2886	3.6060	1.0408	3.4646	73 54	73.9
16.2	16 12	0.2790	0.9603	0.2905	3.5843	1.0413	3.4420	73 48	73.8
16.3	16 18	0.2807	0.9598	0.2924	3.5629	1.0419	3.4197	73 42	73.7
16.4	16 24	0.2823	0.9593	0.2943	3.5418	1.0424	3.3977	73 36	73.6
16.5	16 30	0.2840	0.9588	0.2962	3.5209	1.0429	3.3759	73 30	73.5
16.6	16 36	0.2857	0.9583	0.2981	3.5003	1.0435	3.3544	73 24	74.4
16.7	16 42	0.2874	0.9578	0.3000	3.4800	1.0440	3.3332	73 18	73.3
16.8	16 48	0.2890	0.9573	0.3019	3.4598	1.0446	3.3122	73 12	73.2
16.9	16 54	0.2907	0.9568	0.3038	3.4399	1.0451	3.2914	73 06	73.1
17.0°	17 00′	0.2924	0.9563	0.3057	3.4203	1.0457	3.2709	73°00′	73.0°
17.1	17 06	0.2940	0.9558	0.3076	3.4009	1.0463	3.2506	72 54	72.9
17.2	17 12	0.2957	0.9553	0.3096	3.3817	1.0468	3.2305	72 48	72.8
17.3	17 18	0.2974	0.9548	0.3115	3.3628	1.0474	3.2106	72 42	72.7
17.4	17 24	0.2990	0.9542	0.3134	3.3440	1.0480	3.1910	72 36	72.6
17.5	17 30	0.3007	0.9537	0.3153	3.3255	1.0485	3.1716	72 30	72.5
17.6	17 36	0.3024	0.9532	0.3172	3.3072	1.0491	3.1524	72 24	72.4
17.7	17 42	0.3040	0.9527	0.3191	3.2891	1.0497	3.1334	72 18	72.3
17.8	17 48	0.3057	0.9521	0.3211	3.2712	1.0503	3.1146	72 12	72.2
17.9	17 54	0.3074	0.9516	0.3230	3.2536	1.0509	3.0961	72 06	72.1
18.0°	18°00′	0.3090	0.9511	0.3249	3.2361	1.0515	3.0777	72°00′	72.0°
18.1	18 06	0.3107	0.9505	0.3268	3.2188	1.0521	3.0595	71 54	71.9
18.2	18 12	0.3123	0.9500	0.3288	3.2017	1.0527	3.0415	71 48	71.8
18.3	18 18	0.3140	0.9494	0.3307	3.1848	1.0533	3.0237	71 42	71.7
18.4	18 24	0.3156	0.9489	0.3327	3.1681	1.0539	3.0061	71 36	71.6
18.5	18 30	0.3173	0.9483	0.3346	3.1515	1.0545	2.9887	71 30	71.5
18.6	18 36	0.3190	0.9478	0.3365	3.1352	1.0551	2.9714	71 24	71.4
18.7	18 42	0.3206	0.9472	0.3385	3.1190	1.0557	2.9544	71 18	71.3
18.8	18 48	0.3223	0.9466	0.3404	3.1030	1.0564	2.9375	71 12	71.2
18.9	18 54	0.3239	0.9461	0.3424	3.0872	1.0570	2.9208	71 06	71.1
19.0°	19°00′	0.3256	0.9455	0.3443	3.0716	1.0576	2.9042	71°00′	71.0°
19.1	19 06	0.3272	0.9449	0.3463	3.0561	1.0583	2.8878	70 54	70.9
19.2	19 12	0.3289	0.9444	0.3482	3.0407	1.0589	2.8716	70 48	70.8
19.3	19 18	0.3305	0.9438	0.3502	3.0256	1.0595	2.8556	70 42	70.7
19.4	19 24	0.3322	0.9432	0.3522	3.0106	1.0602	2.8397	70 36	70.6
19.5	19 30	0.3338	0.9426	0.3541	2.9957	1.0608	2.8239	70 30	70.5
19.6	19 36	0.3355	0.9421	0.3561	2.9811	1.0615	2.8083	70 24	70.4
19.7	19 42	0.3371	0.9415	0.3581	2.9665	1.0622	2.7929	70 18	70.3
19.8	19 48	0.3387	0.9409	0.3600	2.9521	1.0628	2.7776	70 12	70.2
19.9	19 54	0.3404	0.9403	0.3620	2.9379	1.0635	2.7625	70°06′	70.1°
		cos θ	sin θ	cot θ	sec θ	csc θ	tan θ	θ deg-min	θ deg

Table VI (continued)

θ deg	θ deg-min	$\sin \theta$	$\cos \theta$	$\tan \theta$	$\csc \theta$	$\sec \theta$	$\cot \theta$		
20.0°	20°00′	0.3420	0.9397	0.3640	2.9238	1.0642	2.7475	70°00′	70.0°
20.1	20 06	0.3437	0.9391	0.3659	2.9099	1.0649	2.7326	69 54	69.9
20.2	20 12	0.3453	0.9385	0.3679	2.8960	1.0655	2.7179	69 48	69.8
20.3	20 18	0.3469	0.9379	0.3699	2.8824	1.0662	2.7034	69 42	69.7
20.4	20 24	0.3486	0.9373	0.3719	2.8688	1.0669	2.6889	69 36	69.6
20.5	20 30	0.3502	0.9367	0.3739	2.8555	1.0676	2.6746	69 30	69.5
20.6	20 36	0.3518	0.9361	0.3759	2.8422	1.0683	2.6605	69 24	69.4
20.7	20 42	0.3535	0.9354	0.3779	2.8291	1.0690	2.6464	69 18	69.3
20.8	20 48	0.3551	0.9348	0.3799	2.8161	1.0697	2.6325	69 12	69.2
20.9	20 54	0.3567	0.9342	0.3819	2.8032	1.0704	2.6187	69 06	69.1
21.0°	21°00′	0.3584	0.9336	0.3839	2.7904	1.0711	2.6051	69°00′	69.0°
21.1	21 06	0.3600	0.9330	0.3859	2.7778	1.0719	2.5916	68 54	68.9
21.2	21 12	0.3616	0.9323	0.3879	2.7653	1.0726	2.5782	68 48	68.8
21.3	21 18	0.3633	0.9317	0.3899	2.7529	1.0733	2.5649	68 42	68.7
21.4	21 24	0.3649	0.9311	0.3919	2.7407	1.0740	2.5517	68 36	68.6
21.5	21 30	0.3665	0.9304	0.3939	2.7285	1.0748	2.5386	68 30	68.5
21.6	21 36	0.3681	0.9298	0.3959	2.7165	1.0755	2.5257	68 24	68.4
21.7	21 42	0.3697	0.9291	0.3979	2.7046	1.0763	2.5129	68 18	68.3
21.8	21 48	0.3714	0.9285	0.4000	2.6927	1.0770	2.5002	68 12	68.2
21.9	21 54	0.3730	0.9278	0.4020	2.6811	1.0778	2.4876	68 06	68.1
22.0°	22°00′	0.3746	0.9272	0.4040	2.6695	1.0785	2.4751	68°00′	68.0°
22.1	22 06	0.3762	0.9265	0.4061	2.6580	1.0793	2.4627	67 54	67.9
22.2	22 12	0.3778	0.9259	0.4081	2.6466	1.0801	2.4504	67 48	67.8
22.3	22 18	0.3795	0.9252	0.4101	2.6354	1.0808	2.4383	67 42	67.7
22.4	22 24	0.3811	0.9245	0.4122	2.6242	1.0816	2.4262	67 36	67.6
22.5	22 30	0.3827	0.9239	0.4142	2.6131	1.0824	2.4142	67 30	67.5
22.6	22 36	0.3843	0.9232	0.4163	2.6022	1.0832	2.4023	67 24	67.4
22.7	22 42	0.3859	0.9225	0.4183	2.5913	1.0840	2.3906	67 18	67.3
22.8	22 48	0.3875	0.9219	0.4204	2.5805	1.0848	2.3789	67 12	67.2
22.9	22 54	0.3891	0.9212	0.4224	2.5699	1.0856	2.3673	67 06	67.1
23.0°	23°00′	0.3907	0.9205	0.4245	2.5593	1.0864	2.3559	67°00′	67.0°
23.1	23 06	0.3923	0.9198	0.4265	2.5488	1.0872	2.3445	66 54	66.9
23.2	23 12	0.3939	0.9191	0.4286	2.5384	1.0880	2.3332	66 48	66.8
23.3	23 18	0.3955	0.9184	0.4307	2.5282	1.0888	2.3220	66 42	66.7
23.4	23 24	0.3971	0.9178	0.4327	2.5180	1.0896	2.3109	66 36	66.6
23.5	23 30	0.3987	0.9171	0.4348	2.5078	1.0904	2.2998	66 30	66.5
23.6	23 36	0.4003	0.9164	0.4369	2.4978	1.0913	2.2889	66 24	66.4
23.7	23 42	0.4019	0.9157	0.4390	2.4879	1.0921	2.2781	66 18	66.3
23.8	23 48	0.4035	0.9150	0.4411	2.4780	1.0929	2.2673	66 12	66.2
23.9	23 54	0.4051	0.9143	0.4431	2.4683	1.0938	2.2566	66 06	66.1
24.0°	24°00′	0.4067	0.9135	0.4452	2.4586	1.0946	2.2460	66°00′	66.0°
24.1	24 06	0.4083	0.9128	0.4473	2.4490	1.0955	2.2355	65 54	65.9
24.2	24 12	0.4099	0.9121	0.4494	2.4395	1.0963	2.2251	65 48	65.8
24.3	24 18	0.4115	0.9114	0.4515	2.4301	1.0972	2.2148	65 42	65.7
24.4	24 24	0.4131	0.9107	0.4536	2.4207	1.0981	2.2045	65 36	65.6
24.5	24 30	0.4147	0.9100	0.4557	2.4114	1.0989	2.1943	65 30	65.5
24.6	24 36	0.4163	0.9092	0.4578	2.4022	1.0998	2.1842	65 24	65.4
24.7	24 42	0.4179	0.9085	0.4599	2.3931	1.1007	2.1742	65 18	65.3
24.8	24 48	0.4195	0.9078	0.4621	2.3841	1.1016	2.1642	65 12	65.2
24.9	24 54	0.4210	0.9070	0.4642	2.3751	1.1025	2.1543	65°06′	65.1°
		$\cos \theta$	$\sin \theta$	$\cot \theta$	$\sec \theta$	$\csc \theta$	$\tan \theta$	θ deg-min	θ deg

Table VI (*continued*)

θ deg	deg-min	sin θ	cos θ	tan θ	csc θ	sec θ	cot θ		
25.0°	25°00′	0.4226	0.9063	0.4663	2.3662	1.1034	2.1445	65°00′	65.0°
25.1	25 06	0.4242	0.9056	0.4684	2.3574	1.1043	2.1348	64 54	64.9
25.2	25 12	0.4258	0.9048	0.4706	2.3486	1.1052	2.1251	64 48	64.8
25.3	25 18	0.4274	0.9041	0.4727	2.3400	1.1061	2.1155	64 42	64.7
25.4	25 24	0.4289	0.9033	0.4748	2.3314	1.1070	2.1060	64 36	64.6
25.5	25 30	0.4305	0.9026	0.4770	2.3228	1.1079	2.0965	64 30	64.5
25.6	25 36	0.4321	0.9018	0.4791	2.3144	1.1089	2.0872	64 24	64.4
25.7	25 42	0.4337	0.9011	0.4813	2.3060	1.1098	2.0778	64 18	64.3
25.8	25 48	0.4352	0.9003	0.4834	2.2976	1.1107	2.0686	64 12	64.2
25.9	25 54	0.4368	0.8996	0.4856	2.2894	1.1117	2.0594	64 06	64.1
26.0°	26°00′	0.4384	0.8988	0.4877	2.2812	1.1126	2.0503	64°00′	64.0°
26.1	26 06	0.4399	0.8980	0.4899	2.2730	1.1136	2.0413	63 54	63.9
26.2	26 12	0.4415	0.8973	0.4921	2.2650	1.1145	2.0323	63 48	63.8
26.3	26 18	0.4431	0.8965	0.4942	2.2570	1.1155	2.0233	63 42	63.7
26.4	26 24	0.4446	0.8957	0.4964	2.2490	1.1164	2.0145	63 36	63.6
26.5	26 30	0.4462	0.8949	0.4986	2.2412	1.1174	2.0057	63 30	63.5
26.6	26 36	0.4478	0.8942	0.5008	2.2333	1.1184	1.9970	63 24	63.4
26.7	26 42	0.4493	0.8934	0.5029	2.2256	1.1194	1.9883	63 18	63.3
26.8	26 48	0.4509	0.8926	0.5051	2.2179	1.1203	1.9797	63 12	63.2
26.9	26 54	0.4524	0.8918	0.5073	2.2103	1.1213	1.9711	63 06	63.1
27.0°	27°00′	0.4540	0.8910	0.5095	2.2027	1.1223	1.9626	63°00′	63.0°
27.1	27 06	0.4555	0.8902	0.5117	2.1952	1.1233	1.9542	62 54	62.9
27.2	27 12	0.4571	0.8894	0.5139	2.1877	1.1243	1.9458	62 48	62.8
27.3	27 18	0.4586	0.8886	0.5161	2.1803	1.1253	1.9375	62 42	62.7
27.4	27 24	0.4602	0.8878	0.5184	2.1730	1.1264	1.9292	62 36	62.6
27.5	27 30	0.4617	0.8870	0.5206	2.1657	1.1274	1.9210	62 30	62.5
27.6	27 36	0.4633	0.8862	0.5228	2.1584	1.1284	1.9128	62 24	62.4
27.7	27 42	0.4648	0.8854	0.5250	2.1513	1.1294	1.9047	62 18	62.3
27.8	27 48	0.4664	0.8846	0.5272	2.1441	1.1305	1.8967	62 12	62.2
27.9	27 54	0.4679	0.8838	0.5295	2.1371	1.1315	1.8887	62 06	62.1
28.0°	28°00′	0.4695	0.8829	0.5317	2.1301	1.1326	1.8807	62°00′	62.0°
28.1	28 06	0.4710	0.8821	0.5339	2.1231	1.1336	1.8728	61 54	61.9
28.2	28 12	0.4726	0.8813	0.5362	2.1162	1.1347	1.8650	61 48	61.8
28.3	28 18	0.4741	0.8805	0.5384	2.1093	1.1357	1.8572	61 42	61.7
28.4	28 24	0.4756	0.8796	0.5407	2.1025	1.1368	1.8495	61 36	61.6
28.5	28 30	0.4772	0.8788	0.5430	2.0957	1.1379	1.8418	61 30	61.5
28.6	28 36	0.4787	0.8780	0.5452	2.0890	1.1390	1.8341	61 24	61.4
28.7	28 42	0.4802	0.8771	0.5475	2.0824	1.1401	1.8265	61 18	61.3
28.8	28 48	0.4818	0.8763	0.5498	2.0758	1.1412	1.8190	61 12	61.2
28.9	28 54	0.4833	0.8755	0.5520	2.0692	1.1423	1.8115	61 06	61.1
29.0°	29°00′	0.4848	0.8746	0.5543	2.0627	1.1434	1.8040	61°00′	61.0°
29.1	29 06	0.4863	0.8738	0.5566	2.0562	1.1445	1.7966	60 54	60.9
29.2	29 12	0.4879	0.8729	0.5589	2.0598	1.1456	1.7893	60 48	60.8
29.3	29 18	0.4894	0.8721	0.5612	2.0434	1.1467	1.7820	60 42	60.7
29.4	29 24	0.4909	0.8712	0.5635	2.0371	1.1478	1.7747	60 36	60.6
29.5	29 30	0.4924	0.8704	0.5658	2.0308	1.1490	1.7675	60 30	60.5
29.6	29 36	0.4939	0.8695	0.5681	2.0245	1.1501	1.7603	60 24	60.4
29.7	29 42	0.4955	0.8686	0.5704	2.0183	1.1512	1.7532	60 18	60.3
29.8	29 48	0.4970	0.8678	0.5727	2.0122	1.1524	1.7461	60 12	60.2
29.9	29 54	0.4985	0.8669	0.5750	2.0061	1.1535	1.7391	60°06′	60.1°
		cos θ	sin θ	cot θ	sec θ	csc θ	tan θ	θ deg-min	θ deg

Table VI (continued)

θ deg	θ deg-min	sin θ	cos θ	tan θ	csc θ	sec θ	cot θ		
30.0°	30°00′	0.5000	0.8660	0.5774	2.0000	1.1547	1.7321	60°00′	60.0°
30.1	30 06	0.5015	0.8652	0.5797	1.9940	1.1559	1.7251	59 54	59.9
30.2	30 12	0.5030	0.8643	0.5820	1.9880	1.1570	1.7182	59 48	59.8
30.3	30 18	0.5045	0.8634	0.5844	1.9821	1.1582	1.7113	59 42	59.7
30.4	30 24	0.5060	0.8625	0.5867	1.9762	1.1594	1.7045	59 36	59.6
30.5	30 30	0.5075	0.8616	0.5890	1.9703	1.1606	1.6977	59 30	59.5
30.6	30 36	0.5090	0.8607	0.5914	1.9645	1.1618	1.6909	59 24	59.4
30.7	30 42	0.5105	0.8599	0.5938	1.9587	1.1630	1.6842	59 18	59.3
30.8	30 48	0.5120	0.8590	0.5961	1.9530	1.1642	1.6775	59 12	59.2
30.9	30 54	0.5135	0.8581	0.5985	1.9473	1.1654	1.6709	59 06	59.1
31.0°	31°00′	0.5150	0.8572	0.6009	1.9416	1.1666	1.6643	59°00′	59.0°
31.1	31 06	0.5165	0.8563	0.6032	1.9360	1.1679	1.6577	58 54	58.9
31.2	31 12	0.5180	0.8554	0.6056	1.9304	1.1691	1.6512	58 48	58.8
31.3	31 18	0.5195	0.8545	0.6080	1.9249	1.1703	1.6447	58 42	58.7
31.4	31 24	0.5210	0.8536	0.6104	1.9194	1.1716	1.6383	58 36	58.6
31.5	31 30	0.5225	0.8526	0.6128	1.9139	1.1728	1.6319	58 30	58.5
31.6	31 36	0.5240	0.8517	0.6152	1.9084	1.1741	1.6255	58 24	58.4
31.7	31 42	0.5255	0.8508	0.6176	1.9031	1.1753	1.6191	58 18	58.3
31.8	31 48	0.5270	0.8499	0.6200	1.8977	1.1766	1.6128	58 12	58.2
31.9	31 54	0.5284	0.8490	0.6224	1.8924	1.1779	1.6066	58 06	58.1
32.0°	32°00′	0.5299	0.8480	0.6249	1.8871	1.1792	1.6003	58°00′	58.0°
32.1	32 06	0.5314	0.8471	0.6273	1.8818	1.1805	1.5941	57 54	57.9
32.2	32 12	0.5329	0.8462	0.6297	1.8766	1.1818	1.5880	57 48	57.8
32.3	32 18	0.5344	0.8453	0.6322	1.8714	1.1831	1.5818	57 42	57.7
32.4	32 24	0.5358	0.8443	0.6346	1.8663	1.1844	1.5757	57 36	57.6
32.5	32 30	0.5373	0.8434	0.6371	1.8612	1.1857	1.5697	57 30	57.5
32.6	32 36	0.5388	0.8425	0.6395	1.8561	1.1870	1.5637	57 24	57.4
32.7	32 42	0.5402	0.8415	0.6420	1.8510	1.1883	1.5577	57 18	57.3
32.8	32 48	0.5417	0.8406	0.6445	1.8460	1.1897	1.5517	57 12	57.2
32.9	32 54	0.5432	0.8396	0.6469	1.8410	1.1910	1.5458	57 06	57.1
33.0°	33°00′	0.5446	0.8387	0.6494	1.8361	1.1924	1.5399	57°00′	57.0°
33.1	33 06	0.5461	0.8377	0.6519	1.8312	1.1937	1.5340	56 54	56.9
33.2	33 12	0.5476	0.8368	0.6544	1.8263	1.1951	1.5282	56 48	56.8
33.3	33 18	0.5490	0.8358	0.6569	1.8214	1.1964	1.5224	56 42	56.7
33.4	33 24	0.5505	0.8348	0.6594	1.8166	1.1978	1.5166	56 36	56.6
33.5	33 30	0.5519	0.8339	0.6619	1.8118	1.1992	1.5108	56 30	56.5
33.6	33 36	0.5534	0.8329	0.6644	1.8070	1.2006	1.5051	56 24	56.4
33.7	33 42	0.5548	0.8320	0.6669	1.8023	1.2020	1.4994	56 18	56.3
33.8	33 48	0.5563	0.8310	0.6694	1.7976	1.2034	1.4938	56 12	56.2
33.9	33 54	0.5577	0.8300	0.6720	1.7929	1.2048	1.4882	56 06	56.1
34.0°	34°00′	0.5592	0.8290	0.6745	1.7883	1.2062	1.4826	56°00′	56.0°
34.1	34 06	0.5606	0.8281	0.6771	1.7837	1.2076	1.4770	55 54	55.9
34.2	34 12	0.5621	0.8271	0.6796	1.7791	1.2091	1.4715	55 48	55.8
34.3	34 18	0.5635	0.8261	0.6822	1.7745	1.2105	1.4659	55 42	55.7
34.4	34 24	0.5650	0.8251	0.6847	1.7700	1.2120	1.4605	55 36	55.6
34.5	34 30	0.5664	0.8241	0.6873	1.7655	1.2134	1.4550	55 30	55.5
34.6	34 36	0.5678	0.8231	0.6899	1.7610	1.2149	1.4496	55 24	55.4
34.7	34 42	0.5693	0.8221	0.6924	1.7566	1.2163	1.4442	55 18	55.3
34.8	34 48	0.5707	0.8211	0.6950	1.7522	1.2178	1.4388	55 12	55.2
34.9	34 54	0.5721	0.8202	0.6976	1.7478	1.2193	1.4335	55°06′	55.1°
		cos θ	sin θ	cot θ	sec θ	csc θ	tan θ	θ deg-min	θ deg

Table VI (continued)

θ deg	θ deg-min	sin θ	cos θ	tan θ	csc θ	sec θ	cot θ		
35.0°	35°00′	0.5736	0.8192	0.7002	1.7434	1.2208	1.4281	55°00′	55.0°
35.1	35 06	0.5750	0.8181	0.7028	1.7391	1.2223	1.4229	54 54	54.9
35.2	35 12	0.5764	0.8171	0.7054	1.7348	1.2238	1.4176	54 48	54.8
35.3	35 18	0.5779	0.8161	0.7080	1.7305	1.2253	1.4124	54 42	54.7
35.4	35 24	0.5793	0.8151	0.7107	1.7263	1.2268	1.4071	54 36	54.6
35.5	35 30	0.5807	0.8141	0.7133	1.7221	1.2283	1.4019	54 30	54.5
35.6	35 36	0.5821	0.8131	0.7159	1.7179	1.2299	1.3968	54 24	54.4
35.7	35 42	0.5835	0.8121	0.7186	1.7137	1.2314	1.3916	54 18	54.3
35.8	35 48	0.5850	0.8111	0.7212	1.7095	1.2329	1.3865	54 12	54.2
35.9	35 54	0.5864	0.8100	0.7239	1.7054	1.2345	1.3814	54 06	54.1
36.0°	36°00′	0.5878	0.8090	0.7265	1.7013	1.2361	1.3764	54°00′	54.0°
36.1	36 06	0.5892	0.8080	0.7292	1.6972	1.2376	1.3713	53 54	53.9
36.2	36 12	0.5906	0.8070	0.7319	1.6932	1.2392	1.3663	53 48	53.8
36.3	36 18	0.5920	0.8059	0.7346	1.6892	1.2408	1.3613	53 42	53.7
36.4	36 24	0.5934	0.8049	0.7373	1.6852	1.2424	1.3564	53 36	53.6
36.5	36 30	0.5948	0.8039	0.7400	1.6812	1.2440	1.3514	53 30	53.5
36.6	36 36	0.5962	0.8028	0.7427	1.6772	1.2456	1.3465	53 24	53.4
36.7	36 42	0.5976	0.8018	0.7454	1.6733	1.2472	1.3416	53 18	53.3
36.8	36 48	0.5990	0.8007	0.7481	1.6694	1.2489	1.3367	53 12	53.2
36.9	36 54	0.6004	0.7997	0.7508	1.6655	1.2505	1.3319	53 06	53.1
37.0°	37°00′	0.6018	0.7986	0.7536	1.6616	1.2521	1.3270	53°00′	53.0°
37.1	37 06	0.6032	0.7976	0.7563	1.6578	1.2538	1.3222	52 54	52.9
37.2	37 12	0.6046	0.7965	0.7590	1.6540	1.2554	1.3175	52 48	52.8
37.3	37 18	0.6060	0.7955	0.7618	1.6502	1.2571	1.3127	52 42	52.7
37.4	37 24	0.6074	0.7944	0.7646	1.6464	1.2588	1.3079	52 36	52.6
37.5	37 30	0.6088	0.7934	0.7673	1.6427	1.2605	1.3032	52 30	52.5
37.6	37 36	0.6101	0.7923	0.7701	1.6390	1.2622	1.2985	52 24	52.4
37.7	37 42	0.6115	0.7912	0.7729	1.6353	1.2639	1.2938	52 18	52.3
37.8	37 48	0.6129	0.7902	0.7757	1.6316	1.2656	1.2892	52 12	52.2
37.9	37 54	0.6143	0.7891	0.7785	1.6279	1.2673	1.2846	52 06	52.1
38.0°	38°00′	0.6157	0.7880	0.7813	1.6243	1.2690	1.2799	52°00′	52.0°
38.1	38 06	0.6170	0.7869	0.7841	1.6207	1.2708	1.2753	51 54	51.9
38.2	38 12	0.6184	0.7859	0.7869	1.6171	1.2725	1.2708	51 48	51.8
38.3	38 18	0.6198	0.7848	0.7898	1.6135	1.2742	1.2662	51 42	51.7
38.4	38 24	0.6211	0.7837	0.7926	1.6099	1.2760	1.2617	51 36	51.6
38.5	38 30	0.6225	0.7826	0.7954	1.6064	1.2778	1.2572	51 30	51.5
38.6	38 36	0.6239	0.7815	0.7983	1.6029	1.2796	1.2527	51 24	51.4
38.7	38 42	0.6252	0.7804	0.8012	1.5994	1.2813	1.2482	51 18	51.3
38.8	38 48	0.6266	0.7793	0.8040	1.5959	1.2831	1.2437	51 12	51.2
38.9	38 54	0.6280	0.7782	0.8069	1.5925	1.2849	1.2393	51 06	51.1
39.0°	39°00′	0.6293	0.7771	0.8098	1.5890	1.2868	1.2349	51°00′	51.0°
39.1	39 06	0.6307	0.7760	0.8127	1.5856	1.2886	1.2305	50 54	50.9
39.2	39 12	0.6320	0.7749	0.8156	1.5822	1.2904	1.2261	50 48	50.8
39.3	39 18	0.6334	0.7738	0.8185	1.5788	1.2923	1.2218	50 42	50.7
39.4	39 24	0.6347	0.7727	0.8214	1.5755	1.2941	1.2174	50 36	50.6
39.5	39 30	0.6361	0.7716	0.8243	1.5721	1.2960	1.2131	50 30	50.5
39.6	39 36	0.6374	0.7705	0.8273	1.5688	1.2978	1.2088	50 24	50.4
39.7	39 42	0.6388	0.7694	0.8302	1.5655	1.2997	1.2045	50 18	50.3
39.8	39 48	0.6401	0.7683	0.8332	1.5622	1.3016	1.2002	50 12	50.2
39.9	39 54	0.6414	0.7672	0.8361	1.5590	1.3035	1.1960	50°06′	50.1°
		cos θ	sin θ	cot θ	sec θ	csc θ	tan θ	θ deg-min	θ deg

Table VI (continued)

θ deg	θ deg-min	sin θ	cos θ	tan θ	csc θ	sec θ	cot θ		
40.0°	40°00′	0.6428	0.7660	0.8391	1.5557	1.3054	1.1918	50°00′	50.0°
40.1	40 06	0.6441	0.7649	0.8421	1.5525	1.3073	1.1875	49 54	49.9
40.2	40 12	0.6455	0.7638	0.8451	1.5493	1.3092	1.1833	49 48	49.8
40.3	40 18	0.6468	0.7627	0.8481	1.5461	1.3112	1.1792	49 42	49.7
40.4	40 24	0.6481	0.7615	0.8511	1.5429	1.3131	1.1750	49 36	49.6
40.5	40 30	0.6494	0.7604	0.8541	1.5398	1.3151	1.1708	49 30	49.5
40.6	40 36	0.6508	0.7593	0.8571	1.5366	1.3171	1.1667	49 24	49.4
40.7	40 42	0.6521	0.7581	0.8601	1.5335	1.3190	1.1626	49 18	49.3
40.8	40 48	0.6534	0.7570	0.8632	1.5304	1.3210	1.1585	49 12	49.2
40.9	40 54	0.6547	0.7559	0.8662	1.5273	1.3230	1.1544	49 06	49.1
41.0°	41°00′	0.6561	0.7547	0.8693	1.5243	1.3250	1.1504	49°00′	49.0°
41.1	41 06	0.6574	0.7536	0.8724	1.5212	1.3270	1.1463	48 54	48.9
41.2	41 12	0.6587	0.7524	0.8754	1.5182	1.3291	1.1423	48 48	48.8
41.3	41.18	0.6600	0.7513	0.8785	1.5151	1.3311	1.1383	48 42	48.7
41.4	41 24	0.6613	0.7501	0.8816	1.5121	1.3331	1.1343	48 36	48.6
41.5	41 30	0.6626	0.7490	0.8847	1.5092	1.3352	1.1303	48 30	48.5
41.6	41 36	0.6639	0.7478	0.8878	1.5062	1.3373	1.1263	48 24	48.4
41.7	41 42	0.6652	0.7466	0.8910	1.5032	1.3393	1.1224	48 18	48.3
41.8	41 48	0.6665	0.7455	0.8941	1.5003	1.3414	1.1184	48 12	48.2
41.9	41 54	0.6678	0.7443	0.8972	1.4974	1.3435	1.1145	48 06	48.1
42.0°	42°00′	0.6691	0.7431	0.9004	1.4945	1.3456	1.1106	48°00′	48.0°
42.1	42 06	0.6704	0.7420	0.9036	1.4916	1.3478	1.1067	47 54	47.9
42.2	42 12	0.6717	0.7408	0.9067	1.4887	1.3499	1.1028	47 48	47.8
42.3	42 18	0.6730	0.7396	0.9099	1.4859	1.3520	1.0990	47 42	47.7
42.4	42 24	0.6743	0.7385	0.9131	1.4830	1.3542	1.0951	47 36	47.6
42.5	42 30	0.6756	0.7373	0.9163	1.4802	1.3563	1.0913	47 30	47.5
42.6	42 36	0.6769	0.7361	0.9195	1.4774	1.3585	1.0875	47 24	47.4
42.7	42 42	0.6782	0.7349	0.9228	1.4746	1.3607	1.0837	47 18	47.3
42.8	42 48	0.6794	0.7337	0.9260	1.4718	1.3629	1.0799	47 12	47.2
42.9	42 54	0.6807	0.7325	0.9293	1.4690	1.3651	1.0761	47 06	47.1
43.0°	43°00′	0.6820	0.7314	0.9325	1.4663	1.3673	1.0724	47°00′	47.0°
43.1	43 06	0.6833	0.7302	0.9358	1.4635	1.3696	1.0686	46 54	46.9
43.2	43 12	0.6845	0.7290	0.9391	1.4608	1.3718	1.0649	46 48	46.8
43.3	43 18	0.6858	0.7278	0.9424	1.4581	1.3741	1.0612	46 42	46.7
43.4	43 24	0.6871	0.7266	0.9457	1.4554	1.3763	1.0575	46 36	46.6
43.5	43 30	0.6884	0.7254	0.9490	1.4527	1.3786	1.0538	46 30	46.5
43.6	43 36	0.6896	0.7242	0.9523	1.4501	1.3809	1.0501	46 24	46.4
43.7	43 42	0.6909	0.7230	0.9556	1.4474	1.3832	1.0464	46 18	46.3
43.8	43 48	0.6921	0.7218	0.9590	1.4448	1.3855	1.0428	46 12	46.2
43.9	43 54	0.6934	0.7206	0.9623	1.4422	1.3878	1.0392	46 06	46.1
44.0°	44°00′	0.6947	0.7193	0.9657	1.4396	1.3902	1.0355	46°00′	46.0°
44.1	44 06	0.6959	0.7181	0.9691	1.4370	1.3925	1.0319	45 54	45.9
44.2	44 12	0.6972	0.7169	0.9725	1.4344	1.3949	1.0283	45 48	45.8
44.3	44 18	0.6984	0.7157	0.9759	1.4318	1.3972	1.0247	45 42	45.7
44.4	44 24	0.6997	0.7145	0.9793	1.4293	1.3996	1.0212	45 36	45.6
44.5	44 30	0.7009	0.7133	0.9827	1.4267	1.4020	1.0176	45 30	45.5
44.6	44 36	0.7022	0.7120	0.9861	1.4242	1.4044	1.0141	45 24	45.4
44.7	44 42	0.7034	0.7108	0.9896	1.4217	1.4069	1.0105	45 18	45.3
44.8	44 48	0.7046	0.7096	0.9930	1.4192	1.4093	1.0070	45 12	45.2
44.9	44 54	0.7059	0.7083	0.9965	1.4167	1.4118	1.0035	45 06	45.1
45.0°	45°00′	0.7071	0.7071	1.0000	1.4142	1.4142	1.0000	45°00′	45.0°
		cos θ	sin θ	cot θ	sec θ	csc θ	tan θ	θ deg-min	θ deg

Odd-Numbered Answers

Exercise 1.1 (page 4)   **1.** {3, 4, 5, 6, 7, 8, 9, 10, 11}   **3.** {a, c, e, h, i, m, s, t}
5. {Monday, Tuesday, Wednesday, Thursday, Friday, Saturday, Sunday}
7. $\neq$ **9.** $=$ **11.** $\not\subset$ **13.** $\subset$ **15.** $\in$ **17.** $\notin$ **19.** {1, 2, 3, 4, 6, 8}   **21.** {6, 8}
23. {5, 6, 7, 8}   **25.** {1, 3, 5, 6, 7, 8}   **27.** {5, 7}   **29.** {1, 3, 5, 6, 7, 8}   **31.** {5, 7}
33. {1, 3}   **35.** {1, 2, 3, 4}   **37.** $\{x \mid x = 2n - 1, n \in N\}$ **39.** $\{x \mid x = 3n, n \in N\}$
41. $\{x \mid x < 100, x \in N\}$ **43.** $B \cap C$ **45.** $A \cap (B \cup C)$ **47.** $A \cap (B \cup C)'$

Exercise 1.2 (page 9)   **1.** False   **3.** True   **5.** True   **7.** False   **9.** False   **11.** False
13. E-2   **15.** E-3   **17.** E-4   **19.** E-1   **21.** F-2   **23.** F-10   **25.** F-3   **27.** F-7
29. F-11   **31.** F-8   **33.** F-4   **35.** F-11   **37.** $ab + a$ **39.** $pq + pr$ **41.** $d^2 + 3d$
43. $(de + g)f$ **45.** $13r + qr$ **47.** $s(t + u + v)$ **49.** $3x + 3y + 3z$

Exercise 1.3 (page 13)   **1.** Theorem 1.11-II   **3.** Theorem 1.1 or 1.4   **5.** Theorem 1.7
7. Theorem 1.11-VII   **9.** Theorem 1.10   **11.** Theorem 1.9   **13.** Theorem 1.11-IV
15. Theorem 1.2   **17.** Theorem 1.9   **19.** Theorem 1.11-IV

Exercise 1.4 (page 17)   **1.** Part II   **3.** Part I   **5.** Part III   **7.** Part IV   **9.** $x > 3$
11. $x \leq 3$ **13.** $x \geq 3$ **15.** $1 \leq x \leq 3$ **17.** $1 \leq x < 3$ **19.** 11   **21.** 11   **23.** -11
25. $\begin{cases} x & \text{if } x \geq 0 \\ -x & \text{if } x < 0 \end{cases}$ **27.** $\begin{cases} 1 - x & \text{if } x \geq 1 \\ x - 1 & \text{if } x < 1 \end{cases}$ **29.** $\begin{cases} 1 + x & \text{if } x \geq -1 \\ -(1 + x) & \text{if } x < -1 \end{cases}$
31. $|-2|$ **33.** 4   **35.** $|-1|$

Exercise 1.5 (page 21)   **1.** $1 - i$ **3.** $9 - i$ **5.** $4 + \frac{5}{2}i$ **7.** $-2i$ **9.** $-\frac{2}{3} + 3i$
11. $-2 + 6i$ **13.** $7 + i$ **15.** $\frac{1}{2} + \frac{1}{2}i$ **17.** $6 + 7i$ **19.** $\frac{2}{5} - \frac{9}{5}i$ **21.** $-1 + 3i$
23. $\frac{1}{10} + \frac{3}{10}i$ **25.** $4 + (3 - \sqrt{2})i$ **27.** $(3 + 3\sqrt{2}) + (9 - \sqrt{2})i$
29. $\frac{3 - 3\sqrt{2}}{11} + \frac{9 + \sqrt{2}}{11}i$ **31.** $(4 + 2\sqrt{3})i$ **33.** 1   **35.** $-i$ **37.** 1   **39.** $-i$

Odd-Numbered Answers

Chapter 1 Review (page 23) 1. $\{1, 4, 6, 8, 9, 10, 12\}$ 2. $\{1, 2, 4, 5, 6, 7, 8, 9, 10, 11, 12\}$
3. $\{2, 3, 5, 6, 7, 9, 11, 12\}$ 4. $\{2, 5, 7, 11\}$ 5. $\{3\}$ 6. $\{1, 4, 6, 8, 9, 10, 12\}$
7. True 8. True 9. False 10. False 11. False 12. True 13. $\{2\}$
14. $\{-2, 0, 2\}$ 15. $\left\{-2, 0, 2, -\frac{1}{2}, \frac{1}{2}\right\}$ 16. $\left\{\sqrt{2}, -\sqrt{2}, \frac{1}{\sqrt{2}}\right\}$ 17. $\left\{\sqrt{-2}, \frac{1}{\sqrt{-2}}\right\}$
18. $\left\{2, \frac{1}{2}, \sqrt{2}, \frac{1}{\sqrt{2}}\right\}$ 19. F-11 20. F-5 21. F-10 22. F-10 23. Theorem 1.1
24. Theorem 1.8-IV 25. Theorem 1.11-VII 26. Theorem 1.8-II 27. Theorem 1.11-IV
28. Theorem 1.3-II 29. Part III 30. Part IV 31. Part II 32. Part III
33. $a < b$ 34. $3 \le a \le 7$ 35. $a \not< b$ or $a \ge b$ 36. $a \ge 2$ 37. 5
38. $\begin{cases} x - 5 & \text{if } x \ge 5 \\ 5 - x & \text{if } x < 5 \end{cases}$ 39. $4 - 2i$ 40. $2 + 4i$ 41. $6 - 8i$ 42. i 43. $2i$
44. $1 - 4i$ 45. i 46. $\frac{2}{3}$

Exercise 2.1 (page 28) 1. $x^2 + x + 1$ 3. $-4x^2 - 4x - 13$ 5. $-x^3 + 2x^2 - x$
7. $\frac{1}{4}x^4 + \frac{1}{6}x^3 + 2x^2 - x + 1$ 9. $2x^2 + 8x + 4$ 11. $5x - 8$ 13. $3x^2 - 4x$
15. $-2x + 1$ 17. $5x^2y$ 19. $-2x^2yz + 2xy^2z - x^3y$ 21. $-x^2 - x - 4$ 23. $-4x - 8$
25. $-4x - 8$ 27. $-2x^2 + 6x + 8$ 29. 2, 2, 8, 0 31. $-1, 0, 1, 2$ 33. 7, 4
35. 25, 19 37. $n, n, n - 2$, no degree

Exercise 2.2 (page 32) 1. $-2x^3y^4$ 3. $-6x^5y^6$ 5. $3x^{2n+2}$ 7. $-2x^{n+2}y^{n+1}$
9. $5x - 10$ 11. $x^2 + 5x + 6$ 13. $3x^2 + 8x - 3$ 15. $6x^2 + 3x - 3$ 17. $-x^2 + 1$
19. $-2x^2 - 5x + 3$ 21. $2x^2 - 4x$ 23. $x^4 + x^2 + x$ 25. $x^3 - 1$
27. $3x^3 + 5x^2 + 2x + 8$ 29. $2x^3 + 5x^2 + 4x + 1$ 31. $x^4 + 2x^3 + 2x^2 + x$
33. $x^4 + 2x^2 + x + 2$ 35. $c^2 + c - 2$, $c^2 + 2ch + h^2 + c + h - 2$, $2ch + h^2 + h$
37. $x^2 + 4x + 5$, $x^4 + 1$, $x^4 + 2x^2 + 1$ 39. $m + n$

Exercise 2.3 (page 36) 1. $2x^2y(xy + 4 + 8y)$ 3. $(x - 4)(x + 4)$ 5. $3x^2(y^2 + 2)(y^2 - 2)$
7. $2(x + 3)^2$ 9. $3y^3(x + 1)^2$ 11. $5x(xyz + 2yz + 3)$ 13. $(x + 5)(x - 2)$
15. $(x - 5)(x - 3)$ 17. $3(x - 1)(x + 5)$ 19. $x(y + 7)(y - 2)$ 21. $(y + 3)(x - 1)$
23. $2(x - 2)(y + 3)$ 25. $(x + 3)(x^2 - 3x + 9)$ 27. $(x + 2)(x^2 + x + 1)$
29. $8(2x - y)(4x^2 + 2xy + y^2)$ 31. $(x^n - 1)(x^n + 1)$ 33. $(x^{2n} - y^n)(x^{2n} + y^n)$
35. $x^{2n}(x^{2n} + 1)^2$ 37. $(x^{2n} + 1)(x^{4n} - x^{2n} + 1)$ 39. $(x^2 + y)(x + 3)$ 41. $(x + 6)(x + 5)$
43. $(x^2 + 1)(y^2 + 3)$ 45. $x(x^2 + 1)(x - 1)(x + 1)$ 47. $(x^2 - xy + y^2)(x^2 + xy + y^2)$
49. $(x^2 - x + 1)(x^2 + x + 1)$

Exercise 2.4 (page 40) 1. $5y^3z$ 3. $6x^3$ 5. $2x^3 + x$ 7. $3x - 2y + 1$
9. $x + 2 + \frac{1}{x}$ 11. $2x^2 - 1 + \frac{3}{x}$ 13. $x^3 + 2x + \frac{x - 4}{x^2}$ 15. $x + 1$

17. $2x^2 + 3x + \frac{7}{2} + \frac{\frac{33}{2}}{2x - 3}$ **19.** $x^4 + 4x^3 + 18x^2 + 72x + 289 + \frac{1152}{x - 4}$

21. $x^2 + 3 + \frac{4x - 4}{x^3 - x}$ **23.** $2x^2 + \frac{3}{2}x - \frac{\frac{1}{2}x}{2x^2 + 1}$ **25.** $x^4 - x^3 + x^2 - x + 1$

Exercise 2.5 (page 43)   **1.** $x + 7$ **3.** $x - 5 - \frac{5}{x + 5}$ **5.** $2x^2 - 4x + 9$

7. $3x^3 + 7x^2 + 12x + 24 + \frac{54}{x - 2}$ **9.** $x^5 + x^4 + x^3 + x^2 + x + 1$

11. $4x^4 + 13x^3 + 37x^2 + 112x + 336 + \frac{1008}{x - 3}$ **13.** $x^2 - 4x + 7 - \frac{8}{x + 1}$

15. $x^3 - 3x^2 + 9x - 26 + \frac{64}{x + 3}$ **17.** $x^2 - 5x + 25 - \frac{114}{x + 5}$

19. $x^4 + 3x^2 + 3x - 6 + \frac{5}{x + 2}$

Exercise 2.6 (page 46)   **1.** $4xy^3$; $x, y \neq 0$ **3.** $2x$, $x \neq -\frac{1}{2}$ **5.** $x + 1$, $x \neq 1$

7. $a - b$, $a \neq -b$ **9.** $b^2(1 - ab)$; $a, b \neq 0$; $ab \neq -1$ **11.** $x - 1$, $x \neq 1$

13. $\frac{x - 4}{x - 1}$; $x \neq 1, -1$ **15.** $\frac{x^2 - x + 1}{(x + 1)^2}$, $x \neq -1$ **17.** $\frac{x^2 + 2x + 4}{x - 3}$, $x \neq 2, 3$

19. $\frac{x^2 - 1}{2}$ **21.** 1, $x \neq 1$ **23.** $\frac{1 + 3x}{4}$, $x \neq \frac{1}{3}$ **25.** $\frac{x^2 - 7}{x}$, $x \neq 0$

27. $\frac{b - 2c}{a + c}$; $a \neq -c$, $b \neq -2c$ **29.** 21 **31.** $x^2y^2z^2$; $x, y, z \neq 0$ **33.** $x^2 + 2x + 1$, $x \neq 1, -1$

35. $x^2 - x + 1$, $x \neq -1$ **37.** No, fails for $x = 1, 2$ **39.** $a < b$, $a > b$

Exercise 2.7 (page 50)   **1.** 1890 **3.** 336 **5.** $x^3(x - 1)(x + 4)$ **7.** $\frac{x}{(x + 1)}$

9. $\frac{y^2 + 3y - 2}{4y^2}$ **11.** $\frac{2}{(x - 1)(x + 1)}$ **13.** $\frac{1 + x}{2 - x}$ **15.** $\frac{2w - 1}{(w - 1)(w - 2)(w + 1)}$

17. $\frac{x^2 - 3x - 1}{(x - 2)(x - 3)^2}$ **19.** $\frac{x^2 - 7x + 8}{(x + 1)(x - 3)^2}$ **21.** $\frac{x^2 + 2x - 2}{x^2 - 1}$ **23.** $\frac{z^3 - z - 1}{z^2}$

25. $\frac{x^2 - 2}{x^2 - 1}$ **27.** $\frac{4}{(x + 1)(x + 3)}$ **29.** $\frac{x - ax + a^2}{(x - a)(x + a)}$

Exercise 2.8 (page 54)   **1.** $\frac{acx}{y}$ **3.** $\frac{20bz}{9y}$ **5.** $\frac{x}{x - 1}$ **7.** $\frac{x^2 + x - 2}{x^2 - x - 2}$ **9.** $\frac{a^2b^2}{a + b}$

11. $\dfrac{z^2-1}{z^2}$ **13.** $\dfrac{1}{x+5}$ **15.** $\dfrac{x-2}{x+3}$ **17.** $\dfrac{3}{2x+2}$ **19.** $\dfrac{x^2+2x-1}{x^2-1}$ **21.** $\dfrac{a+1}{a+3}$
23. $\dfrac{x-3}{x+3}$ **25.** $\dfrac{x+y}{x-y}$

Exercise 2.9 (page 58) **1.** $\dfrac{4}{x}-\dfrac{4}{x+1}$ **3.** $\dfrac{-1}{x+1}+\dfrac{2}{x+2}$ **5.** $\dfrac{-4}{x+2}+\dfrac{5}{x+3}$

7. $\dfrac{\frac{1}{2}}{x}+\dfrac{1}{x+1}+\dfrac{-\frac{3}{2}}{x+2}$ **9.** $\dfrac{\frac{1}{2}}{x}+\dfrac{-3}{x+1}+\dfrac{\frac{7}{2}}{x+2}$ **11.** $\dfrac{\frac{1}{2}}{x}+\dfrac{-1}{x+1}+\dfrac{\frac{1}{2}}{x+2}$

13. $\dfrac{-\frac{3}{4}}{x}+\dfrac{-\frac{1}{2}}{x^2}+\dfrac{\frac{3}{4}}{x-2}$ **15.** $\dfrac{\frac{1}{2}}{x-1}+\dfrac{\frac{1}{4}}{(x-1)^2}+\dfrac{\frac{1}{2}}{x+1}+\dfrac{-\frac{1}{4}}{(x+1)^2}$

17. $\dfrac{-1}{x}+\dfrac{-1}{x^2}+\dfrac{1}{x-1}$ **19.** $x^2+\dfrac{2}{x}+\dfrac{-1}{x+1}$ **21.** $x+2+\dfrac{3}{x-1}+\dfrac{1}{(x-1)^2}$

Chapter Review 2 (page 61) **1.** $2x^2-2x-1$ **2.** $6x^2$ **3.** 13 **4.** -3
5. $6x^2+10x+4$ **6.** $3x^3+7x^2+3x+2$ **7.** $(y-5)(y-3)$ **8.** $x(x+2)^2$
9. $(x^2+4)(x-2)(x+2)$ **10.** $(z-4)(z^2+4z+16)$ **11.** $6xy^2z$ **12.** $6y^2+7y+1$
13. $2x-5$ **14.** $3x+2$ **15.** x^3+2x^2-x+5 **16.** $3x^2-4+\dfrac{3}{x+2}$
17. $2y+11z;\ y,z\neq 0$ **18.** $x+7,\ x\neq -3$ **19.** $\dfrac{x+7}{x+3};\ x\neq 3,-3$
20. $\dfrac{x+2}{x-2};\ x\neq 2,-2$ **21.** $\dfrac{7x+3}{12}$ **22.** $\dfrac{6x}{x^2+x-2}$ **23.** $\dfrac{x^2-2x+1}{x+1}$
24. $\dfrac{y^2-4y+8}{y^2-y-2}$ **25.** $\dfrac{x^4 z}{y^2}$ **26.** $\dfrac{x^2+8x+7}{x+5}$ **27.** $\dfrac{x^2+x-2}{x+3}$ **28.** $\dfrac{x-3}{x+1}$
29. $\dfrac{2}{2x-3}-\dfrac{1}{x+1}$ **30.** $\dfrac{-1}{x}+\dfrac{1}{x-1}+\dfrac{1}{x+1}$

Exercise 3.1 (page 66) **1.** $\dfrac{1}{3}$ **3.** $-\dfrac{1}{8}$ **5.** 4 **7.** $\dfrac{1}{12}$ **9.** 3 **11.** $\dfrac{1}{2}$ **13.** x^3
15. $\dfrac{1}{x^2 y}$ **17.** $x^{10}y^5$ **19.** $\dfrac{x^6}{8y^3}$ **21.** $\dfrac{4x^2 z^2}{y^4}$ **23.** $\dfrac{z^8}{x^3 y^4}$ **25.** $\dfrac{y^2+x}{xy^2}$ **27.** $\dfrac{x^2-y^2}{xy}$
29. $\dfrac{x^3 y-1}{xy^2}$ **31.** $y+x$ **33.** $x+y$ **35.** $\dfrac{x^4 y^2}{x^4+2x^2 y+y^2}$ **37.** x^{2n-3} **39.** y^2
41. y^{5n+1} **43.** x^{2n+2} **45.** $x^{n+1}y^{n+3}$ **47.** $\dfrac{x^{n-1}}{y}$ **49.** $\dfrac{1-x^2}{x^2}$ **51.** $\dfrac{x^2+2xy+y^2}{xy}$
53. $\dfrac{2x^5-x^3+2x^2-1}{x^3}$ **55.** $\dfrac{y^3+4x^4}{x^3 y}$ **57.** $\dfrac{3x+4y}{x^4 y^4}$ **59.** $\dfrac{xy^3 z^6-x^2}{yz^3}$

61. 1.9711×10^4 **63.** 1.976×10^3 **65.** 5.86×10^{-2} **67.** 1.001×10^{-2}

Exercise 3.2 (page 71)   **1.** 3   **3.** $\frac{1}{8}$ **5.** 4   **7.** 4   **9.** $\frac{1}{4}$ **11.** $\frac{27}{8}$ **13.** $x^{5/6}$
15. $y^{4/3}$ **17.** $y^{1/3}$ **19.** $4x^{3/4}$ **21.** $x^{16}y^3$ **23.** $\frac{5}{3}x^3y$ **25.** $9x^2y^4$ **27.** $x^{3n/2}y^{3n+3}$
29. a^{5n-3} **31.** $x^{3/2} + x^2$ **33.** $x - x^{1/2}$ **35.** $y^{5/3} - y$ **37.** $x - y$
39. $x + y - (x+y)^{3/2}$ **41.** $x^{1/2} + 1$ **43.** $1 + x$ **45.** $(x+1) - 1$ **47.** 9
49. $3|x|$ **51.** $\frac{1}{x^2(1-x)^{1/2}}$

Exercise 3.3 (page 74)   **1.** $\sqrt{5}$ **3.** $2\sqrt[3]{x}$ **5.** $3\sqrt{y}$ **7.** $\sqrt[5]{x^3}$ **9.** $x\sqrt[3]{y}$ **11.** $\sqrt[3]{xy}$
13. $\sqrt{x^2+y^2}$ **15.** $\frac{1}{\sqrt[4]{(x^2-y^2)^3}}$ **17.** $2x^{4/3}$ **19.** $x^{1/2}y^{3/4}$ **21.** $(x+y)^{3/2}$
23. $(x^2+1)^{1/3}$ **25.** -3 **27.** -2 **29.** x^3y **31.** $3xy^2$ **33.** $x^3\sqrt{x}$
35. $-3x\sqrt[3]{x}$ **37.** $x\sqrt{y}$ **39.** $4x^4$ **41.** x **43.** $2x\sqrt[3]{y}$ **45.** $\frac{\sqrt{2}}{2}$ **47.** $\frac{2\sqrt{2x}}{x}$
49. $\frac{\sqrt[3]{2}}{2}$ **51.** $\frac{\sqrt[3]{y}}{y}$ **53.** $\frac{1}{\sqrt{2}}$ **55.** $\frac{x}{y\sqrt[3]{x^2}}$ **57.** $\sqrt[3]{9}$ **59.** $2\sqrt{x}$ **61.** $\sqrt{2x}$
63. $(x-1)^2$

Exercise 3.4 (page 78)   **1.** $7\sqrt{5}$ **3.** $5\sqrt{3}$ **5.** $-\sqrt{3}$ **7.** $8\sqrt{2}$ **9.** $7\sqrt[3]{2}$
11. $-\sqrt[4]{2}$ **13.** $6 - 3\sqrt{3}$ **15.** 2   **17.** $-3 - \sqrt{5}$ **19.** $3x\sqrt{2}$ **21.** $2x - 3 + \sqrt{x}$
23. $3 - x$ **25.** $\frac{1+\sqrt{3}}{2}$ **27.** $\frac{x + x\sqrt{x}}{1-x}$ **29.** $\frac{3x\sqrt{x}+6x}{x-4}$ **31.** $\frac{\sqrt{x+1}+1}{x}$
33. $\frac{\sqrt{xy}+y}{x-y}$ **35.** $2x + 1 + 2\sqrt{x^2+x}$ **37.** $\frac{1}{\sqrt{3}-1}$
39. $\frac{1-x}{x\sqrt{2}-x\sqrt{x+1}}$ **41.** $\frac{x-y}{x+y-2\sqrt{xy}}$

Chapter 3 Review (page 80)   **1.** $x^{12}y^8$ **2.** $\frac{y^3}{x^3}$ **3.** $\frac{y}{x^2}$ **4.** x^3 **5.** $\frac{x^2+1}{x^3}$
6. $xy^2 + x^2$ **7.** x^{-1} **8.** $x^n y^{2-n}$ **9.** 3.51×10^4 **10.** 1.8×10^{-4} **11.** $x^{7/6}$
12. $\frac{y}{x^{1/2}}$ **13.** $\frac{1}{x^n y^{2n}}$ **14.** $\frac{1}{xy}$ **15.** $y^{4/3} + y$ **16.** $4y^2 - y$ **17.** $1 + x^{3/4}$
18. $x^{4/3} - 1$ **19.** $2x^2 y\sqrt{xy}$ **20.** $6x\sqrt{y}$ **21.** $x\sqrt{3}$ **22.** $3x$ **23.** $\frac{\sqrt[3]{xy^2}}{y}$
24. $\frac{1}{2\sqrt{3y}}$ **25.** $\sqrt{5}$ **26.** $\sqrt{3xy}$ **27.** $-3\sqrt{2}$ **28.** $8\sqrt[3]{5}$ **29.** 1   **30.** $2x - 5 - 3\sqrt{x}$

31. $\dfrac{3+\sqrt{2}}{7}$ **32.** $\dfrac{2x+1+3\sqrt{x}}{4x-1}$ **33.** $\dfrac{1}{\sqrt{3}-1}$ **34.** $\dfrac{x-1}{x+2+3\sqrt{x}}$

Exercise 4.1 (page 85) **1.** $\{11\}$ **3.** $\{5\}$ **5.** $\{2\}$ **7.** $\{9\}$ **9.** $\{0\}$ **11.** $\left\{\dfrac{1}{2}\right\}$

13. Any x; $x \neq 1, -1$ **15.** $\{0\}$ **17.** $k = v - gt$ **19.** $c = \dfrac{2A - bh}{h}$, $h \neq 0$

21. $n = \dfrac{l - a + d}{d}$, $d \neq 0$ **23.** $y' = \dfrac{1 + 3x}{x^2 - 2y^3}$, $x^2 \neq 2y^3$

25. $x_1 = \dfrac{x_4}{x_2 - 2x_3}$, $x_2 \neq 2x_3$ **27.** $y = 6x - 6x_1 + y_1$, $x \neq x_1$ **29.** $k = \dfrac{-3}{5}$

Exercise 4.2 (page 89) **1.** $\{-2, 1\}$ **3.** $\{0, 5\}$ **5.** $\{1\}$ **7.** $\left\{-\dfrac{1}{2}, -\dfrac{1}{3}\right\}$

9. $\left\{-\dfrac{3}{2}, -3\right\}$ **11.** $\{1, 3\}$ **13.** $\{1, -3\}$ **15.** $\{3i, -3i\}$ **17.** $\{1 + i, 1 - i\}$

19. $\{1, -9\}$ **21.** $\{-6, -5\}$ **23.** $\left\{2, -\dfrac{1}{3}\right\}$ **25.** $\left\{\dfrac{1}{2} + \dfrac{\sqrt{17}}{2}, \dfrac{1}{2} - \dfrac{\sqrt{17}}{2}\right\}$

27. $\left\{\dfrac{3}{2} + \dfrac{\sqrt{3}i}{2}, \dfrac{3}{2} - \dfrac{\sqrt{3}i}{2}\right\}$ **29.** $\left\{\dfrac{3i}{2}, \dfrac{-3i}{2}\right\}$ **31.** $(x - 2)^2 + (y - 2)^2 = 25$

33. $(x + 3)^2 + (y - 1)^2 = 4$ **35.** $\left(x - \dfrac{1}{2}\right)^2 + (y + 1)^2 = 4$ **37.** $y = (x - 1)^2 + 4$

39. $y = (x + 4)^2 - 12$ **41.** $y = \left(x - \dfrac{3}{2}\right)^2 + \dfrac{11}{4}$ **43.** $x = 1$ yields $y = 4$

45. $x = -4$ yields $y = -12$ **47.** $x = \dfrac{3}{2}$ yields $y = \dfrac{11}{4}$ **49.** $x^2 - 5x + 6 = 0$

51. $6x^2 + x - 2 = 0$

Exercise 4.3 (page 93) **1.** $\{3, 4\}$ **3.** $\left\{-2, \dfrac{3}{4}\right\}$ **5.** $\left\{-1, -\dfrac{1}{2}\right\}$

7. $\left\{\dfrac{-1 + \sqrt{13}}{2}, \dfrac{-1 - \sqrt{13}}{2}\right\}$ **9.** $\left\{-1, -\dfrac{1}{2}\right\}$ **11.** $\left\{\dfrac{3 + \sqrt{3}}{2}, \dfrac{3 - \sqrt{3}}{2}\right\}$

13. $\left\{1, \dfrac{1}{2}\right\}$ **15.** $\{-4 + i, -4 - i\}$ **17.** $\left\{\dfrac{-3 + i}{2}, \dfrac{-3 - i}{2}\right\}$

19. $\left\{\dfrac{-3 + \sqrt{11}i}{2}, \dfrac{-3 - \sqrt{11}i}{2}\right\}$ **21.** $\left\{\dfrac{1 + \sqrt{7}i}{4}, \dfrac{1 - \sqrt{7}i}{4}\right\}$

23. Let $r_1 = \dfrac{-b + \sqrt{b^2 - 4ac}}{2a}$ and $r_2 = \dfrac{-b - \sqrt{b^2 - 4ac}}{2a}$, then simplify $r_1 + r_2$ and $r_1 \cdot r_2$

25. $-\dfrac{3}{2}, -3$ **27.** $3, -\dfrac{5}{3}$

Exercise 4.4 (page 97) 1. $\{64\}$ 3. $\{-7\}$ 5. $\{4\}$ 7. $\{-25\}$ 9. $\{16\}$ 11. $\{1\}$ 13. $\{5\}$ 15. $\{16\}$ 17. $\{4\}$ 19. $\{1, 3\}$ 21. $A = \pi r^2$ 23. $y = \dfrac{1}{x^3}$ 25. $y = \pm \sqrt{a^2 - x^2}$

Exercise 4.5 (page 98) 1. $\{1, -1, 3, -3\}$ 3. $\{2, -2, \sqrt{3}, -\sqrt{3}\}$ 5. $\{2, -9\}$ 7. $\{1\}$ 9. $\{216, -27\}$ 11. $\left\{\dfrac{1}{9}, -\dfrac{1}{2}\right\}$ 13. 15. $\{16, 256\}$ 17. $\{87\}$ 19. $\left\{-\dfrac{25}{8}, -\dfrac{11}{4}\right\}$ 21. $\{0, -2, -1 + \sqrt{2}, -1 - \sqrt{2}\}$

Exercise 4.6 (page 102) 1. $(-6, +\infty)$ 3. $(-\infty, 3]$

5. $[-6, +\infty)$ 7. $\left[\dfrac{5}{2}, +\infty\right)$

9. $(-\infty, -1)$ 11. $[-2, +\infty)$

13. $[-6, 1]$ 15. $\left(-\dfrac{22}{3}, -4\right]$

17. $[-25, -16]$ 19. $(0, 4]$

21. $(0, 1]$ 23. The intersection is the empty set.

25. 27.

29.

Odd-Numbered Answers

Exercise 4.7 (page 107) **1.** $(-\infty, -1) \cup (5, +\infty)$ **3.** $\left[-\frac{1}{2}, \frac{8}{3}\right]$ **5.** $(-\infty, -3) \cup (2, +\infty)$
7. $(-1, 1)$ **9.** $(-\infty, 2 - \sqrt{3}] \cup [2 + \sqrt{3}, +\infty)$ **11.** No solution
13. $\left(-\infty, -\frac{1}{2}\right] \cup (0, +\infty)$ **15.** $(-\infty, 1)$ **17.** $(-2, 0)$ **19.** $(-1, 1) \cup (3, +\infty)$
21. $(-\infty, -1) \cup (0, 2)$ **23.** $[-3, -1] \cup [1, +\infty)$ **25.** $(-\infty, -4) \cup (-2, 0) \cup (2, +\infty)$

Exercise 4.8 (page 111) **1.** $\{-5, 1\}$ **3.** $\{-1, 7\}$ **5.** $\{1, -2\}$ **7.** $(-3, 7)$
9. $\left(-\frac{1}{6}, \frac{7}{6}\right)$ **11.** $(-\infty, 3) \cup (5, +\infty)$ **13.** $(-\infty, -10) \cup (4, +\infty)$ **15.** $[3, 5]$
17. $[0, 3]$ **19.** $\left(-\infty, -\frac{3}{2}\right] \cup \left[\frac{5}{2}, +\infty\right)$ **21.** $\left(-\infty, -\frac{2}{3}\right] \cup \left[\frac{4}{3}, +\infty\right)$ **23.** $[-3, 3]$
25. $|x| \leq 3$ **27.** $|x - 2| < 4$ **29.** $|x + 5| \leq 3$ **31.** $\left|x + \frac{1}{3}\right| < 1$

Exercise 4.9 (page 113) **1.** 15 dimes, 17 nickels **3.** 15 pounds of 32% silver alloy
5. $1200 at 6%, $800 at 7% **7.** $11,000 in 4%, $33,000 in 6%
9. Rate of automobile is 60 miles per hour, rate of plane is 180 miles per hour **11.** 400 miles
13. 11 and 14 **15.** 11 and 12; -11 and -12 **17.** $\sqrt{13,000}$ miles **19.** $\frac{1}{2}$ second, $3\frac{1}{2}$ seconds
21. 6 days for the father alone, 12 days for the son alone **23.** A grade of at least 94%

Chapter 4 Review (page 117) **1.** $\left\{-11\frac{1}{2}\right\}$ **2.** $\{3\}$ **3.** $\left\{-\frac{5}{2}\right\}$ **4.** $\left\{\frac{3}{5}\right\}$ **5.** $y = \frac{x}{4}$
6. $x = 4y$ **7.** $\{3, -2\}$ **8.** $\{-7, 3\}$ **9.** $\{\sqrt{5}, -\sqrt{5}\}$ **10.** $\{5 + \sqrt{2}, 5 - \sqrt{2}\}$
11. $\left\{\frac{-5}{2} + \frac{\sqrt{33}}{2}, \frac{-5}{2} - \frac{\sqrt{33}}{2}\right\}$ **12.** $\left\{\frac{1}{2}, -1\right\}$ **13.** $(x - 4)^2 + (y + 3)^2 = 37$
14. $y = \left(x - \frac{b}{2}\right)^2 + \left(2 - \frac{b^2}{4}\right)$ **15.** $\frac{1 \pm i\sqrt{15}}{4}$ **16.** $\{-1 \pm \sqrt{5}\}$ **17.** $\left\{\frac{3 \pm \sqrt{9 - 4k}}{2k}\right\}$
18. $\left\{\frac{-k \pm \sqrt{k^2 + 16}}{2}\right\}$ **19.** $\{4, 9\}$ **20.** $\{8\}$ **21.** $\{1, -1, \sqrt{2}, -\sqrt{2}\}$ **22.** $\left\{\frac{1}{7}, -\frac{1}{6}\right\}$
23. $[38, \infty)$ **24.** $\left(-\infty, -\frac{6}{5}\right)$ **25.** ⟵┼┼┼┼┼●●⟶ **26.** ⟵●┼┼┼┼┼●⟶
27. $(-5, 2)$ **28.** $\left(\frac{1}{3}, 1\right)$ **29.** $\left\{-2, \frac{4}{3}\right\}$ **30.** $\left\{-1, \frac{7}{3}\right\}$
31. $(-\infty, -1) \cup (7, \infty)$ ⟵┼┼●┼┼┼┼●┼┼⟶ **32.** $|x + 1| \leq 4$ **33.** 2113 and 2263 votes
34. 6 centimeters **35.** 6 miles per hour going; 4 miles per hour returning

Exercise 5.1 (page 124)   **1.** $\{(0, 6), (1, 4), (2, 2), (-3, 12), (\frac{2}{3}, \frac{14}{3})\}$

3. $\{(0, 0), (1, -3), (2, 3), (-3, -\frac{9}{7}), (\frac{2}{3}, -\frac{9}{7})\}$

5. $\{(0, \sqrt{11}), (1, \sqrt{14}), (2, \sqrt{17}), (-3, \sqrt{2}), (\frac{2}{3}, \sqrt{13})\}$ **7. a.** domain: $\{2, 5, 7\}$ **b.** function

9. a. domain: $\{2, 3\}$ **b.** not a function   **11. a.** domain: $\{5, 6, 7\}$ **b.** function   **13.** domain: $\{x \mid x \in R\}$
15. domain: $\{x \mid x \in R\}$ **17.** domain: $\{x \mid x \in R, x \neq 2\}$ **19.** domain: $\{x \mid x \in R, x \geq 0\}$
21. domain: $\{x \mid x \in R, -2 \leq x \leq 2\}$ **23.** domain: $\{x \mid x \in R, x \neq 0, 1\}$ **25.** 2   **27.** -1 **29.** $a + 2$
31. $2a + 2$ **33.** 9   **35.** 4   **37.** $a^2 - 4a + 4$ **39.** $\dfrac{a^2 - 4a + 4}{4}$ **41.** 1   **43.** $a - 2$
45. ± 1 **47.** ± 3 **49. a.** 2   **b.** 0   **c.** 0   **d.** -12 **51.** 3   **53.** 1
55. No. Both $(1, 1)$ and $(1, -1)$ are solutions.   **57. a.** even   **b.** odd

Exercise 5.2 (page 130)

1.   **3.**   **5.**   **7.**

9. 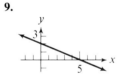   **11.**   **13.**   **15.** Range: $\{y \mid -1 \leq y \leq 3\}$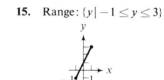

17. Distance, 5; slope, $\dfrac{4}{3}$ **19.** Distance, 13; slope $\dfrac{12}{5}$ **21.** Distance, $\sqrt{2}$; slope, 1

23. Distance, $3\sqrt{5}$; slope, $\dfrac{1}{2}$ **25.** Distance, 5; slope 0   **27.** Distance, 10; slope, undefined

29. 7, $\sqrt{68}$, $\sqrt{89}$ **31.** 10, 21, 17

33. From Exercise 30, the lengths of the sides of the triangle are 15, $3\sqrt{5}$, and $6\sqrt{5}$, respectively; since $15^2 = 225$, $(3\sqrt{5})^2 = 45$, and $(6\sqrt{5})^2 = 180$, it follows that $15^2 = (3\sqrt{5})^2 + (6\sqrt{5})^2$, and the converse of the Pythogorean theorem applies. Hence, the triangle is a right triangle.   **35.** $x - 2y = 8$

37. This follows from the fact that the midpoint of the segment PR is $\left(\dfrac{x_1 + x_2}{2}, y_1\right)$ and the midpoint of QR is $\left(x_2, \dfrac{y_1 + y_2}{2}\right)$. Then, since $\triangle POM$ is similar to $\triangle PQR$, $PO = \dfrac{1}{2} PQ$. Hence, $\left(\dfrac{x_1 + x_2}{2}, \dfrac{y_1 + y_2}{2}\right)$ is the midpoint of PQ.

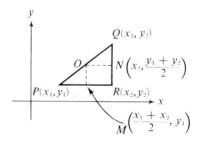

Odd-Numbered Answers

Exercise 5.3 (page 134)   **1.** $4x - y - 7 = 0$ **3.** $x + y - 10 = 0$ **5.** $3x - y = 0$
7. $x + 2y + 2 = 0$ **9.** $3x + 4y + 14 = 0$ **11.** $y - 2 = 0$
13. $y = -x + 3$; slope, -1; intercept, 3   **15.** $y = -\frac{3}{2}x + \frac{1}{2}$; slope, $-\frac{3}{2}$; intercept, $\frac{1}{2}$
17. $y = \frac{1}{3}x - \frac{2}{3}$; slope, $\frac{1}{3}$; intercept, $-\frac{2}{3}$ **19.** $3x + 2y - 6 = 0$ **21.** $5x + 2y + 10 = 0$
23. $6x - 2y + 3 = 0$ **25.** $3x - y - 5 = 0$ **27.** $2x - 3y + 19 = 0$ **29.** $x - 2 = 0$
31. Substituting $\frac{y_2 - y_1}{x_2 - x_1}$ for m in the formula $y - y_1 = m(x - x_1)$ gives the desired result.
33. $y = 0$ **35.** $x - y = 0$

Exercise 5.4 (page 140)

1. $y, 4$; x, 1 and 4; $\left(\frac{5}{2}, -\frac{9}{4}\right)$, minimum   **3.** $y, -7$; x, -1 and 7; $(3, -16)$, minimum

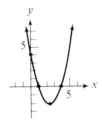
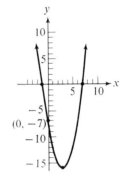

5. $y, -4$; x, 1 and 4; $\left(\frac{5}{2}, \frac{9}{4}\right)$, maximum   **7.** $y, 2$; no x; $(0, 2)$, minimum

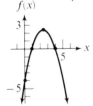

9.

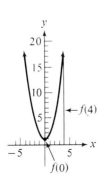

11. 4; 4

13. Varying k has the effect of translating the graph along the y-axis.

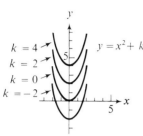

15. a. Parabola
b. No; there are two values of y associated with each value of x, except $x = 0$.
c. No

17.

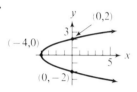

19.

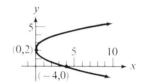

Exercise 5.5 (page 144)   **1.** $3; 17; 47$ **3.** $56; 12; 326$ **5.** $-685; 95; 719$

7. 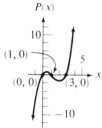   **9.** 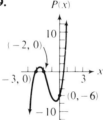   **11.** 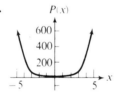   **13.**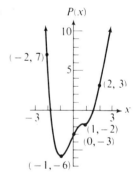

Exercise 5.6 (page 149)   **1.** $x = 3$ **3.** $x = 3; x = -2$ **5.** $x = -1; x = -4$ **7.** $x = -1$

9. 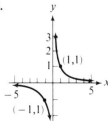   **11.** 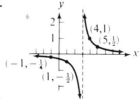   **13.** 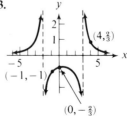   **15.**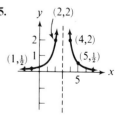

Odd-Numbered Answers   495

17. $x = 2$; $x = -2$; $y = 0$ **19.** $x = 4$; $y = x + 4$ **21.** $x = 4$; $x = -1$; $y = 1$

23. 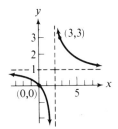   **25.** 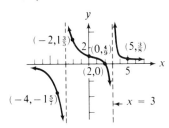   **27.**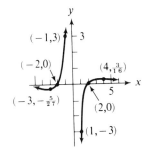

29. 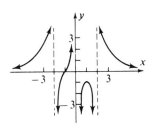   **31.**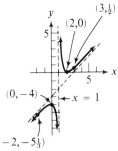

33. Behaves like $y = x^2$ as $|x|$ becomes large.

35. No, because if $x = a$ is a vertical asymptote, then the function is not defined at $x = a$; thus there is no y so that the point (a, y) is on the graph.

37. $(1, 3)$

Chapter 5 Review (page 151)   **1.** $\{4, 2, 3\}$ **2.** $\{1\}$ **3.** $\{x \mid x \neq -4\}$ **4.** $\{x \mid x \geq 6\}$ **5.** 13
6. -5 **7.** $x - h - 3$ **8.** $x^2 + 2xh + h^2 + 4$ **9.** $\sqrt{41}$, $m = -\dfrac{4}{5}$ **10.** $\sqrt{5}$, $m = -\dfrac{1}{2}$
11. $4x - y - 15 = 0$ **12.** $x - 2y + 12 = 0$ **13.** $m = -4$, y-intercept $= 6$
14. $m = \dfrac{3}{2}$, y-intercept $= -8$ **15.** $2x - 3y + 6 = 0$ **16.** $12x - y - 4 = 0$
17. $4x - 3y + 11 = 0$ **18.** $x + 6y + 21 = 0$
19. x-intercepts $3, -2$; axis of symmetry $x = \dfrac{1}{2}$; minimum pt. $\left(\dfrac{1}{2}, \dfrac{-25}{4}\right)$
20. x-intercepts $5, 2$; axis of symmetry $x = \dfrac{7}{2}$; maximum pt. $\left(\dfrac{7}{2}, \dfrac{9}{4}\right)$

21. 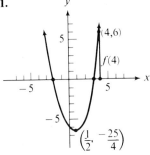   **22.** 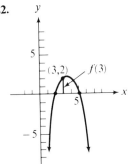   **23.** 1   **24.** 53

25. $(-4, -10), (-3, 0), (-2, 0), (-1, -4), (0, -6), (1, 0), (2, 20), (3, 60), (4, 126)$

26.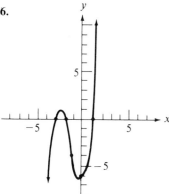

27.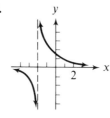

vertical asymptote: $x = -2$
horizontal asymptote: $y = 0$

28.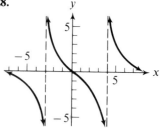

vertical asymptotes: $x = -3; x = 4$
horizontal asymptote: $y = 0$

29.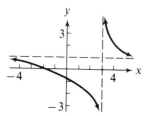

vertical asymptote: $x = 3$
horizontal asymptote: $y = 1$

30.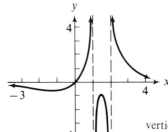

vertical asymptotes: $x = 1; x = 2$
horizontal asymptote: $y = 0$

Exercise 6.1 (page 156)

1.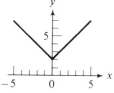

3.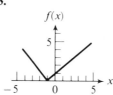

5.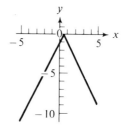

7.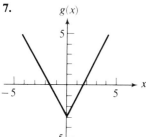

9.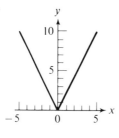

11.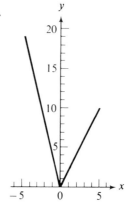

Odd-Numbered Answers

13.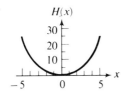

15.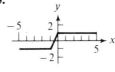

17.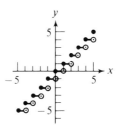

19.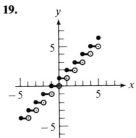

21.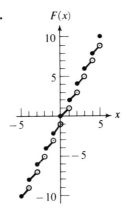

23.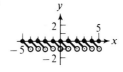

25.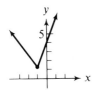

27.

29. Cost in cents: $C = -c[-x]$, $x > 0$

Exercise 6.2 (page 159)

1.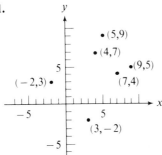

3.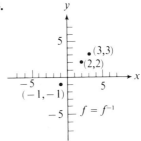

5.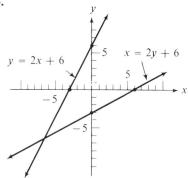

7.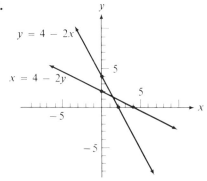

9.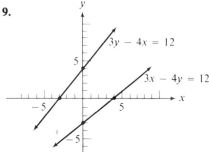

11.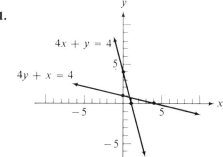

13.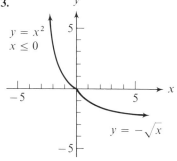

15. $F(x) = x$, $F^{-1}(x) = x$; $F[F^{-1}(x)] = F(x) = x$, $F^{-1}[F(x)] = F^{-1}(x) = x$

17. $F(x) = 4 - 2x$, $F^{-1}(x) = \dfrac{4-x}{2}$;

$F[F^{-1}(x)] = F\left[\dfrac{4-x}{2}\right] = 4 - 2\left(\dfrac{4-x}{2}\right)$
$= 4 - 4 + x = x$,

$F^{-1}[F(x)] = F^{-1}[4 - 2x] = \dfrac{4 - (4-2x)}{2}$
$= \dfrac{4 - 4 + 2x}{2} = \dfrac{2x}{2} = x$

19. $F(x) = \dfrac{3x - 12}{4}$, $F^{-1}(x) = \dfrac{4x + 12}{3}$;

$F[F^{-1}(x)] = F\left[\dfrac{4x+12}{3}\right] = \dfrac{(3)\left(\dfrac{4x+12}{3}\right) - 12}{4} = \dfrac{4x + 12 - 12}{4} = \dfrac{4x}{4} = x$,

$F^{-1}[F(x)] = F^{-1}\left[\dfrac{3x-12}{4}\right] = \dfrac{(4)\left(\dfrac{3x-12}{4}\right) + 12}{3} = \dfrac{3x - 12 + 12}{3} = \dfrac{3x}{3} = x$

Odd-Numbered Answers

21. $F(x) = x^2 + 2, \ x \geq 0; \ F^{-1}(x) = \sqrt{x-2}$
$F[F^{-1}(x)] = (\sqrt{x-2})^2 + 2 = x - 2 + 2 = x$
$F^{-1}[F(x)] = \sqrt{(x^2+2) - 2} = \sqrt{x^2} = x$ (since $x \geq 0$)

Exercise 6.3 (page 163)   **1.** 1, 16   **3.** 400 feet   **5.** 160 pounds per square foot
7. 1,687.5 pounds   **9.** 16   **11.** 160 pounds per square foot   **13.** 1,687.5 pounds

15. Let c_1 and d_1 be the circumference and diameter of one circle and c_2 and d_2 be corresponding parts in a second circle; then $c_2 = \pi d_2$ and $\pi = \dfrac{c_2}{d_2}$; similarly $c_1 = \pi d_1$, and substituting for π yields $c_1 = \dfrac{c_2}{d_2} d_1$; and since $c_2 \neq 0$, multiplying both sides by $\dfrac{1}{c_2}$ yields $\dfrac{c_1}{c_2} = \dfrac{d_1}{d_2}$.

17.   **19.** "Increases" the curvature 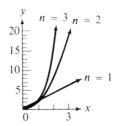   **21.** The curves decrease faster.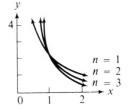

Exercise 6.4 (page 171)   **1. a.** $y = \pm\sqrt{4-x^2}$ **b.** $y = \sqrt{4-x^2}; \ y = -\sqrt{4-x^2}$
c. Domain $= \{x \mid -2 \leq x \leq 2\}$ **3. a.** $y = \pm 3\sqrt{4-x^2}$ **b.** $y = 3\sqrt{4-x^2}; \ y = -3\sqrt{4-x^2}$
c. Domain $= \{x \mid -2 \leq x \leq 2\}$ **5. a.** $y = \pm\dfrac{1}{2}\sqrt{16-x^2}$ **b.** $y = \dfrac{1}{2}\sqrt{16-x^2}; \ y = -\dfrac{1}{2}\sqrt{16-x^2}$
c. Domain $= \{x \mid -4 \leq x \leq 4\}$ **7. a.** $y = \pm\dfrac{1}{\sqrt{3}}\sqrt{24-2x^2}$
b. $y = \dfrac{1}{\sqrt{3}}\sqrt{24-2x^2}; \ y = \dfrac{-1}{\sqrt{3}}\sqrt{24-2x^2}$ **c.** Domain $= \{x \mid -\sqrt{12} \leq x \leq \sqrt{12}\}$
9. a. $y = \pm\sqrt{x^2-1}$ **b.** $y = \sqrt{x^2-1}; \ y = -\sqrt{x^2-1}$ **c.** Domain $= \{x \mid x \geq 1 \text{ or } \leq -1\}$
11. a. $y = \pm\sqrt{9+x^2}$ **b.** $y = \sqrt{9+x^2}; \ y = -\sqrt{9+x^2}$ **c.** Domain $= \{x \mid x \in R\}$

13. 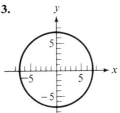   **15.** 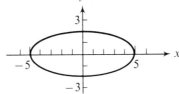   **17.**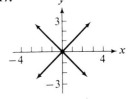

19. 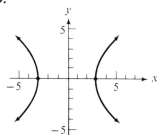   **21.** 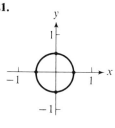   **23.**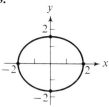

25. Pair of straight lines intersecting at the origin   **27.** A line, the y-axis

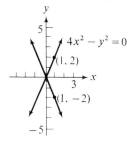   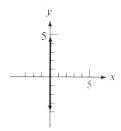

29. Since $x^2 + y^2 \geq 0$ for all $x, y \in R$, and $-1 < 0$, it follows that $x^2 + y^2 \neq -1$ for any $x, y \in R$.

31. Let $P(x, y)$ be a point at distance 4 from $(2, -3)$; then by the distance formula, $\sqrt{(x-2)^2 + (y+3)^2} = 4$; squaring and simplifying yields $(x-2)^2 + (y+3)^2 = 16$.

33. Given $Ax^2 + By^2 = C$ with intercepts $(a, 0)$ and $(0, b)$. If $y = 0$, $x = a$ and $x^2 = \dfrac{C}{A}$; hence $a^2 = \dfrac{C}{A}$. If $x = 0$, $y = b$ and $y^2 = \dfrac{C}{B}$; hence $b^2 = \dfrac{C}{B}$. Because $Ax^2 + By^2 = C$, $\dfrac{x^2}{C/A} + \dfrac{y^2}{C/B} = 1$ and $\dfrac{x^2}{a^2} + \dfrac{y^2}{b^2} = 1$.

35. As $|x|$ increases, so does the denominator Ax^2 of the fraction $\dfrac{C}{Ax^2}$; the value of this fraction approaches zero, and hence, $\sqrt{1 - \dfrac{C}{Ax^2}}$ approaches 1, and $y = \pm\sqrt{\dfrac{A}{B}}\,|x|\left(\sqrt{1 - \dfrac{C}{Ax^2}}\right)$ approaches $y = \pm\sqrt{\dfrac{A}{B}}\,|x|$.

Exercise 6.5 (page 175)

1.   **3.**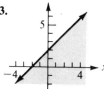

5. 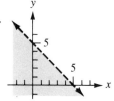   **7.**

Odd-Numbered Answers

9.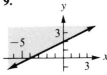
11.
13.
15.

y-axis not part of the solution

17.
19.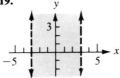
21.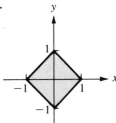

23.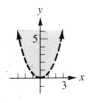
25.
27.

29.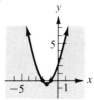
31.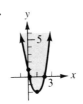
33.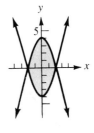

Chapter 6 Review (page 176)

1.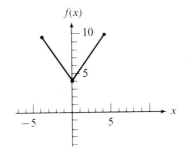
2.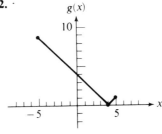

3. 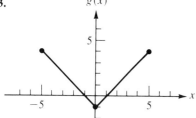   **4.**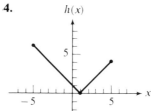

5. 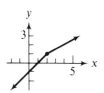   **6.**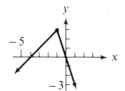

7. 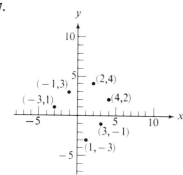   **8.**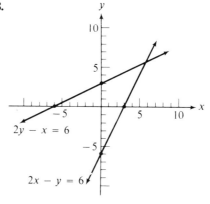

9. 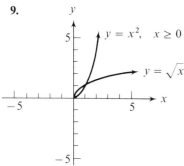   **10.**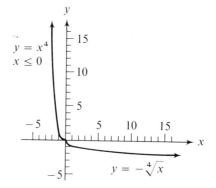

11. $y = 125$ **12.** $r = \dfrac{128}{243}$ **13.** 64 posts   **14.** 432 rpm

Odd-Numbered Answers

15. 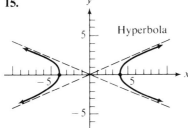 Hyperbola

16. 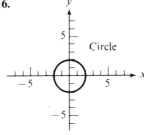 Circle

17. 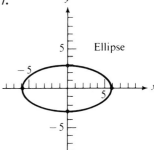 Ellipse

18. 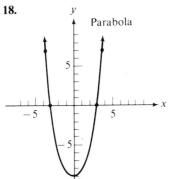 Parabola

19. 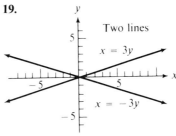 Two lines
$x = 3y$
$x = -3y$

20. 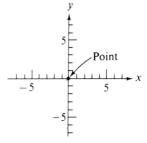 Point

21.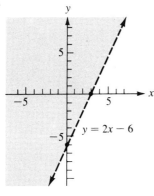
$y = 2x - 6$

22.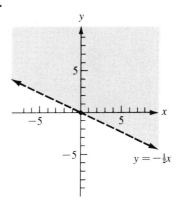
$y = -\frac{1}{2}x$

23.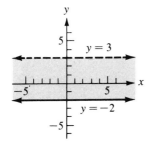
$y = 3$
$y = -2$

24.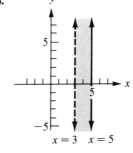
$x = 3 \quad x = 5$

25.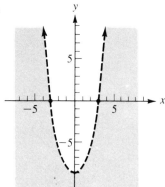

26.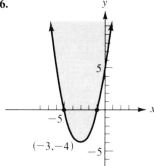

Exercise 7.1 (page 179) **1.** $(0, 1), (1, 3), (2, 9)$ **3.** $(0, -1), (1, -5), (2, -25)$ **5.** $(-3, 8), (0, 1), \left(3, \dfrac{1}{8}\right)$ **7.** $\left(-2, \dfrac{1}{100}\right), (1, 10), (0, 1)$

9. 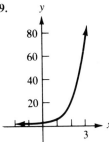 **11.** 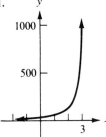 **13.** 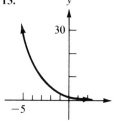 **15.**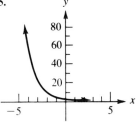

17. 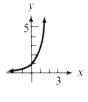 **19.** 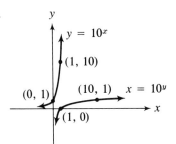 **21.** $\{-2\}$

23. $\left\{\dfrac{3}{4}\right\}$ **25.** 2 **27.** -2 **29.**

Exercise 7.2 (page 183) **1.** $\log_4 16 = 2$ **3.** $\log_3 27 = 3$ **5.** $\log_{1/2} \dfrac{1}{4} = 2$ **7.** $\log_8 \dfrac{1}{2} = -\dfrac{1}{3}$

Odd-Numbered Answers

9. $\log_{10} 100 = 2$ **11.** $\log_{10}(0.1) = -1$ **13.** $2^6 = 64$ **15.** $3^2 = 9$ **17.** $\left(\dfrac{1}{3}\right)^{-2} = 9$ **19.** $10^3 = 1000$

21. 2   **23.** 3   **25.** -1 **27.** 1   **29.** 2   **31.** -1 **33.** $\{2\}$ **35.** $\{2\}$ **37.** $\{64\}$ **39.** $\{-3\}$

41. $\{100\}$ **43.** $\{4\}$ **45.** $\log_b x + \log_b y$ **47.** $\log_b x - \log_b y$ **49.** $5 \log_b x$ **51.** $\dfrac{1}{3} \log_b x$

53. $\dfrac{1}{2}(\log_b x - \log_b z)$ **55.** $\dfrac{1}{3}(\log_{10} x + 2 \log_{10} y - \log_{10} z)$ **57.** $\log_b 2xy^3$ **59.** $\log_b x^{1/2} y^{2/3}$

61. $\log_b \dfrac{x^3 y}{z^2}$ **63.** $\log_{10} \dfrac{x(x-2)}{z^2}$

65. By definition, $\log_b 1$ is a number such that $b^{\log_b 1} = 1$; therefore $\log_b 1 = 0$.

67. Since $x_1 = b^{\log_b x_1}$ and $x_2 = b^{\log_b x_2}$, it follows that

$$\frac{x_2}{x_1} = \frac{b^{\log_b x_2}}{b^{\log_b x_1}} = b^{\log_b x_2 - \log_b x_1},$$

and by definition, $\log_b \left(\dfrac{x_2}{x_1}\right) = \log_b x_2 - \log_b x_1$.

Exercise 7.3 (page 189)   **1.** 2   **3.** -3, or $7 - 10$ **5.** -4, or $6 - 10$ **7.** 4   **9.** 0.8280
11. $9.9101 - 10$ **13.** $8.9031 - 10$ **15.** 2.3945   **17.** 4.10   **19.** 3.67   **21.** 0.0642
23. 5480   **25.** 0.000718   **27.** 1.0986   **29.** 2.8332   **31.** 5.7991   **33.** 6.1093   **35.** 1.6487
37. 29.964   **39.** 1.2586   **41.** 0.8607   **43.** 0.0821   **45.** 0.7788
47. Since $e^N = x$, we have $\log_{10} e^N = \log_{10} x$, thus by Theorem 7.1-III,

$$N \log_{10} e = \log_{10} x.$$

Hence $N = \dfrac{\log_{10} x}{\log_{10} e}.$ **49.** 2.63

Exercise 7.4 (page 193)   **1.** $\left\{\dfrac{\log_{10} 7}{\log_{10} 2}\right\}$, $\{2.8076\}$ **3.** $\left\{\dfrac{\log_{10} 8}{\log_{10} 3} - 1\right\}$, $\{0.8929\}$

5. $\left\{\dfrac{1}{2}\left(\dfrac{\log_{10} 3}{\log_{10} 7} + 1\right)\right\}$, $\{0.7823\}$ **7.** $n = \dfrac{\log_{10} y}{\log_{10} x}$ **9.** $t = \dfrac{\log_{10} y}{k \log_{10} e}$ **11.** $x = 1 + 5^{(y-3)/2}$

13. 4%   **15.** \$7,400, \$7,430   **17.** 6.93%   **19.** 7.7   **21.** 2.51×10^{-6}
23. 30.0 inches; 16.1 inches   **25.** 12.05 grams   **27.** $k = 0.23026$

Chapter 7 Review (page 196)

1.

2.

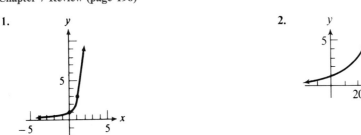

3. $\log_{16} \frac{1}{4} = -\frac{1}{2}$ **4.** $\log_7 343 = 3$ **5.** $2^3 = 8$ **6.** $10^{-4} = 0.0001$ **7.** $y = 4$ **8.** $x = 1{,}000$
9. $\frac{1}{3} \log_{10} x + \frac{2}{3} \log_{10} y$ **10.** $\log_{10} 2 + 3\log_{10} R - \frac{1}{2} \log_{10} P - \frac{1}{2} \log_{10} Q$ **11.** $\log_b \frac{x^2}{\sqrt[3]{y}}$
12. $\log_{10} \frac{\sqrt[3]{x^2 y}}{z^3}$ **13.** 1.6232   **14.** 7.4969 − 10   **15.** 2.8338   **16.** 8.6170 − 10   **17.** 67.4
18. 0.0466   **19.** 2.65   **20.** 0.664   **21.** 1.9459   **22.** 3.1355   **23.** 5.4807   **24.** 6.2344
25. 403.43   **26.** 6.0496   **27.** 0.0302   **28.** 0.6570   **29.** $x = \frac{\log_{10} 2}{\log_{10} 3} \approx 0.6309$
30. $x = \frac{\log_{10} 80}{\log_{10} 3} - 1 \approx 2.9889$ **31.** $t = -\log_e \left(\frac{y - A}{k} \right)$ **32.** $x = e^{y - N/N_0}$ **33.** 4.06   **34.** −1.9188
35. 4%   **36.** 6.31×10^{-7} **37.** 7.13%   **38.** $k = 0.65$

Exercise 8.1 (page 201)   **1.** I   **3.** C   **5.** H   **7.** K   **9.** I   **11.** K   **13.** M   **15.** K
17. positive   **19.** positive   **21.** positive   **23.** positive   **25.** positive
27. $\frac{4}{5}$, I   **29.** $-\frac{12}{13}$, IV   **31.** $\frac{\sqrt{5}}{3}$, IV   **33.** $-\frac{1}{2}$, III   **35.** $\frac{4}{5}$, IV

Exercise 8.2 (page 206)   **1.** $\frac{1}{2}$ **3.** $\frac{\sqrt{2}}{2}$ **5.** 1   **7.** 1   **9.** $\frac{2 + \sqrt{2}}{2}$ **11.** $\frac{1}{2}$ **13.** 0.9460
15. 0.6816   **17.** 0.1601   **19.** 0.9735   **21.** 0.9940   **23.** 0.9490
25. 0.5000   **27.** 0.0045   **29.** 2.4572   **31.** $K = 25$ **33.** $d = 0$
35. $E = 1.2$ **37.** $E = 0.6$ **39.** $E = 0$

Exercise 8.3 (page 212)   *Note:* A calculator in conjunction with a reference arc was used to compute the values in Problems 13–24, 31–36, and 43–48. If you used Table V, your answers may differ slightly from those given below.
1. $-\frac{\sqrt{3}}{2}$ **3.** $-\frac{1}{2}$ **5.** $\frac{\sqrt{2}}{2}$ **7.** $-\frac{\sqrt{3}}{2}$ **9.** $-\frac{\sqrt{2}}{2}$ **11.** $\frac{\sqrt{2}}{2}$
13. −0.5148   **15.** −0.9323   **17.** 0.9916   **19.** 0.9999
21. −0.0200   **23.** −0.9992   **25.** $\frac{1}{2}$ **27.** $-\frac{\sqrt{2}}{2}$ **29.** $-\frac{1}{2}$
31. −0.9608   **33.** 0.4169   **35.** 0.9208   **37.** $\frac{\sqrt{3}}{2}$ **39.** $\frac{\sqrt{2}}{2}$
41. $\frac{1}{2}$ **43.** 0.1197   **45.** 0.9927   **47.** −0.5396

Exercise 8.4 (page 220)

1.
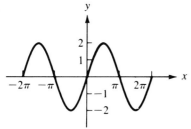
$A = 2$, $p = 2\pi$

3.
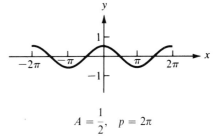
$A = \frac{1}{2}$, $p = 2\pi$

5.
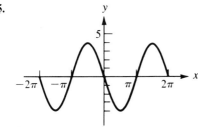
$A = 4$, $p = 2\pi$

7.
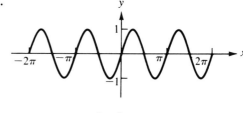
$A = 1$, $p = \pi$

9.
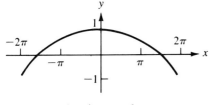
$A = 1$, $p = 6\pi$

11.
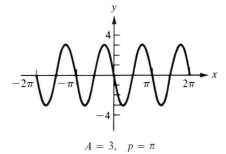
$A = 3$, $p = \pi$

13.
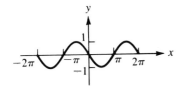
$A = 1$, $p = 2\pi$, leads $y = \sin x$ by π

15.
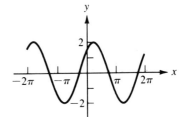
$A = 2$, $p = 2\pi$, lags $y = 2\cos x$ by $\frac{\pi}{4}$

17.

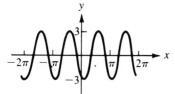

$A = 3$, $p = \pi$, lags $y = 3 \sin 2x$ by $\dfrac{2\pi}{3}$

19.

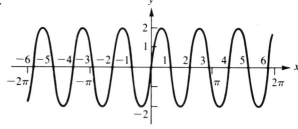

$A = 2$, $p = 2$

21.

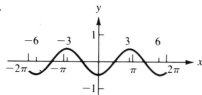

$A = \dfrac{1}{2}$, $p = 6$

23. $\left\{-2\pi, -\dfrac{3\pi}{2}, -\pi, -\dfrac{\pi}{2}, 0, \dfrac{\pi}{2}, \pi, \dfrac{3\pi}{2}, 2\pi\right\}$

Note: long marks, π units; short marks, integers

25. $\left\{-\dfrac{3\pi}{2}, \dfrac{3\pi}{2}\right\}$ **27.** $\{-6, -5, -4, -3, -2, -1, 0, 1, 2, 3, 4, 5, 6\}$

29.

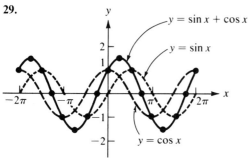

31.

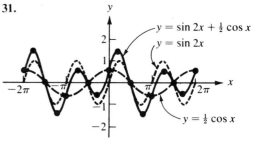

33.

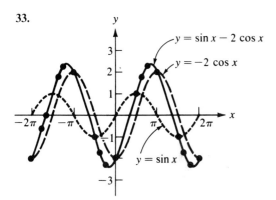

35.

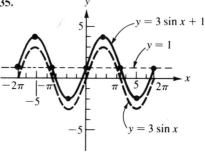

37.

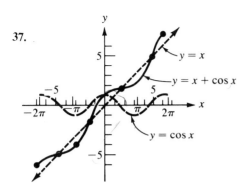

Note: long marks, π units; short marks, integers

39. Since the circumference of the unit circle is exactly 2π, any arc of length $x + 2\pi$ along the unit circle from the point $(1, 0)$ has the same terminal point as an arc of length x along the unit circle from the point $(1, 0)$. Thus,

$$\sin(x + 2\pi) = \sin x.$$

Exercise 8.5 (page 225) *Note:* A calculator in conjunction with a reference arc was used to compute the values in Problems 17–24. If you used Table V, your answers may differ slightly from those given below.

1. 0 **3.** 1 **5.** undefined **7.** undefined **9.** $-\sqrt{3}$ **11.** $\dfrac{\sqrt{3}}{3}$

13. $-\dfrac{\sqrt{3}}{3}$ **15.** undefined **17.** 5.1774 **19.** 2.3600 **21.** -0.0601 **23.** 0.1003

25. 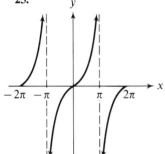 **27.**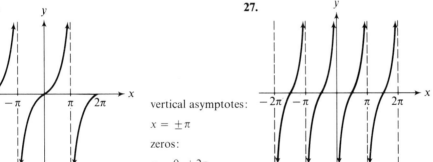

For 25: vertical asymptotes: $x = \pm\pi$; zeros: $x = 0, \pm 2\pi$

For 27: vertical asymptotes: $x = 0, \pm\pi, \pm 2\pi$; zeros: $x = \pm\dfrac{\pi}{2}, \pm\dfrac{3\pi}{2}$

Exercise 8.6 (page 230) *Note:* A calculator in conjunction with a reference arc was used to compute the values in Problems 21–32. If you used Table V, your answers may differ slightly from those given below.

1. 0 **3.** $\sqrt{2}$ **5.** $\dfrac{2\sqrt{3}}{3}$ **7.** undefined **9.** $\sqrt{3}$ **11.** $\sqrt{2}$ **13.** 2

15. undefined **17.** $\dfrac{2\sqrt{3}}{3}$ **19.** -1 **21.** 0.4237 **23.** 2.0098

25. 1.0033 **27.** 0.5870 **29.** -1.1395 **31.** -7.1662

Exercise 8.7 (page 236) *Note:* A calculator in conjunction with a reference arc was used to compute the value in Problem 31. If you used Table V, your answer may differ slightly from the one given below.

1. $\dfrac{\pi}{6}$ **3.** $\dfrac{\pi}{4}$ **5.** $\dfrac{\pi}{3}$ **7.** $-\dfrac{\pi}{6}$ **9.** 0.10 **11.** 0.39 **13.** -1.04 **15.** 0.32

17. $\dfrac{\pi}{4}$ **19.** $\dfrac{\pi}{3}$ **21.** $\dfrac{\sqrt{3}}{2}$ **23.** $\dfrac{\sqrt{3}}{2}$ **25.** 0.4472 **27.** 2.4 **29.** $\dfrac{3\pi}{4}$ **31.** 1

33. $\dfrac{\sqrt{2}}{2}$ **35.** $x = \dfrac{1}{2}\cos\dfrac{y}{3}$, $0 \le y \le 3\pi$ **37.** $x = -\pi + \tan 2y$, $-\dfrac{\pi}{4} < y < \dfrac{\pi}{4}$

Chapter 8 Review (page 238) Note: A calculator (in conjunction with a reference arc where appropriate) was used to compute the values in Problems 11–16, 21–28, 37–40, 50–52, and 59–64. If you used Table V, your answers may differ slightly from those given below.

1. negative **2.** negative **3.** positive **4.** negative

5. $-\dfrac{\sqrt{5}}{3}$, IV **6.** $\dfrac{\sqrt{55}}{8}$, I **7.** $\dfrac{\sqrt{3}}{2}$ **8.** 0 **9.** 0 **10.** $\dfrac{\sqrt{3}}{3}$

11. 0.9581 **12.** -0.2160 **13.** 0.0962 **14.** 3.2842 **15.** $K = -25$

16. $d = 71.4$ **17.** $-\dfrac{\sqrt{2}}{2}$ **18.** $\dfrac{\sqrt{3}}{2}$ **19.** $-\dfrac{\sqrt{2}}{2}$ **20.** $\dfrac{1}{2}$

21. -0.9902 **22.** 0.9856 **23.** 0.5891 **24.** -0.2085 **25.** 0.0408

26. 0.9611 **27.** -0.9967 **28.** 0.8016 **29.** $A = 2, p = \dfrac{2\pi}{3}$; no phase shift

30. $A = \dfrac{3}{2}, p = 2\pi$; lags $y = \dfrac{3}{2}\cos x$ by $\dfrac{\pi}{4}$. **31.** $A = 4, p = \pi$; lags $y = 4\sin 2x$ by $\dfrac{\pi}{2}$.

32. 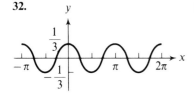 **33.** 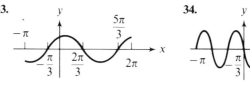 **34.**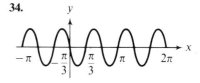

35. 1 **36.** $\sqrt{3}$ **37.** -2.4979 **38.** 0.5594 **39.** 1.4284 **40.** 1.4592

41. 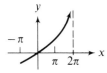 **42.**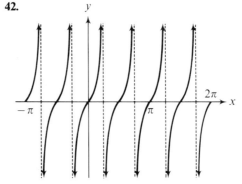

43. $\sqrt{3}$ **44.** $\dfrac{2\sqrt{3}}{3}$ **45.** 1 **46.** undefined **47.** -1 **48.** $\sqrt{2}$

49. -2 **50.** 1.1634 **51.** -1.0099 **52.** -1.0033 **53.** $\dfrac{\pi}{4}$ **54.** 0

55. $\dfrac{\pi}{6}$ **56.** $\dfrac{\pi}{6}$ **57.** $\dfrac{2\pi}{3}$ **58.** $-\dfrac{\pi}{6}$ **59.** 0.95 **60.** 1.20 **61.** 0.30 **62.** 0.90

63. 0.60 **64.** 0.80 **65.** 0.25 **66.** 0 **67.** 1 **68.** $\dfrac{\sqrt{3}}{2}$

Odd-Numbered Answers

Exercise 9.1 (page 244)   **1. a.** $0°$ **b.** $90°$ **c.** $180°$ **d.** $270°$ **e.** $360°$ **3.** $40°$ **5.** $252°$ **7.** $17.2°$
9. $207.5°$ **11.** 0.35 **13.** 2.27 **15.** 7.33 **17.** 13.08 **19.** $57.32°$
21. $s \approx 2.09$ cm   **23.** $s \approx 0.72$ m   **25.** $s \approx 4.71$ m   **27.** $90°$ **29.** $180°$ **31.** $45°$
33. $390°, -330°; 30° + 360°k, k \in J$ **35.** $120°, 480°; -240° + 360°k, k \in J$
37. $60°, -300°; 420° + 360°k, k \in J$ **39.** $30°, 390°; -330° + 360°k, k \in J$

Exercise 9.2 (page 253)   *Note:* A calculator (in conjunction with a reference angle where appropriate) was used to compute the values in Problems 27–56. If you used Table V or Table VI, your answers may differ slightly from those given below.

1. $\sin \alpha = \frac{4}{5}$, $\cos \alpha = \frac{3}{5}$, $\tan \alpha = \frac{4}{3}$, $\cot \alpha = \frac{3}{4}$, $\sec \alpha = \frac{5}{3}$, $\csc \alpha = \frac{5}{4}$

3. $\sin \alpha = -\frac{\sqrt{2}}{2}$, $\cos \alpha = \frac{\sqrt{2}}{2}$, $\tan \alpha = -1$, $\cot \alpha = -1$, $\sec \alpha = \sqrt{2}$, $\csc \alpha = -\sqrt{2}$

5. $\sin \alpha = -\frac{4}{5}$, $\cos \alpha = -\frac{3}{5}$, $\tan \alpha = \frac{4}{3}$, $\cot \alpha = \frac{3}{4}$, $\sec \alpha = -\frac{5}{3}$, $\csc \alpha = -\frac{5}{4}$

7. $\sin \alpha = 1$, $\cos \alpha = 0$, $\tan \alpha$ not defined, $\cot \alpha = 0$, $\sec \alpha$ not defined, $\csc \alpha = 1$

9. IV   **11.** I   **13.** IV   **15.** $-\frac{1}{\sqrt{3}}$ **17.** $-\frac{1}{\sqrt{2}}$ **19.** $-\frac{1}{2}$ **21.** $-\frac{1}{2}$ **23.** $\frac{\sqrt{3}}{2}$

25. $\sqrt{3}$ **27.** 0.5299 **29.** 0.6494 **31.** 1.0457 **33.** 0.8080 **35.** 0.1519 **37.** 1.0148
39. 0.7431 **41.** -0.8391 **43.** -0.6428 **45.** 0.8772 **47.** 1.8270 **49.** -0.0608
51. a. $\frac{\pi^R}{2}$ **b.** $90°$ **53. a.** $\frac{\pi^R}{4}$ **b.** $45°$ **55. a.** 0.19^R **b.** $11.0°$

57. If the coordinates are associated with lengths of line segments as indicated in the figure, then by the Pythagorean theorem, $OP_1 = \sqrt{x_1^2 + y_1^2}$ and $OP_2 = \sqrt{x_2^2 + y_2^2}$. Since $\triangle OAP_1$ and $\triangle OBP_2$ are similar, their corresponding sides are proportional and the desired results follow by substituting in the following ratios:

$$\frac{OA}{AP_1} = \frac{OB}{BP_2}; \quad \frac{OA}{OP_1} = \frac{OB}{OP_2}; \quad \text{and} \quad \frac{AP_1}{OP_1} = \frac{BP_2}{OP_2}$$

Exercise 9.3 (page 258)   **1.** $c = 10$, $A = 36.9°$, $B = 53.1°$
3. $c = 148.3$ $a = 87.2$, $A = 36°$ **5.** $a = 6.2$, $b = 14.8$, $B = 67.3°$ **7.** 16
9. $\cos \theta = \frac{-\sqrt{3}}{2}$, $\tan \theta = \frac{1}{\sqrt{3}}$, $\cot \theta = \sqrt{3}$, $\csc \theta = -2$, $\sec \theta = \frac{-2}{\sqrt{3}}$
11. $\sin \theta = \frac{2\sqrt{10}}{7}$, $\tan \theta = -\frac{2\sqrt{10}}{3}$, $\cot \theta = -\frac{3}{2\sqrt{10}}$, $\sec \theta = -\frac{7}{3}$, $\csc \theta = \frac{7}{2\sqrt{10}}$
13. $\frac{\sqrt{2}}{4}$ **15.** $\frac{3\sqrt{13}}{13}$ **17.** $\sqrt{1-x^2}$ **19.** $\frac{y}{\sqrt{1-y^2}}$

21. Let $\alpha = \text{Arcsin}\left(\frac{2}{5}\right)$. Then from the illustrated triangle we read $\tan \alpha = \frac{2}{\sqrt{21}}$. Thus, $\text{Arctan}\left(\frac{2}{\sqrt{21}}\right) = \alpha = \text{Arcsin}\frac{2}{5}$.

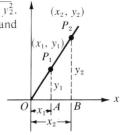

23. $\dfrac{\sqrt{2}+3}{2}$ **25.** 20.5 **27.** 38.7° and 51.3° **29.** 24.0 meters **31.** 927.2 meters
33. 10,585.6 meters

Exercise 9.4 (page 264) **1.** $\gamma = 70°$, $c = 18.8$, $a = 19.7$ **3.** $\beta = 58.7°$, $b = 74.8$, $a = 74.3$
5. $B = 47.9°$, $b = 66.5$, $c = 33.7$ **7.** one triangle possible **9.** one triangle possible
11. two triangles possible **13.** a right triangle **15.** $\gamma = 48.4°$, $\beta = 18.6°$, $b = 1.7$
17. $\gamma = 28.6°$, $\alpha = 108.7°$, $a = 8.7$
19. $\beta = 139.9°$, $\alpha = 7.6°$, $a = 0.4$ and $\beta = 40.1°$, $\alpha = 107.4°$, $a = 2.7$
21. no solution **23.** $B = 25.7°$, $C = 94.3°$, $c = 15.9$
25. $C = 53.1°$, $A = 93.5°$, $a = 767.8$ and $C = 126.9°$, $A = 19.7°$, $a = 259.3$
27. 197 feet **29.** 2,235.5 meters **31.** 6.9

Exercise 9.5 (page 268) **1.** $c = 6.9$, $\alpha = 131.9°$, $\beta = 17.5°$ **3.** $b = 10.3$, $\alpha = 23.8°$, $\gamma = 33.8°$
5. $a = 14.9$, $B = 75°$, $C = 24.3°$ **7.** $C = 60°$, $B = 81.8°$, $A = 38.2°$ **9.** 71.7° **11.** 77.7
13. 279.3 yards **15.** 23.2° west of north or on a heading of 336.8°
17. Set the triangle on a coordinate system as shown in the adjoining figure, then $a^2 = (b - x)^2 + y^2$. But notice $c^2 = x^2 + y^2$ and thus

$$a^2 = (b - x)^2 + c^2 - x^2$$
$$= b^2 - 2bx + x^2 + c^2 - x^2$$
$$= b^2 + c^2 - 2bx.$$

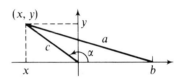

Since $\cos \alpha = \dfrac{x}{c}$, we have $a^2 = b^2 + c^2 - 2bc \cos \alpha$.

Exercise 9.6 (page 274) **1.** 1; 90° **3.** 2; $-25°$ **5.**

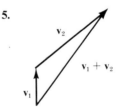

7.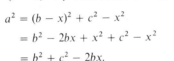

9.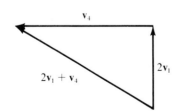

11. $\|v_x\| = 3$, $\|v_y\| = 3$ **13.** $\|v_x\| = \dfrac{7\sqrt{3}}{2}$, $\|v_y\| = \dfrac{7}{2}$

15. $R \approx 8.60$ lbs; $\alpha \approx 54.5°$ **17.** $R \approx 7.2$ lbs; $\alpha \approx 106.1°$

19. $\sqrt{520{,}000}$ mi ≈ 721.1 mi **21.** 80 lbs

Odd-Numbered Answers

Chapter 9 Review (page 275) *Note:* A calculator (in conjunction with a reference angle where appropriate) was used to calculate the values in Problems 17–24 and 26. If you used Table V or Table VI, your answers may differ slightly from those given below.

1. $67.5°$ 2. $499.9°$ 3. 0.70^R 4. 10.64^R 5. $5.0''$ 6. $4.2'$ 7. $30°, 390°$
8. $-202°, 158°$ 9. III 10. III 11. $-\dfrac{\sqrt{3}}{2}$ 12. $-\dfrac{1}{2}$ 13. 1
14. $-\dfrac{1}{2}$ 15. $\dfrac{1}{2}$ 16. $\dfrac{1}{\sqrt{2}}$ 17. 0.9205 18. 2.9119 19. 0.8660 20. -0.7071
21. -0.2079 22. 1.8807 23. -0.8577 24. 0.9168 25. a. $\dfrac{\pi^R}{6}$ b. $30°$
26. a. 0.09^R b. $4.9°$ 27. $c = 15$, $A = 36.9°$, $B = 53.1°$ 28. $a = 7.2$, $b = 8.3$, $\beta = 49°$
29. $c = 24$, $b = 23.5$, $B = 78°$ 30. $b = 9.8$, $B = 63°$, $A = 27°$
31. $\cos\theta = \dfrac{4}{5}$, $\tan\theta = \dfrac{3}{4}$, $\cot\theta = \dfrac{4}{3}$, $\sec\theta = \dfrac{5}{4}$, $\csc\theta = \dfrac{5}{3}$
32. $\sin\theta = \dfrac{5}{13}$, $\cos\theta = \dfrac{12}{13}$, $\cot\theta = \dfrac{12}{5}$, $\sec\theta = \dfrac{13}{12}$, $\csc\theta = \dfrac{13}{5}$ 33. 101.45 meters
34. 4.3 meters
35. $a = 45.7$, $c = 46.6$, $\gamma = 80°$ 36. $b = 22.4$, $\beta = 111.3°$, $\gamma = 38.7°$ $b' = 3.6$, $\beta' = 8.7°$, $\gamma' = 141.3°$
37. $a = 15.2$, $c = 19.8$, $A = 50°$ 38. $a = 36$, $b = 31.7$, $B = 60°$ 39. 538.2 feet
40. 4.2 kilometers 41. $c = 9.4$, $\beta = 7.8°$, $\alpha = 147.2°$
42. $a = 14.1$, $\beta = 63.6°$, $\gamma = 35.7°$ 43. $C = 58.8°$, $B = 71.8°$, $A = 49.5°$
44. $C = 36.9°$, $B = 53.1°$, $A = 90°$ 45. $A = 40.2°$, $B = 19.6°$, $C = 120.3°$
46. 329.3 meters 47. $\|\mathbf{v}_x\| = 6\sqrt{3}$, $\|\mathbf{v}_y\| = 6$ 48. $\|\mathbf{v}_x\| = 11.88$, $\|\mathbf{v}_y\| = 11.88$
49. $\|\mathbf{v}_3\| \approx 13.16$; $\alpha \approx 46.3°$ 50. 2.56 mph

Exercise 10.1 (page 282) 1. Recall $\tan x = \dfrac{\sin x}{\cos x}$; hence $\cos x \tan x = \cos x \left(\dfrac{\sin x}{\cos x}\right) = \sin x$.

3. Recall $\cot x = \dfrac{\cos x}{\sin x}$; then $\cot^2 x = \dfrac{\cos^2 x}{\sin^2 x}$; hence $\sin^2 x \cot^2 x = \sin^2 x \left(\dfrac{\cos^2 x}{\sin^2 x}\right) = \cos^2 x$.

5. By Equation (2), $1 + \tan^2 \alpha = \sec^2 \alpha$; also $\sec \alpha = \dfrac{1}{\cos \alpha}$. Then $\sec^2 \alpha = \dfrac{1}{\cos^2 \alpha}$;

hence $(\cos^2 \alpha)(1 + \tan^2 \alpha) = (\cos^2 \alpha)(\sec^2 \alpha) = (\cos^2 \alpha)\left(\dfrac{1}{\cos^2 \alpha}\right) = 1$.

7. By Equation (1), $\sin^2 \alpha + \cos^2 \alpha = 1$; hence $1 - \cos^2 \alpha = \sin^2 \alpha$. Using the definitions of secant, cosecant, and tangent yields $\sec \alpha \csc \alpha - \cot \alpha = \left(\dfrac{1}{\cos \alpha}\right)\left(\dfrac{1}{\sin \alpha}\right) - \dfrac{\cos \alpha}{\sin \alpha} = \dfrac{1 - \cos^2 \alpha}{\cos \alpha \sin \alpha} = \dfrac{\sin^2 \alpha}{\cos \alpha \sin \alpha} = \dfrac{\sin \alpha}{\cos \alpha} = \tan \alpha$.

9. By the definitions of secant and tangent, $\dfrac{\sin \theta \sec \theta}{\tan \theta} = \dfrac{(\sin \theta)(1/\cos \theta)}{(\sin \theta/\cos \theta)} = \dfrac{\sin \theta}{\sin \theta} = 1$.

11. By Equation (2), $\tan^2 \theta + 1 = \sec^2 \theta$; hence $\sec^2 \theta - 1 = \tan^2 \theta$; also, by Equation (1), $\sin^2 \theta + \cos^2 \theta = 1$, and if $\sin^2 \theta \neq 0$, then $\dfrac{\sin^2 \theta}{\sin^2 \theta} + \dfrac{\cos^2 \theta}{\sin^2 \theta} = \dfrac{1}{\sin^2 \theta}$, or $1 + \cot^2 \theta = \csc^2 \theta$; hence $\csc^2 \theta - 1 = \cot^2 \theta$. Using these

results with the fact that $\cot x = \dfrac{1}{\tan x}$ yields $(\sec^2 \theta - 1)(\csc^2 \theta - 1) = (\tan^2 \theta)(\cot^2 \theta) = (\tan^2 \theta)\left(\dfrac{1}{\tan^2 \theta}\right) = 1$.

13. From Equation (1), since $\sin^2 \theta + \cos^2 \theta = 1$, $1 - \sin^2 \theta = \cos^2 \theta$. From this result and the definition of secant, $\dfrac{1}{1 + \sin \theta} + \dfrac{1}{1 - \sin \theta} = \dfrac{1 - \sin \theta + 1 + \sin \theta}{(1 + \sin \theta)(1 - \sin \theta)} = \dfrac{2}{1 - \sin^2 \theta} = \dfrac{2}{\cos^2 \theta} = 2\left(\dfrac{1}{\cos \theta}\right)^2 = 2 \sec^2 \theta$.

15. From the definition of cotangent, $\sin x \cot x = \sin x \, \dfrac{\cos x}{\sin x} = \cos x$.

17. Recall that $\tan \theta = \dfrac{\sin \theta}{\cos \theta}$; hence $\dfrac{\sin \theta}{\tan \theta} = \dfrac{\sin \theta}{\sin \theta / \cos \theta} = \cos \theta$. By Equation (5), $\cos(-\theta) = \cos \theta$; hence $\dfrac{\sin \theta}{\tan \theta} = \cos(-\theta)$.

19. Recall that $\tan x = \dfrac{\sin x}{\cos x}$; hence, $\tan^2 x \cos^2 x = \dfrac{\sin^2 x}{\cos^2 x} \cos^2 x = \sin^2 x$.

21. By Equation (1), $1 - \sin^2 \theta = \cos^2 \theta$; hence

$$\dfrac{(1 + \sin \theta)(1 - \sin \theta)}{\cos \theta} = \dfrac{1 - \sin^2 \theta}{\cos \theta} = \dfrac{\cos^2 \theta}{\cos \theta} = \cos \theta.$$

23. From the definition of secant and tangent,

$$\sec x - \cos x = \dfrac{1}{\cos x} - \cos x = \dfrac{1 - \cos^2 x}{\cos x} = \dfrac{\sin^2 x}{\cos x} = \sin x \left(\dfrac{\sin x}{\cos x}\right) = \sin x \tan x.$$

25. By Equation (3), $1 + \cot^2 x = \csc^2 x$; also, $\cot x = \dfrac{1}{\tan x}$; hence $\tan x \cot x = 1$ and $\tan x = \dfrac{1}{\cot x}$, $\cot x \neq 0$. Then

$$\dfrac{1 + \tan^2 x}{\tan^2 x} = \dfrac{1}{\tan^2 x} + 1 = \cot^2 x + 1 = \csc^2 x.$$

27. From the definition of tangent and Equation (1),

$$\tan^2 \alpha - \sin^2 \alpha = \dfrac{\sin^2 \alpha}{\cos^2 \alpha} - \sin^2 \alpha = \dfrac{\sin^2 \alpha - \sin^2 \alpha \cos^2 \alpha}{\cos^2 \alpha} = \dfrac{\sin^2 \alpha(1 - \cos^2 \alpha)}{\cos^2 \alpha}$$

$$= \left(\dfrac{\sin \alpha}{\cos \alpha}\right)^2 (1 - \cos^2 \alpha) = \tan^2 \alpha \sin^2 \alpha = \sin^2 \alpha \tan^2 \alpha.$$

29. From Equation (2),

$$\dfrac{1}{\sec \alpha - \tan \alpha} \cdot \dfrac{\sec \alpha + \tan \alpha}{\sec \alpha + \tan \alpha} = \dfrac{\sec \alpha + \tan \alpha}{\sec^2 \alpha - \tan^2 \alpha} = \dfrac{\sec \alpha + \tan \alpha}{(\tan^2 \alpha + 1) - \tan^2 \alpha} = \sec \alpha + \tan \alpha.$$

31. Recall $\cot x = \dfrac{\cos x}{\sin x}$; hence $\dfrac{\cos x + \sin x}{\sin x} = \dfrac{\cos x}{\sin x} + \dfrac{\sin x}{\sin x} = \cot x + 1$.

33. Adding the fractions in the left-hand member, we have $\dfrac{1}{1 + \sin x} + \dfrac{1}{1 - \sin x} = \dfrac{2}{1 - \sin^2 x}$. By the definition of secant and Equation (1), $\dfrac{1}{1 + \sin x} + \dfrac{1}{1 - \sin x} = \dfrac{2}{1 - \sin^2 x} = \dfrac{2}{\cos^2 x} = 2 \sec^2 x$.

35. Recall $\sec\theta = \dfrac{1}{\cos\theta}$; hence $\dfrac{1+\sec\theta}{\sec\theta} = \dfrac{1+(1/\cos\theta)}{1/\cos\theta} = \dfrac{(\cos\theta+1)/\cos\theta}{1/\cos\theta} = \cos\theta + 1$.

Now, multiplying by $\dfrac{1-\cos\theta}{1-\cos\theta}$, we have $\dfrac{1+\sec\theta}{\sec\theta} = \dfrac{(1+\cos\theta)(1-\cos\theta)}{1-\cos\theta} = \dfrac{1-\cos^2\theta}{1-\cos\theta}$.

By Equation (1), $1 - \cos^2\theta = \sin^2\theta$; hence, $\dfrac{1+\sec\theta}{\sec\theta} = \dfrac{\sin^2\theta}{1-\cos\theta}$.

37. Multiplying the left-hand member by $\dfrac{1-\sin\alpha}{1-\sin\alpha}$, we have

$$\dfrac{1+\sin\alpha}{\cos\alpha} = \dfrac{(1+\sin\alpha)(1-\sin\alpha)}{\cos\alpha(1-\sin\alpha)} = \dfrac{1-\sin^2\alpha}{\cos\alpha(1-\sin\alpha)}.$$

By Equation (1), $1 - \sin^2\alpha = \cos^2\alpha$; hence $\dfrac{1+\sin\alpha}{\cos\alpha} = \dfrac{\cos^2\alpha}{\cos\alpha(1-\sin\alpha)} = \dfrac{\cos\alpha}{1-\sin\alpha}$.

39. Multiplying the left-hand member by $\dfrac{\sec\alpha}{\sec\alpha}$, we have

$$\dfrac{1-\cos\alpha}{1+\cos\alpha} = \dfrac{(1-\cos\alpha)\sec\alpha}{(1+\cos\alpha)\sec\alpha} = \dfrac{\sec\alpha - 1}{\sec\alpha + 1}.$$

41. By Equation (2), $\tan^2\theta = \sec^2\theta - 1$;

hence $\tan^4\theta + \tan^2\theta = (\sec^2\theta - 1)^2 + (\sec^2\theta - 1) = \sec^4\theta - 2\sec^2\theta + 1 + \sec^2\theta - 1 = \sec^4\theta - \sec^2\theta$.

43. $\tan^2 45° + \sec^2 45° = 1 + 2 = 3 \neq 1$.   **45.** $\sin 45° + \tan 45° = 1 + \dfrac{\sqrt{2}}{2} \neq \dfrac{\sqrt{2}}{2} = \cos 45°$.

47. Expanding the terms of the left-hand member, we have $(a\sin\alpha + b\cos\alpha)^2 + (a\cos\alpha - b\sin\alpha)^2$
$= a^2\sin^2\alpha + 2ab\sin\alpha\cos\alpha + b^2\cos^2\alpha + a^2\cos^2\alpha - 2ab\sin\alpha\cos\alpha + b^2\sin^2\alpha$
$= a^2(\sin^2\alpha + \cos^2\alpha) + b^2(\sin^2\alpha + \cos^2\alpha)$. By Equation (1), $\sin^2\alpha + \cos^2\alpha = 1$;
hence $(a\sin\alpha + b\cos\alpha)^2 + (a\cos\alpha - b\sin\alpha)^2 = a^2 + b^2$.

49. Convert the terms in the left-hand member into terms involving only sines and cosines and simplify to get

$$\left(\dfrac{\sec\theta - 1}{\sec\theta + 1}\right)\sin^2\theta + \sin^2\theta - 2 = \left(\dfrac{1/\cos\theta - 1}{1/\cos\theta + 1}\right)\sin^2\theta + \sin^2\theta - 2 = \left(\dfrac{1-\cos\theta}{1+\cos\theta}\right)\sin^2\theta + \sin^2\theta - 2.$$

By Equation (1), $\sin^2\theta = 1 - \cos^2\theta$; hence $\left(\dfrac{\sec\theta - 1}{\sec\theta + 1}\right)\sin^2\theta + \sin^2\theta - 2 =$

$\dfrac{(1-\cos\theta)(1-\cos^2\theta)}{1+\cos\theta} + (1-\cos^2\theta) - 2 = (1-\cos\theta)(1-\cos\theta) + 1 - \cos^2\theta - 2 = -2\cos\theta$.

Exercise 10.2 (page 286)   **1.** $\dfrac{-\sqrt{2}(1+\sqrt{3})}{4}$ **3.** $\dfrac{\sqrt{2}(1-\sqrt{3})}{4}$

5. $\dfrac{\sqrt{2}(\sqrt{3}-1)}{4}$ **7. a.** $\dfrac{-7}{25}$ **b.** $\dfrac{\sqrt{5}}{5}$

9. $\cos(\pi + x) = \cos\pi\cos x - \sin\pi\sin x = -\cos x$

11. $\cos\left(\dfrac{\pi}{2} + x\right) = \cos\dfrac{\pi}{2}\cos x - \sin\dfrac{\pi}{2}\sin x = -\sin x$

13. $\cos\left(\dfrac{3\pi}{2} + x\right) = \cos\dfrac{3\pi}{2}\cos x - \sin\dfrac{3\pi}{2}\sin x = \sin x$

15. By Equation (4), $\cos 2x = 1 - 2\sin^2 x$; hence $\dfrac{1 - \cos 2x}{2} = \dfrac{1 - (1 - 2\sin^2 x)}{2} = \sin^2 x$.

17. Write the terms of the right-hand member in terms of sine and cosine to obtain

$$\dfrac{1 - \tan^2\theta}{1 + \tan^2\theta} = \dfrac{1 - \dfrac{\sin^2\theta}{\cos^2\theta}}{1 + \dfrac{\sin^2\theta}{\cos^2\theta}} = \dfrac{\cos^2\theta - \sin^2\theta}{\cos^2\theta + \sin^2\theta} = \dfrac{\cos^2\theta - \sin^2\theta}{\cos^2\theta + \sin^2\theta}.$$

By Equation (1) of Section 9.1 and Equation (3), we have

$$\dfrac{1 - \tan^2\theta}{1 + \tan^2\theta} = \cos 2\theta.$$

19. $\cos 3x = \cos(x + 2x) = \cos x \cos 2x - \sin x \sin 2x$
$= \cos x(2\cos^2 x - 1) - \sqrt{1 - \cos^2 x}\sqrt{1 - \cos^2 2x}$
$= \cos x(2\cos^2 x - 1) - \sqrt{1 - \cos^2 x}\sqrt{1 - (2\cos^2 x - 1)^2}$
$= \cos x(2\cos^2 x - 1) - \sqrt{1 - \cos^2 x}\sqrt{1 - 4\cos^4 x + 4\cos^2 x - 1}$
$= \cos x(2\cos^2 x - 1) - \sqrt{1 - \cos^2 x}\sqrt{4\cos^2 x(1 - \cos^2 x)}$
$= \cos x(2\cos^2 x - 1) - 2\cos x(1 - \cos^2 x)$
$= 4\cos^3 x - 3\cos x$

21. $\dfrac{\cos(x+h) - \cos x}{h} = \dfrac{\cos x \cos h - \sin x \sin h - \cos x}{h} = \cos x\left(\dfrac{\cos h - 1}{h}\right) - \dfrac{\sin h}{h}\sin x$

Exercise 10.3 (page 290) 1. $\dfrac{\sqrt{2}(\sqrt{3} - 1)}{4}$ 3. $\dfrac{-\sqrt{2}(1 + \sqrt{3})}{4}$

5. $\dfrac{\sqrt{2}(-\sqrt{3} + 1)}{4}$ 7. a. $\dfrac{-2\sqrt{5}}{5}$ b. $\dfrac{24}{25}$

9. $\sin(\pi + x) = \sin\pi\cos x + \sin x\cos\pi = -\sin x$

11. $\sin\left(\dfrac{\pi}{2} + x\right) = \sin\dfrac{\pi}{2}\cos x + \sin x\cos\dfrac{\pi}{2} = \cos x$

13. $\sin\left(\dfrac{3\pi}{2} + x\right) = \sin\dfrac{3\pi}{2}\cos x + \sin x\cos\dfrac{3\pi}{2} = -\cos x$

15. By Equation (5), $\sin 2\alpha = (2\sin\alpha\cos\alpha)\left(\dfrac{\cos\alpha}{\cos\alpha}\right) = 2\tan\alpha\cos^2\alpha = \dfrac{2\tan\alpha}{\sec^2\alpha} = \dfrac{2\tan\alpha}{1 + \tan^2\alpha}$.

17. By Equation (5) of Section 10.2 and Equation (5),

$$\dfrac{1 + \cos 2x}{\sin 2x} = \dfrac{1 + 2\cos^2 x - 1}{2\sin x\cos x} = \dfrac{\cos^2 x}{\sin x\cos x} = \dfrac{\cos x}{\sin x} = \cot x.$$

19. $\sin 3x = \sin(2x + x) = \sin 2x\cos x + \sin x\cos 2x$
$= 2\sin x\cos^2 x + \sin x(1 - 2\sin^2 x) = 2\sin x - 2\sin^3 x + \sin x - 2\sin^3 x = 3\sin x - 4\sin^3 x$

Odd-Numbered Answers

21. $\dfrac{\sin(x+h) - \sin x}{h} = \dfrac{\sin x \cos h + \cos x \sin h - \sin x}{h} = \sin x \left(\dfrac{\cos h - 1}{h}\right) + \cos x \dfrac{\sin h}{h}$

Exercise 10.4 (page 294) **1.** $\dfrac{\sqrt{3}+1}{1-\sqrt{3}}$ **3.** $\sqrt{2}-1$ **5.** $\sqrt{2}+1$ **7. a.** $\dfrac{24}{7}$ **b.** $\dfrac{1}{3}$

9. $\tan(\pi - x) = \dfrac{\tan \pi - \tan x}{1 + \tan \pi \tan x} = -\tan x$ **11.** $\tan(2\pi - x) = \dfrac{\tan 2\pi - \tan x}{1 + \tan 2\pi \tan x} = -\tan x$

13. By Identity 19, $\cot \theta = \dfrac{1}{\tan \theta}$ and $\cot 2\theta = \dfrac{1}{\tan 2\theta}$; by Identity 22, $\tan 2\theta = \dfrac{2\tan\theta}{1-\tan^2\theta}$;

hence $\cot\theta - \cot 2\theta = \dfrac{1}{\tan\theta} - \dfrac{1-\tan^2\theta}{2\tan\theta} = \dfrac{2-1+\tan^2\theta}{2\tan\theta} = \dfrac{1+\tan^2\theta}{2\tan\theta}$. By Identity 14,

$1 + \tan^2\theta = \sec^2\theta$; hence $\cot\theta - \cot 2\theta = \dfrac{\sec^2\theta}{2\tan\theta}$.

15. $\tan 3x = \tan(2x + x) = \dfrac{\tan 2x + \tan x}{1 - \tan 2x \tan x}$

$= \dfrac{\dfrac{2\tan x}{1-\tan^2 x} + \tan x}{1 - \dfrac{2\tan x}{1-\tan^2 x} \cdot \tan x} = \dfrac{\dfrac{2\tan x + \tan x - \tan^3 x}{1-\tan^2 x}}{\dfrac{1-\tan^2 x - 2\tan^2 x}{1-\tan^2 x}} = \dfrac{3\tan x - \tan^3 x}{1 - 3\tan^2 x}$

17. $2\tan 4x = 2\tan 2(2x) = 2\dfrac{2\tan 2x}{1-\tan^2 2x} = \dfrac{4\tan 2x}{1-\tan^2 2x}$

19. By Exercise 19 of Sections 10.2 and 10.3, $\cos 3x = 4\cos^3 x - 3\cos x$ and $\sin 3x = 3\sin x - 4\sin^3 x$; hence $\tan\dfrac{3x}{2} = \dfrac{1-\cos 3x}{\sin 3x} = \dfrac{1 - 4\cos^3 x + 3\cos x}{3\sin x - 4\sin^3 x}$

21. $\tan\left(\dfrac{\pi}{2} - x\right) = \dfrac{\sin\left(\dfrac{\pi}{2} - x\right)}{\cos\left(\dfrac{\pi}{2} - x\right)} = \dfrac{\cos x}{\sin x} = \cot x$ **23.** $\tan\left(\dfrac{3\pi}{2} - x\right) = \dfrac{\sin\left(\dfrac{3\pi}{2} - x\right)}{\cos\left(\dfrac{3\pi}{2} - x\right)} = \dfrac{-\cos x}{-\sin x} = \cot x$

Exercise 10.5 (page 297)

1. a. $\{(60 + 360k)°\} \cup \{(300 + 360k)°\}, k \in J$ **b.** $\left\{\left(\dfrac{\pi}{3} + k \cdot 2\pi\right)^R\right\} \cup \left\{\left(\dfrac{5\pi}{3} + k \cdot 2\pi\right)^R\right\}, k \in J$

3. a. $\{(60 + 180k)°\}, k \in J$ **b.** $\left\{\left(\dfrac{\pi}{3} + k\pi\right)^R\right\}, k \in J$

5. a. $\{(45 + 360k)°\} \cup \{(315 + 360k)°\}, k \in J$ **b.** $\left\{\left(\dfrac{\pi}{4} + k \cdot 2\pi\right)^R\right\} \cup \left\{\left(\dfrac{7\pi}{4} + k \cdot 2\pi\right)^R\right\}, k \in J$

7. a. $\{14.5° + 360°k\} \cup \{163.5° + 360°k\}, k \in J$ **b.** $\{(0.25 + k \cdot 2\pi)^R\} \cup \{(2.89 + k \cdot 2\pi)^R\}, k \in J$

9. a. $\{71.6° + 180°k\}, k \in J$ **b.** $\{(1.25 + k\pi)^R\}, k \in J$

11. a. $\{66.4° + 360°k\} \cup \{293.6° + 360°k\}, k \in J$ **b.** $\{(1.16 + k \cdot 2\pi)^R\} \cup \{(5.12 + k \cdot 2\pi)^R\}, k \in J$

13. $\left\{x \mid x = \dfrac{\pi}{3} + k\pi\right\}, k \in J$ **15.** $\left\{x \mid x = \dfrac{\pi}{4} + \dfrac{k\pi}{2}\right\}, k \in J$ **17.** $\left\{x \mid x = \dfrac{\pi}{2} + k\pi\right\}, k \in J$

19. {30°, 45°, 135°, 150°, 225°, 315°}   **21.** {0°, 120°, 180°, 240°}   **23.** $\left\{\dfrac{\pi^R}{4}, \dfrac{5\pi^R}{4}\right\}$
25. {0^R}   **27.** {1.11^R, 1.77^R, 4.25^R, 4.91^R}   **29.** {0.67^R, 2.47^R}
31. {1.9, 0, −1.9}

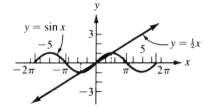

Note: long marks, π units; short marks, integers

Exercise 10.6 (page 301)   **1.** {22.5°, 157.5°, 202.5°, 337.5°}   **3.** {60°, 300°}
5. {0°, 60°, 120°, 180°, 240°, 300°}   **7.** {45°, 225°}   **9.** {67.5°, 157.5°, 247.5°, 337.5°}
11. {30°, 90°, 150°, 210°, 270°, 330°}   **13.** $\left\{\dfrac{\pi}{2} + k\pi, \dfrac{\pi}{6} + 2k\pi, \dfrac{5\pi}{6} + 2k\pi\right\}, k \in J$
15. $\left\{\dfrac{\pi}{2} + k\pi, \dfrac{\pi}{3} + 2k\pi, \dfrac{5\pi}{3} + 2k\pi\right\}, k \in J$ **17.** $\left\{k\pi, \dfrac{\pi}{2} + k\pi\right\}, k \in J$ **19.** $\left\{\dfrac{k\pi}{2}\right\}, k \in J$

Chapter 10 Review (page 301)   **1.** $\tan x = \dfrac{\sin x}{\cos x} = \sin x \sec x$

2. $\cos x - \sin x = (\cos x - \sin x)\dfrac{\cos x}{\cos x} = \left(\dfrac{\cos x}{\cos x} - \dfrac{\sin x}{\cos x}\right)\cos x = (1 - \tan x)\cos x$

3. $\dfrac{1 - \tan^2 \alpha}{\tan \alpha} = \dfrac{1}{\tan \alpha} - \dfrac{\tan^2 \alpha}{\tan \alpha} = \cot \alpha - \tan \alpha$

4. $\dfrac{\cos^2 \alpha}{1 - \sin \alpha} = \dfrac{\cos \alpha \cdot \dfrac{1}{\sec \alpha}}{1 - \sin \alpha} = \dfrac{\cos \alpha}{(1 - \sin \alpha)(\sec \alpha)} = \dfrac{\cos \alpha}{\sec \alpha - \dfrac{\sin \alpha}{\cos \alpha}} = \dfrac{\cos \alpha}{\sec \alpha - \tan \alpha}$

5. $\dfrac{1}{\cot^2 \theta} - \dfrac{1}{\csc^2 \theta} = \dfrac{\csc^2 \theta - \cot^2 \theta}{\csc^2 \theta \cot^2 \theta} = \dfrac{1}{\csc^2 \theta \cot^2 \theta}$

6. $\dfrac{2}{1 - \sin \theta} + \dfrac{2}{1 + \sin \theta} = \dfrac{2(1 + \sin \theta) + 2(1 - \sin \theta)}{1 - \sin^2 \theta} = \dfrac{4}{\cos^2 \theta}$

7. $\dfrac{1 + \sin x}{\cos x} = \dfrac{1}{\cos x} + \dfrac{\sin x}{\cos x} = \sec x + \tan x = \dfrac{\sec^2 x - \tan^2 x}{\sec x - \tan x} = \dfrac{1}{\sec x - \tan x}$

8. $\dfrac{1}{\cos^2 x} + \dfrac{1}{\sin^2 x} = \dfrac{\sin^2 x + \cos^2 x}{\sin^2 x \cos^2 x} = \dfrac{1}{\sin^2 x \cos^2 x}$

9. $\dfrac{-2\sqrt{3} - \sqrt{5}}{6}$ **10.** $\dfrac{-2 - \sqrt{15}}{6}$ **11.** $\dfrac{-2\sqrt{3} + \sqrt{5}}{6}$ **12.** $\dfrac{-2 + \sqrt{15}}{6}$ **13.** $-\dfrac{1}{9}$ **14.** $\dfrac{\sqrt{6}}{6}$

15. $\dfrac{-\sqrt{5} + 2\sqrt{3}}{6}$ **16.** $\dfrac{-\sqrt{15} + 2}{6}$ **17.** $\dfrac{-\sqrt{5} - 2\sqrt{3}}{6}$

Odd-Numbered Answers

18. $\dfrac{-\sqrt{15}-2}{6}$ 19. $\dfrac{-4\sqrt{5}}{9}$ 20. $\sqrt{\dfrac{3+\sqrt{5}}{6}}$ 21. $\dfrac{-6\sqrt{5}+5\sqrt{3}}{15+2\sqrt{15}}$

22. $\dfrac{5\sqrt{3}-2\sqrt{5}}{5+2\sqrt{15}}$ 23. $\dfrac{5\sqrt{3}+6\sqrt{5}}{15-2\sqrt{15}}$ 24. $\dfrac{5\sqrt{3}+2\sqrt{5}}{5-2\sqrt{15}}$ 25. $-4\sqrt{5}$ 26. $\dfrac{3+\sqrt{5}}{2}$

27. $\{0°, 180°, 360°\}$ 28. $\{30°, 90°, 150°, 270°\}$ 29. $\left\{\dfrac{\pi^R}{2}\right\}$ 30. $\left\{\dfrac{\pi^R}{6}+k\pi^R\right\}, k \in J$ 31. $\{20°, 40°\}$

32. $\{90°\}$ 33. $\left\{\left(\dfrac{\pi}{8}+\dfrac{k}{2}\pi\right)\right\}, k \in J$ 34. $\left\{\left(\dfrac{\pi}{6}+\dfrac{k}{2}\pi\right)\right\} \cup \left\{\left(\dfrac{\pi}{3}+\dfrac{k}{2}\pi\right)\right\}, k \in J$

Exercise 11.1 (page 308) 1. $\{(3, 2)\}$ 3. $\{(2, 1)\}$ 5. $\{(-5, 4)\}$ 7. $\left\{\left(0, \dfrac{3}{2}\right)\right\}$ 9. $\left\{\left(\dfrac{2}{3}, -1\right)\right\}$

11. $\{(1, 2)\}$ 13. $\left\{\left(-\dfrac{19}{5}, -\dfrac{18}{5}\right)\right\}$ 15. $a = 1, b = -1$ 17. $y = -\dfrac{10}{3}x + 2$

19. $C = \dfrac{5}{9}(F - 32)$ 21. 32 pounds   23. \$7200 at 4%; \$8200 at 5%   25. 32 and 64 miles per hour

27. **a.** From $\dfrac{a_1}{a_2} = \dfrac{b_1}{b_2}$, it follows that $\dfrac{a_1}{b_1} = \dfrac{a_2}{b_2}$. Hence the slopes are equal; the two lines are the same or they are parallel. The equations are either consistent or inconsistent, respectively. Assume that the equations are consistent and the two lines are the same. Then, the y-intercepts $\dfrac{-c_1}{b_1} = \dfrac{-c_2}{b_2}$, from which $\dfrac{b_1}{b_2} = \dfrac{c_1}{c_2}$; but this contradicts the hypothesis $\dfrac{b_1}{b_2} \neq \dfrac{c_1}{c_2}$; hence the equations are inconsistent.   **b.** from (a) above, $\dfrac{a_1}{b_1} = \dfrac{a_2}{b_2}$, and it follows that $\dfrac{a_1}{a_2} = \dfrac{b_1}{b_2}$. Now, either $\dfrac{b_1}{b_2} = \dfrac{c_1}{c_2}$ or $\dfrac{b_1}{b_2} \neq \dfrac{c_1}{c_2}$. Assume $\dfrac{b_1}{b_2} = \dfrac{c_1}{c_2}$; then $\dfrac{c_1}{b_1} = \dfrac{c_2}{b_2}$; the graphs would be the same straight line, and the equations would be consistent. This contradicts the hypothesis that states that the equations are inconsistent; hence $\dfrac{b_1}{b_2} \neq \dfrac{c_1}{c^2}$.

Exercise 11.2 (page 314) 1. $\{(1, 2, -1)\}$ 3. $\{(2, -2, 0)\}$ 5. $\{(2, 2, 1)\}$ 7. $\{(0, 1, 2)\}$

9. Dependent   11. $\{(4, -2, 2)\}$ 13. 3, 6, 6   15. 60 nickels, 20 dimes, 5 quarters

17. Thirty-six 1's, thirty-four 5's, twenty-four 10's   19. $x^2 + y^2 + 2x + 2y - 23 = 0$

21. $a = \dfrac{1}{4}, b = \dfrac{1}{2}, c = 0$ 23. $\left\{\left(\dfrac{5}{3}, \dfrac{1}{3}, 0\right), (2, 2, -1)\right\}$

25. Since the solution set for (a), (b), (c) is $\{(1, 1, 1)\}$ and this is not a solution of (d) since $2 - 1 + 2 \neq 4$, the solution set for (a), (b), (c), (d) is $\emptyset$.

Exercise 11.3 (page 320) 1. $\{(-1, -4), (5, 20)\}$ 3. $\{(2, 3), (3, 2)\}$ 5. $\{(4, -3), (-3, 4)\}$

7. $\{(-1, 3), (-1, -3), (1, 3), (1, -3)\}$ 9. $\{(-3, \sqrt{2}), (-3, -\sqrt{2}), (3, \sqrt{2}), (3, -\sqrt{2})\}$

11. $\{(\sqrt{3}, 4), (\sqrt{3}, -4), (-\sqrt{3}, 4), (-\sqrt{3}, -4)\}$ 13. $\{(1, -2), (-1, 2), (2, -1), (-2, 1)\}$

15. $\{(3, 1), (-3, -1), (-2\sqrt{7}, \sqrt{7}), (2\sqrt{7}, -\sqrt{7})\}$

17. {(0, 2)}

19. $\left\{(1, 0), \left(\dfrac{3}{2}, \dfrac{1}{2}\right)\right\}$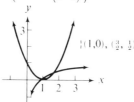

21. {(1, 0)}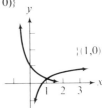

23. {(1, 0)} (see graph Exercise 21, where $y_1 = 10^{-x}$ and $y_2 = \log_{10} x$).

25. a. 1 **b.** 2 **c.** 4 **27.** $\dfrac{7}{2}, \dfrac{5}{2}$ **29.** 6 pounds per square inch; 5 cubic inches

31. {(2,2), (−2, −2)}

Exercise 11.4 (page 323)

1. 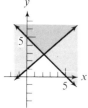 **3.**   **5.**   **7.**

9. 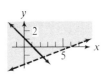 **11.** 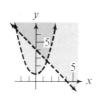 **13.**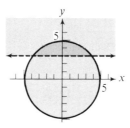

15. 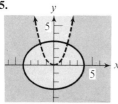 **17.** 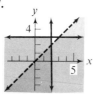 **19.**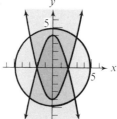

21. 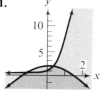 **23.**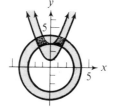

Exercise 11.5 (page 326)

1. 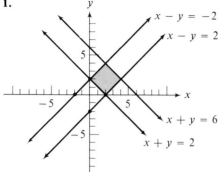   **3.**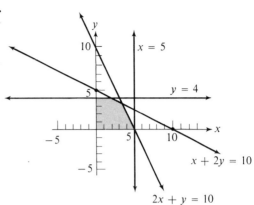

7. No   **9.** Yes

5.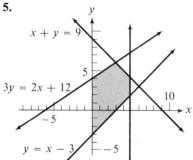

Exercise 11.6 (page 328)
1. Maximum: 14; Minimum: 2   **3.** Maximum: 80/3; Minimum: 0
5. Maximum: 17; Minimum: -16 **7.** Maximum profit: $1860   **9.** Maximum profit: $770
11. 200 standard models and 400 deluxe models

Chapter 11 Review (page 329)

1. $\left\{\left(\frac{1}{2}, \frac{7}{2}\right)\right\}$ **2.** $\{(1, 2)\}$ **3.** $\left\{\left(\frac{37}{13}, -\frac{10}{13}\right)\right\}$ **4.** $a = 2, b = 5$
5. $\{(2, 0, -1)\}$ **6.** $\{(2, 1, -1)\}$ **7.** $\{(2, -1, 3)\}$ **8.** $a = 2, b = -3, c = 4$
9. $\{(1, 2), (4, -13)\}$ **10.** $\left\{\left(\frac{6}{7}, -\frac{37}{7}\right), (2, -3)\right\}$ **11.** $\{(2, 3), (2, -3), (-2, 3), (-2, -3)\}$
12. $\left\{\left(\frac{7}{8}, \frac{17}{8}\right)\right\}$ **13.** 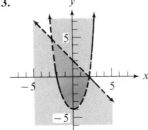   **14.**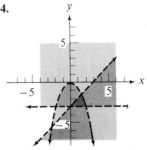

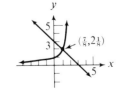

15. 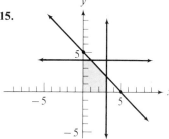 **16.**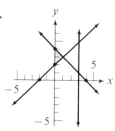

17. Minimum: 0;   Maximum: 8   **18.** Minimum: 0;   Maximum: 11
19. Maximum profit: $1100   **20.** Maximum profit: $210

Exercise 12.1 (page 335) 1. 2×2, $\begin{bmatrix} 6 & 2 \\ -1 & 3 \end{bmatrix}$ **3.** 2×3, $\begin{bmatrix} 2 & 1 \\ -7 & 4 \\ 3 & 0 \end{bmatrix}$ **5.** 3×3, $\begin{bmatrix} 2 & 4 & -2 \\ 3 & 0 & 3 \\ -1 & 1 & 1 \end{bmatrix}$

7. 2×4, $\begin{bmatrix} 4 & 2 \\ -3 & 1 \\ -1 & 1 \\ 0 & 6 \end{bmatrix}$ **9.** $\begin{bmatrix} 3 & 1 \\ 3 & 9 \end{bmatrix}$ **11.** $\begin{bmatrix} 9 & -1 & -1 \\ 2 & 3 & 6 \end{bmatrix}$ **13.** $\begin{bmatrix} -2 & -3 \\ 4 & 0 \end{bmatrix}$

15. $\begin{bmatrix} 2 & -9 & -13 \\ 7 & -4 & 1 \\ 9 & 0 & 0 \end{bmatrix}$ **17.** $\begin{bmatrix} 10 \\ 3 \\ -3 \end{bmatrix}$ **19.** $\begin{bmatrix} 2 & 3 & 4 \\ -1 & 6 & 2 \\ 1 & 0 & 3 \end{bmatrix}$

21. Let X, $X + A$, and B have each entry x_{ij}, $x_{ij} + a_{ij}$, and b_{ij}, respectively. Then since $X + A = B$, $x_{ij} + a_{ij} = b_{ij}$ and since x_{ij}, a_{ij}, and b_{ij} are real numbers, $x_{ij} = b_{ij} - a_{ij}$ (cancellation law of addition); thus, $X = B - A$.

23. $\begin{bmatrix} 2 & 2 \\ -1 & 1 \end{bmatrix}$ **25.** $\begin{bmatrix} 1 & 1 \\ -4 & 3 \end{bmatrix}$

27. The i, j entry of $(A + B)_{m \times n} = a_{ij} + b_{ij}$; and since a_{ij} and b_{ij} are real numbers, $a_{ij} + b_{ij}$ is also a real number (closure law of addition); thus, $(A + B)_{m \times n}$ is a matrix with real-number entries.

29. The i, j entry of $(A + B)_{m \times n}$ is $a_{ij} + b_{ij}$, and similarly of $(B + A)_{m \times n}$ is $b_{ij} + a_{ij}$; and since a_{ij} and b_{ij} are real numbers, $a_{ij} + b_{ij} = b_{ij} + a_{ij}$ (commutative law of addition); hence, $(A + B)_{m \times n} = (B + A)_{m \times n}$, by Definition 12.1.

Exercise 12.2 (page 340) 1. $\begin{bmatrix} 0 & -5 & 5 \\ -15 & 5 & -10 \end{bmatrix}$ **3.** $[-1]$ **5.** $\begin{bmatrix} 1 & -13 \\ 4 & -7 \end{bmatrix}$ **7.** $\begin{bmatrix} 30 & -39 \\ 29 & 14 \end{bmatrix}$

9. $\begin{bmatrix} -5 & -1 \\ 8 & -1 \end{bmatrix}$ **11.** $\begin{bmatrix} -1 & 0 & -2 \\ 1 & 2 & 8 \\ 0 & 1 & 3 \end{bmatrix}$ **13.** $\begin{bmatrix} 1 & 0 & 0 \\ 0 & 1 & 0 \\ 0 & 0 & 1 \end{bmatrix}$ **15.** $\begin{bmatrix} 1 & 0 \\ -1 & 2 \end{bmatrix}$ **17.** $\begin{bmatrix} 1 & -2 \\ 1 & 2 \end{bmatrix}$

19. $\begin{bmatrix} -2 & 3 \\ 2 & -4 \end{bmatrix}$ **21.** $\begin{bmatrix} -1 & 1 \\ -1 & -1 \end{bmatrix}$ **23.** $\begin{bmatrix} 1 & -1 \\ 1 & 0 \end{bmatrix}$

25. Since $A + B = \begin{bmatrix} 0 & 2 \\ -1 & 3 \end{bmatrix}$, $A - B = \begin{bmatrix} -2 & 2 \\ 1 & -1 \end{bmatrix}$, $A^2 = \begin{bmatrix} 1 & 0 \\ 0 & 1 \end{bmatrix}$, $B^2 = \begin{bmatrix} 1 & 0 \\ -3 & 4 \end{bmatrix}$

Odd-Numbered Answers

and $AB = \begin{bmatrix} -3 & 4 \\ -1 & 2 \end{bmatrix}$, then (a) $(A+B)(A+B) = \begin{bmatrix} 0 & 2 \\ -1 & 3 \end{bmatrix} \cdot \begin{bmatrix} 0 & 2 \\ -1 & 3 \end{bmatrix} = \begin{bmatrix} -2 & -6 \\ -3 & 7 \end{bmatrix}$

and $A^2 + 2AB + B^2 = \begin{bmatrix} 1 & 0 \\ 0 & 1 \end{bmatrix} + 2\begin{bmatrix} -3 & 4 \\ -1 & 2 \end{bmatrix} + \begin{bmatrix} 1 & 0 \\ -3 & 4 \end{bmatrix} = \begin{bmatrix} -4 & 8 \\ -5 & 9 \end{bmatrix}$;

hence $(A+B)(A+B) \neq A^2 + 2AB + B^2$. (b) $(A+B)(A-B) = \begin{bmatrix} 0 & 2 \\ -1 & 3 \end{bmatrix} \cdot \begin{bmatrix} -2 & 2 \\ 1 & -1 \end{bmatrix} = \begin{bmatrix} 2 & -2 \\ 5 & -5 \end{bmatrix}$

and $A^2 - B^2 = \begin{bmatrix} 1 & 0 \\ 0 & 1 \end{bmatrix} - \begin{bmatrix} 1 & 0 \\ -3 & 4 \end{bmatrix} = \begin{bmatrix} 0 & 0 \\ 3 & -3 \end{bmatrix}$; hence $(A+B)(A-B) \neq A^2 - B^2$.

For Exercises 27 and 29, let $A = \begin{bmatrix} a_{11} & a_{12} \\ a_{21} & a_{22} \end{bmatrix}$, $B = \begin{bmatrix} b_{11} & b_{12} \\ b_{21} & b_{22} \end{bmatrix}$, and $C = \begin{bmatrix} c_{11} & c_{12} \\ c_{21} & c_{22} \end{bmatrix}$.

27. Since $AB = \begin{bmatrix} a_{11}b_{11} + a_{12}b_{21} & a_{11}b_{12} + a_{12}b_{22} \\ a_{21}b_{11} + a_{22}b_{21} & a_{21}b_{12} + a_{22}b_{22} \end{bmatrix}$,

$(AB)C = \begin{bmatrix} (a_{11}b_{11} + a_{12}b_{21})c_{11} + (a_{11}b_{12} + a_{12}b_{22})c_{21} & (a_{11}b_{11} + a_{12}b_{21})c_{12} + (a_{11}b_{12} + a_{12}b_{22})c_{22} \\ (a_{21}b_{11} + a_{22}b_{21})c_{11} + (a_{21}b_{12} + a_{22}b_{22})c_{21} & (a_{21}b_{11} + a_{22}b_{21})c_{12} + (a_{21}b_{12} + a_{22}b_{22})c_{22} \end{bmatrix}$.

The element in the first row, first column is given by $a_{11}b_{11}c_{11} + a_{12}b_{21}c_{11} + a_{11}b_{12}c_{21} + a_{12}b_{22}c_{21} =$

$(a_{11}b_{11}c_{11} + a_{11}b_{12}c_{21}) + (a_{12}b_{21}c_{11} + a_{12}b_{22}c_{21}) = a_{11}(b_{11}c_{11} + b_{12}c_{21}) + a_{12}(b_{21}c_{11} + b_{22}c_{21})$.

When the remaining elements are treated in a similar manner, $(AB)C =$

$\begin{bmatrix} a_{11}(b_{11}c_{11} + b_{12}c_{21}) + a_{12}(b_{21}c_{11} + b_{22}c_{21}) & a_{11}(b_{11}c_{12} + b_{12}c_{22}) + a_{12}(b_{21}c_{12} + b_{22}c_{22}) \\ a_{21}(b_{11}c_{11} + b_{12}c_{21}) + a_{22}(b_{21}c_{11} + b_{22}c_{21}) & a_{21}(b_{11}c_{12} + b_{12}c_{22}) + a_{22}(b_{21}c_{12} + b_{22}c_{22}) \end{bmatrix} =$

$\begin{bmatrix} a_{11} & a_{12} \\ a_{21} & a_{22} \end{bmatrix} \cdot \begin{bmatrix} b_{11}c_{11} + b_{12}c_{21} & b_{11}c_{12} + b_{12}c_{22} \\ b_{21}c_{11} + b_{22}c_{21} & b_{21}c_{12} + b_{22}c_{22} \end{bmatrix} = \begin{bmatrix} a_{11} & a_{12} \\ a_{21} & a_{22} \end{bmatrix} \cdot \left(\begin{bmatrix} b_{11} & b_{12} \\ b_{21} & b_{22} \end{bmatrix} \cdot \begin{bmatrix} c_{11} & c_{12} \\ c_{21} & c_{22} \end{bmatrix} \right) = A(BC)$.

29. $B + C = \begin{bmatrix} b_{11} + c_{11} & b_{12} + c_{12} \\ b_{21} + c_{21} & b_{22} + c_{22} \end{bmatrix}$,

hence $(B+C)A = \begin{bmatrix} (b_{11} + c_{11})a_{11} + (b_{12} + c_{12})a_{21} & (b_{11} + c_{11})a_{12} + (b_{12} + c_{12})a_{22} \\ (b_{21} + c_{22})a_{11} + (b_{22} + c_{22})a_{21} & (b_{21} + c_{22})a_{12} + (b_{22} + c_{22})a_{22} \end{bmatrix}$.

The element in the first row, first column is given by

$b_{11}a_{11} + c_{11}a_{21} + b_{12}a_{21} + c_{12}a_{21} = (b_{11}a_{11} + b_{12}a_{21}) + (c_{11}a_{21} + c_{12}a_{21})$.

When the remaining elements are treated in a similar manner,

$(B+C)A = \begin{bmatrix} (b_{11}a_{11} + b_{12}a_{21}) + (c_{11}a_{11} + c_{12}a_{21}) & (b_{11}a_{12} + b_{12}a_{22}) + (c_{11}a_{12} + c_{12}a_{22}) \\ (b_{21}a_{11} + b_{22}a_{21}) + (c_{21}a_{11} + c_{22}a_{21}) & (b_{21}a_{12} + b_{22}a_{22}) + (c_{21}a_{12} + c_{22}a_{22}) \end{bmatrix}$

$= \begin{bmatrix} b_{11}a_{11} + b_{12}a_{21} & b_{11}a_{12} + b_{12}a_{22} \\ b_{21}a_{11} + b_{22}a_{21} & b_{21}a_{12} + b_{22}a_{22} \end{bmatrix} + \begin{bmatrix} c_{11}a_{11} + c_{12}a_{21} & c_{11}a_{12} + c_{12}a_{22} \\ c_{21}a_{11} + c_{22}a_{21} & c_{21}a_{12} + c_{22}a_{22} \end{bmatrix}$

$= \begin{bmatrix} b_{11} & b_{12} \\ b_{21} & b_{22} \end{bmatrix} \cdot \begin{bmatrix} a_{11} & a_{12} \\ a_{21} & a_{22} \end{bmatrix} + \begin{bmatrix} c_{11} & c_{12} \\ c_{21} & c_{22} \end{bmatrix} \cdot \begin{bmatrix} a_{11} & a_{12} \\ a_{21} & a_{22} \end{bmatrix} = BC + CA$.

31. $(A_{2 \times 2} \cdot B_{2 \times 2}) = \begin{bmatrix} a_{11}b_{11} + a_{12}b_{21} & a_{11}b_{12} + a_{12}b_{22} \\ a_{21}b_{11} + a_{22}b_{21} & a_{21}b_{12} + a_{22}b_{22} \end{bmatrix}$.

Hence, $(A_{2 \times 2} \cdot B_{2 \times 2})^t = \begin{bmatrix} a_{11}b_{11} + a_{12}b_{21} & a_{21}b_{11} + a_{22}b_{21} \\ a_{11}b_{12} + a_{12}b_{22} & a_{21}b_{12} + a_{22}b_{22} \end{bmatrix}$.

By the commutative property of multiplication for real numbers,

$$(A_{2\times 2}\cdot B_{2\times 2})^t = \begin{bmatrix} b_{11}a_{11}+b_{21}a_{12} & b_{11}a_{21}+b_{21}a_{22} \\ b_{12}a_{11}+b_{22}a_{12} & b_{12}a_{21}+b_{22}a_{22} \end{bmatrix} = \begin{bmatrix} b_{11} & b_{21} \\ b_{12} & b_{22} \end{bmatrix} \cdot \begin{bmatrix} a_{11} & a_{21} \\ a_{12} & a_{22} \end{bmatrix} = B_{2\times 2}^t \cdot A_{2\times 2}^t.$$

33. Let $A = \begin{bmatrix} a_{11} & a_{12} \\ a_{21} & a_{22} \end{bmatrix}$ and c be a scalar; then by Definition 12.6 and the closure property of multiplication of real numbers, $c\begin{bmatrix} a_{11} & a_{12} \\ a_{21} & a_{22} \end{bmatrix} = \begin{bmatrix} ca_{11} & ca_{12} \\ ca_{21} & ca_{22} \end{bmatrix}$, which is an $m \times n$ matrix.

35. By Definition 12.6 and the distributive property for real numbers, and by Definition 12.3,

$$(c+d)\begin{bmatrix} a_{11} & a_{12} \\ a_{21} & a_{22} \end{bmatrix} = \begin{bmatrix} (c+d)a_{11} & (c+d)a_{12} \\ (c+d)a_{21} & (c+d)a_{22} \end{bmatrix} = \begin{bmatrix} ca_{11}+da_{11} & ca_{12}+da_{12} \\ ca_{21}+da_{21} & ca_{22}+da_{22} \end{bmatrix}$$

$$= \begin{bmatrix} ca_{11} & ca_{12} \\ ca_{21} & ca_{22} \end{bmatrix} + \begin{bmatrix} da_{11} & da_{12} \\ da_{21} & da_{22} \end{bmatrix} = c\begin{bmatrix} a_{11} & a_{12} \\ a_{21} & a_{22} \end{bmatrix} + d\begin{bmatrix} a_{11} & a_{12} \\ a_{21} & a_{22} \end{bmatrix} = cA + dA.$$

37. By Definitions 12.6 and 12.5, $(-1)A = -1\begin{bmatrix} a_{11} & a_{12} \\ a_{21} & a_{22} \end{bmatrix} = \begin{bmatrix} -a_{11} & -a_{12} \\ -a_{21} & -a_{22} \end{bmatrix} = -A.$

39. $0_{2\times 2} = \begin{bmatrix} 0 & 0 \\ 0 & 0 \end{bmatrix}$; hence, by Definition 12.6, $c \cdot 0_{2\times 2} = c\begin{bmatrix} 0 & 0 \\ 0 & 0 \end{bmatrix} = \begin{bmatrix} c\cdot 0 & c\cdot 0 \\ c\cdot 0 & c\cdot 0 \end{bmatrix} = \begin{bmatrix} 0 & 0 \\ 0 & 0 \end{bmatrix} = 0_{2\times 2}.$

Exercise 12.3 (page 347)   **1.** $\{(2, -1)\}$ **3.** $\{(5, 1)\}$ **5.** $\{(8, 1)\}$ **7.** $\{(-2, 2, 0)\}$

9. $\left\{\left(\dfrac{5}{4}, \dfrac{5}{2}, -\dfrac{1}{2}\right)\right\}$ **11.** $\left\{\left(-\dfrac{77}{27}, -\dfrac{8}{27}, \dfrac{29}{27}\right)\right\}$

13. $\{(1, 1-z, z)\}$ **15.** $(3, 3, -1)$ **17.** No solution   **19.** No solution

21. $\begin{bmatrix} k & 0 \\ 0 & 1 \end{bmatrix}\begin{bmatrix} a & b \\ c & d \end{bmatrix} = \begin{bmatrix} k\cdot a+0\cdot c & k\cdot b+0\cdot d \\ 0\cdot a+1\cdot c & 0\cdot b+1\cdot d \end{bmatrix} = \begin{bmatrix} ka & kb \\ c & d \end{bmatrix}$

23. $\begin{bmatrix} 0 & 1 \\ 1 & 0 \end{bmatrix}\begin{bmatrix} a & b \\ c & d \end{bmatrix} = \begin{bmatrix} 0\cdot a+1\cdot c & 0\cdot b+1\cdot d \\ 1\cdot a+0\cdot c & 1\cdot b+0\cdot d \end{bmatrix} = \begin{bmatrix} c & d \\ a & b \end{bmatrix}$

25. $\begin{bmatrix} 1 & 0 \\ k & 1 \end{bmatrix}\begin{bmatrix} a & b \\ c & d \end{bmatrix} = \begin{bmatrix} 1\cdot a+0\cdot c & 1\cdot b+0\cdot d \\ k\cdot a+1\cdot c & k\cdot b+1\cdot d \end{bmatrix} = \begin{bmatrix} a & b \\ ka+c & kb+d \end{bmatrix}$

Exercise 12.4 (page 353)   **1.** 0   **3.** -6 **5.** -2

7. $M_{11} = \begin{vmatrix} 0 & 3 & -1 \\ 1 & 2 & 2 \\ -1 & 3 & 1 \end{vmatrix}$, $A_{11} = \begin{vmatrix} 0 & 3 & -1 \\ 1 & 2 & 2 \\ -1 & 3 & 1 \end{vmatrix}$

9. $M_{23} = \begin{vmatrix} 2 & 1 & 0 \\ -2 & 1 & 2 \\ 1 & -1 & 1 \end{vmatrix}$, $A_{23} = -\begin{vmatrix} 2 & 1 & 0 \\ -2 & 1 & 2 \\ 1 & -1 & 1 \end{vmatrix}$

11. $M_{31} = \begin{vmatrix} 1 & -2 & 0 \\ 0 & 3 & -1 \\ -1 & 3 & 1 \end{vmatrix}$, $A_{31} = \begin{vmatrix} 1 & -2 & 0 \\ 0 & 3 & -1 \\ -1 & 3 & 1 \end{vmatrix}$

Odd-Numbered Answers

13. $M_{44} = \begin{vmatrix} 2 & 1 & -2 \\ 1 & 0 & 3 \\ -2 & 1 & 2 \end{vmatrix}$, $A_{44} = \begin{vmatrix} 2 & 1 & -2 \\ 1 & 0 & 3 \\ -2 & 1 & 2 \end{vmatrix}$

15. 1   17. 0   19. -30 21. $3x + 1$ 23. 3   25. 0   27. -1 29. 0

31. x^3 33. $x = 5$ 35. $x = 3$

37. Expanding by elements in the first row yields $\begin{vmatrix} 0 & 1 & 0 & 0 \\ 1 & 0 & 3 & 2 \\ 5 & -1 & 2 & 1 \\ 1 & 0 & 1 & 1 \end{vmatrix} = -\begin{vmatrix} 1 & 3 & 2 \\ 5 & 2 & 1 \\ 1 & 1 & 1 \end{vmatrix}$;

expanding by elements in the third row yields $-\begin{vmatrix} 3 & 2 \\ 2 & 1 \end{vmatrix} + \begin{vmatrix} 1 & 2 \\ 5 & 1 \end{vmatrix} - \begin{vmatrix} 1 & 3 \\ 5 & 2 \end{vmatrix} = 1 - 9 + 13 = 5.$

39. 2, 6, 24

41. Let $A = \begin{bmatrix} a_{11} & a_{12} \\ a_{21} & a_{22} \end{bmatrix}$; by Definition 12.6 and the fact that $\delta(A) = a_{11}a_{22} - a_{12}a_{21}$, it follows that

$aA = \begin{bmatrix} aa_{11} & aa_{12} \\ aa_{21} & aa_{22} \end{bmatrix}$ and $\delta(aA) = a^2 a_{11}a_{22} - a^2 a_{12}a_{21} = a^2(a_{11}a_{22} - a_{12}a_{21}) = a^2 \delta(A)$.

43. Let $A = \begin{bmatrix} a_{11} & a_{12} \\ a_{21} & a_{22} \end{bmatrix}$ and $B = \begin{bmatrix} b_{11} & b_{12} \\ b_{21} & b_{22} \end{bmatrix}$; then $\delta(A) = a_{11}a_{22} - a_{12}a_{21}$

and $\delta(B) = b_{11}b_{22} - b_{12}b_{21}$, $AB = \begin{bmatrix} a_{11}b_{11} + a_{12}b_{21} & a_{11}b_{12} + a_{12}b_{22} \\ a_{21}b_{11} + a_{22}b_{21} & a_{21}b_{12} + a_{22}b_{22} \end{bmatrix}$

and $\delta(AB) = (a_{11}b_{11} + a_{12}b_{21})(a_{21}b_{12} + a_{22}b_{22}) - (a_{11}b_{12} + a_{12}b_{22})(a_{21}b_{11} + a_{22}b_{21})$, which simplifies to $a_{11}b_{11}a_{22}b_{22} - a_{11}b_{12}a_{22}b_{21} - a_{12}b_{22}a_{21}b_{11} + a_{12}b_{21}a_{21}b_{12}$
$= a_{11}a_{22}(b_{11}b_{22} - b_{12}b_{21}) - a_{12}a_{21}(b_{11}b_{22} - b_{12}b_{21}) = (a_{11}a_{22} - a_{12}a_{21})(b_{11}b_{22} - b_{12}b_{21}) = \delta(A) \cdot \delta(B)$.

Exercise 12.5 (page 361)   1. Theorem 12.7   3. Theorem 12.9   5. Theorem 12.10
7. Theorem 12.10   9. Theorem 12.12   11. Theorem 12.12

13. $\begin{vmatrix} 1 & 3 \\ 0 & -4 \end{vmatrix}$ 15. $\begin{vmatrix} 1 & -2 & 1 \\ 0 & 7 & 1 \\ 0 & 2 & 1 \end{vmatrix}$ 17. $\begin{vmatrix} 0 & 1 & -3 & -2 \\ 0 & 2 & 1 & 2 \\ 1 & 1 & 2 & 3 \\ 0 & 1 & 1 & 1 \end{vmatrix}$

19. $-1 \begin{vmatrix} 2 & 1 \\ -1 & 2 \end{vmatrix} = -5$ 21. $\begin{vmatrix} -1 & -5 \\ 2 & -2 \end{vmatrix} = 12$ 23. $\begin{vmatrix} 4 & 4 \\ 3 & 7 \end{vmatrix} = 16$

25. $\begin{vmatrix} -1 & 1 \\ 4 & -1 \end{vmatrix} = -3$ 27. $\begin{vmatrix} 3 & 1 \\ 5 & 4 \end{vmatrix} = 7$ 29. $\begin{vmatrix} 3 & 6 \\ 0 & 0 \end{vmatrix} = 0$

31. $\begin{vmatrix} 6 & 1 \\ 0 & 3 \end{vmatrix} = 18$ 33. $-16 \begin{vmatrix} 1 & 2 \\ 2 & 3 \end{vmatrix} = 16$ 35. $\begin{vmatrix} 4 & -4 \\ 3 & -9 \end{vmatrix} = -24$

37. Let $A = \begin{vmatrix} x & y & 1 \\ x_1 & y_1 & 1 \\ x_2 & y_2 & 1 \end{vmatrix} = 0$.

Expanding about the first row gives $\delta(A) = x\begin{vmatrix} y_1 & 1 \\ y_2 & 1 \end{vmatrix} - y\begin{vmatrix} x_1 & 1 \\ x_2 & 1 \end{vmatrix} + 1\begin{vmatrix} x_1 & y_1 \\ x_2 & y_2 \end{vmatrix} = 0$;

hence $(y_1 - y_2)x + (x_2 - x_1)y + (x_1y_2 - y_1x_2) = 0$. Further, y_1, y_2, x_1, x_2 are real numbers; hence there exist real numbers, a, b, c such that $(y_1 - y_2) = a$, $(x_2 - x_1) = b$, and $(x_1y_2 - y_1x_2) = c$ with a and b not both 0 since $(x_1, y_1) \neq (x_2, y_2)$. Substituting yields $ax + by + c = 0$, which is the equation of a straight line. By Theorem 12.9, the line passes through (x_1, y_1) and (x_2, y_2).

39. Multiply column 1 by $(-a)$ and add result to column 2; also, multiply column 1 by $(-a^2)$ and add result to column 3, to obtain $\begin{vmatrix} 1 & a & a^2 \\ 1 & b & b^2 \\ 1 & c & c^2 \end{vmatrix} = \begin{vmatrix} 1 & 0 & 0 \\ 1 & b-a & b^2-a^2 \\ 1 & c-a & c^2-a^2 \end{vmatrix}$. Expand about the first row to obtain

$$1 \begin{vmatrix} b-a & b^2-a^2 \\ c-a & c^2-a^2 \end{vmatrix} = (b-a)[c^2-a^2] - (c-a)[b^2-a^2]$$
$$= (b-a)[(c-a)(c+a)] - (c-a) \cdot [(b-a)(b+a)]$$
$$= -(a-b)[(c-a)(c+a)] + (c-a)[(a-b)(a+b)]$$
$$= (a-b)(c-a)[-(c+a) \cdot (a+b)]$$
$$= (a-b)(c-a)(b-c) = (b-c)(c-a)(a-b).$$

Exercise 12.6 (page 367) **1.** $\begin{bmatrix} 3 & -2 \\ -1 & 1 \end{bmatrix}$ **3.** $\dfrac{1}{5}\begin{bmatrix} 1 & 3 \\ -1 & 2 \end{bmatrix}$ **5.** $|A| = 0$; no inverse

7. $-1\begin{bmatrix} 4 & -7 \\ -3 & 5 \end{bmatrix}$ **9.** $\dfrac{1}{2}\begin{bmatrix} -2 & -4 \\ 4 & 7 \end{bmatrix}$ **11.** $|A| = 0$; no inverse

13. $\dfrac{1}{6}\begin{bmatrix} 2 & 2 & -5 \\ -4 & 2 & 1 \\ 0 & 0 & 3 \end{bmatrix}$ **15.** $\dfrac{1}{3}\begin{bmatrix} -2 & 3 & -1 \\ -1 & 0 & 1 \\ 6 & -6 & 3 \end{bmatrix}$ **17.** $|A| = 0$; no inverse

19. $|A| = 0$; no inverse **21.** $\begin{bmatrix} 2 & -3 & 11 \\ -2 & 4 & -13 \\ 1 & -2 & 7 \end{bmatrix}$ **23.** $-1\begin{bmatrix} 0 & 0 & -1 \\ 0 & -1 & 0 \\ -1 & 0 & 0 \end{bmatrix}$

25. Let $A \cdot B = \begin{bmatrix} 2 & 3 \\ 1 & -1 \end{bmatrix} \cdot \begin{bmatrix} 0 & 1 \\ 3 & 1 \end{bmatrix}$; $A \cdot B = \begin{bmatrix} 9 & 5 \\ -3 & 0 \end{bmatrix}$ and $\delta(AB) = 15$,

so $(A \cdot B)^{-1} = \dfrac{1}{15}\begin{bmatrix} 0 & -5 \\ 3 & 9 \end{bmatrix}$; also, since $\delta(A) = -5$ and $\delta(B) = -3$,

$B^{-1} = -\dfrac{1}{3}\begin{bmatrix} 1 & -1 \\ -3 & 0 \end{bmatrix}$ and $A^{-1} = \dfrac{1}{5}\begin{bmatrix} -1 & -3 \\ -1 & 2 \end{bmatrix}$; $B^{-1} \cdot A^{-1} = \dfrac{1}{15}\begin{bmatrix} 0 & -5 \\ 3 & 9 \end{bmatrix}$;

hence $(A \cdot B)^{-1} = B^{-1} \cdot A^{-1}$.

27. Let $A = \begin{bmatrix} 3 & 0 & 1 \\ 2 & 1 & 0 \\ 0 & 1 & 2 \end{bmatrix}$; then $\delta(A) = 8$ and $A^{-1} = \dfrac{1}{8}\begin{bmatrix} 2 & 1 & -1 \\ -4 & 6 & 2 \\ 2 & -3 & 3 \end{bmatrix}$.

Let $B = \begin{bmatrix} 2 & 1 & 0 \\ 1 & 1 & 2 \\ 0 & 1 & 0 \end{bmatrix}$; then $\delta(B) = -\dfrac{1}{4}$ and $B^{-1} = -\dfrac{1}{4}\begin{bmatrix} -2 & 0 & 2 \\ 0 & 0 & -4 \\ 1 & -2 & 1 \end{bmatrix}$;

$A \cdot B = \begin{bmatrix} 6 & 4 & 0 \\ 5 & 3 & 2 \\ 1 & 3 & 2 \end{bmatrix}$, $\delta(A \cdot B) = -32$, and $(A \cdot B)^{-1} = -\dfrac{1}{32}\begin{bmatrix} 0 & -8 & 8 \\ -8 & 12 & -12 \\ 12 & -14 & -2 \end{bmatrix}$;

$B^{-1} \cdot A^{-1} = -\dfrac{1}{4}\begin{bmatrix} -2 & 0 & 2 \\ 0 & 0 & -4 \\ 1 & -2 & 1 \end{bmatrix} \cdot \dfrac{1}{8}\begin{bmatrix} 2 & 1 & -1 \\ -4 & 6 & 2 \\ 2 & -3 & 3 \end{bmatrix} = -\dfrac{1}{32}\begin{bmatrix} 0 & -8 & 8 \\ -8 & 12 & -12 \\ 12 & -14 & -2 \end{bmatrix}$;

hence $(A \cdot B)^{-1} = B^{-1} \cdot A^{-1}$.

Odd-Numbered Answers

29. Let $A = \begin{bmatrix} a_{11} & a_{12} \\ a_{21} & a_{22} \end{bmatrix}$; then $\delta(A) = (a_{11}a_{22} - a_{12}a_{21}) \neq 0$, since A is nonsingular, and hence $\dfrac{1}{\delta(A)}$ is defined and A^{-1} exists.

$$A^{-1} = \frac{1}{\delta(A)} \begin{bmatrix} a_{22} & -a_{12} \\ -a_{21} & a_{11} \end{bmatrix} = \begin{bmatrix} \dfrac{a_{22}}{\delta(A)} & \dfrac{-a_{12}}{\delta(A)} \\ \dfrac{-a_{21}}{\delta(A)} & \dfrac{a_{11}}{\delta(A)} \end{bmatrix}$$

and $\delta(A^{-1}) = \dfrac{a_{22}a_{11}}{[\delta(A)]^2} - \dfrac{a_{12}a_{21}}{[\delta(A)]^2} = \dfrac{a_{22}a_{11} - a_{12}a_{21}}{[\delta(A)]^2} = \dfrac{\delta(A)}{[\delta(A)]^2} = \dfrac{1}{\delta(A)}$.

31. From the results of Exercise 12.4–43, and Exercise 29 above,

$$\delta[B^{-1}AB] = \delta[B^{-1}(AB)] = \delta(B^{-1}) \cdot \delta(AB) = \delta(B^{-1}) \cdot \delta(A) \cdot \delta(B) = \frac{1}{\delta(B)} \cdot \delta(A) \cdot \delta(B) = \delta(A).$$

Exercise 12.7 (page 371) **1.** $\{(1, 1)\}$ **3.** $\{(2, 2)\}$ **5.** $\{(6, 4)\}$ **7.** $\{(1, 1, 1)\}$ **9.** $\{(1, 1, 0)\}$
11. $\{(1, -2, 3)\}$ **13.** $\{(3, -1, -2)\}$

Exercise 12.8 (page 376) **1.** $\left\{\left(\dfrac{13}{5}, \dfrac{3}{5}\right)\right\}$ **3.** $\left\{\left(\dfrac{22}{7}, \dfrac{20}{7}\right)\right\}$ **5.** $\{(6, 4)\}$ **7.** Inconsistent
9. $\{(4, 1)\}$ **11.** $\left\{\left(\dfrac{1}{a+b}, \dfrac{1}{a+b}\right)\right\}$ $(a \neq -b)$ **13.** $\{(1, 1, 0)\}$ **15.** $\{(1, -2, 3)\}$
17. $\{(3, -1, -2)\}$ **19.** $\left\{\left(-\dfrac{1}{3}, -\dfrac{25}{24}, -\dfrac{5}{8}\right)\right\}$ **21.** $\left\{\left(1, -\dfrac{1}{3}, \dfrac{1}{2}\right)\right\}$
23. $\{(w, x, y, z)\} = \{(2, -1, 1, 0)\}$ **25.** $A = \begin{vmatrix} a_1 & b_1 \\ a_2 & b_2 \end{vmatrix}$ and $\delta(A) = a_1b_2 - a_2b_1$;

$A_y = \begin{vmatrix} a_1 & c_1 \\ a_2 & c_2 \end{vmatrix}$ and $\delta(A_y) = a_1c_2 - a_2c_1 = 0$, so $a_1c_2 = a_2c_1$;

$A_x = \begin{vmatrix} c_1 & b_1 \\ c_2 & b_2 \end{vmatrix}$ and $\delta(A_x) = b_2c_1 - b_1c_2 = 0$, so $b_1c_2 = b_2c_1$;

hence, $\dfrac{a_1c_2}{b_1c_2} = \dfrac{a_2c_1}{b_2c_1}$; $\dfrac{a_1}{b_1} = \dfrac{a_2}{b_2}$ and $a_1b_2 = a_2b_1$, so $a_1b_2 - a_2b_1 = 0$; therefore $\delta(A) = 0$.

Chapter 12 Review (page 377) **1.** $\begin{bmatrix} 1 & -1 \\ 1 & 1 \end{bmatrix}$ **2.** $\begin{bmatrix} 2 & 5 & -2 \\ 14 & -1 & 12 \end{bmatrix}$ **3.** $\begin{bmatrix} -2 & -3 & 1 \\ 3 & -7 & 6 \\ -5 & 2 & 5 \end{bmatrix}$

4. $\begin{bmatrix} -8 & -3 & -6 \\ 6 & -3 & 0 \\ -9 & 5 & 5 \end{bmatrix}$ **5.** $\begin{bmatrix} -21 & 7 \\ -14 & 0 \\ -7 & -7 \end{bmatrix}$ **6.** $[13]$ **7.** $\begin{bmatrix} -13 & 3 \\ -19 & 27 \end{bmatrix}$ **8.** $\begin{bmatrix} 18 & 7 & 25 \\ 8 & -1 & 11 \\ 3 & 0 & 3 \end{bmatrix}$

9. $\{(2, -1)\}$ **10.** $\{(2, 1, 1)\}$ **11.** -3 **12.** 7 **13.** -3 **14.** 14 **15.** 15

16. -1578 **17.** $\dfrac{1}{34}\begin{bmatrix} -3 & 2 \\ 11 & 4 \end{bmatrix}$ **18.** $\dfrac{1}{6}\begin{bmatrix} 1 & 3 & -2 \\ -3 & -3 & 6 \\ 1 & -3 & 4 \end{bmatrix}$ **19.** $\{(-2, 1)\}$ **20.** $\{(3, -1, 1)\}$

21. $\left\{\left(-\frac{16}{7}, -\frac{13}{7}\right)\right\}$ **22.** $\{(2, -1, 0)\}$

Exercise 13.1 (page 381)   **1.** $-4, -3, -2, -1$ **3.** $-\frac{1}{2}, 1, \frac{7}{2}, 7$ **5.** $2, \frac{3}{2}, \frac{4}{3}, \frac{5}{4}$ **7.** $0, 1, 3, 6$

9. $-1, 1, -1, 1$ **11.** $1, 0, -\frac{1}{3}, \frac{1}{2}$ **13.** $11, 15, 19$ **15.** $x + 2, x + 3, x + 4$

17. $2x + 7, 2x + 10, 2x + 13$ **19.** $32, 128, 512, 2048$ **21.** $\frac{8}{3}, \frac{16}{3}, \frac{32}{3}, \frac{64}{3}$

23. $\frac{x}{a}, -\frac{x^2}{a^2}, \frac{x^3}{a^3}, -\frac{x^4}{a^4}$ **25.** $4n + 3, 31$ **27.** $-5n + 8, -92$ **29.** $48(2)^{n-1}, 1536$

31. $-\frac{1}{3}(-3)^{n-1}, -243$ **33.** $2; 3; 41$ **35.** 28th   **37.** 3

Exercise 13.2 (page 387)   **1.** $1 + 4 + 9 + 16$ **3.** $-\frac{1}{2} + \frac{1}{4} - \frac{1}{8}$ **5.** $1 + \frac{1}{2} + \frac{1}{4} + \cdots$

7. $\sum_{j=1}^{4} x^{2j-1}$ **9.** $\sum_{j=1}^{5} j^2$ **11.** $\sum_{j=1}^{\infty} j(j+1)$ **13.** $\sum_{j=1}^{\infty} \frac{j+1}{j}$ **15.** 63   **17.** 806   **19.** -6

21. 1092   **23.** $\frac{31}{32}$ **25.** $\frac{364}{729}$ **27.** 168   **29.** 196   **31.** $\frac{3}{4}, \frac{7}{8}, \frac{15}{16}, \frac{31}{32}; 1$

33. $p = 4, q = -3$

35. Let the two sequences be given by $s_n = a + (n - 1)d$ and $s'_n = a' + (n - 1)d'$. Then the sequence formed by adding corresponding terms has nth term

$$t_n = s_n + s'_n = a + (n - 1)d + a' + (n - 1)d'$$
$$= a + a' + (n - 1)(d + d').$$

Therefore t_n defines an arithmetic sequence with first term $a + a'$ and common difference $d + d'$.

Exercise 13.3 (page 394)   **1.** $\lim_{n \to \infty} s_n = 0$ **3.** $\lim_{n \to \infty} s_n = 1$ **5.** $\lim_{n \to \infty} s_n$ is undefined   **7.** $\lim_{n \to \infty} s_n = 0$

9. Convergent, $\lim_{n \to \infty} \left| 0 - \frac{1}{2^n} \right| = 0$ **11.** Divergent, $\lim_{n \to \infty} n$ is undefined

13. Convergent, $\lim_{n \to \infty} \left| 0 - (-1)^{n+1} \frac{1}{2^{n-1}} \right| = 0$ **15.** 24   **17.** No sum   **19.** 2   **21.** $\frac{31}{99}$

23. $\frac{2408}{999}$ **25.** $\frac{29}{225}$ **27.** 20 cm

Exercise 13.4 (page 399)   **1.** $8 \cdot 7 \cdot 6 \cdot 5 \cdot 4 \cdot 3 \cdot 2 \cdot 1$ **3.** $6 \cdot 5 \cdot 4 \cdot 3 \cdot 2 \cdot 1$

5. $5 \cdot 4 \cdot 3 \cdot 2 \cdot 1 = 120$ **7.** $\frac{9 \cdot 8 \cdot 7!}{7!} = 72$ **9.** $\frac{5 \cdot 4 \cdot 3 \cdot 2 \cdot 1 \cdot 7!}{8 \cdot 7!} = 15$ **11.** $\frac{8 \cdot 7 \cdot 6!}{2 \cdot 1 \cdot 6!} = 28$

13. $3!$ **15.** $\frac{6!}{2!}$ **17.** $\frac{8!}{5!}$ **19.** $\frac{6!}{5! \, 1!} = 6$ **21.** $\frac{3!}{3! \, 0!} = 1$ **23.** $\frac{7!}{0! \, 7!} = 1$ **25.** $\frac{5!}{2! \, 3!} = 10$

27. $(n)(n - 1)(n - 2) \cdot \cdots \cdot 3 \cdot 2 \cdot 1$ **29.** $(3n)(3n - 1)(3n - 2) \cdot \cdots \cdot 3 \cdot 2 \cdot 1$

Odd-Numbered Answers

31. $(n-2)(n-3)(n-4) \cdot \cdots \cdot 3 \cdot 2 \cdot 1$ **33.** $(n+2)(n+1)$ **35.** $\dfrac{n+1}{n+3}$ **37.** $\dfrac{2n-1}{2n-2}$

39. $x^5 + 15x^4 + 90x^3 + 270x^2 + 405x + 243$ **41.** $x^4 - 12x^3 + 54x^2 - 108x + 81$

43. $8x^3 - 6x^2y + \dfrac{3}{2}xy^2 - \dfrac{1}{8}y^3$ **45.** $\dfrac{1}{64}x^6 + \dfrac{3}{8}x^5 + \dfrac{15}{4}x^4 + 20x^3 + 60x^2 + 96x + 64$

47. $x^{20} + 20x^{19}y + \dfrac{20 \cdot 19}{2!}x^{18}y^2 + \dfrac{20 \cdot 19 \cdot 18}{3!}x^{17}y^3,$ or

$\binom{20}{0}x^{20} + \binom{20}{1}x^{19}y + \binom{20}{2}x^{18}y^2 + \binom{20}{3}x^{17}y^3$

49. $a^{12} + 12a^{11}(-2b) + \dfrac{12 \cdot 11}{2!}a^{10}(-2b)^2 + \dfrac{12 \cdot 11 \cdot 10}{3!}a^9(-2b)^3,$ or

$\binom{12}{0}a^{12} + \binom{12}{1}a^{11}(-2b) + \binom{12}{2}a^{10}(-2b)^2 + \binom{12}{3}a^9(-2b)^3$

51. $x^{10} + 10x^9(-\sqrt{2}) + \dfrac{10 \cdot 9}{2!}x^8(-\sqrt{2})^2 + \dfrac{10 \cdot 9 \cdot 8}{3!}x^7(-\sqrt{2})^3,$ or

$\binom{10}{0}x^{10} + \binom{10}{1}x^9(-\sqrt{2}) + \binom{10}{2}x^8(-\sqrt{2})^2 + \binom{10}{3}x^7(-\sqrt{2})^3$ **53.** 1.22   **55.** $1480

57. $-3003a^{10}b^5$ **59.** $3360x^6y^4$ **61. a.** $1 - x + x^2 - x^3 + \cdots$ **b.** $1 - x + x^2 - x^3 + \cdots$

Exercise 13.5 (page 406)   **1. a.** For $n = 1$: $\dfrac{n}{2} = \dfrac{1}{2}$; $\dfrac{n(n+1)}{4} = \dfrac{1(1+1)}{4} = \dfrac{1}{2}$.

b. For $n = k$: $\dfrac{1}{2} + \dfrac{2}{2} + \dfrac{3}{2} + \cdots + \dfrac{k}{2} = \dfrac{k(k+1)}{4}$ and $(k+1)$th term $= \dfrac{k+1}{2}$;

hence $\dfrac{1}{2} + \dfrac{2}{2} + \dfrac{3}{2} + \cdots + \dfrac{k}{2} + \dfrac{k+1}{2} = \dfrac{k(k+1)}{4} + \dfrac{k+1}{2} = \dfrac{k^2 + k + 2k + 2}{4} = \dfrac{k^2 + 3k + 2}{4} = \dfrac{(k+1)(k+2)}{4}$.

3. a. For $n = 1$: $2n = 2(1) = 2$; $n(n+1) = 1(1+1) = 2$.

b. For $n = k$: $2 + 4 + 6 + \cdots + 2k = k(k+1)$ and $(k+1)$th term is $2(k+1)$;

hence $2 + 4 + 6 + \cdots + 2k + 2(k+1) = k(k+1) + 2(k+1) = (k+1)(k+2)$.

5. a. For $n = 1$: $n^2 = 1^2 = 1$; $\dfrac{n(n+1)(2n+1)}{6} = \dfrac{1(2)(3)}{6} = 1$.

b. For $n = k$: $1^2 + 2^2 + 3^2 + \cdots + k^2 = \dfrac{k(k+1)(2k+1)}{6}$ and $(k+1)$th term is $(k+1)^2$;

hence $1^2 + 2^2 + 3^2 + \cdots + k^2 + (k+1)^2 = \dfrac{k(k+1)(2k+1)}{6} + (k+1)^2 = \dfrac{k(k+1)(2k+1) + 6(k+1)^2}{6}$

$= \dfrac{(k+1)[k(2k+1) + 6(k+1)]}{6} = \dfrac{(k+1)(2k^2 + 7k + 6)}{6} = \dfrac{(k+1)(k+2)(2k+3)}{6}$

$= \dfrac{(k+1)[(k+1) + 1][2(k+1) + 1]}{6}$.

7. a. For $n = 1$: $(2n-1)^3 = (2-1)^3 = 1^3 = 1$; $n^2(2n^2 - 1) = 1(2 - 1) = 1(1) = 1$.

b. For $n = k$: $1^3 + 3^3 + 5^3 + \cdots + (2k-1)^3 = k^2(2k^2 - 1)$ and the $(k+1)$th term is
$[2(k+1) - 1]^3 = (2k+1)^3$; hence $1^3 + 3^3 + 5^3 + \cdots + (2k-1)^3 + (2k+1)^3 = k^2(2k^2 - 1) + (2k-1)^3 =$
$2k^4 + 8k^3 + 11k^2 + 6k + 1$; by use of the factor theorem and synthetic division (see Section A.1),
$2k^4 + 8k^3 + 11k^2 + 6k + 1 = (k+1) \times (k+1)(2k^2 + 4k + 1)$;
also, $2k^2 + 4k + 1 = 2(k^2 + 2k + 1) - 2 + 1 = 2(k+1)^2 - 1$;
hence $2k^4 + 8k^3 + 11k^2 + 6k + 1 = (k+1)^2[2(k+1)^2 - 1]$.

9. a. For $n = 1$: $n(n+1) = 1(2) = 2$; $\dfrac{n(n+1)(n+2)}{3} = \dfrac{1(2)(3)}{3} = 2$.

   b. For $n = k$: $1 \cdot 2 + 2 \cdot 3 + 3 \cdot 4 + \cdots + k(k+1) = \dfrac{k(k+1)(k+2)}{3}$ and the $(k+1)$th term is $(k+1)[(k+1)+1] = (k+1)(k+2)$; hence $1 \cdot 2 + 2 \cdot 3 + 3 \cdot 4 + \cdots + k(k+1) + [(k+1)(k+2)]$
$= \dfrac{k(k+1)(k+2)}{3} + (k+1)(k+2) = \dfrac{[k(k+1)(k+2)] + [3(k+1)(k+2)]}{3}$
$= \dfrac{(k+1)(k+2)(k+3)}{3} = \dfrac{(k+1)[(k+1)+1][(k+1)+2]}{3}$.

11. a. For $n = 1$: $1^3 + 2 \cdot 1 = 3$; 3 is divisible by 3.

   b. For $n = k$: assume $k^3 + 2k$ is divisible by 3; for $n = k+1$:
$$(k+1)^3 + 2(k+1) = k^3 + 3k^2 + 3k + 1 + 2k + 2$$
$$= (k^3 + 2k) + (3k^2 + 3k + 3).$$

Since by hypothesis $k^3 + 2k$ is divisible by 3 and since $3k^2 + 3k + 3$ is divisible by 3, then $(k+1)^3 + 2(k+1)$ is divisible by 3.

13. For $n = k$: $2 + 4 + 6 + \cdots + 2k = k(k+1) + 2$ and $(k+1)$th term is $2(k+1)$;
hence $2 + 4 + 6 + \cdots + 2k + 2(k+1) = k(k+1) + 2 + 2(k+1) = (k^2 + 3k + 2) + 2 = (k+1)(k+2) + 2$
$= (k+1)[(k+1)+1] + 2$.
However, for $n = 1$: $2n = 2(1) = 2$; $n(n+1) + 2 = 1(2) + 2 = 4$. Hence not true for every $n \in N$.

15. Since $s_1 = a$ and $s_{n+1} = s_n + d$, it follows that

   a. For $n = 1$: $s_n = s_1 = a = a + 0 \cdot d = a = (n-1)d$.

   b. For $n = k$: $s_k = a + (k-1)d$; now
$$s_{k+1} = s_k + d = a + (k-1)d + d = a + kd = a + (k+1-1)d.$$

17. Since $s_n = ar^{n-1}$ and $S_{n+1} = S_n + s_{n+1}$, it follows that

   a. For $n = 1$: $S_1 = s_1 = a = a\dfrac{1-r}{1-r} = \dfrac{a - ar^1}{1-r}$.

   b. For $n = k$: $S_k = \dfrac{a - ar^k}{1-r}$ and $s_{k+1} = ar^k$; now

$$S_{k+1} = S_k + s_{k+1} = \dfrac{a - ar^k}{1-r} + ar^k = \dfrac{a - ar^k + ar^k - ar^{k+1}}{1-r} = \dfrac{a - ar^{k+1}}{1-r}.$$

Exercise 13.6 (page 411) **1.** 1, 5, 8 **3.** 2, 6, 16 **5.** 2, 2, 4 **7.** 4 **9.** 24 **11.** 16

13. 64 **15.** 24 **17.** 216 **19.** 375 **21.** 30 **23.** 10 **25.** 48 **27.** $\dfrac{5!}{2!}$, or 60

29. $\dfrac{8!}{3!}$, or 6720 **31.** $P_{5,3} = \dfrac{5!}{2!} = \dfrac{5 \cdot 4!}{2!} = 5\left(\dfrac{4!}{2!}\right) = 5(P_{4,2})$

33. $P_{n,3} = \dfrac{n!}{(n-3)!} = \dfrac{n(n-1)!}{(n-3)!} = n\left[\dfrac{(n-1)!}{(n-3)!}\right] = n(P_{n-1,2})$ **35.** 9 **37.** 24 **39.** 48

Exercise 13.7 (page 415) **1.** 7 **3.** 15 **5.** $\binom{52}{5}$ **7.** $\binom{13}{5} \cdot \binom{13}{5} \cdot \binom{13}{3}$ **9.** $4 \cdot \binom{13}{5}$

11. 164 **13.** 10 **15.** 210 **17.** 12

Odd-Numbered Answers

Chapter 13 Review (page 417) 1. 2, 5, 10   2. $\frac{1}{2}, \frac{1}{3}, \frac{1}{4}$ 3. 13, 16, 19   4. $a-4, a-6, a-8$
5. $-18, 54, -162$ 6. $\frac{3}{2}, \frac{9}{4}, \frac{27}{8}$ 7. $s_n = 5n - 8$; $s_7 = 27$ 8. $s_n = (-2)\left(\frac{-1}{3}\right)^{n-1}$; $s_5 = \frac{-2}{81}$
9. 25   10. 6th term   11. $2 + 6 + 12 + 20$ 12. $\sum_{k=1}^{\infty} x^{k+1}$ 13. 119   14. $\frac{121}{243}$
15. 3   16. $\frac{8}{3}$ 17. $\frac{1}{2}$ 18. $\frac{4}{9}$ 19. $5 \cdot (2 \cdot 1)$ 20. 48   21. 21   22. $\frac{1}{n(n+1)!}$
23. $x^{10} - 20x^9 y + 180x^8 y^2 - 960x^7 y^3$ 24. $-15{,}360 x^3 y^7$
25. Formula holds for 1. Assume formula holds for n; test for $n + 1$:
$$3 + 6 + 9 + \cdots + 3n + 3(n+1) = \frac{3n(n+1)}{2} + 3(n+1).$$
Right-hand member is equivalent to:
$$\frac{3n^2 + 3n}{2} + \frac{6(n+1)}{2} = \frac{3n^2 + 9n + 6}{2} = \frac{3(n^2 + 3n + 2)}{2} = \frac{3(n+1)(n+2)}{2} = \frac{3(n+1)((n+1)+1)}{2}.$$
26. Formula holds for 1. Assume formula holds for n; test for $n + 1$:
$$\frac{1}{2} + \frac{1}{4} + \frac{1}{8} + \cdots + \frac{1}{2^n} + \frac{1}{2^{n+1}} = 1 - \frac{1}{2^n} + \frac{1}{2^{n+1}}.$$
Right-hand member is equivalent to: $\frac{2^{n+1} - 2 + 1}{2^{n+1}} = \frac{2^{n+1} - 1}{2^{n+1}} = 1 - \frac{1}{2^{n+1}}.$
27. 16   28. 64   29. 128   30. 360   31. 792   32. 1,033,885,600   33. 200   34. 84

Exercise 14.1 (page 420)   1. 40   3. $-41 + 33i$ 5. 36   7. $-11 + 40i$ 9. $-2 - 7i$
11. $Q = 2x^2 + 5x + 18$; $R = 58$ 13. $Q = x^2 + (1 + 2i)x - 7 + 2i$; $R = -1 - 14i$
15. $Q = x^3 - ix^2 - 4x + 1 + 4i$; $R = 3 - i$ 17. $Q = x^2 + (i - 1)x - 1$; $R = -i$
19. Divide $x^3 + 2x^2 - 5x - 6$ by $x - 2$. Remainder is 0, hence $x - 2$ is a factor.
21. $(-2i)^3 - (-2i)^2 + 4(-2i) - 4 = 0$. Hence $-2i$ is a root.   23. $Q(x) = x^2 + (-3 + i)x - 3i$
25. 80   27. -22

Exercise 14.2 (page 423)   1. $-2i$ 3. $-i$ 5. $i, -2i$ 7. $3i, \frac{-3 + \sqrt{29}}{2}, \frac{-3 - \sqrt{29}}{2}$
9. $1 - i$; $x^3 - 2x + 4 = 0$ 11. $P(x) = (2x - 3)(x - 2 + 2i)(x - 2 - 2i)$ 13. $4 + i, 1 + i, 1 - i$
15. Let $P(x) = 0$ be a polynomial equation with real coefficients of degree n, where n is odd; then, $P(x) = 0$ must have at least n zeros and, since $P(x)$ has only real coefficients, any complex zeros must occur as conjugate pairs; since there must always be at least one real zero, then $P(x) = 0$ must have at least one real root.
17. $-1, \frac{1 + \sqrt{3}\,i}{2}, \frac{1 - \sqrt{3}\,i}{2}$

Exercise 14.3 (page 429)   1. 2 positive real, 2 negative real; or 0 positive real, 2 negative real, 2 imaginary; or 0 positive real, 0 negative real, 4 imaginary; or 2 positive real, 0 negative real, and 2 imaginary.

3. 0 positive real, 1 negative real, 4 complex   5. 0 positive real, 0 negative real, 4 complex
7. Upper bound 3, lower bound -4 9. Upper bound 4, lower bound -3
11. Upper bound 2, lower bound -3 13. Upper bound 1, lower bound -2
15. $f(0) = 1$ and $f(1) = -1$ 17. $g(-3) = 10$ and $g(-2) = -33$
19. $P(-3) = 2$ and $P(-2) = -4$; $P(0) = -4$ and $P(1) = 2$

Exercise 14.4 (page 431)   1. $\{4\}$ 3. $\{1, -2\}$ 5. None   7. $\left\{-\dfrac{3}{2}\right\}$ 9. $\left\{2, -\dfrac{7}{4}\right\}$ 11. $\left\{\dfrac{3}{2}\right\}$
13. $\left\{-\dfrac{1}{3}, 1 + \sqrt{5}, 1 - \sqrt{5}\right\}$ 15. $\left\{-1, \dfrac{1}{2}, \dfrac{-1 + i\sqrt{3}}{2}, \dfrac{-1 - i\sqrt{3}}{2}\right\}$ 17. $\left\{\dfrac{3}{4}, -\dfrac{4}{3}, i, -i\right\}$
19. $(2x + 3)(x - 1)(x + 1)$
21. Consider $x^2 - 3 = 0$. If the equation has real zeros, then they are either rational or irrational. If rational, by Theorem 14.11, the only possibilities are ± 1 and ± 3; however, direct substitution shows none of these satisfies the equation. Now, $\sqrt{3}$ is a real number, and since $(\sqrt{3})^2 - 3 = 3 - 3 = 0$, $\sqrt{3}$ is a real zero of the equation, and since it is not among the rational zeros, it must be irrational.

Chapter 14 Review (page 432)   1. -132 2. $-13 + 18i$
3. $x^2 + (-2 + i)x + (2 - 2i)$; remainder: $-3 + 2i$
4. Upon division by $x - i$, the remainder is 0.   5. $\{2, 2i, -2i\}$ 6. $\left\{\dfrac{3}{2}, 2 + 2i, 2 - 2i\right\}$
7. There are two sign changes and therefore the equation has two or no positive real solutions. The equation also has two or no negative real solutions.
8. Integral least upper bound: 3; integral greatest lower bound: -4
9. $P(1) = -3$ and $P(2) = 10$; hence there is a zero between 1 and 2.   10. $\{4\}$ 11. $\left\{-\dfrac{1}{2}, 3\right\}$
12. $\{-2, i\sqrt{2}, -i\sqrt{2}\}$

Exercise A.1 (page 439)

1. 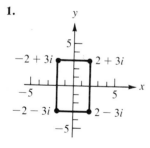   3. 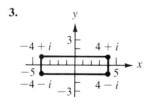   5.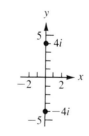

7. 4   9. $\sqrt{13}$ 11. 2   13. $\sqrt{5}$ 15. $3\sqrt{2}$ cis $45°$ 17. 5 cis $0°$ 19. 4 cis $330°$
21. $-2 - 2\sqrt{3}\,i$ 23. $3\sqrt{3} - 3i$ 25. $6 + 6\sqrt{3}i$ 27. a. $-3 + 3i$ b. $\dfrac{3}{2} + \dfrac{3}{2}i$
29. a. 108   b. $\dfrac{1}{6} - \dfrac{\sqrt{3}}{6}i$ 31. a. $10 + 10i$ b. $\dfrac{1}{10} - \dfrac{7}{10}i$
33. Let $z = a + bi$; then $\bar{z} = a - bi$. $(a + bi) + (a - bi) = (a + a) + (b - b)i = 2a + 0i = 2a \in R$; $(a + bi) \cdot (a - bi) = (a^2 + b^2) + (ab - ab)i = (a^2 + b^2) + 0i = (a^2 + b^2) \in R$.

Odd-Numbered Answers

35. From the result of Exercise 34, and from Th. A.1-I

$$(a+bi)^3 = (a+bi)^2(a+bi) = (r^2 \operatorname{cis} 2\theta)(r \operatorname{cis} \theta) = r^3 \operatorname{cis}(2\theta + \theta) = r^3 \operatorname{cis} 3\theta.$$

Exercise A.2 (page 443) **1.** $-64\sqrt{3} + 64i$ **3.** $1 + 0i$ **5.** $\dfrac{729}{2} + \dfrac{729\sqrt{3}}{2}i$ **7.** $\dfrac{-\sqrt{3}}{64} + \dfrac{1}{64}i$

9. $\dfrac{\sqrt{3}-1}{64} - \dfrac{\sqrt{3}+1}{64}i$ **11.** $-\dfrac{1}{2} - \dfrac{1}{2}i$

13. 2 cis 9°; 2 cis 81°; 2 cis 153°; 2 cis 225°; 2 cis 297°
15. 2 cis 30°; 2 cis 102°; 2 cis 174°; 2 cis 246°; 2 cis 318°
17. cis 45°; cis 105°; cis 165°; cis 225°; cis 285°; cis 345°
19. 2 cis 60°; 2 cis 132°; 2 cis 204°; 2 cis 276°; 2 cis 348°
21. cis $25\frac{5}{7}°$; cis $77\frac{1}{7}°$; cis $128\frac{4}{7}°$; cis 180°; cis $231\frac{3}{7}°$; cis $282\frac{6}{7}°$; cis $334\frac{2}{7}°$
23. $(x - 2 \operatorname{cis} 45°)(x - 2 \operatorname{cis} 135°)(x - 2 \operatorname{cis} 225°)(x - 2 \operatorname{cis} 315°)$
25. The four roots are $-1, 1, i, -i$; hence their sum is $0 + 0i$.

Exercise A.3 (page 447) **1.** $(6, 125°), (6, -235°), (-6, 305°), (-6, -55°)$
3. $(-2, -90°), (-2, 270°), (2, 90°), (2, -270°)$ **5.** $(6, -240°), (6, 120°), (-6, -60°), (-6, 300°)$
7. $\left(\dfrac{5}{\sqrt{2}}, \dfrac{5}{\sqrt{2}}\right)$ **9.** $\left(\dfrac{\sqrt{3}}{4}, -\dfrac{1}{4}\right)$ **11.** $\left(\dfrac{-10}{\sqrt{2}}, \dfrac{-10}{\sqrt{2}}\right)$ **13.** $(6, 45°), (6, -315°)$
15. $(2, 240°), (2, -120°)$ **17.** $(0, \theta°), (0, -\theta°)$, for any $\theta > 0$ **19.** $r = 5$ **21.** $r \sin \theta = -4$
23. $r^2(\cos^2 \theta + 9 \sin^2 \theta) = 9$ **25.** $x^2 + y^2 = 25$ **27.** $x^2 + y^2 - 9x = 0$ **29.** $y^2 = 4x + 4$
31. $\sec(\theta/2) = \dfrac{1}{\cos(\theta/2)}$ and $\cos^2(\theta/2) = \dfrac{1 + \cos\theta}{2}$; hence $r = \sec^2(\theta/2) = \dfrac{1}{\cos^2(\theta/2)} = \dfrac{2}{1 + \cos\theta}$.

Now $\cos\theta = \dfrac{x}{r}$ and $r^2 = x^2 + y^2$; so $r = \dfrac{2}{1 + \dfrac{x}{r}}$ and $r\left(1 + \dfrac{x}{r}\right) = 2$; hence $r + x = 2$, or

or $r = 2 - x$. Squaring each member gives $r^2 = 4 - 4x + x^2$ and substituting for r^2 gives $x = 1 - \dfrac{1}{4}y^2$, whose graph is a parabola.

33.
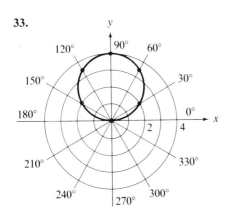

35.
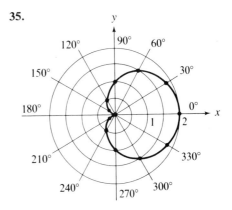

37.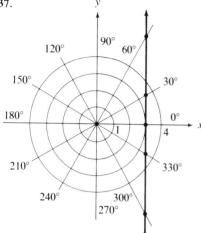

39.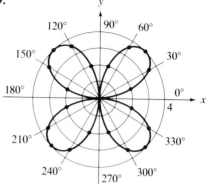

41.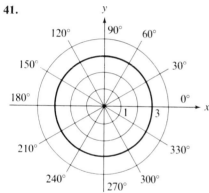

Index

a^0, 64
a^{-n}, 64
Abscissa, 120
Absolute value:
   in equation, 109*ff*
   in inequalities, 109*ff*
   of a complex number, 437
   of a real number, 16
Addition:
   associative law of:
      for matrices, 334
      for real numbers, 8
   closure for, of matrices, 334
      of real numbers, 8
   commutative law of:
      for matrices, 334
      for real numbers, 8
   identity element for:
      matrices, 334
      real numbers, 8
   law of:
      for equality, 11
      of matrices, 331
      of quotients, 12
      of vectors, 271
   parallelogram law for, of vectors, 271
Additive inverse:
   of a matrix, 334
   of a real number, 8
   uniqueness of, 11
Algebraic expression, 25
Alternating sequence, 391
Ambiguous case for law of sines, 262
Amplitude:
   of a complex number, 437
   of a sine wave, 216
Angle(s):
   coterminal, 244

Angle(s) *(continued)*:
   definition of, 241
   in a quadrant, 241
   in standard position, 241
   initial side of, 241
   measure of an, using a circle, 242
   reference, 251
   sides of, 241
   symbols for measure, 242
   terminal side of, 241
   vertex of, 241
Antilogarithm, 188
Arc(s):
   reference, 208
Arccosecant function, 233
Arccosine function, 233
Arccosine relation, 232
Arccotangent function, 233
Arcsecant function, 233
Arcsine:
   function, 233
   relation, 232
Arctangent function, 233
Arctangent relation, 232
Argand plane, 436
Argument of a complex number, 437
Arithmetic progression:
   common difference of, 380
   meaning of, 380
   nth term of, 381
   sum of n terms, 385
Associative law:
   of addition of matrices, 334
   of addition of real numbers, 8
   of multiplication of a matrix by a real number, 337
   of multiplication of real numbers, 8

Asymptote(s):
  horizontal, 146
  oblique, 148
  of graph of a hyperbola, 168
  of graph of a rational function, 145*ff*
  vertical, 145
Augmented matrix, 344
Axioms, 6
Axis:
  imaginary, 436
  polar, 444
  real, 436
Axis, of symmetry, 137

Base of a logarithm, 181
Base of a power, 25
Binary operation, 7
Binomial:
  coefficient of tth term in expansion of, 397
  definition, 26
  expansion of, 397
  products, 32
  theorem, 396*ff*

Cartesian coordinate system, 120
  graph of an ordered pair on, 120
  in three dimensions, 312
Cartesian product, 120
Characteristic of a logarithm, 187
Circle, 166, 169
Circular functions, 231*ff*
Closure law:
  for addition of matrices, 334
  for addition of positive real numbers, 15
  for addition of real numbers, 8
  for multiplication of positive real numbers, 15
  for multiplication of real numbers, 8
Coefficient(s):
  detached, 42
  leading, of a polynomial, 27
  numerical, 26
  of terms in a binomial expansion, 397
Coefficient matrix of a system of linear equations, 344
Cofactor, of an element in a determinant, 351
Column matrix, 331
Column vector, 331
Combination(s):
  definition of, 413
  of n things taken r at a time, 414
  related to coefficients in binomial expansion, 415
Common difference of an arithmetic progression, 380
Common logarithms (*see also* Logarithms), 185
Common ratio of a geometric progression, 381

Commutative law:
  of addition of matrices, 334
  of addition of real numbers, 8
  of multiplication of real numbers, 8
Completely factored polynomial, 34
Completing the square, 89
Complex fractions, 52
Complex number(s):
  absolute value of, 437
  addition of, 19
  amplitude of, 437
  argument of, 437
  as zeros of a polynomial function, 421
  conjugate of, 21
  definition of set of, 18, 19
  division of, 21
  equality of, 18
  factor theorem over, 420
  graph of, 436
  imaginary part of, 436
  modulus of, 437
  multiplication of, 19
  notation for, 18
  polar form of, 437
  powers of, 441
  pure imaginary, 20
  real numbers related to, 20
  real part of, 436
  roots of, 443
  subtraction of, 20
  synthetic division over, 419
  trigonometric form of, 436*ff*
Complex plane, 436
Components of an ordered pair, 120
Composite number, 429
Conditional equation(s), 295*ff*
Conformable matrices, 338
Conics, 168
Conic sections, 165*ff*
Conjugate(s):
  definition, 77
  of a complex number, 21
Conjugate complex roots, 421
Consistent equations, 304, 313
Constant, 3
Constant function, 127
Constant of variation, 161
Convergent sequence, 391
Conversion formulas for angle measure, 243
Convex polygon, 324
Convex sets, 324
  intersection of, 325
Coordinate(s):
  of a point in the plane, 120
  of a point on the number line, 14
  of points on unit circle, 198

Cosecant:
  definition, 229, 248
  domain of function, 229
  graph of function, 230
  period, 229
  range of function, 229
Cosine(s):
  definition, 198, 248
  domain of function, 200
  double-angle formula for, 285
  function, 198
  graph of function, 214ff
  half-angle formula for, 285
  law of, 266ff
  period, 215
  range of function, 200
  sum and difference formulas, 285
  table of values of, 205
Cotangent:
  definition, 227, 248
  domain of function, 227
  graph of function, 228
  period, 227
  range of function, 227
Coterminal angles, 244
Counting properties, 407
Cramer's rule, 372ff
Critical numbers, 105
Cube root, 69
Cycle of a periodic function, 216
Cycle of a sine wave, 216

Decreasing function, 179
Degree as angle measure, 242
Degree of a monomial, 26
Degree of a polynomial, 26
De Moivre's theorem, 441
Denominator of a fraction, 44
Descartes' rule of signs, 425
Detached coefficients, 42
Determinant(s):
  cofactor of an element of, 351
  definition of, 349, 350, 353
  expansion of, 350, 353
  function, 349ff
  minor of an element of, 351
  notation, 349
  properties of, 355ff
  use of, in solutions of systems, 373
  value of, 350
Diagonal matrix, 339
Difference:
  definition of, 9
  of complex numbers, 20
  of matrices, 333
  of quotients, 13

Difference (continued):
  of real numbers, 9
Difference formula:
  for cosine function, 285
  for sine function, 288
  for tangent function, 291
Dimension of a matrix, 331
Direct variation, 160
Direction angle of a vector, 270
Discriminant of a quadratic equation, 93
Disjoint sets, 3
Distance formula, 129
Distributive law:
  for $n \times n$ square matrices, 339
  for real numbers, 8, 31
  for scalar multiplication over matrix addition, 337
Divergent sequence, 392
Division:
  of complex numbers, 21
  of polynomials, 37ff
  of quotients, 13
  of real numbers, 9
  synthetic, 41
Domain:
  of a logarithmic function, 181
  of a relation or function, 121
  of cosecant function, 229
  of cosine function, 200
  of cotangent function, 227
  of secant function, 228
  of sine function, 200
  of tangent function, 222
Double-angle formula:
  for cosine function, 285
  for sine function, 289
  for tangent function, 292

e, 188
Element:
  of a matrix, 331
  of a set, 1
Elementary transformation:
  in terms of matrices, 342
  of an equation, 83
  of an inequality, 99
Ellipse, 167, 169
Empty set, 2
Entry of a matrix, 331
Equality:
  addition law for, 11
  multiplication law for, 11
  of complex numbers, 18
  of like powers, 95
  of matrices, 332
  of quotients, 12

Equality *(continued)*:
  of sets, 2
  of vectors, 271
  postulates, 450
  properties of, 7
Equation(s):
  conditional, 295*ff*
  consistent, 304, 313
  elementary transformation of an, 83
  equivalent, 83
  first-degree, in one variable, 82
  first-degree, in two variables, 126
  graph of a linear, in two variables, 126
  in one variable, 82*ff*
  inconsistent, 304, 313
  intercept form, 133
  in two variables, 121
  linear, in one variable, 84
  linear, in two variables, 126
  linearly dependent, 304
  linearly independent, 304
  nonlinear, 316
  point-slope form, 133
  polar form, 447
  quadratic, in one variable, 87
  radical, 95*ff*
  root of an, 82
  second-degree, in one variable, 87
  slope-intercept form, 133
  solution of an, 82*ff*
  solution set of an, 82
  two-point form, 135
Equivalence axioms, 7
Equivalence symbol, 2
Equivalent expressions, 25
Equivalent rational expressions, 44
Equivalent systems, 304
Even function, 126
Expansion:
  binomial, 397
  of a determinant, 350
Exponent(s):
  definition, 25
  laws of integral, 64
  laws of natural number, 31, 38
  laws of rational, 68*ff*
  logarithms as, 180
Exponential function(s):
  definition of an, 178
  graph of an, 179
Extraction of roots, 88
Extraneous solutions, 95

Factor(s):
  by grouping, 35
  in a domain, 34

Factor(s) *(continued)*:
  of quadratics, 35
  zero as a, 12
Factor theorem, 419
Factorial notation, 396
Factoring polynomials, 34*ff*
Field:
  definition of a, 7
  ordered, 15
  postulates, 7, 450
  properties of a, 11*ff*
  set of real numbers as a, 7
Finite sequences, 379
Finite sets, 2
First-degree equation(s):
  in one variable, 84
  in two variables, 126*ff*
  intercept form of a, in two variables, 133
  point-slope form for, in two variables, 133
  slope-intercept form for, in two variables, 133
  solution of systems of, by Cramer's rule, 372
  standard form for, in two variables, 126
Fractions:
  equivalent, 44
  fundamental principle of, 12
  lowest terms, 45
  partial, 55
  reducing, 45
  use of diagonal lines, 45
Function(s):
  absolute value, 153
  bracket, 155
  circular, 231*ff*
  constant, 127
  cosecant, 229, 248
  cosine, 198
  cotangent, 227, 248
  counting, for sets, 407
  decreasing, 179
  definition of a, 123
  determinant, 349*ff*
  even, 126
  exponential, 178*ff*
  increasing, 179
  inverse, 156*ff*
  inverse of an exponential, 180
  inverse of a trigonometric, 233
  linear, 126*ff*
  logarithmic, 180*ff*
  nonlinear, 136
  notation, 123
  odd, 126
  one-to-one, 157
  period of, 214
  periodic, 214*ff*
  polynomial, 141

Index

Functions *(continued)*:
  quadratic, 136*ff*
  rational, 145*ff*
  real-valued, 123
  secant, 228, 264
  sequence, 379
  sine, 198
  tangent, 222, 263
  trigonometric, 248*ff*
  zeros of, 139, 421
Fundamental period, 214
  of cosecant function, 229
  of cosine function, 215
  of cotangent function, 217
  of secant function, 228
  of sine function, 215
  of tangent function, 223
Fundamental principle of fractions, 12
Fundamental theorem of algebra, 422
Fundamental theorem of arithmetic, 429

Gauss plane, 437
Geometric plane, 120
Geometric progression(s):
  common ratio of, 381
  definition of, 381
  infinite, 393
  nth term of, 381
  sum of an infinite, 393
  sum of n terms of, 386
Geometric vector, 269*ff*
Graph(s):
  in Argand plane, 436
  in three dimensions, 312
  line, 101
  maximum point of, 137
  minimum point of, 137
  of an absolute value function, 153
  of a conic section, 170
  of an exponential function, 179
  of a first-degree equation, 126
  of a first-degree relation, 173
  of an inequality in two variables, 173
  of a logarithmic function, 181
  of a point on the number line, 14
  of a polynomial function, 141
  of a quadratic function, 136
  of a rational function, 145
  of cosecant function, 230
  of cosine function, 214*ff*
  of cotangent function, 228
  of secant function, 229
  of sine function, 214*ff*
  of tangent function, 224
Greater than, 15

Half-angle formula:
  for cosine function, 285
  for sine function, 289
  for tangent function, 292
Half-plane, 174, 324
Horizontal:
  asymptote, 146
Hyperbola, 168, 169
  asymptotes of, 168

i, 19
Identities:
  definition of, 279
  proof of, 280
  summary of, 293
Identity element:
  for addition of matrices, 334
  for addition of real numbers, 8
  for multiplication of matrices, 340
  for multiplication of real numbers, 8
If and only if, 2
Imaginary:
  number, 20
  part of complex number, 436
  pure, 20
Inclination, 128
Inconsistent equations, 304
Increasing function, 179
Index of a radical, 73
Index of summation, 387
Induction, mathematical, 403*ff*
Inequality (Inequalities):
  absolute, 99
  as open sentences, 99
  conditional, 99
  elementary transformations of, 99
  equivalent, 99
  graph of, in one variable, 101
  graph of, in two variables, 173
  in one variable, 82*ff*
  involving absolute values, 109*ff*
  second-degree, in one variable, 103*ff*
  second-degree, in two variables, 173*ff*
  solution of, 99
  solution set, 99
  unconditional, 99
Infinite series, 392
Infinite sets, 2
Initial side of an angle, 241
Integers, set of, 6
Intercept form for a linear equation, 133
Intercept of a graph, 126
Intersection of sets, 2, 325
Interval notation, 102
Inverse:
  additive, 8

Inverse *(continued)*:
   multiplicative, 8
   of a circular function, 231
   of a function, 156*ff*
   of a square matrix, 363*ff*
   of a trigonometric function, 233
   of an exponential function, 180
Inverse function, 156*ff*, 233
Inverse operations, 9
Inverse relation, 156, 231
Inverse variation, 161
Inversions, 350
Irrational numbers, set of, 6

Joint variation, 162

Lag, in phase of periodic function, 219
Law of cosines, 266*ff*
Law of sines, 260*ff*
Laws of signs, 12
Lead, in phase of periodic function, 219
Least common denominator, 49
Least common multiple, 49
Length of a line segment, 128
Less than, 15
Limit:
   as sum of an infinite series, 391
   of a sequence, 390
Line segment:
   inclination of a, 128
   length of a, 128
   slope of a, 129
Linear equation (*See* First-degree equation)
Linear function, 126*ff*
Linear programming, 327
Linearly dependent equations, 304
Linearly independent equations, 304
Location theorem, 427
Logarithm(s):
   applications of, 191
   base *e*, 188
   base 10, 185*ff*
   characteristic of, 187
   common, 185*ff*
   computations with, 191*ff*
   laws of, 182
   mantissa of, 187
   natural, 188
   reading tables of, 185
Logarithmic function, 180*ff*
Long-division algorithm, 39
Lower bound:
   for zeros of real polynomial function, 426
Lowest terms, 45, 430

Magnitude of vector, 270
Mantissa of a logarithm, 187

Mathematical induction:
   principle of, 403*ff*
   proof by, 403*ff*
Mathematical structure, 450*ff*
Mathematical systems, 463
Matrix (Matrices):
   addition of, 331
   additive inverse of a, 334
   associative law of addition for, 334
   augmented, 344
   coefficient, of a system of linear equations, 344
   column, 331
   commutative law of addition for, 334
   conformable, for multiplication, 338
   diagonal, 339
   difference of, 333
   dimension of, 331
   equality of, 332
   identity element for addition of, 334
   identity element for multiplication of, 340
   inverse of a square, 363*ff*
   meaning of, 331
   negative of, 333
   noncommutativity of, products, 338
   nonsingular, 343
   notation, 331
   order of, 331
   principal diagonal of, 339
   product of, and a scalar, 337
   product of two, 337
   properties of products, 337
   properties of sums, 334
   rank, 375
   row, 331
   row-equivalent, 343, 347
   solution of linear systems, 344, 347
   square, 339
   sum of two, 333
   transpose of a, 333
   zero, 333
Maximum point:
   on a graph, 141
   on a parabola, 137
Member of a set, 1
Minimum point:
   on a graph, 141
   on a parabola, 137
Minor of an element in a determinant, 351
Modulus of a complex number, 437
Monomial, 26
Multiplication:
   associative law of, for matrices, 337
   associative law of, for real numbers, 8
   closure under, of set of real numbers, 8
   commutative law of, for real numbers, 8
   identity element for, in set of matrices, 340

Multiplication *(continued)*:
   identity element for, in set of real numbers, 8
   law of, for equality, 11
   of complex numbers, 19
   of a matrix by a scalar, 337
   of a vector by a scalar, 271
   of matrices, 337
Multiplicative identity, 8
Mutliplicative inverse, 8

Natural number, 1, 5
Negative:
   of a matrix, 333
   of a real number, 8
   of a vector, 271
Negative number(s):
   product of a, with a positive, 12
   product of two, 12
   real, 8, 15
Norm of a vector, 270
Null set, 2
Number(s):
   lowest terms for a rational, 45
   negative of a, 8
   properties of, 451
   relatively prime, 45
Number line, 14*ff*
Numerator of a fraction, 44

Oblique asymptote, 148
Odd function, 126
One-to-one correspondence, 14
Operation, binary, 7
Order:
   of a matrix, 331
   postulates for real numbers, 15, 451
   properties of, 15
   trichotomy law of, 15
Ordered field:
   rational numbers as, 15
   real numbers as, 15
Ordered pair(s):
   components of, 120
   definition of, 120
   in Cartesian products, 120
Ordinate, 120
Origin, on a number line, 14

Parabola, 137, 168
   axis of a, 137
   lowest or highest point, 137
   vertex of a, 137
Parallelogram law for vector sums, 271
Partial fractions, 55
Partial sums, 392
Perfect square, 89

Period:
   fundamental, of a function, 214
   of a function, 214
Periodic function, 214*ff*
Permutation(s):
   circular, 413
   definition of a, 408
   of n things taken n at a time, 409
   of n things taken r at a time, 409
   with some identical objects, 410
Phase shift, 219
Point-slope form of a linear equation, 133
Polar axis, 444
Polar coordinates, 444*ff*
Pole, 444
Polygon, convex, 324
Polygonal regions, 324
Polynomial(s):
   addition of, 26
   continuity of a, function, 425
   definition of, 26
   degree of, 26
   Descartes' rule of signs for, 425
   division of, 37*ff*
   exactly divisible, 41
   function, 141
   graph of, function, 141
   in a real variable, 27
   leading coefficient of, 27
   leading term, 27
   location theorem for zeros of, function, 427
   monic, 27
   multiplication of, 31*ff*
   over R, 27
   rational zeros, 429*ff*
   real, 27
   real zeros of, function, 424*ff*
   remainder theorem for, 143, 419
   simplification of, 27
   value of, 28
Polynomial equation(s) (*see also* Polynomial function):
   rational roots of, 429*ff*
Polynomial function(s):
   continuity of, 425
   graph of, 141
   isolating zeros of real, 425
   rational zeros of, 429*ff*
   real zeros of, 424*ff*
   zeros of, 421*ff*
Positive real number(s):
   closure of set of, 15
   product of a, with a negative real number, 12
Postulate, 6*ff*
Power(s):
   definition, 25
   integral, 63

Power(s) *(continued)*:
  of complex numbers, 441
  rational, of a real number, 68
  real, of a real number, 178
Prime number, 429
Prime polynomial, 34
Principal diagonal of a matrix, 339
Product(s):
  Cartesian, 120
  of complex numbers, 19
  of matrices, 337
  of matrix and scalar, 337
  of monomials, 31
  of negative real numbers, 12
  of polynomials, 31
  of positive and negative real numbers, 12
  of powers, 31
  of quotients, 13
  of vector and scalar, 271
Progression:
  arithmetic, 380
  geometric, 381
Proof:
  by mathematical induction, 403
Proportion, 162
Pure imaginary number, 20

Quadratic equation(s) (*see* Second-degree equations)
Quadratic formula, 93
  discriminant of, 93
Quadratic function(s), 136*ff*
  graph of, 136
Quadratic inequalities, 103
Quotient(s):
  addition of, 13
  definition of, 9
  difference of two, 13
  equality of, 12
  of complex numbers, 21
  of polynomials, 38*ff*
  of powers, 38
  of quotients, 13
  of real numbers, 9
  products of, 13
  subtraction of, 13
  sums of, 13

Radian measure of an angle, 242
Radical expressions:
  definition of, 73*ff*
  operations on, 77*ff*
  products of, 73
  properties of, 73
  simplest form, 74
  standard form, 74
  sums of, 77

Radicand, 73
Range:
  of an exponential function, 179
  of a logarithmic function, 181
  of a relation or function, 121
  of cosecant function, 229
  of cosine function, 200
  of cotangent function, 227
  of secant function, 228
  of sine function, 200
  of summation, 387
  of tangent function, 222
Rank of a matrix, 375
Rational expressions:
  definition of, 44
  differences of, 48*ff*
  products of, 51*ff*
  quotients of, 51*ff*
  sums of, 48*ff*
Rational functions, 145*ff*
Rational numbers:
  as an ordered field, 15
  set of, 6
Rational powers, properties of, 68
Ratios, trigonometric, 256
Real number(s):
  absolute value of a, 16
  equality of, 7
  field properties of, 7
  negative, 8, 15
  negative of, 8, 15
  order in set of, 15
  positive, 18
  postulates for, 450
  set of, 6, 7
  set of, as ordered field, 15
Real part of a complex number, 436
Real polynomial, 27
Real-valued function, 123
Reciprocal, 8
Rectangular coordinate system, 120
Recursive property of a factorial, 396
Reference angle, 251
Reference arc, 208
Reflexive law of equality, 7
Regions, polygonal, 324
Relation(s):
  definition of, 121
  domain of, 121
  inverse, 156, 231
  nonlinear, 136
  range of, 121
Remainder theorem, 143, 419
Replacement set, 3
Right triangles, solution of, 256
Root(s):
  cube, 69

Index

Root(s) *(continued)*:
    extraction of, 88
    of a complex number, 443
    of a polynomial equation, 421*ff*
    of a quadratic equation, 88
    of real numbers, 69
    rational, of a polynomial equation, 429*ff*
    square, 69
Row-equivalent matrices, 343
Row matrix, 331
Row vector, 339

Scalar:
    in vector algebra, 271
    product of a matrix by, 337
    product of a vector by, 271
Scalar product of vectors, 271
Secant:
    definition, 228, 248
    domain of, 228
    fundamental period of, 228
    graph of, 229
    period, 228
    range of, 228
Second-degree equation(s):
    definition of, 87
    discriminant of, 93
    in one variable, 87*ff*
    number of solutions, 88
    solution of, by completing the square, 89
    solution of, by factoring, 87
    solution of, by formula, 93
    standard form, 87
Second-degree inequalities, 103*ff*
Sentence(s):
    open, 82
Sequence:
    alternating, 391
    convergent, 391
    definition of, 379
    divergent, 392
    finite, 379
    function, 379
    limit of, 390
    notation, 380
    nth term, 381
Series:
    definition of, 384
    infinite, 392
    sigma notation for, 386
Set(s):
    Cartesian product of, 120
    closed convex polygonal, 324
    convex, 324
    counting properties of, 407
    definition, 1
    designation of a, 1

Set(s) *(continued)*:
    disjoint, 3
    domain of, 3
    element of a, 1
    empty, 2
    equality of, 2
    equivalence of, 2
    finite, 2
    infinite, 2
    intersection, 2
    member of a, 1
    notation, 1, 3
    null, 2
    of complex numbers, 20, 21
    of imaginary numbers, 20
    of integers, 6
    of irrational numbers, 6
    of natural numbers, 1, 5
    of rational numbers, 6
    of real numbers, 6, 7
    polygonal, 325
    replacement, 3
    union of, 2
    universal, 3
Set-builder notation, 4
Sigma notation, 386
Sine:
    definition, 198, 248
    domain of function, 200
    double-angle formula for, 289
    graph of, 214*ff*
    half-angle formula for, 289
    period, 215
    range of, 200
    sum and difference formulas, 288
    table of values of, 205
Sine waves, 216
Sines, law of, 260*ff*
Sinusoid, 216
Slope:
    of a line, 129
    of a line segment, 129
Slope-intercept form for a linear equation, 133
Solution(s):
    extraneous, 95
    matrix, of linear systems, 344, 369
    of a system, 303*ff*
    of an equation in one variable, 82
    of an equation in two variables, 121
    of linear inequalities, 99
    of nonlinear inequalities, 103
    of triangles, 256*ff*
Solution set, 82
Square matrix (*see* Matrix)
Square root, definition, 69
Standard form:
    for first-degree equations, 126

Standard form *(continued)*:
   for quadratic equations, 87
Standard position of an angle, 241
Statement, 82
Subset:
   definition of, 2
Substitution in solving equations, 97
Substitution law for equality, 7
Subtraction:
   of complex numbers, 20
   of quotients, 13
   of real numbers, 9
Sum(s):
   of a geometric progression, 386
   of an arithmetic progression, 385
   of an infinite geometric progression, 393
   of an infinite series, 392
   of complex numbers, 19
   of matrices, 333
   of polynomials, 26
   of quotients, 13
   of real numbers, 7*ff*
   partial, 392
Sum formula:
   for cosine function, 285
   for sine function, 288
   for tangent function, 291
Summation notation, 386
   index of, 387
   range of, 387
Symmetric law of equality, 7
Synthetic division, 41, 419
Systems, equivalent, 304
Systems of inequalities, 322
Systems of linear equations:
   in three variables, 310*ff*
   in two variables, 303*ff*
Systems of linear equations, solution by:
   addition, 304
   determinants, 372
   linear combinations, 304
   matrices, 342*ff*
   substitution, 306
Systems of nonlinear equations, 316
   graphs of, 319

Tangent:
   definition, 222, 248
   domain of, 222
   double-angle, 292
   graph of, 224
   half-angle, 292
   period, 223
   range of, 222
   sum and difference formulas, 291

Term, 26
Terminal side of an angle, 241
Theorem:
   definition of, 10
Transformation(s):
   elementary, 83, 342
Transitive law for inequalities, 16
Transitive law of equality, 7
Transpose of a matrix, 333
Triangle:
   right, 255*ff*
   solution of, 256*ff*
Trichotomy law, 15
Trigonometric functions, 248*ff*
Trigonometric ratios, 256
Trinomial, 26
Turning point of a graph, 141
Two-dimensional vector, 269

Union of sets, 2
Unique-factorization theorem, 429
Uniqueness:
   of additive inverse, 11
   of multiplicative inverse, 11
Unit circle, 243
Universal set, 3
Upper bound:
   for zeros of real polynomial function, 426

Value:
   absolute, 16
   of a determinant, 350
Variable, 3
Variation:
   constant of, 161
   direct, 160
   inverse, 161
   joint, 162
Variation in sign of coefficients, 425
Vector(s):
   applications, 272*ff*
   column, 331
   direction angle of, 270
   equality of, 271
   equivalence of, 271
   geometric, 269*ff*
   magnitude of, 270
   negative of, 271
   norm of, 270
   notation for, 269
   parallelogram law of addition for, 271
   resultant, 271
   row, 331
   scalar product of, 271
   sum, 271

Vector(s) *(continued)*:
    two-dimensional, 269
    with opposite direction, 272
    with same direction, 272
    zero, 272
Vertex of an angle, 241
Vertical:
    asymptote, 145

Zero(s):
    as a factor in a product, 12

Zero(s) *(continued)*:
    bounds for, of real polynomial function, 426
    complex, of a polynomial function, 421*ff*
    location theorem for, 427
    of a function, 139
    rational, of polynomial function, 429*ff*
    real, of polynomial function, 424*ff*
    test for rational, 431
Zero matrix, 333
Zero vector, 272

[8.1]	$\cos x$	element in the range of the cosine function
	$\sin x$	element in the range of the sine function
[8.3]	$\bar{s}$ or $\bar{x}$	length of reference arc corresponding to arc of length s or x
[8.5]	$\tan x$	element in the range of the tangent function
[8.6]	$\cot x$	element in the range of the cotangent function
	$\sec x$	element in the range of the secant function
	$\csc x$	element in the range of the cosecant function
[8.7]	$\operatorname{Sin}^{-1} x$, $\operatorname{Cos}^{-1} x$, etc.	elements in the range of the principal-value inverse circular functions; Arcsin x, Arccos x, etc.
[9.1]	$\alpha \cong \beta$	α is congruent to β
	$\overrightarrow{AB}$	the ray AB
	$^\circ$	degree measure
	^{R}	radian measure
[9.2]	$\bar{\alpha}$	the reference angle corresponding to the angle α
	$T(\alpha)$	any one of the trigonometric function values of α
[9.6]	$\mathbf{v}$	vector
	$\|\mathbf{v}\|$	the norm, or magnitude, of $\mathbf{v}$
	$\mathbf{0}$	the zero vector
[11.2]	(x, y, z)	the ordered triple of numbers whose first component is x, second component is y, and third component is z
[11.5]	$\mathcal{P}, \mathcal{S}$, etc.	a set of points
[12.1]	$\begin{bmatrix} a_1 & b_1 & c_1 \\ a_2 & b_2 & c_2 \end{bmatrix}$, etc.	matrix
	$A_{m \times n}$	m by n matrix
	$a_{i,j}$	the element in the ith row and jth column of the matrix A
	A^t	the transpose of the matrix A
	$0_{m \times n}$	the m by n zero matrix
	$-A_{m \times n}$	the negative of $A_{m \times n}$
[12.2]	$I_{n \times n}$	the n by n identity matrix